AF443761

DL 8241171 9

History of Geomorphology and Quaternary Geology

The Geological Society of London
Books Editorial Committee

Chief Editor

BOB PANKHURST (UK)

Society Books Editors

JOHN GREGORY (UK)
JIM GRIFFITHS (UK)
JOHN HOWE (UK)
PHIL LEAT (UK)
NICK ROBINS (UK)
JONATHAN TURNER (UK)

Society Books Advisors

MIKE BROWN (USA)
ERIC BUFFETAUT (FRANCE)
JONATHAN CRAIG (ITALY)
RETO GIERÉ (GERMANY)
TOM MCCANN (GERMANY)
DOUG STEAD (CANADA)
RANDELL STEPHENSON (NETHERLANDS)

IUGS/GSL publishing agreement

This volume is published under an agreement between the International Union of Geological Sciences and the Geological Society of London and arises from IUGS International Commission on the History of Geological Sciences (INHIGEO).

GSL is the publisher of choice for books related to IUGS activities, and the IUGS receives a royalty for all books published under this agreement.

Books published under this agreement are subject to the Society's standard rigorous proposal and manuscript review procedures.

It is recommended that reference to all or part of this book should be made in one of the following ways:

GRAPES, R. H., OLDROYD, D. & GRIGELIS, A. (eds) 2008. *History of Geomorphology and Quaternary Geology*. Geological Society, London, Special Publications, **301**.

BAKER, V. R. 2008. The Spokane Flood debates: historical background and philosophical perspective. *In*: GRAPES, R. H., OLDROYD, D. & GRIGELIS, A. (eds) *History of Geomorphology and Quaternary Geology*. Geological Society, London, Special Publications, **301**, 33–50.

GEOLOGICAL SOCIETY SPECIAL PUBLICATION NO. 301

History of Geomorphology and Quaternary Geology

EDITED BY

R. H. GRAPES
Korea University, South Korea

D. OLDROYD
The University of New South Wales, Australia

and

A. GRIGELIS
Lithuanian Academy of Sciences, Lithuania

2008
Published by
The Geological Society
London

THE GEOLOGICAL SOCIETY

The Geological Society of London (GSL) was founded in 1807. It is the oldest national geological society in the world and the largest in Europe. It was incorporated under Royal Charter in 1825 and is Registered Charity 210161.

The Society is the UK national learned and professional society for geology with a worldwide Fellowship (FGS) of over 9000. The Society has the power to confer Chartered status on suitably qualified Fellows, and about 2000 of the Fellowship carry the title (CGeol). Chartered Geologists may also obtain the equivalent European title, European Geologist (EurGeol). One fifth of the Society's fellowship resides outside the UK. To find out more about the Society, log on to www.geolsoc.org.uk.

The Geological Society Publishing House (Bath, UK) produces the Society's international journals and books, and acts as European distributor for selected publications of the American Association of Petroleum Geologists (AAPG), the Indonesian Petroleum Association (IPA), the Geological Society of America (GSA), the Society for Sedimentary Geology (SEPM) and the Geologists' Association (GA). Joint marketing agreements ensure that GSL Fellows may purchase these societies' publications at a discount. The Society's online bookshop (accessible from www.geolsoc.org.uk) offers secure book purchasing with your credit or debit card.

To find out about joining the Society and benefiting from substantial discounts on publications of GSL and other societies worldwide, consult www.geolsoc.org.uk, or contact the Fellowship Department at: The Geological Society, Burlington House, Piccadilly, London W1J 0BG: Tel. +44 (0)20 7434 9944; Fax +44 (0)20 7439 8975; E-mail: enquiries@geolsoc.org.uk.

For information about the Society's meetings, consult *Events* on www.geolsoc.org.uk. To find out more about the Society's Corporate Affiliates Scheme, write to enquiries@geolsoc.org.uk.

Published by The Geological Society from:
The Geological Society Publishing House, Unit 7, Brassmill Enterprise Centre, Brassmill Lane, Bath BA1 3JN, UK

(*Orders*:   Tel. +44 (0)1225 445046, Fax +44 (0)1225 442836)
Online bookshop: www.geolsoc.org.uk/bookshop

The publishers make no representation, express or implied, with regard to the accuracy of the information contained in this book and cannot accept any legal responsibility for any errors or omissions that may be made.

British Library Cataloguing in Publication Data

A catalogue record for this book is available from the British Library.

ISBN 978-1-86239-255-7

Typeset by Techset Composition Ltd., Salisbury, UK

Printed by Cromwell Press, Trowbridge, UK

Distributors

North America
For trade and institutional orders:
The Geological Society, c/o AIDC, 82 Winter Sport Lane, Williston, VT 05495, USA
Orders: Tel +1 800-972-9892
             Fax +1 802-864-7626
             Email gsl.orders@aidcvt.com

For individual and corporate orders:
AAPG Bookstore, PO Box 979, Tulsa, OK 74101-0979, USA
Orders: Tel +1 918-584-2555
             Fax +1 918-560-2652
             Email bookstore@aapg.org
             Website http://bookstore.aapg.org

India
Affiliated East-West Press Private Ltd, Marketing Division, G-1/16 Ansari Road, Darya Ganj, New Delhi 110 002, India
Orders: Tel +91 11 2327-9113/2326-4180
             Fax +91 11 2326-0538
             Email affiliat@vsnl.com

Contents

Preface

The Baltic States – Lithuania, Latvia and Estonia – form a region that experienced substantial glaciation during the Pleistocene which left a cover of up to 160 m of glacial sediments and reached a thicknesses of up to 310 m in buried palaeovalleys. The subsequent deglaciation and development of river networks during the Holocene gave rise to the relief that we see today, producing particularly interesting geomorphological features. Given this environment, the International Commission on the History of Geological Sciences (INHIGEO) chose the theme 'History of Quaternary Geology and Geomorphology' for its annual conference, held in the Baltic States in 2006.

It was the first time that the Commission had met in this region of eastern Europe. The main part of the meeting took place in Vilnius – the ancient and beautiful capital of Lithuania – and was followed by a field excursion through all three Baltic States. The presentation of the papers in Vilnius and the discussions during the field excursion allowed participants to examine the geological and geomorphological phenomena of the three countries, and their relationship to human history.

The Quaternary Period is no exception to the idea that different conditions prevailed at different times in different parts of the world, leading to variations in the geological record, as was stated by Leopold von Buch in the early nineteenth century. Thus, for example, when, 16–10 ka ago, ice sheets covered northern Europe, the Tamala Limestone, containing marine fossils, was being deposited in warm shallow seas in the region of Western Australia. With the amelioration of climate following the 'Ice Ages', and the land elevation of Scandinavia, the present relief and river networks of the Baltic States were formed. All round the world, rising sea levels produced changes in coastlines and estuaries.

The conference papers considered the histories of Quaternary geology and geomorphology in different parts of the world, with emphasis on the pioneers of these branches of geoscience in central and eastern Europe. It helped participants to improve their understanding of how Quaternary and land-surfaces research originated and has subsequently been developed, as well as understanding the numerous particular problems associated with Quaternary geology, compared with other parts of the stratigraphic column.

The conference also provided a valuable opportunity for participants from countries other than those of eastern Europe to get to know something of the history, geology and culture of a region that has been part of European civilization for about a thousand years. It also offered a chance for Lithuania and her sister states to open their doors to the world and display the geohistorical work that has been going on there for some considerable time, rather little noticed by outsiders.

I am convinced that the conference generated useful information on the themes discussed, which will serve it as a worthy Special Publication of the Geological Society of London, providing valuable insights into the histories of geomorphology and Quaternary geology in many parts of the world. This volume should be of value to all those interested in these two important branches of Earth science.

Let me wish this edition good fortune to survive in the Recent Era.

Algimantas Grigelis
Convener, INHIGEO Conference Vilnius 2006
1 July 2007

Contributions to the history of geomorphology and Quaternary geology: an introduction

DAVID R. OLDROYD[1] & RODNEY H. GRAPES[2]

[1]School of History and Philosophy, The University of New South Wales, Sydney, NSW 2052, Australia (e-mail: doldroyd@optushome.com.au)

[2]Department of Earth and Environmental Sciences, Korea University, Seoul, 136-701, Korea (e-mail: grapes@korea.ac.kr)

This Special Publication deals with various aspects of the histories of geomorphology and Quaternary geology in different parts of the world. Geomorphology is the study of landforms and the processes that shape them, past and present. Quaternary geology studies the sediments and associated materials that have come to mantle much of Earth's surface during the relatively recent Pleistocene and Holocene epochs. Geomorphology, with its concern for Earth's surface features and processes, deals with information that is much more amenable to observation and measurement than is the case for most geological work. Quaternary geology focuses mostly, but not exclusively, on the Earth's surficial sedimentary cover, which is usually more accessible than the harder rocks of the deeper past.

Institutionally, geomorphology is usually situated alongside, or within, academic departments of geology or geography. In most English-speaking countries, its links are more likely to be with geography; but in the United States these connections are usually shared between geography and geology, although rarely in the same institution. In leading institutions everywhere, strong links exist between geomorphology and such cognate disciplines as soil science, hydrology, oceanography and civil engineering. Although nominally part of geology, Quaternary geology also has strong links with geography and with those disciplines, such as climatology, botany, zoology and archaeology, concerned with environmental change through the relatively recent past.

Given that geomorphology concerns the study of the Earth's surface (i.e. landforms, and their origin, evolution and the processes that shape them) and that the uppermost strata are in many cases of Pleistocene and Holocene age, it is unsurprising that this Special Publication should deal 'promiscuously' with topics in both geomorphology and Quaternary studies. This particular selection has been developed from a nucleus of papers presented at a conference on the histories of geomorphology and Quaternary geology held in the Baltic States in 2006, where a great deal of what the geologist sees consists of Quaternary sediments. However, much of the Earth's surface is not formed of these sediments but of older rocks exposed at the surface by erosion and structural displacement. Here, geomorphology can seek answers to questions regarding the past histories of these rocks, their subsequent erosion, and present location and form. Geomorphology also raises questions, and may provide answers, regarding tectonic issues, for example from deformed marine terraces and offset fault systems. In all these instances, the history of geological and geomorphological investigations can serve to illustrate both the progress and pitfalls involved in the scientific understanding of the Earth's surface and recent geological history.

There are relatively few books but a growing number of research papers on the history of geomorphology. For readers of English, there is a short book by Tinkler (1985) and a collection edited by the same writer (Tinkler 1989), an elegantly written volume on British geomorphology from the sixteenth to the nineteenth century by Davies (1969), and a series of essays by Kennedy (2006). But towering over all other writings are three volumes: those by Chorley et al. (1964) on geomorphology up to the time of the American, William Morris Davis (1850–1934); by Chorley et al. (1973) dealing exclusively with Davis; and by Beckinsale & Chorley (1991) on some aspects of work after Davis. As envisaged by Chorley and Beckinsale, who died in 2002 and 1999, respectively, a fourth volume by other authors is soon to emerge (Burt et al. 2008). A series of essays edited by Stoddart (1997) on *Process and Form in Geomorphology* (1997) also contains valuable historical material, while papers edited by Walker & Grabau (1993) discuss the development of geomorphology in different countries, of which Australia, China, Estonia, Iceland, Japan, Lithuania, New Zealand, The Netherlands, the USA and the USSR are specifically mentioned in the present volume.

From: GRAPES, R. H., OLDROYD, D. & GRIGELIS, A. (eds) *History of Geomorphology and Quaternary Geology*. Geological Society, London, Special Publications, **301**, 1–17.
DOI: 10.1144/SP301.1   0305-8719/08/$15.00 © The Geological Society of London 2008.

A framework for geomorphology

Connections between geomorphology and geology go back to the early days of Earth science, but it is to developments in the later eighteenth century that we often attribute the foundations of modern links between the disciplines, notably to scholars such as Giovanni Targioni-Tarzetti (1712–1783) in Italy, Jean-Etienne Guettard (1715–1786) in France, Mikhail Lomonosov (1711–1765) in Russia and James Hutton (1726–1797) in Scotland. Hutton gave much thought to extended Earth time, and to the processes whereby soil and rock are eroded from the land to the sea. In 1802, Hutton's friend and biographer, John Playfair (1748–1810), not only rescued Hutton's ideas from relative obscurity but contributed original ideas on the nature and behaviour of river systems. However, the intellectual climate of the time worked against the ready acceptance of their views.

Following the leads provided by Hutton and Playfair, Charles Lyell (1797–1875) also addressed questions of extended Earth time and of erosion in his well-known and influential three-volume treatise *Principles of Geology* (Lyell 1830–1833). He emphasized the differential erosive powers that rivers or the sea could have on strata of different hardness, and discussed cases where river systems did *not* divide simply, like the branches of a tree, but cut through higher ground or occupied the eroded axes of anticlines. The latter phenomenon could be explained by supposing that folding had fractured the rocks at an anticlinal crest so that they became more prone to erosion, with the result that 'reversal' of drainage might occur. But Lyell realized that most of the rivers draining the Weald of SE England did not follow the main axis of the Wealden anticline but often cut through the North or South Downs that formed the flanks of the fold. He attributed such anomalous configurations to fractures that cut across the Wealden axis and to the interaction of Earth movements and fluvial erosion. Thus, Lyell invoked geomorphological and tectonic considerations in order to develop a geological history of a region.

A name that often emerges in the present collection of papers is that of W.M. Davis, with his theory of a cycle of erosion that was constructed in part on the work of his compatriots John Wesley Powell (1834–1902), Clarence Edward Dutton (1841–1912) and Grove Karl Gilbert (1843–1918) (Davis 1889, 1899, 1912). And one may reiterate that Davis's work was considered by Chorley *et al.* (1973) to be so influential as to warrant an entire volume of their comprehensive historical study of geomorphology.

Davis's initial cyclic ideas were encapsulated in the hypothesis that, following initial structural uplift, landforms shaped by rivers pass through different stages of development, which he dubbed 'youth', 'maturity' and 'old age', until they are reduced to a nearly level surface or 'peneplain'. The peneplain, for which he found evidence in the Appalachians, could later be 'rejuvenated' by uplift, thereby initiating a new cycle of erosion. This model led to studies of 'denudation chronology', or the reconstruction of landscape histories based of the recognition of erosion cycles and peneplains in various stages of development. Without a clear understanding of the processes and time involved, however, 'reading a landscape' through the lens of Davisian doctrine, or elucidating its 'denudation chronology', became an art form, rather than a rigorous science. Davis's geomorphic model was essentially qualitative and difficult to test but, as Charles Darwin famously wrote regarding his notion of natural selection, 'here then I had at last a theory by which to work' (Darwin 1887, p. 83).

Davis's ideas were challenged in his own time, particularly by German geomorphologists such as Albrecht Penck (1858–1945), Professor of Physical Geography at the University of Vienna and later of Geography at Berlin, and more particularly his son Walther Penck (1888–1923). Before the World War I, the Pencks and Davis were on good terms, but they subsequently drifted apart, partly owing to world politics and partly owing to Walther's rejection of the idealized character of Davis's theory along with disagreements as to the relationship between Earth movements and erosion. The Pencks objected to the notion of discrete upward Earth movements as the cause of topographic rejuvenation and also argued that erosion wears *back* a surface just as much as *down*. However, Walther Penck's proposed model of slope retreat would eventually yield a gently sloping surface resembling a Davisian peneplain (Penck 1924, 1953). Penck also envisaged an empirical relationship between tectonic activity and slope development, owing to the changing rates of river incision as the land itself was raised at varying rates. This idea was rejected vigorously by some in the English-speaking community, with Douglas Johnson (1878–1944), for example, describing it as 'one of the most fantastic ideas ever introduced into geomorphology'! (Johnson 1940, p. 231).

Ultimately, the differences between Davis and Penck lay in their different objectives and scientific approaches. Davis regarded geomorphology as a branch of geography, with geomorphic processes furnishing the topography upon which geography 'resided'. He, together with a number of like-minded geologists, geomorphologists and natural scientists, founded the Association of American Geographers in 1904, in part as a forum for his views (Orme 2005). Penck, in contrast, saw the field as being one that could elucidate problems of crustal movements and he was apparently less concerned with process and time (Hubbard 1940). It may be noted, though, that in his old age Davis accepted the idea of parallel slope retreat, such as is usually associated with the name of Walter Penck. Davis's changed views were given in lectures delivered at the University of Texas in 1929 but were not published until as late as 1980 (King & Schumm 1980).

Another major figure in the modern formulation of ideas on landscape evolution was the South African geomorphologist Lester C. King (1907–1989). Imbued with Davisian ideas and the triad of process, time and structure, as a graduate student of Charles Cotton (1885–1970) in New Zealand, King nevertheless went on to challenge much of Davisian theory. While still invoking the cyclic concept, like Penck he emphasized the importance of surficial processes, particularly in relation to the role of scarp retreat and pediment formation, and the considerable antiquity (e.g. Cretaceous) of some erosion surfaces. Given the structure of his adopted homeland in Africa, with its extensive flat-lying strata and thus many potential cap rocks, it is not surprising that King interpreted landscapes primarily in terms of scarp recession with consistency of slope form and inclination in any area and structural setting indicating parallel retreat. He thought that steep slopes are shaped by gravity and turbulent water flow (e.g. in gullying), whereas pediments, the typical landform of erosional plains, are the result of surface water flow (sheet wash), capable of transporting sediment and 'smoothing' the bedrock (King 1953). Pediments or pediplains persisted until another cycle of river incision or change in base level occurs, causing further slope retreat.

Thus, although King concluded that the evolution of landscapes by the action of running water would occur everywhere, except in glacial and desert areas, his ideas stemmed from observations in a semi-arid South Africa with limited river action, where weathering and rockfall predominate, and where scarp retreat, which occurs everywhere, is closely linked to pedimentation, which is of limited importance. King's recognition of a Mesozoic (or Gondwana) surface on the Drakensberg gave support to the idea that not all landforms are necessarily Late Cenozoic in age, as postulated in other theories of landscape evolution (e.g. Hack 1960). Mesozoic or Early Tertiary palaeosurface remnants have been identified in many other cratonic and old orogenic areas (e.g. China and Australia; see articles by Branagan (2008) and Zhang (2008), respectively, in this Special Publication), and their persistence raises fundamental questions about the complex interaction of surface-shaping processes such as erosion, the effects of climate change, tectonic uplift and deformation, etc., the duration of erosion 'cycles', or rock composition and structure.

Davis's erosion model was imbued with ideas drawn from Darwinian biology and his interests in entomology, and his diction was full of evolutionary metaphors. He was also interested in the pragmatic philosophy of Charles Peirce, as has been remarked by Baker (1996). By contrast, an awareness of recent developments in thermodynamics manifested itself in Gilbert's geomorphology through notions of dynamic equilibrium, grade and feedback loops. Gilbert's concept of 'negative feedback' in stream systems leading to 'graded rivers' occurred some 7 years before Henri Le Chatelier (1850–1936) enunciated his well-known principle as a general feature of chemical systems. Gilbert wrote:

Let us suppose that a stream endowed with a constant volume of water is at some point continuously supplied with as great a load as it is capable of carrying. For so great a distance as its velocity remains the same, it will neither corrade (downward) nor deposit, but will leave the grade of its bed unchanged. But if in its progress it reaches a place where a less declivity of bed gives a diminished velocity, its capacity for transportation will become less than the load and part of the load will be deposited. Or if in its progress it reaches a place where a greater declivity of bed gives an increased velocity, the capacity for transportation will become greater than the load and there will be corrasion of the bed. In this way a stream which has a supply of *débris* equal to its capacity, tends to build up the gentler slopes of its bed and cut away the steeper. It tends to establish a single uniform grade.
(Gilbert 1877, p. 112)

In the same publication, Gilbert also enunciated 'laws' for the formation of uniform slopes, structure and divides, and the concept of planation. In vegetated areas, he believed that the 'law of divides' was likely to prevail; in arid regions, he favoured the 'law of structure'. Thus, the 'laws' were not universal, in the style of Newton's laws. Nevertheless, Gilbert's work marked a significant advance in the search for geomorphological principles and thereby a step towards the establishment of geomorphology as a physical science (rather than an historical 'art'!). By contrast, Davis's 'evolutionary geomorphology', although attractive to his contemporaries and through much of the first half of the twentieth century, has now been largely or wholly superseded.

But Gilbert's concept of 'grade' also presents problems. It is supposedly a situation of balance between the transport of material in a river and the widening or deepening of the river bed by corrasion. According to Davis, for a 'mature' river 'a balanced condition is brought about by changes in the capacity of a river to do work, and in the quantity of work that the river has to do' (Davis 1902, p. 86). This assumes that for a given rate of river flow, there is a limit to the load that it can carry, and that the energy available can be used for either transport or corrasion. But these cannot just be summed, so that for a given stream flow if there is an increased load there is an equivalent decrease in corrasion. But this is simply not the case: halving the load does not double the corrasion (Wooldridge 1953, p. 168).

Walther Penck's interest in the relationships between Earth movements and landforms was shared by the French geomorphologist Henri Baulig (1877–1962), who was a student of Davis for 6 years at Harvard. Baulig's main area of research was France's Massif Central for which he tried to synthesize the ideas of Davis and those of the notable Austrian geologist Eduard Suess (1831–1914) (Baulig 1928). During the later nineteenth century, Suess had sought a global understanding of geological phenomena in terms of the increasingly questionable notion of a cooling and contracting Earth, which led to lateral compressive forces that produced great orogenies. With each large-scale collapse of Earth's crust, Suess believed that there was a concomitant global lowering of sea level as well as elevation of mountain ranges. Worldwide erosion and sedimentation would follow, and the ocean basins would receive sediment, leading to global marine transgressions. These would supposedly account for the correlations that might be made worldwide for different parts of the stratigraphic column. In 1888, the global changes in sea level, arising from spasmodic tectonic episodes, were called 'eustatic movements' by Suess (English translation 1906, p. 538); and, thus, there emerged the concept of global 'eustasy', based on intelligible (albeit mistaken) explanatory principles. Suess's ideas were attractive in Baulig's earlier years as the basis of a general geological theory and it is therefore unsurprising that Baulig sought to link them to his geomorphological studies.

In considering the relative levels of land and sea (globally), one could consider epeirogenic movements, isostasy and eustasy (the latter being due to epeirogeny/diastrophism or the waxing or waning of glaciation, which could also generate isostatic responses). And if a land surface is reduced by erosion there will also be an isostatic response. Despite these complexities, Baulig favoured global eustasy as the main source of the formation of planar erosion surfaces. This opened the prospect of worldwide temporal correlation of peneplains:

[R]egions, widely-spaced and totally independent from a structural viewpoint, show perfectly clearly an exactly similar geomorphological development since the Upper Pleistocene. This similarity, in the present state of ideas and knowledge, admits of only one explanation: that it is eustasy pure and simple.

(Baulig 1928, p. 543; translation from Beckinsale & Chorley 1991, p. 268)

Of course, it was easy to conflate or confuse glacial and Suessian eustasy. Nevertheless, Baulig reiterated his ideas in 1935, extending his claims of uniformity of marine terraces at distant locations back into the Pliocene (Baulig 1935). But this line of inquiry led to confusion as much as to understanding.

The search for guiding principles, or 'laws' as they were (or are) sometimes mistakenly called, has been a recurrent feature of the history of geomorphology. As early as 1802, Playfair enunciated a general principle that, despite many exceptions, became known as 'Playfair's Law', thus:

Every river appears to consist of a main trunk, fed from a variety of branches, each running in a valley proportional to its size, and all them together forming a system of vallies, communicating with one another, and having such a nice adjustment of their declivities, that none of them join the principal valley, either on too high or too low a level; a circumstance which would be infinitely improbable, if each of these valleys were not the work of the stream that flows in it.

(Playfair 1802, p. 102)

As the field of geomorphology developed, the search for so-called laws among drainage networks continued to interest scholars. For example, is there any pattern, any law-like behaviour in such networks? Can a mathematical model of stream branching be discerned? Very early on, Leonardo da Vinci (1452–1519) had noted the similarity of branching in trees and stream systems (Shepherd & Ellis 1977). Later, following the physician James Keill's (1673–1719) (1708) early anatomical work on arterial trees, known to Hutton and Playfair, Julian Jackson (1790–1853) (1833) addressed the notion of 'stream order' in 1834, and Harry Gravelius (1861–1938) of the Dresden Technical Institute later expanded on these ideas (Gravelius 1914). The largest or stem stream was designated as being of Order 1; the first tributary was of Order 2; and so on back to the unbranched 'fingertip' tributaries.

This nomenclature (or taxonomy) prevailed in Europe for a considerable time. But the US Geological Survey hydrologist Robert Horton (1875–1945) reversed the terminology so that 'un-branched tributaries are of 1st order; streams that receive 1st-order tributaries, but these only, are of the 2nd order; third order streams receive 2nd- or 1st- and 2nd-order tributaries; and so on, until, finally, the main stream is of the highest order and characterizes the order of the drainage basin' (i.e. the highest order stream extends from source to outlet) (Horton 1945, p. 277). Subsequently, the American geologist Arthur Strahler (1918–2002) proposed an alternative system of stream ordering designed to give an idea of the relative power of the different waterways in a drainage system (the higher the order the higher the power) (Strahler 1952).

However, the Horton and Strahler schemes were misleading, and mathematically cumbrous, in that they ignored downstream links with streams of lower order. Later schemes by Adrian Scheidegger (b. 1925) (1965) and Ronald Shreve (1966) resolved this problem. Although Horton's early system allowed some interesting 'laws' to be defined and for drainage densities to be calculated, by the 1970s it had come to be realized that such relationships were in large measure a consequence of the ordering systems used and the topological randomness of such networks. Nevertheless, stream-ordering systems continue to be used for ranking purposes by drainage basin specialists.

Among Strahler's students at Columbia University, Stanley Schumm (b. 1927), who spent most of his career with the US Geological Survey and at Colorado State University, and Mark Melton (b. 1930) were particularly prominent. Melton did his PhD at Columbia and moved from there to the University of Chicago, where he was given to understand that strongly mathematical and statistical work would be appreciated. Schumm (1956) measured and analysed both the surface and subsurface processes involved in slope development in order to provide a theoretical analysis of fluvial erosion. Melton's mathematically sophisticated work used a systems approach and ergodic reasoning for the analysis of geomorphological problems (Melton 1958). Perhaps unsurprisingly, he demonstrated that channel frequency was a function of drainage density. In these and other ways, Playfair's early insight on the form and interrelationships of drainage networks was given mathematical expression during the so-called quantitative revolution in geomorphology during the mid-twentieth century.

Prior to the 'quantitative revolution', the time-dependent models of Davis and Penck had incorporated into geomorphology the notion of uniformitarianism: the assumption of gradual change through time based on the principle that 'the present is the key to the past' (Geikie 1962, p. 299). In contrast, the earlier work of Gilbert, based on the dynamic interactions of landform processes, was more suited to a time-independent approach, although he did not develop a comprehensive model. And in 1960 such a model, based on the study of humid temperate drainage basins, was proposed by John T. Hack (1913–1991) of the US Geological Survey. Hack's model revealed conflicts with erosion-cycle concepts and presented time-independent equilibrium as an alternative to the Davisian system. Instead of attributing it to age, landscape variability was considered to result from interacting contemporary processes wherein a state of balance, or dynamic equilibrium, existed between fluctuating inputs and outputs of material and energy. According to Hack, landforms were open systems so that similar landforms could have different origins. For example, accordant summit heights, invoked by Davis as evidence of former peneplains, may originate in rocks with similar hardness, structure and drainage density. Where rocks differ in resistance, there is the possibility of different levels of accordant summits, which are not necessarily explained in terms of multiple erosion cycles. Implicit in Hack's work was the assumption that there is a uniform lowering of the landscape with little obvious change in rate and process unless there is a change in climate, tectonism or geology. Thus, Hack's model was in broad agreement with Penck and, provided that uplift was slow enough to balance the rate of erosion, a steady-state relief would result. Sudden uplift would produce transitional relief, with relict landforms disappearing as a new equilibrium state was approached.

The advantages of this dynamic equilibrium approach to landscape evolution was that it was not constrained by, or dependent on, a Davisian stage, and it provided a convenient entry point (current conditions) for understanding the system because the past is usually poorly known. The idea of dynamic equilibrium relies on the notion that landscape systems near equilibrium change slowly (time-dependent) and those that are far from equilibrium change rapidly (time-independent). The concept thus unites two viewpoints. But debate continues, with arguments that equilibrium probably never exists in the multivariant, often chaotic and non-linear nature of Earth processes, and that they more probably reflect disequilibrium (e.g. Phillips 1999).

The concept of time is important to geomorphology, but it was not until the twentieth century that the traditional preoccupation with time-dependent landscape evolution could be tested. The long-established foundations of the geological timescale, such as the principles of stratigraphy, could not readily be applied to landforms undergoing denudation but for which there were no residual deposits. For relatively short-term geomorphic events, time might be measured directly during a particular event or period of measurement, or by reference to records over periods of recorded time. By contrast, for studies of landscape change over longer periods, say from thousands to millions of years, some means of establishing the time frame is necessary and it is only within the past 100 years or so, and often more recently, that such methods have become available. They include absolute dating techniques, such as dendrochronology, thermoluminescence and radiometric dating, and the development in recent decades of a wide range of surface-exposure dating techniques that can provide ages for eroded rocks and surfaces. For a review of such methods, see Walker & Lowe (2007).

A geomorphic division of time based on whether the variables of landscape evolution are independent, semi-independent or dependent was proposed by Schumm & Lichty (1965) in terms of cyclic time (hundreds of thousands to millions of years that would cover the duration of a Davisian cycle), graded time (thousands to hundreds of years) and steady time (a few days). In the first case, time is the most important independent variable and all others, such as climate, initial relief and geology, are dependent on it. Graded time is a short segment of cyclic time during which a graded condition exists that involves a fluctuating dynamic equilibrium as the reduction of relief approaches a steady state. Steady time represents a brief period during which some parts of the system remain unchanged (e.g. uniform stream-flow or channel form) and hence are time-independent.

The reintroduction of *process* to geomorphology in the 1950s brought about an inquiry into the effects of processes of different frequency and magnitude, encapsulated in the benchmark studies of Wolman & Miller (1960). In general, frequency and magnitude are inversely related. Although relatively infrequent, large-magnitude events such as great earthquakes or major floods can have catastrophic geomorphic consequences. Wolman & Miller's work sought to show that most changes in the landscape are carried out by frequent events of moderate magnitude, for example by peak annual stream flows. They suggested that such events do most of the work in geomorphic systems, but their model is not well supported in areas such as Mediterranean-type regions and desert margins, where rare high-magnitude events clearly do most of the work, and it has subsequently been modified (e.g. Baker 1977; Wolman & Gerson 1978).

The foregoing gives some indications of the background to work in modern geomorphology, against which to evaluate many of the historical essays in this book. In recent years, geomorphology has continued to grow as a discipline with emphasis on quantitative data, experimentation, predictive modelling (e.g. Wilcock & Iverson 2003), tectonic geomorphology (e.g. Burbank & Anderson 2001), and the understanding of links between process and form. Much recent work is driven by the need for hazard prediction and landscape management in a world that is becoming ever more crowded.

A framework for Quaternary geology

The Quaternary is the shortest and most recent of the geological periods recognized in Earth history, defined here as the last 2.6 Ma. It is, perhaps, the most important period of time because, despite its brevity relative to earlier periods, its materials cover much of the present landscape and provide soils and resources for agriculture and other human activities. It has been a period of pronounced climate change with all that implies for Earth's land surface, hydrosphere and biosphere. It has also witnessed the later evolution of hominids and the emergence of *Homo sapiens*. In short, it is a period of Earth time of great intrinsic and practical interest. It is also, as we shall show, the focus of much controversy.

Recognition of the peculiar properties of the Quaternary Period was slow to emerge, and debate continues as to the precise nature of the changes and forces involved. A major issue for geoscientists in the early–middle nineteenth century was the origin of the extensive surficial deposits, from clay to boulders, found across northern Eurasia and North America. The deposits were mostly poorly consolidated, unsorted, poorly structured and devoid of guide fossils. Following the dominant biblical beliefs of the time, these deposits had often been ascribed to materials deposited by the Noachian Deluge. Later, several catastrophic episodes were thought to have interrupted geological history from time to time, as supposed by Georges Cuvier (1769–1832) (1813). The deposits supposedly derived from the Deluge were termed 'Diluvium' and were distinguished from 'Alluvium', which was still to be seen being laid down by rivers (Buckland 1819, pp. 532–533).

Roderick Murchison (1839, vol. 1, p. 509) preferred the term 'drift' to Diluvium, as that word did not have any connotations of the 'Deluge of Holy Writ' and might be applied to deposits of similar character from different locations and of different ages, many of them attributable to marine currents. The term 'drift' caught on and, despite its 'archaic' implications, has survived to the present in many areas. In describing his observations on coastal exposures in Norfolk in 1839, Lyell (1840, p. 176) deployed Murchison's term 'drift' as a substitute for 'Diluvium' and added the suggestion that erratic boulders and the like were emplaced as drop-stones by floating icebergs at a time of reduced global temperature, rather than by exceptionally violent marine currents. Thus, where Lyell found such materials onshore, the land surface must have been lower than at present. The 'iceberg theory' accorded with Lyell's objections to catastrophic floods as geological agents but it initiated the unhelpful notion of 'glacial submergence': the idea that epochs of cold in regions presently mantled in drift deposits coincided with marine transgressions. Yet, the involvement of the sea in drift deposits seemed helpful in seeking to explain the occurrence of stratified layers in some 'tills' (the Scottish term that Lyell also employed).

The 'glacial theory' (or 'land-ice theory') emerged through the work of Ignace Venetz (1788–1859) and Jean de Charpentier (1786–1855), among others, in the European Alps, and was popularized by Louis Agassiz (1807–1873) with his publication of *Etudes sur les glaciers* (1840). It was Agassiz who brought the Swiss land-ice theory to Britain at the Glasgow meeting of the British Association in 1840; and then to America with his appointment to Harvard University in 1846. Yet, for a time the unhelpful 'glacial submergence' theory remained popular, at least in insular Britain, as it gave an attractive explanation of the presence of erratics over much of the low-lying ground of northern Europe. In his *Antiquity of Man*, Lyell (1863) seemingly accepted the 'land-ice theory' but he later reverted to his 'iceberg theory'. Meanwhile, stimulated by writings on climate change by the astronomer John Herschel (1792–1871) (1830) and Joseph Adhémar (1797–1862) (1842), the Scotsman James Croll (1821–1890) developed the then remarkable explanation for glacial epochs in terms of an astronomical theory (Croll 1867) – a forerunner of the early twentieth-century work of the Serbian mathematician Milutin Milanković (1879–1958) (synthesized in 1941 and republished in English in 1998).

Despite, or perhaps because of, Agassiz's advocacy of an extreme monoglacial 'land-ice theory' and Lyell's focus on his 'iceberg theory', the concept of widespread continental glaciation made slow progress in the mid-nineteenth century, and some opposition persisted to the close of the century (Orme 2002). Eventually, the publication of *The Great Ice Age* by James Geikie (1839–1915) (1874) and *The Ice Age in North America* by George Frederick Wright (1838–1921) (1889) did much to confirm the theory. Geikie's book was especially influential because he moved among influential scientists, including Thomas C. Chamberlin (1843–1928) in North America and Otto Torell (1828–1900) in Sweden, who early recognized the evidence for multiple glaciations. By then, evidence for extensive non-glacial deposits of Quaternary age, such as loess and pluvial lake deposits, was also emerging.

The term 'Quaternary' ('*Quaternaire*') was first proposed by Jules Desnoyers (1801–1887) (1829) as an 'extra' to the Primary, Secondary and Tertiary subdivisions of the stratigraphic column that had been proposed in Italy in the eighteenth century by Giovanni Arduino (1760). The term 'Pleistocene' ('*Pléistocène* tirée du grec pleiston, plus kainos, *recent*') was introduced by Lyell in 1839 in the Appendix to the French edition of his *Elements of Geology* (1839, p. 622), as an alternative name for his previous term 'Newer Pliocene', proposed in his *Principles of Geology* (1833) on palaeontological grounds. (We thank Dr G. Gohau for checking this reference in a Paris library.) Lyell (1833, vol. 3, p. 61) referred to post-Tertiary sediments as 'Recent' and stated that 'some authors' used the term for 'formations which have originated during the human epoch'; but he did not favour that definition (Lyell 1833, vol. 3, pp. 52–53). The name '*Holocène*' (= 'wholly recent') was subsequently suggested by Paul Gervais (1867, vol. 2, p. 32), and the term 'Holocene' as a synonym for 'Recent' was ratified at the Third International Geological Congress in London in 1885. For Gervais, the Quaternary was made up of the Pleistocene and the Holocene. The latter was estimated as being some 8–10 ka in duration. Moreover, with the general acceptance of the idea of a 'Great Ice Age', the term Pleistocene came to be used to represent the period of time when glaciation was widespread in the northern hemisphere (as suggested by Edward Forbes (1846, p. 403)). Lyell, however, did not use the term 'Quaternary'.

So where should the base of the Pleistocene be located, and how should it be related to the Quaternary? Maurice Gignoux (1910) suggested that the base of the Quaternary should be defined by a site in Calabria in southern Italy, where sediment containing cold-water fossils (especially *Cyprina* (*Arctica*) *islandica*) was seen to overlie sediments containing fossils indicative of a relatively warm climate. This event was not well suited for international correlation but was nevertheless accepted at the Eighteenth International Geological Congress in London in 1948 (King & Oakley 1950). Later, Hays & Berggren (1971) showed that the Calabrian deposit coincided quite closely with the top of the so-called 'Olduvai Normal Event', a short episode within the Matuyama Reversed Epoch of the geomagnetic polarity timescale, which occurred about 1.8 Ma. This geomagnetic marker offered the possibility of unambiguous worldwide correlation, and hence became widely accepted (Haq *et al.* 1977). A proposal for a Global Stratotype Section and Point (GSSP) for the boundary at Vrica in Calabria was ratified by the International Commission on Stratigraphy in 1983 and at the Moscow International Geological Congress in 1984 (Aguirre & Passini 1985) (initially it was set at 1.64 Ma and later charged to 1.81 Ma).

This choice was based chiefly on 'classical' biostratigraphic criteria. But many students of Quaternary geology were not happy with the decision, as there was substantial evidence of earlier glaciations in other parts of the world (see, for example, Milanovsky 2008, published in this Special Publication). Moreover, the term 'Quaternary' was considered by many to be outmoded, being out of line with other stratigraphic terminology, given that Primary and Secondary had long been obsolete and that the Cenozoic had been divided into Palaeogene and Neogene, with the consequent demise of the Tertiary (Fig. 1). The 'issue of the Neogene' has been an important additional factor confounding discussions of the Quaternary and the Pleistocene. The term was introduced by the Austrian Moritz Hörnes (1815–1868) (1853). For background on the Neogene and Palaeogene, see Berggren (1998).

Given the evidence for late Cenozoic glaciation in some parts of the world earlier than 1.8 Ma, and the fact that the Vrica GSSP was based on neither clear-cut bioevents nor climatic criteria (Partridge 1997, p. 8), an earlier date for the boundary was sought, particularly in the light of the discovery of the arrival of organisms indicative of a cold climate in the Mediterranean region prior to 1.8 Ma. Thus, the Gauss–Matuyama reversal at 2.6 Ma has been suggested as a suitable boundary (Pillans & Naish 2004). Although glaciation is known to have occurred in some parts of the world earlier than 2.6 Ma, this geomagnetic reversal meshes with determinable biostratigraphic changes indicative of climate change, a clearly identifiable event in oxygen-isotope stratigraphy (Shackleton 1997), and changes in grain size in Chinese loess deposits. The chosen golden spike, in Sicily, for the bottom of the new Gelasian Stage of the Upper Pliocene is thought to be only about 20 ka younger than the Gauss–Matuyama reversal (Rio *et al.* 1998, p. 85), so the fit is good. Moreover, the climatic deterioration could be related to a change from the dominance of orbital precession to that of the obliquity of the ecliptic, according to Milanković theory (Lourens & Hilgen 1997; Maslin *et al.* 1998).

But to place the base of the Pleistocene at 2.6 Ma would take the Quaternary down to the bottom of the uppermost (Gelasian) stage of the Pliocene (Fig. 1). Alternatively, it would require a decoupling of the Pleistocene from the Quaternary; yet, both have long been associated in geoscientists' minds with the 'glacial epoch'. In consequence, some authorities have proposed that the term 'Quaternary' should be dropped from the stratigraphic column (e.g. Berggren 1998). In Berggren's view, it should (or would or could) only survive 'for geopolitical purposes'! The issue has been particularly sensitive because Quaternary studies are such a well-established branch of geoscience, with the International Union for Quaternary Research (INQUA)

having been established back in 1928. An attempt in 1998 to have the Pliocene–Pleistocene boundary placed at the bottom of the Gelasian by the Commission for Neogene Stratigraphy of the ICS was unsuccessful (Ogg 2004, p. 125).

Given this messy situation, various proposals have been put forward (e.g. Suc *et al.* 1997; Pillans 2004; Gibbard *et al.* 2005; Suguio *et al.* 2005). But none of these provided a neat and *consensual* solution. Suc *et al.* (1997) favoured the move of the base of the Pleistocene to the bottom of the Gelasian and disuse of the term 'Quaternary'. Pillans (2004) believed that the Quaternary should be preserved and should run from the bottom of the Gelasian to the present (there being no Holocene unit as the Neogene also runs through to the present); but the Pliocene–Pleistocene boundary should continue to be located at the *top* of the Gelasian; and there should be a Holocene above the Pleistocene. This was anything but tidy. Details of the comings and goings of the debate may be found at www.quaternary.stratigraphy. org.uk/meetings/Quat_TaskGroup_25Aug05.doc: Definition and geochronologic/chronostratigraphic rank of the term *Quaternary*. Recommendations of the Quaternary Task Group jointly of the International Commission on Stratigraphy (ICS, of the International Union of Geological Sciences, IUGS) and the International Union for Quaternary Research (INQUA). The issue had to do with the problem of synthesizing different dating methods and the various overt institutional and invisible networks of members of different research fields.

The debates were not confined to the Anglophone community. The Chinese Association for Quaternary Research (2005), based on the significance of Chinese loess deposits, supported the INQUA position, and maintained that the Quaternary 'should be a formal unit with full Period/System status in geological time work' with a base at 2.6 Ma. In 2007, the Chinese President of the IUGS (Zhang Hongren) wrote to the bickering chairs of the relevant ICS subcommissions, as well as the ICS Executive, telling them, in effect, to co-ordinate their activities and formulate a solution ready for ratification by the IGC in Oslo in 2008. (The letter, and many other relevant documents, may be viewed at http://www.quaternary.stratigraphy.org. uk.) And as it appears at the time of writing, the Quaternary will survive as a received stratigraphic unit (Period or System) with its base at 2.6 Ma, the Pleistocene and Holocene being its constituent Series or Epochs (Bowen & Gibbard 2007) (Fig. 1).

Future changes notwithstanding, the term Quaternary is used in this Special Publication to refer to the Pleistocene and the Holocene, the latter denoting the last 10 000 radiocarbon years (11.5 ka calendar years), which is approximately equivalent

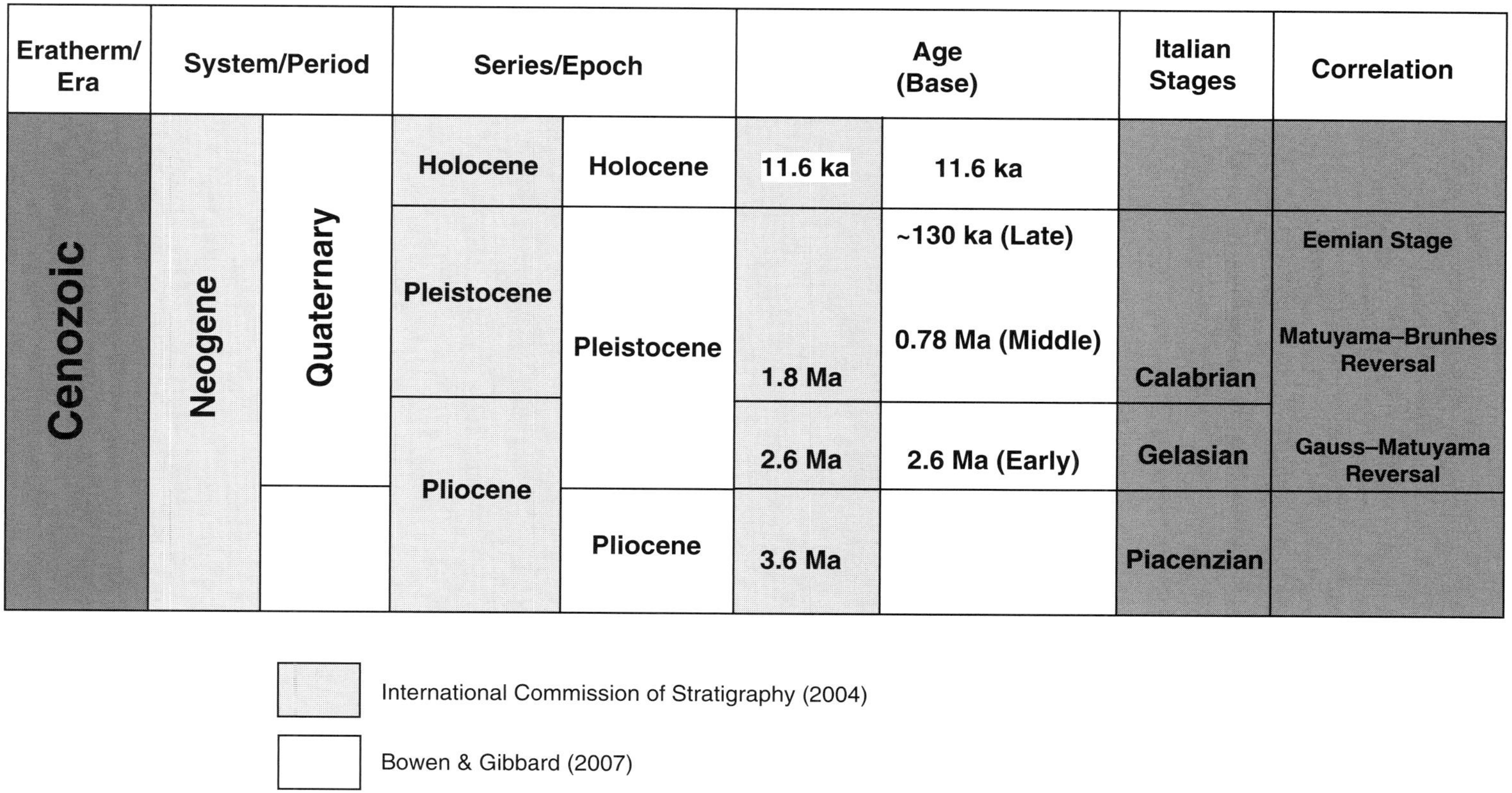

Eratherm/Era	System/Period	Series/Epoch		Age (Base)		Italian Stages	Correlation
Cenozoic	Quaternary	Holocene	Holocene	11.6 ka	11.6 ka		
		Pleistocene	Pleistocene	1.8 Ma	~130 ka (Late)	Calabrian	Eemian Stage
					0.78 Ma (Middle)		Matuyama–Brunhes Reversal
	Neogene	Pliocene		2.6 Ma	2.6 Ma (Early)	Gelasian	Gauss–Matuyama Reversal
			Pliocene	3.6 Ma		Piacenzian	

International Commission of Stratigraphy (2004)

Bowen & Gibbard (2007)

Fig. 1. Age of base of the Pleistocene and status of the Neogene–Quaternary periods after the International Commission of Stratigraphy (2004) (Gradstein *et al.* 2004) and Bowen & Gibbard (2007).

to the end of the last ice age and the emergence of civilized humans.

Stratigraphically, then, the Quaternary is subdivided on the basis of climate into cold (glacial) and warm (interglacial) stages. Global landscapes cannot be explained without reference to these alternations of cold and warmth associated with the repeated alternations of ice advance into, and retreat from, middle latitudes of the northern hemisphere. For example, alternations of wetter and drier conditions during the Pleistocene in the present deserts of the northern hemisphere reflect the pervading influence of climate changes that led, farther north, to glacial and interglacial conditions. Similar conditions occurred in the southern hemisphere, but there the potential impacts of climate change have been cushioned by the dominance of oceans, such that during cold stages nival and aeolian effects often take precedence over actual glaciation, although multiple glaciations have been identified in Tasmania and Patagonia. The development of ice sheets in high latitudes also caused the migration and compression of other climatic zones towards the equator so that nival or periglacial conditions extended well beyond their present limits. The climate changes were such that even in low latitudes many glaciers developed on the mountains of Papua New Guinea, Borneo, the tropical Andes and East Africa. Also, glacio-eustatic changes in sea level are global events. Thus, the climatic changes of the Quaternary had worldwide impacts.

Nevertheless, remnants of older landscapes persisted in regions beyond the reach of glaciers, and even in glaciated areas major preglacial features survived the passage of ice sheets, which, although scraping away loose debris, made only minor modifications to bedrock forms.

The environmental changes of the Quaternary have been influenced in part by the continued migration of Earth's continental and oceanic plates, albeit at much slower rates than climate change, and by episodes of greater or lesser volcanicity, which often ejected large quantities of climate-changing tephra into the atmosphere. Nevertheless, for most of the climate changes experienced in the Quaternary, there is now broad consensus that changes in Earth–Sun relations are broadly responsible, accompanied and aided by complex feedback mechanisms including ocean–atmosphere and biosphere–atmosphere linkages. Thus, variations in the eccentricity of Earth's orbit, the obliquity of its axis relative to the plane of the ecliptic and the precession of the equinoxes have caused cyclic changes, within periods of about 100, 41 and 21 ka, respectively, onto which shorter, more rapid fluctuations have been superimposed.

Modern ice sheets and glaciers are but shrunken remnants of their former areal extents, but the interpretation of the deposits left behind by the ice and the landforms shaped by the moving ice has been greatly facilitated by research on modern glaciers and by imaginative theoretical ideas. The first systematic explorations and studies of the Quaternary Period were carried out in Europe and North America, because by happy coincidence these were glaciated lands where the geological sciences initially blossomed and where social conditions encouraged travel and research. The knowledge of glacier motion and the recognition of glacial and interglacial deposits that flowed from this research have proved invaluable to the understanding of Quaternary events elsewhere, for example in Africa, Australia and South America. Similarly, early observations in the northern hemisphere regarding changing sea levels, now attributed to a variety of climatic and tectonic forces, paved the way for sophisticated studies of Quaternary land–sea relationships elsewhere in the world.

The Quaternary Period is important for many reasons. For the geologist, it affords insight into relatively recent events that also contribute to the understanding of the more distant past. For the climatologist, it presents evidence of past climates in a world beset with problems associated with continuing climate change. For biologists and soil scientists, events during the Quaternary help to explain much of the Earth's present plant and animal distributions, and its soils, which are so important to agriculture.

The Quaternary has also seen the continued evolution of *Homo sapiens*. Some human impacts have been deleterious. As with climatic changes, their importance can be exaggerated but cannot be ignored. Without the advent of the human mind there would be no Quaternary, for the System (like all the units in the stratigraphic column) is a human construct – in this case the latest of several known periods of climatic aberration.

The papers

The papers contained in the present volume do not pretend to cover every aspect of the innumerable components of geomorphology or Quaternary geology, but they touch upon many and in a number of cases deal with topics that are not generally known to those interested in these branches of geoscience.

For the first paper in the present collection, **Klemun**, an historian of science at the Institute of History, Vienna University, Austria, explores the contributions of Adolphe Morlot to the establishment of the term 'Quaternary'. Both he and Jules

Desnoyers have been credited with introducing the term, but, in fact, they did so independently, at different times and places. Desnoyers simply wanted to give a name to distinctive strata that lay above those that Arduino had named Tertiary back in the eighteenth century. Morlot's proposal came later and for quite different reasons. He found evidence in Switzerland for two distinct glacial episodes, separated by deposits that suggested a milder climate: an interglacial as we would say. These three sets of sediments, together with those of recent times, gave four 'layers'. Hence, the name Quaternary. Professor Klemun's work is a result of her interest in the historical development of scientific classificatory terms.

The papers by **Baker** (Department of Hydrology and Water Resources, University of Arizona, Tucson, USA) and **Orme** (Department of Geography, University of California, Los Angeles, USA) discuss issues relating to the Quaternary geology and geomorphology of large parts of western North America, of which both authors have detailed knowledge. Readers are likely to be familiar with the 'Great Scablands Debate' (Gould 1978), which had to do with certain features of the topography of eastern Washington where gigantic ripple marks, too large to be discerned from a single vantage point on the ground, suggested to the Chicago geologist J Harlen Bretz (1882–1981) that they had been produced by 'catastrophic' flooding. Joseph T. Pardee (1871–1960) had previously suggested large-scale flooding produced by the unblocking of a lake ponded by a glacier (Pardee 1910). Bretz's fieldwork in the 1920s suggested that there had, indeed, been some such 'catastrophic' cause; and, in fact, the hypothesis that was eventually accepted was somewhat akin to the one suggested long before in the nineteenth century, by Leopold von Buch, to account for the movement of erratics in the Alpine valleys, which was also later adopted by the 'uniformitarian' Charles Lyell. Yet (ironically), Lyell's 'uniformitarian' programme became so successful that any seeming 'catastrophist' theory, such as appeared to be suggested by Bretz's observations, was dismissed for a long time.

The intricacies of this interesting debate are carefully traced by Victor Baker, who used to know Bretz personally and so knows his side of the story well. Baker's account also includes reference to the philosophical issues involved: the reluctance of geologists in America in Bretz's time to countenance anything that hinted at catastrophism; and the American preference at that time for what might be called naive empiricism. For Baker, however, a theory that 'binds together' a number of distinctive facts (giving a consilience of inductions) has much to recommend it. He (and we) are devotees of the use of 'coherence' as a criterion in the pragmatic pursuit of truth, the virtues of which approach are well illustrated by the Great Scablands Debate. Baker was the keynote speaker at the Vilnius meeting.

The paper by Antony Orme on the Quaternary pluvial lakes of the American West is detailed, but at the same time offers a wide historical sweep, from the observations of the early European explorers in the region to the issues arising from the use of modern dating methods in the study of these lakes. Formerly, the study of the history of such lakes may have seemed a somewhat esoteric undertaking, but in more recent years it has come to be a matter of considerable practical concern, both from the perspective of water supplies and of data relevant to current problems relating to climate change. Here, then, is a case where the study of the history of geology turns out to have practical value to contemporary problems. The problem of understanding the history of these pluvial lakes is complex because there have been several independent causal factors involved in their history, such as rainfall, temperature, erosion and crustal deformation. Orme's paper is also valuable, for the purposes of the present collection, in that it attends to issues relating to dating in the Quaternary.

The theory of continental glaciation, beginning from observations of 'living glaciers' and areas previously occupied by ice in the Swiss Alps, was worked out independently by Otto Torell (1828–1900) (in 1872), Piotr Kropotkin (1842–1921) (between 1862 and 1876) and others in northern Europe, to which workers in the Baltic States made significant contributions. **Raukas** at the Institute of Geology at Tallinn University of Technology, Estonia, describes contributions to the evolution of the continental glaciation theory by (amongst others): Karl Eduard Eichwald (1795–1876), the first person in the Baltic provinces to consider the possibility of the wide distribution of ice in lowland areas in 1853; Friedrich Schmidt (1832–1908), who studied the Quaternary deposits of Estonia, demonstrating a correlation with those in Sweden in 1865; and the Estonian stratigrapher Gregor Helmersen (or Gelmerson) (1803–1885) (who became Director of the St Petersburg Mining Academy), who rejected the drift hypothesis in favour of a continental ice sheet in 1869 on the basis of the distribution of erratic boulders, glacial clay and striations. It was Schmidt, in 1871, who proved that during the Pleistocene the Scandinavian ice sheet covered the Baltic Sea depression and surrounding territories, and thereby resolved the early controversy over continental glaciation in northern Europe.

Milanovsky is one of the 'grand old men' of Russian geology and a former head of department at the Moscow State University. He comes from a family of distinguished geologists and has led a remarkable life, including fighting with a tank unit from Moscow to Berlin during the Great Patriotic War! He has recounted that, when he was a young man, he did not know whether he wished to study art or geology but in the end he opted for geology. Yet, his facility for drawing has been retained through his long career, and in this paper we see numerous examples of his quick sketches made during the course of his fieldwork. His special area of interest has been neotectonics; and for the purposes of this Special Publication volume his 'proof' of the occurrence of glaciation in the Pliocene is particularly interesting, as it contributed evidence that was relevant to the changes of definition of the Quaternary and the Pleistocene that were discussed earlier in this introduction. Milanovsky's use of the term 'Eopleistocene' in his stratigraphic table should be noted, as that has been the preferred term in Russia, as opposed to Lower Pleistocene. The present paper is to be understood as an *autobiographical contribution* to the study of the history of Quaternary geology.

Ivanova & Markin, from the Earth Science Museum of Moscow State University, Russia, give an overview of a work by the famous anarchist philosopher, Prince Piotr Kropotkin: *Researches on the Glacial Period* (1876). This book is virtually unknown to Anglophones and there appears to be no copy of it in either the Library of Congress or the British Library. It is, we are told, a rare volume, even in Russia. Kropotkin's book *Mutual Aid* (1902) is well known in the West and makes reference to his travels in Siberia. Early in his career he travelled in that part of the world as a military surveyor, and later he went to Scandinavia, as a result of which he became interested in the 'land-ice theory'. His recognition of the evidence for glaciation was not remarkable at the time of his travels, but his studies of eskers were noteworthy and original. *Researches on the Glacial Period* was written when Kropotkin was in prison in St Petersburg on account of his subversive political views, before he made his dramatic escape to the West. It is pleasing that his scientific investigations, so different from his well-known philosophical and political writings, should be recognized here. Kropotkin's work also shows that there was not a huge difference between 'Eastern' and 'Western' science in his day. Unfortunately, not a great deal is known in the West about early Russian geology and the commonalities are not sufficiently recognized.

Ideas associated with the development of Quaternary research in the Baltic countries are also chronicled by **Gaigalas** from the Department of Geology and Mineralogy at Vilnius University, Lithuania, in terms of their geopolitical position, economic and social conditions, establishment of science centres, progress of geological thought, and the natural environment. Quaternary geological investigations in the Baltic region were, and continue to be, an important link between the East and West in northern Europe. Among the lineage of those involved in Quaternary research, the contributions of the Lithuanian geologist Česlovas Pakuckas (1898–1956) to glaciomorphological investigations of the Baltic marginal highlands in Lithuania and Poland are documented by the paper by Gaigalas *et al.* Pakuckas, who is regarded as a pioneer of modern glaciomorphological investigations, concluded that during the last glaciation the continental glacier was not a single ice sheet but consisted of a number of flows, each dependent on topography and each with its specific glacial centre. A section of organic-rich sediments that he discovered on the banks of the Nemunas River has recently become the stratotype of the Eemian Interglacial in Lithuania.

Kondratienė & Stančikaitė at the Institute of Geology and Geography at Vilnius, Lithuania, evaluate the contributions of another Lithuanian scientist, Valerija Čepulytė (1904–1987). During 46 years of research, she deciphered different aspects of Quaternary stratigraphy, palaeogeography, the extent of the last glacial event (Weichselian) and deglaciation in Lithuania, studies that provided the framework for her 1968 doctorate degree in geology. Čepulytė's primary interests were concerned with the development of geomorphological terminology and the methodology of geomorphological mapping in relation to the last glaciation – aspects that have influenced Quaternary research both in Lithuania and elsewhere.

A former petroleum geologist, **van Veen** at the Department of Earth Sciences, Technical University Delft, The Netherlands, writes about early ideas on glaciation in The Netherlands, using as his 'spyglass' the several essays on ideas about glaciation that were submitted as entries to prize competitions organized by the Hollandsche Maatschappij der Wetenschappen (Holland Society of Sciences) and the Teyler Genootschap (Teyler Society). In the nineteenth century, such prize competitions were popular, with questions being posed about controversial scientific problems, for which there were no known or definite answers. Several prize topics had to do with glacial theory, and eventually they yielded a submission from the Swedish geologist Otto Torell that clinched the land-ice theory for geologists in Germany and northern Europe, although he did not originate this idea. van Veen's paper is helpful in that he delves into a corner of the history of geoscience that can only be studied by someone fluent in Dutch and with access to rare publications in that language.

Zhang, Associate Professor in the Department of Earth Sciences, Sun Yat-sen University, Guangzhou, China, is involved in research in tectonic geomorphology, particularly in the Ordos Plateau area of north China. He documents the punctuated progress of 100 years of investigation of Tertiary (Neogene) planation surfaces in China. This began with the pioneering work of Bailey Willis in 1903–1904, who introduced the Davisian idea of erosion cycles and recognized peneplain remnants in north China. Zhang documents studies of planation surfaces in that part of the world between 1910 and the 1940s, in which scientists arrived at different and similar conclusions as to their ages, and the discovery of similar surface remnants in south China, prompted by the inland migration caused by the southward advance of Japanese forces prior to and during World War II. Since the 1970s, restudy of planation surfaces over the whole of China has been undertaken with the contemporary focus being on uplift of the Tibetan Plateau and its impact on the environment.

Thick marine Quaternary sediments underlying the Kanto Plain of Honshu, Japan are regarded as having been deposited in a 'Palaeo-Tokyo Bay', a concept first proposed by Hisakatsu Yabe (1878–1969) in 1913 and 1914, based on molluscan fossils that record changing sea levels during the glacial and interglacial periods. **Yajima**, from the Tokyo Medical and Dental University, describes how this now-accepted idea came about in conjunction with resolution of the question of Pleistocene glaciation in Japan and a change from Palaeo-Tokyo Bay being open to the east and, at the time of the high sea-level phase, and perhaps also to the south, to modern Tokyo Bay which opens to the south.

Branagan, at the School of Geosciences, University of Sydney, Australia, is the doyen of historians of geology in Australia. His paper gives a comprehensive account of the earliest studies of the rocks of the Australian coastline and, subsequently, in inland regions as explorers pushed into the continent's interior. There they encountered many hardships in the desert regions, but were also able to make numerous observations of Quaternary deposits in well-exposed country. In the first stages, all the information was transferred to 'centres of calculation' (cf. Latour 1987, chap. 6) in Europe and published there. Subsequently, geologists began to settle in Australia and publish there, even though they had mostly been trained in Europe. Thus, Branagan's paper illustrates the first two stages of 'Colonial Science', as envisaged by George Basalla in his threefold classification of the stages of development of science in European colonies or settlements (Basalla 1967). In the early work described by Branagan, much of

Australian geology was, as might be expected, seen through European eyes.

Twidale is a geomorphologist of international standing and Honorary Visiting Fellow in Geology and Geophysics at the School of Earth and Environmental Sciences at Adelaide University, Australia. He is the author of numerous books on geomorphology and has a special interest is desert landscapes, particularly in Australia. His paper provides much information about the character of Australian deserts (which are notably different from those in North Africa or China, say, and can at times provide brief nutriment for abundant life forms) and the early history of exploration in these mostly inhospitable regions. He focuses on the history of the study of the dunes of the Australian deserts, and those of South Australia more particularly. Like Orme, Twidale also discusses dating problems – but as regards individual sand grains using thermoluminescence methods as opposed to the radiometric studies of materials chiefly used in studying the pluvial lakes of the American West. Some of the Australian dunes are only a few thousand years old and thus presumably Holocene in age. The dating of the lake-fill and alluviation of the Australian desert lakes has furnished information about the histories of past climates, which is relevant to biogeography and contemporary problems of climate change.

Oldroyd, Honorary Visiting Professor at the University of New South Wales, Sydney, Australia, is a historian of science who has written chiefly on topics to do with the Earth sciences. In his contribution, he examines the history of ideas about the landforms of the Sydney Basin, with special reference to the patterns of the area's rivers, and considers two important early twentieth-century Australian geologists: Ernest Andrews and Griffith Taylor. Both men were keen disciples of the evolutionary ideas about landforms developed by Davis in the United States, whose ideas were enthusiastically applied in Australasia. From the appearance of a geological map, the geology of the Sydney area appears very simple, but detailed work has revealed numerous difficulties and complexities. Early studies of the area relied heavily on geomorphological considerations, which were used to try to probe the tectonic history of the area. Unfortunately, surfaces that were construed by Taylor and Andrews as being the relics of Davisian peneplains were (we think) depositional surfaces or ones that were generated according to the relative hardnesses of nearly horizontal sedimentary strata. Oldroyd shows that, even now, there is no full consensus about the history of the Sydney Basin, and that debates on the matter still depend to a considerable degree on geomorphological evidence.

Mayer, at the Department of Earth and Marine Sciences, Australian National University in Canberra, is an authority on early scientific voyages to Australia, especially those conducted by French explorers. His paper relates to the Tamala Limestone, a well-known Quaternary formation that crops out extensively along the coast of Western Australia. Work continues on the study of this unit, which presents problems with respect to sea-level and climate change, and the role of tectonic deformations of the crust in that part of the world. It is noteworthy that the early investigators were struck by the indications of changes in the relative levels of land and sea, which was a 'hot topic' in the early nineteenth century. Other matters of interest were the discovery of the 'living fossil', *Trigonia*, which gave comfort to those opposed to the notion of transmutation because it indicated the possibility that various organisms no longer extant in Europe might turn up one day in remote parts of the world; and to certain enigmatic calcareous structures that might have had various origins. It was Darwin who first suggested that the Tamala Limestone was of aeolian origin.

From New Zealand, a biography of the internationally known geomorphologist Sir Charles Cotton (1885–1970) is given by **Grapes**, a New Zealander, now Professor at Korea University, Seoul, and formerly at Sun Yat-sen University, China, and Victoria University, Wellington, New Zealand, where he had first-hand access to and experience of Cotton's legacy. Cotton is perhaps best known for his textbooks, the most influential being *Geomorphology of New Zealand*, first published in 1922, as well as numerous pioneering papers on a great variety of subjects in geomorphology. Although much of Cotton's earlier work followed the ideas of W. M. Davis in terms of an explanatory description of landforms (structure, process, form), he also emphasized, in a qualitative way, the importance of climate change and tectonic movements in landscape-forming processes. Cotton could not have been in a better place to develop these ideas than in New Zealand, a tectonically active country where the continuing relationship between geologically recent Earth movements, denudation, erosion and associated deposition is particularly obvious.

Brook, from the School of People, Environment and Planning, Massey University in the North Island of New Zealand, a geomorphologist interested in geomorphic controls of landscape evolution, writes on the role of an amateur in geology. His paper recounts the pioneering studies of George Leslie Adkin (1888–1964), a farmer, amateur geologist and well-known ethnologist, whose observations provided evidence for localized glaciation and uplift in the axial Tararua Range of the North Island. Adkin's work was published in 1911 and helped resolve a controversy over the extent of glaciation in New Zealand, although his ideas, in this instance, ran counter to those of Charles Cotton. Later studies have broadly vindicated Adkin's work. It has been postulated that cooling of some 4 °C could have occurred during the Pleistocene, resulting in a lowering of the glacier equilibrium line by approximately 670 m, substantially below the crest of the Tararua Range.

We wish to express our sincere thanks to the valuable comments on a draft of this Introduction made by the contributors of this volume. We are particularly grateful to R. Twidale, A. Orme and V. Baker for their considerable input.

References

ADHÉMAR, J. 1842. *Evolutions de la mer: déluges périodiques*. Carilian-Goery & Dalmont, Paris.

AGASSIZ, L. 1840. *Etudes sur les glaciers; par L. Agassiz. Ouvrage accompagné d'un atlas de 32 planches*. Jent et Gassmann, Neuchâtel.

AGUIRRE, E. & PASSINI, G. 1985. The Pliocene–Pleistocene boundary. *Episodes*, **8**, 116–120.

ARDUINO, G. 1760. Due lettere del Sig. Giovanni Arduino sopra varie sue osservazioni naturali. Al chiaris. Sig. Antonio Vallisneri … Vicenza. *Nuovo raccolto d'opusculi scientifici e filologici*, **6**, 99–180.

BAKER, V. R. 1977. Stream channel responses to floods with examples from central Texas. *Geological Society of America Bulletin*, **89**, 198–208.

BAKER, V. R. 1996. The pragmatic roots of American Quaternary geology and geomorphology. *Geomorphology*, **16**, 197–221.

BASALLA, G. 1967. The spread of Western science. *Science*, **156**, 611–621.

BAULIG, H. 1928. *Le Plateau Central de la France et sa bordure Méditerranéan: étude morphologique*. Colin, Paris.

BAULIG, H. 1935. *The Changing Sea Level*. George Philip & Son, London; Philip, Son & Nephew, Liverpool (available in full at: http://www.archive.org/details/changingsealevel033566mpb).

BECKINSALE, R. P. & CHORLEY, R. J. 1991. *The History of the Study of Landforms or The Development of Geomorphology, Volume 3: Historical and Regional Geomorphology 1890–1950*. Routledge, London.

BERGGREN, W. A. 1998. The Cenozoic Era: Lyellian (crono)stratigraphy and nomenclatural reform at the Millenium. *In*: BLUNDELL, D. J. & SCOTT, A. C. (eds) *Lyell: The Past is the Key to the Present*. Geological Society, London, 111–132.

BOWEN, D. Q. & GIBBARD, P. L. 2007. The Quaternary is here to stay. *Journal of Quaternary Science*, **22**, 3–8.

BRANAGAN, D. 2008. Australia – a Cenozoic history. *In*: GRAPES, R. H., OLDROYD, D. R. & GRIGELIS, A. (eds) *Histroy of Geomorphology and Quaternary*

Geology. Geological Society, London, Special Publications, **301**, 189–213.

BUCKLAND, W. 1819. Description of the Quartz Rock of the Lickey Hill in Worcestershire, and the strata immediately surrounding it; with considerations on the evidence of a recent Deluge, afforded by the gravel beds, and state of the plains and vallies of Warwickshire, and the north of Oxfordshire; and of the valley of the Thames from Oxford downwards to London; and an Appendix, containing analogous proofs of diluvian action. Collected from various authorities. *Transactions of the Geological Society, London*, **5**, 516–544 & plates.

BURBANK, D. & ANDERSON, R. S. 2001. *Tectonic Geomorphology*. Blackwell, Malden, MA.

BURT, T. P., CHORLEY, R. J., BRUNSDEN, D., COX, N. J. & GOUDIE, A. S. 2008. *The History of the Study of Landforms or The Development of Geomorphology volume 4: Quaternary and Recent Processes and Forms (1890–1965) and the Mid-Century Revolutions*. Geological Society, London.

CHORLEY, R. J., BECKINSALE, R. P. & DUNN, A. J. 1973. *The History of the Study of Landforms or The Development of Geomorphology, Volume 2: The Life and Work of William Morris Davis*. Methuen, London.

CHORLEY, R. J., DUNN, A. J. & BECKINSALE, R. P. 1964. *The History of the Study of Landforms or The Development of Geomorphology, Volume 1: Geomorphology Before Davis*. Methuen, London.

COTTON, C. A. 1922. *Geomorphology of New Zealand: An Introduction to the Study of Land-Forms. Part 1: Systematic*. Wellington: Dominion Museum.

CROLL, J. 1867. On the excentricity of the Earth's orbit. *Philosophical Magazine*, **33**, (Series 4), 121–137.

CUVIER, J. L. N. F. [G.] & BARON, 1813. *Essay on the Theory of the Earth. Translated from the French of M. Cuvier . . . by Robert Kerr . . . with Mineralogical Notes, and an Account of Cuvier's Geological Discoveries by Professor Jameson*. William Blackwood, Edinburgh; John Murray, London; Robert Baldwin, London.

DARWIN, C. 1887. *The Life and Letters of Charles Darwin, Including an Autobiographical Chapter. Edited by his Son, Francis Darwin. In Three Volumes:– Volume 1*. John Murray, London.

DAVIES, G. L. [HERRIES]. 1969. *The Earth in Decay: A History of British Geomorphology 1578–1878*. London: Macdonald Technical and Scientific.

DAVIS, W. M. 1889. The rivers and valleys of Pennsylvania. *National Geographic Magazine*, **1**, 183–253.

DAVIS, W. M. 1899. The geographical cycle. *Geographical Journal*, **14**, 481–504.

DAVIS, W. M. 1902. Base-level, grade, and peneplain. *Journal of Geology*, **10**, 77–111.

DAVIS, W. M. 1912. *Die erklärende Beschreibung der Landformen [An Explanatory Description of Landforms]*. B.G. Teubner, Berlin.

DESNOYERS, J. P. F. S. 1829. Observations sur un ensemble de dépôts marins plus récents que les terrains tertiaires du bassin de la Seine, et constituant une formation géologique distincte; précédées d'un aperçu de la nonsimultanéité des bassins tertiaires. *Annales des sciences naturelles*, **16**, 171–214, 402–491.

FORBES, E. 1846. On the connexion between the distribution of the existing fauna and flora of the British Isles, and the geological changes which have affected the area, especially during the epoch of the Northern Drift. *Memoirs of the Geological Survey of Great Britain, and the Museum of Economic Geology in London*, **1**, 336–403.

GEIKIE, A. 1962. *The Founders of Geology*. Dover, New York (1st edn 1897).

GEIKIE, J. 1874. *The Great Ice Age and its Relation to the Antiquity of Man*. Appleton, New York.

GERVAIS, P. 1867. *Recherches sur l'ancienneté de l'homme et la période quaternaire*. A. Bertrand, Paris.

GIBBARD, P. L., SMITH, A. G. ET AL. 2005. What status for the Quaternary? *Boreas*, **34**, 1–6.

GIGNOUX, M. 1910. Sur la classification du Pliocène et du Quaternaire de l'Italie du Sud. *Comptes rendus de l'Académie des Sciences, Paris*, **150**, 841–844.

GILBERT, G. K. 1877. *Report on the Geology of the Henry Mountains*. United States of America Geographical and Geological Survey of the Rocky Mountain Region. Government Printing Office, Washington, DC.

GOULD, S. J. 1978. The great scablands debate. *Natural History*, **87**, 12–18.

GRADSTEIN, F., SMITH, A. G. & OGG, J. G. 2004. *A Geologic Time Scale 2004*. Cambridge University Press, Cambridge.

GRAVELIUS, H. 1914. *Flusskunde: Grundriss der gesamten Gewässerkunde* [Rivers: A Sketch of General Hydrology]. G. J. Goschenesche, Berlin.

HACK, J. T. 1960. Interpretation of erosional topography in humid temperate regions. *American Journal of Science*, **258(A)**, 80–97.

HAQ, B. U., BERGGREN, W. A. & VAN COUVERING, J. A. 1977. Corrected age of the Plio/Pleistocene boundary. *Nature*, **269**, 483–488.

HAYS, J. & BERGGREN, W. 1971. Quaternary boundaries and correlations. *In*: FUNNELL, B. & RIEDEL, W. (eds) *The Micropaleontology of Oceans*. Cambridge University Press, Cambridge, 669–691.

HERSCHEL, J. 1830. On the astronomical causes which may influence geological phenomena. *Proceedings of the Heological Society of London*, **1**, 244–245.

HÖRNES, M. 1853. Mittheilung an Professor Bronn gerichtet, Wien. *Neues Jahrbuch für Mineralogie, Geologie, Geognosie und Petrefaktenkunde*, 806–810.

HORTON, R. E. 1945. Erosional development of streams and their drainage basins: hydrophysical approach to quantitative morphology. *Geological Society of America, Bulletin*, **56**, 275–370.

HUBBARD, G. D. 1940. Major objectives of Penck and of Davis in geomorphic studies. *Annals of the Association of American Geographers*, **30**, 237–240.

JACKSON, J. R. 1833. *Observations on Lakes, Being an Attempt to Explain the Laws of Nature Regarding Them*. Bossange, Barthels, Barthels, and Lowell, London.

JOHNSON, D. W. 1940. Memorandum by Douglas Johnson, Columbia University, on the mimeographed outline of the proposed symposium on the geomorphic ideas of Davis and Walther Penck. *Annals of the Association of American Geographers*, **30**, 228–233.

KEILL, J. 1708. *An Account of Animal Secretion, the Quantity of Blood in the Humane Body, and Muscular Motion*. George Strahan, London.

KENNEDY, B. A. 2006. *Inventing the Earth: Ideas on Landscape Development Since 1740.* Blackwell, Oxford.

KING, L. C. 1953. Canons of landscape evolution. *Geological Society of America Bulletin,* **64**, 721–752.

KING, P. B. & SCHUMM, S. A. 1980. *The Physical Geology (Geomorphology) of William Morris Davis: Compiled, Illustrated, Edited and Annotated by Philip B. King and Stanley A. Schumm.* Geo Abstracts (Geo Books Series), Norwich.

KING, W. R. & OAKLEY, K. P. 1950. Report on the temporary commission on the Pliocene–Pleistocene boundary. *In: 18th International Geological Congress, Great Britain, Part 1.* International Geological Congress, London, 213–214.

KROPOTKIN, P. A. 1902. *Mutual Aid: A Factor of Evolution.* Heinemann, London.

LATOUR, B. 1987. *Science in Action: How to Follow Scientists and Engineers Through Society.* Open University Press, Milton Keynes.

LOURENS, L. J. & HILGEN, F. J. 1997. Long-term variations in the Earth's obliquity and their relation to third-order eustatic cycles and late Neogene glaciations. *Quaternary International,* **40**, 43–52.

LYELL, C. 1830–1833. *Principles of Geology, Being an Attempt to Explain the Former Changes of the Earth's Surface, by Reference to Causes now in Operation* (3 vols). John Murray, London.

LYELL, C. 1839. *Nouveaux éléments de géologie,* transl. by Tullia Meulien. Pitoit-Levrault et Compagnie, Paris.

LYELL, C. 1840. On the Boulder Formation or Drift, and associated freshwater deposits composing the mud cliffs of Eastern Norfolk. *Proceedings of the Geological Society, London,* **3**, 171–179.

LYELL, C. 1863. *The Antiquity of Man with Remarks on the Origin of Species by Variation,* John Murray, London.

MASLIN, M. A., LI, X. S., LOUTRE, M.-F. & BERGER, A. 1998. The contribution of orbital forcing to the progressive intensification of northern hemisphere glaciation. *Quaternary Science Reviews,* **17**, 411–426.

MELTON, M. A. 1958. Geometric properties of mature drainage systems and their representation in an E_4 phase space. *Journal of Geology,* **66**, 35–56.

MILANKOVIĆ, M. 1998. *Canon of Insolation and the Ice-Age Problem.* Zavod Udžbenike i Nastavna Sredstva, Belgrade.

MILANOVSKY, E. E. 2008. Origin and development of ideas on Pliocene and Quaternary glaciations in northern and eastern Europe, Iceland, Caucasus and Siberia. *In:* GRAPES, R. H., OLDROYD, D. R. & GRIGELIS, A. (eds) *History of Geomorphology and Quaternary Geology.* Geological Society, London, Special Publications, **301**, 87–115.

MURCHISON, R. I. 1839. *The Silurian System, Founded on Geological Researches in the Counties of Salop, Hereford, Radnor, Montgomery, Caermarthen, Brecon, Pembroke, Monmouth, Gloucester, Worcester, and Stafford; With Descriptions of the Coalfields and Overlying Formations* (3 vols). John Murray, London.

OGG, J. 2004. Introduction to concepts and proposed standardization of the term 'Quaternary'. *Episodes,* **27**, 125–126.

ORME, A. R. 2002. Shifting paradigms in geomorphology: the fate of research ideas in an educational context. *Geomorphology,* **47**, 325–342.

ORME, A. R. 2005. American geomorphology at the dawn of the 20th century. *Physical Geography,* **25**, 361–381.

PARDEE, J. T. 1910. The glacial Lake Missoula. *Journal of Geology,* **18**, 376–386.

PARTRIDGE, T. C. 1997. Reassessment of the position of the Plio–Pleistocene boundary: is there a case for lowering it to the Gauss–Matuyama palaeomagnetic reversal? *Quaternary International,* **40**, 5–10.

PENCK, W. 1924. *Die Morphologische Analyse: Ein Kapitel der physikalischen Geologie.* J. Engelhorns, Stuttgart.

PENCK, W. 1953. *Morphological Analysis of Landforms: A Contribution to Physical Geology. Translated by Hella Czech and Katharine Cumming,* St Martin's Press, New York.

PHILLIPS, J. D. 1999. *Earth Surface Systems: Complexity, Order, and Scale.* Blackwell, Maldon, MA.

PILLANS, B. 2004. Proposal to redefine the Quaternary. *Episodes,* **27**, 127.

PILLANS, B. & NAISH, T. 2004. Defining the Quaternary. *Quaternary Science Reviews,* **23**, 2271–2282.

PLAYFAIR, J. 1802. *Illustrations of the Huttonian Theory of the Earth.* Cadell and Davies, London, William Creech, Edinburgh.

RIO, D., SPROVIERI, R., CASTRADORI, D. & DI STEFANO, E. 1998. The Gelasian Stage (Upper Pliocene): a new unit of the global standard chronostratigraphic scale. *Episodes,* **21**, 82–87.

SCHEIDEGGER, A. G. 1965. The algebra of stream-order numbers. *United States Geological Survey, Professional Paper,* **525-B**, 187–189.

SCHUMM, S. A. 1956. The role of rainwash and creep on the retreat of badland slopes. *American Journal of Science,* **254**, 693–706.

SCHUMM, S. A. & LICHTY, R. W. 1965. Time, space, and causality in geomorphology. *American Journal of Science,* **77**, 397–414.

SHACKLETON, N. J. 1997. The deep-sea record and the Pliocene–Pleistocene boundary. *Quaternary International,* **40**, 33–36.

SHEPHERD, R. G. & ELLIS, B. N. 1977. Leonardo da Vinci's tree and the law of channel widths – combining geomorphology and art. *Journal of Geoscience Education,* **45**, 425–427.

SHREVE, R. L. 1966. Statistical law of stream numbers. *Journal of Geology,* **74**, 17–37.

STODDART, D. R. (ed.) 1997. *Process and Form in Geomorphology.* Routledge, London.

STRAHLER, A. N. 1952. Dynamic basis of geomorphology. *Bulletin of the Geological Society of America,* **63**, 923–938.

SUC, J.-P., BERTINI, A., LEROY, S. A. G. & SUBALLYOVA, D. 1997. Towards the lowering of the Pliocene/Pleistocene boundary to the Gauss–Matuyama reversal. *Quaternary International.* **40**, 37–42.

SUESS, E. 1906. *The Face of the Earth (Das Antlitz der Erde)* transl. by Hertha B. C. Sollas, Volume 2. Clarendon Press, Oxford.

SUGUIO, K., SALLUN, A. E. M. & SOARES, E. A. A. 2005. Quaternary: 'Quo vadis'? *Episodes*, **28**, 197–200.

TINKLER, K. J. 1985. *A Short History of Geomorphology*. Croom Helm, London.

TINKLER, K. J. (ed.) 1989. *History of Geomorphology: From Hutton to Hack*. Unwin Hyman, Boston, MA.

WALKER, H. J. & GRABAU, W. E. (eds) 1993. *The Evolution of Geomorphology: A Nation-by-Nation Summary of Development*. John Wiley & Sons, Chichester.

WALKER, M. & LOWE, J. 2007. Quaternary science 2007: a 50-year retrospective. *Journal of the Geological Society*, **164**, 1073–1092.

WILCOCK, P. R. & EVERSON, R. S. (eds) 2003. *Prediction in Geomorphology*. American Geophysical Union, Geographical Monograph, **135**.

WOLMAN, M. G. & GERSON, R. 1978. Relative scales of time and effectiveness of climate in watershed geomorphology. *Earth Surface Processes*, **3**, 189–208.

WOLMAN, M. G. & MILLER, J. P. 1960. Magnitude and frequency of forces in geomorphic processes. *Journal of Geology*, **68**, 54–56.

WOOLDRIDGE, S. W. 1953. The progress of geomorphology. *In*: TAYLOR, T. G. (ed.) *Geography in the Twentieth Century: A Study of Growth, Fields, Techniques, Aims and Trends*. Methuen, London; Philosophical Library, New York, 165–177.

WRIGHT, G. F. 1889. *The Ice Age in North America, and its Bearings upon the Antiquity Of Man, . . . With an Appendix on 'The Probable Cause of Glaciation', by Warren Upham*. D. Appleton, New York.

ZHANG, K. 2008. Planation surfaces in China: one hundred years of investigation. *In*: GRAPES, R. H., OLDROYD, D. R. & GRIGELIS, A. (eds) *History of Geomorphology and Quaternary Geology*. Geological Society, London, Special Publications, **301**, 171–178.

Questions of periodization and Adolphe von Morlot's contribution to the term and the concept '*Quaternär*' (1854)

MARIANNE KLEMUN

Institut für Geschichte der Universität Wien, Dr Karl Lueger-Ring, 1010 Vienna, Austria (e-mail: marianne.klemun@univie.ac.at)

Abstract: Questions concerning periodization in geology are obviously still with us, and the same goes for the relationships of time, change and discontinuity. The fact that such questions are debated repeatedly in both history and geology is illustrated by the extensive discussion in recent years about the use of the term 'Quaternary' as a stratigraphic unit. Thus periodization is not merely a philosophical issue. Neither does it belong solely to the sociology or politics of science. Rather it must be seen as an essential instrument and an integral part of an on-going discussion of fundamental ideas about time in general.

Several texts state that it was Adolphe Morlot (1820–1867) who coined the term 'Quaternary', but in fact there were earlier usages, with different meanings. This paper discusses not so much the 'invention' of the term Quaternary, but its range of meaning during the early phase of its introduction and development, in order to give and appropriate categorization of Morlot's specific contribution and the reason why he introduce the term *Quaternär*. The discussion is based to a considerable extent on correspondence between Morlot and Friedrich Simony (1813–1896) of Vienna University.

Reforms of periodization constitute a vital element in geology as well of history, with a recent controversy being that concerning the subdivision of the Cenozoic Era. The controversies centring on nomenclature and the adoption of certain terms such as Palaeogene and Neogene – in preference to the terms Tertiary, Quaternary and Pleistocene – provide ample illustration (Gibbard *et al.* 2005). As is well known, the terms Quaternary and Pleistocene were coined in the nineteenth century. In the course of time, however, they have been associated with various concepts and meanings. Arising from the present controversies about the term Quaternary as a chronostratigraphic unit, as William Berggren (1998) stressed the necessity of looking at the histories of the terms and analysing their specific connotations and their various contemporary levels of meaning: '[w]hat appears to be absent from the debate is an awareness of the history and variability of the terms in question, that might suggest points of view that are not apparent from the trenches' (Berggren 2005, p. 1). Thus, discussing the invention and establishing of the term Quaternary and its range of meanings in a comparative way seems to be a worthwhile exercise. The focus on the first phase of the introduction and development of the term Quaternary, in the context of broader geological concepts, in order to be able to categorize Morlot's contributions seems justified for it was with Morlot that the conceptualization of the term and the periodization with which it was linked in German-speaking countries began, and priority has been given to him – particularly in German sources (e.g. Wagenbreth 1999, p. 117) – for the use of the term. This paper will seek to assess Morlot's contribution in terms of its wider context.

The beginnings in France: the first appearance of the term 'Quaternary'

The term Quaternary (*Quaternaire ou Tertiaire recent*) was proposed by Jules Pierre François Stanislaus Desnoyers (1800–1887) in 1829 as the fourth, and final, subdivision of the previously proposed threefold subdivision of the geological record (Primary, Secondary and Tertiary) for the rocks in the Loire–Touraine Basin and Languedoc, which were demonstrably younger than those of the Seine–Paris Basin (Desnoyers 1829). Desnoyers was not only a geologist but also an historian and archaeologist. He was Secretary of the Société de l'histoire de France, and in the year after the publication of his paper on the Quaternary he was one of the founding fathers of the Société géologique de France in Paris. Desnoyers coined the term 'Quaternary' to describe a sequence of rocks apparently younger than what were then regarded as the youngest Tertiary deposits. He discovered that the marine sandy deposit (known as 'Faluns'), which was filled with sea shells and coral, and the marls near Tours in the Loire Basin, overlay a freshwater deposit that constituted the highest subdivision of the Paris group and extended without further subdivision over the plateau between the Seine and Loire Basins.

From: GRAPES, R. H., OLDROYD, D. & GRIGELIS, A. (eds) *History of Geomorphology and Quaternary Geology.*
Geological Society, London, Special Publications, **301**, 19–31.
DOI: 10.1144/SP301.2   0305-8719/08/$15.00 © The Geological Society of London 2008.

That was the starting point for the argument that a new stratigraphic category (or unit as we would say) had to be defined; and 'Quaternary' was chosen:

[L]a série des terrains tertiaires s'est prolongée, et même a commencé dans les bassins plus nouveaux, long-temps peut-être après que celui de la Seine a été entièrement comblé, et que ces formations postérieures, *Quaternaires* pour ainsi dire, ne doivent pas plus conserver le nom d'alluvions que les vrais et anciens terrains tertiaires, dont il faut également les distinguer.

(Desnoyers 1929, p. 193)

Desnoyers subdivided this unit into three parts, from younger to older: (3) Recent; (2) Diluvium; (1) 'Faluns de Touraine' (shelly marls of the Touraine), etc. In retrospect, it can be seen that this original definition included what Charles Lyell (1797–1875) was later to include in his Miocene, Newer Pliocene and Recent (Lyell 1830–1833). Subsequently, Lyell proposed the term 'Pleistocene' for his younger Pliocene in an appendix to a French translation of his *Elements of Geology* (Lyell 1839). So Desnoyer's 'Quaternary' was not the same as that which followed and continues into modern times.

In Lyell's writings, the change from a biostratigraphic to a chronostratigraphic unit was completed over many years (Berggren 2005, p. 1). He subsequently modified his chronology by defining the most recent phase as 'Post-Tertiary' and dispensing with the term Pleistocene. But he was, in any case, aware of the significance of Desnoyer's advance and said of it that 'a new chronological system had been founded' (Lyell 1857, p. 146). Lyell judged this method to be analogous to the then much-praised method of establishing species in botany and zoology. However, he emphasized that the difficulty of the determination of species and the obvious artificiality of the system of classification would increase as more facts became known (Lyell 1857, p. 147).

It is interesting to note that Desnoyers used the term 'Quaternairies' very sparingly in his 134-page paper. Nevertheless, the name and meaning of the term were quickly adopted, although subsequently modified. Marcel de Serres (1783–1862), Professor at the University of Montpellier, used the term '*Quaternaire*' in 1830, considering it to be synonymous with the term 'Diluvium', as first proposed by Gideon Mantell (1790–1852) in 1822 and William Buckland (1784–1856) in 1823 for deposits of the Biblical Deluge (Serres 1830; Berggren 2005, p. 7). It was subsequently used in this sense by German geologists. For example, in 1844 Carl Cäsar von Leonhard (1779–1862) discussed whether diluvial debris belonged 'to the last of the great upheavals' and mentioned the fact that there were investigators who classified such upheavals as evidence of different periods'

(Leonhard 1844, pp. 232 and 233). Moreover, the term Quaternary was given a faunal connotation by Henri Reboul (1833) (1763–1839). He distinguished it as containing living species of animals and plants, as opposed to the Tertiary, which was believed to hold mostly, if not exclusively, extinct species (which was not, of course, Lyell's view).

De Serres first 'introduced' prehistoric humans into the concept of Quaternary, suggesting that the first evidence for them was to be found in that period. After the excavations at the Grotto of Aurignac by Edward Lartet (1801–1871) in 1863, this 'archaeological' aspect was reinforced. Such works show that there were various quite different criteria for characterizing the Quaternary in the few years of its existence.

The Viennese palaeontologist Moriz [*sic*] Hörnes (1815–1865), Curator of the Mineralogical Collection of the Royal Museum of Natural History in Vienna, observed in 1853 that Miocene and Pliocene molluscan fauna had closer affinities to one another than to the Eocene faunas. He therefore wanted to refer to them collectively as Neogene. In creating this term, Hörnes referred specifically to the stratigraphic subdivisions of the Molasse Mountains – proposed in 1838 by Heinrich Georg Bronn (1800–1862), Professor of Zoology at Heidelberg – and to the younger 'Alluvial' subdivision (Hörnes 1853). In Hörnes' original definition the Neogene spanned the time interval from the Miocene to Pliocene and up to Recent.

Adolphe Morlot: biography and an outline of the secondary literature

According to the Polish/French cultural historian, Krzysztof Pomian, every period is determined in two ways: by facts and by concepts (Pomian 1984). So far, the criteria for the concepts that we have been discussing have been essentially biostratigraphic or biochronological in nature. However, Adolphe Morlot's concept, which was based on stratigraphy and climatology, also included the then controversial notion of the glacial theory and differed from the biostratigraphic approach.

In several texts, Morlot is mentioned in various contexts as far as his involvement with the initial use of the term Quaternary is concerned. William Sarjeant simply asserted that Morlot was the 'originator of the term "Quaternary"' (Sarjeant 1980, vol. 3, p. 1751), not mentioning others before Morlot at that point in his bibliography. Moreover, Sarjeant failed to state what Morlot *meant* by the term.

In the introductory part of his chapter on Quaternary geology in his *Geschichte der Geologie in Deutschland* [*History of German Geology*],

Otfried Wagenbreth (1999, p. 117) emphasized that the 'most recent layers of the surface of the earth were called Quaternary by the Swiss scholar A. v. Morlot as an extension of the term "Tertiary"' ['Die jüngsten Schichten der Erdoberfläche wurden von dem Schweizer A. v. Morlot in Fortführung des Begriffes "Tertiär" als Quartär bezeichnet']. In the *Geschichte der Geologie und Paläontologie* [*History of Geology and Palaeontology*] by Karl Alfred Zittel (1899, p. 717) it was briefly stated that: 'in 1839, Lyell suggested the term Pleistocene for Buckland's Diluvium; in 1854 Morlot's suggestion was Quaternary' ['Lyell hat 1839 für das Buckland'sche Diluvium den Namen Pleistocän, Morlot 1854 die Bezeichnung Quaternär'].

First, in anticipation of my further investigation, it should not be thought that I am arguing from the vague statements cited above that it was Morlot who gave the term 'Quaternary' an 'official' character or, less dramatically, that he was responsible for coining the name in 1854: the term Quaternary was already accepted in German textbooks (Morlot 1854*a*). In that same year, for example, Carl Vogt (1817–1895), in an overview in his geology textbook, gave the Pleistocene Era a further name, that of 'Quaternary Formations' (tidal deposits, loess, erratic phenomena) (Vogt 1854, p. 624). But Morlot gave the term a new twist – namely a *glacial* connotation. (However, Edward Forbes (1846) had previously given a glacial signification to the term Pleistocene – which Lyell opposed. And if not ascribed to the Deluge, erratics were generally associated with glaciation in some way.)

The differing statements on the importance of Morlot's contribution to the introduction of the Quaternary in German-speaking countries made me curious, especially as there was not only a lack of precision but also different ways of understanding his concept. But through them I became acquainted with Morlot as an interesting personality and his work increasingly attracted my interest.

Adolph von Morlot (1820–1867), the son of a Bernese patrician family, studied mathematics, first in his home town and then in Paris at the Collège St Barbe. His lecture notes in algebra and geometry from 1838 to 1841, which are among his Berne papers, are evidence of his excellent education (*Burgerbibliothek Bern, Manuscripta Historia helvetica* **45**, 23). From Paris, Morlot proceeded to Freiberg in order to study mining. In 1844 Carl Friedrich Naumann (1797–1873) travelled in the area around Rüdersdorf in Saxony, prompted by Carl Bernhard von Cotta (1808–1879), who was fascinated by the glacial striae in the Swiss Jura, and was accompanied by the 24-year old Morlot. While Naumann (1844) was hesitant in ascribing the striae to some previous glaciation, Morlot, who was familiar with such

phenomena and the discussions of them from his home country, Switzerland, went a step further and solved the question by means of a bold hypothesis. To him, findings such as boulders and the so called 'Gletschertöpfe' ('glacier pots'), as well as the scratches on the backs of shells, were evidence of a former glaciation of the entire region. He postulated that a Scandinavian glaciation had extended southwards to central Saxony in historical times. This fact has previously been remarked by Eiszmann (1974), who called Morlot the founder of the theory of inland glaciation. However, this theory was not credited to Morlot in his own time.

The idea and theory of the Ice Age

The impulse for the general theory of an Ice Age started in Switzerland with the work of a construction engineer from Wallis, Ignatz Venetz (1788–1859). He was responsible for repairs to bridges and roads, and during his official travels he found stone rubble and moraine in many places far from modern glaciers (Truffer 1990, pp. 7–27; Roten & Kalbermatter 1990, pp. 33–46). He therefore proposed that they could be related to a former period of glaciation of considerably greater scope than that known in Switzerland at present (Kaiser 1990). Venetz also gave a glacigenic explanation for the existence of erratic blocks. He lectured on his ideas in 1822, but his innovative explanation was only published in 1833 (Venetz 1833). These thoughts were taken up by Jean de Charpentier (1786–1855), Director of the salt mines in Bex. He collected evidence for this theory, such as the accumulation of erratic blocks, and he, too, was soon voicing his ideas, at meetings of the Lucerne Society (Charpentier 1836, 1837), although in the face of opposition or even derision from others in the scientific community.

From 1832 Louis Agassiz (1897–1873) taught natural history at the new college (not then a university) in the little town of Neuchâtel, and developed into the leading authority on fossil fish. When Agassiz visited his former teacher, de Charpentier, in Bex in the summer of 1836 and met up with Venetz, he accepted the invitation to examine the moraines and erratic blocks. Agassiz had become dissatisfied with the older explanations (or non-explanations) and was converted, so to speak, to the group of Ice Age proponents. What was important now was to press on with the collection of the countless indications of glacial activity in the Rhone Valley and in the Jura, in order to have sufficient facts to support the argument. There was soon a vigorous debate between British and Swiss geologists, and joint excursions were undertaken in Scotland and Switzerland, with Buckland being a convert to 'Agassiz's theory' (as it soon became). But Lyell was only finally convinced as late as 1857.

The specific term 'Ice Age' was coined by the German botanist Karl Schimper (1803–1867). Agassiz gave it a 'catastrophist' connotation, which he supported with appropriate linguistic imagery. For example, he wrote: 'Accordingly, the phenomenon of temperature reduction at the end of a geological period could be viewed as analogous to the stiffening [or rigor] that occurs on the death of an individual, and the increase in temperature as parallel to the development of a characteristic warmth as a being comes into existence' (Agassiz 1841, p. 396). This was, in a way, a form of the ancient macrocosm/microcosm analogy, although Agassiz probably only used the comparison as a rhetorical device.

Personal contact between Morlot and de Charpentier can be proved by the exchange of letters between them that are preserved from 1844 (Burgerbibliothek Bern, *Manuscripta Historia helvetica*, **45**, 9/No. 126). With Agassiz, Morlot seems to have started personal contact in 1844, as is documented in his personal correspondence (Burgerbibliothek Bern, *Manuscripta Historia helvetica*, **45**, 9/No. 2).

The theory of (in)land glaciation underwent the same fate as the Ice Age theory: neither was generally accepted at first. However, the theory of inland glaciation subsequently received general approval, especially when it was discussed again in Otto Torell's lecture to the German Geological Association in Berlin in 1875 (Torell 1875) (see van Veen 2008 in this volume). In 1875 Torell also referred to Rüdesheim (Torell 1875, 1880), a site that was important for Morlot in 1844 and which had in the meantime become famous and was visited by Lyell in 1851, and to the Hohburger Berge (Hohburg Mountains) near Wurzen. In that year (1875), if not earlier, the 'drift theory' collapsed. (This theory went back to Murchison and Lyell, and postulated a sea reaching from Scandinavia to central Europe and transporting icebergs southwards, leaving behind boulders.)

In 1854, in a letter to Friedrich Simony (1813–1896) – the leading geographer and glaciologist in Vienna and Professor of Geography at the University of Vienna – Morlot described his version of the development of the outdated theories as a dialogue that took place between de Charpentier and Léonce Élie de Beaumont (1798–1874). This document shows how Morlot ridiculed Lyell's theory of 'drift' or Leopold von Buch's (1774–1853) 'mud theory' (diluvium). Morlot found it amusing that, during an expedition, de Charpentier called out to Élie de Beaumont in jest:

'Watch the blocks swimming', while in reality they were rather rolling over the smaller Polish plains in the rather shallow river, just as the big block of the Pallas statue [Pallas Athene, also *Athene Parthenos*, the Greek goddess] is rolling on metal balls to Petersburg. E. de Beaumont took this joke seriously & the learned orthodox world that rejects the well-founded glacial theory, believes the utter stupidity without a second's thought: that stone can swim in a mixture of the same stone with water!
(Morlot to Simony, 8 January 1854, ÖNB, Vienna, 463/28–17)

While the 'drift' theory had a substantial number of followers in German-speaking countries, at first no one believed in the glacial theory in the form suggested by de Charpentier and Agassiz (Agassiz 1841; Charpentier 1841). This is why Morlot's astute study of 1844 on inland glaciation was quite courageous. And up to the turn of the century (1900), geologists sometimes rejected glacial phenomena, even when they referred to Morlot's correctly interpreted classical site in Saxony.

The young, well-educated and aspiring geologist Morlot was attracted to Vienna after his time in Freiberg (i.e. after 1845), where, together with renowned investigators such as Ami Boué (1794–1881), Wilhelm Haidinger (1795–1871) and Franz von Hauer (1822–1899), he belonged to a group that, in connection with the Mining Museum at the Royal and Imperial Chamber for Minting and Mining, no longer did geological work privately but instead at the public's expense. Thus, the discipline was free to expand in the Habsburg Empire, and Vienna began to develop a favourable climate for the sciences in general and geology in particular, especially after the revolution of 1848, when the university was reformed and several associations and institutions for the natural sciences were founded.

Morlot, who came to Vienna with the recommendation of his teacher Bernhard von Cotta, was soon employed (in 1846) by the Geognostic Mining Society of Inner Austria (=Carinthia, Styria, Carniolia and Istria), and was responsible for geological mapping in the Austrian Alps from 1846 to 1850 in the course of the first pre-state geological mapping of Inner Austria. He soon published a large number of small but important papers (Morlot 1847*a*, 1848, 1850*a–f*) in the course of his employment, and especially a geological map of the Eastern Alps (Morlot 1847*b*). But he was unsuccessful in joining the Geologische Reichsanstalt in Vienna (founded in 1849, the first state-funded Geological Survey of the Habsburg Monarchy). He would undoubtedly have liked to participate in the mapping of the countries belonging to the monarchy. But having lost the opportunity of employment in Vienna, he returned to Switzerland, where he published further geological papers and then, after 1860, dedicated himself to archaeology. Respecting the wishes of his father, who found him a position (a fact revealed by one of Morlot's letters to Friedrich Simony), Morlot first obtained a post as Professor of Geology at the Academy and the College in Lausanne (1851–1853), before he retired from teaching. After 1860, he devoted himself to archaeology (Morlot 1859*a*, 1865).

Morlot and periodization: the connections between geology and archaeology, archaeology and ethnology, geology, and evolution

Just as the discussions on the origin of the glaciers often referred to northern Europe, the origin of mankind was also linked to that region, and archaeology was mainly professionalized in Denmark (Morlot 1860*a*). Perhaps the common root for Morlot's fascination with both types of phenomena can be found there. On his journeys through Carinthia he had already shown an interest in Celtic culture (Morlot to Simony, 19 November 1848, ÖNB, Vienna, 463/28–1). Subsequently, Morlot transferred his knowledge of strata, as basic units for structuring knowledge, from geology to archaeology and pre-history, respectively, and applied this to the first Celtic pile dwellings at the Lake of Zurich (Zürichsee), which were discovered by Ferdinand Keller (1800–1881) (Keller 1865–1866). It was also Keller who, in 1832, used the illustrations of a travel report on a pile-dwelling village in (what is now) Papua New Guinea, called Dorei, for a visual reconstruction of such lake-dwelling cultures (Kauz 2000). Thus, a connection was established between the stages in historic cultures and present-day ethnology. Morlot encapsulated this way of thinking, emphasizing the analogies between physical geography and geology, on the one hand, and ethnology and archaeology, on the other:

Ethnology is to ancient history what physical geography is to geology, namely a set of signposts or a thread to guide us through the complicated realm of the past, or a fixed departure point for any comparative research that has as its goal the understanding of humans and their development. (Morlot 1859*a*, p. 6)

Transferring structurally founded elements from one field of knowledge into another one is an interesting process, and one that is often at work in historical investigations. The connection between geology and archaeology, as well as the *borrowing* from the geological repertoire in order to establish archaeology – something that Morlot called for in his archaeological work (Morlot 1859*a*) – has not received much discussion from historians of science (but see Rudwick 1997 on 'transposed concepts'). But this phenomenon need only be mentioned briefly here, as Morlot was only concerned with a geologically based determination of age. The starting point for his arguments in archaeology or ancient history was the work of Lyell and ethnology, which could be used to demonstrate stages in a culture. As in geology, layered artefacts found their positions according to the geological principle of superposition. This was a useful preliminary induction, but the methods taken over from geology remained rather inaccurate. So Morlot, who – unlike some of his contemporaries – initially kept prehistoric man out of his geological work (Morlot 1854*a*), dedicated the last phase of his life to the periodization of early human history. A letter to Franz Hauer (1822–1899), the first geologist of the Geological Survey in Vienna, demonstrates how deeply interested Morlot was, as early as 1849, in the exact determination of geological and historical periods:

Now we have found the thread (*Faden*) for measuring, or at least estimating, the years of the geological periods. They are obviously connected to the duration of the species. We have the total duration of mankind, i.e. also that of the present period, which has a certain connection and relation to the previous one. – exactly in harmony with the hint I obtained eight years ago from Agassiz. [Nun ist der Faden gefunden zur Messung, oder vorderhand nur wenigstens zur Schätzung in Jahreszahlender geologischen Perioden. Sie hängen offenbar mit der Dauer der Spezies zusammen, die Totaldauer des Menschengeschlechts haben wir, also auch die der gegenwärtigen Periode, die in einem bestimmten Zusammenhang Verhältnis zu der früheren steht, Ganz nach dem Fingerzeig, den ich vor acht Jahren von Agassiz erhielt.]
(Geologische Bundesanstalt Wien, Morlot to Hauer, 3 May 1849)

The exact categorization, which, according to his notes (Burgerbibliothek Bern, *Manuscripta Historia helvetica*, **45**, 21/No.10, fol.1/7) Morlot called *absolute chronology*, and which was also discussed by Lyell, had the function of establishing 'grades' or steps in the unified process of 'perfection of civilization', by analogy with evolution. In this respect, however, Morlot thought along a different track, compared to Lyell, and his ideas about change could be described as directionalist or progressionist. In geology, however, Morlot argued along non-directionalist lines.

Archaeology, together with anthropology and researches in history, had long sought to prove and support the theory of progression. When, no later than the middle of the nineteenth century, the narrow time frame of the history of humans was broken open by finds of fossilized hominids and artefacts, and chronological calculations expanded dramatically, the 'revolution of time' (Stocking 1968) took place. Those pre-historic discoveries (Van Riper 1993) deserved the status of an 'inaugurating moment' along the path to Darwin's notion of the 'origin of species'. Evolutionary theory experienced a breakthrough insofar as the necessary time span was provided by expanding the geological timescale, as was necessary in order to locate the progressing development within the realm of human understanding. Both fields, geology and archaeology, were linked in Morlot's understanding of the question of periodization when accessing the past. He was concerned with the internal differentiation and subdivision of the larger historical epochs and argued successfully for the division of one of the epochs of ancient history, the Hallstatt Period, into two different phases; the older and younger 'Irontime' – or Iron Age (Morlot 1859*a*). This crude division was accepted and has been retained through to the present.

Morlot's connection between stratigraphy and glaciology: the 'double glaciation'

Let me now return to the aspect of periodization and Morlot's research on the Quaternary. As mentioned earlier, when I began my investigation of this topic I read in several places that Morlot had coined the expression 'Quaternary'. But we recall that French works (i.e. the publications of Desnoyers, Serres and Reboul) were clearly ahead of Morlot's suggestion. Morlot, of course, knew this French literature, including the text that I mentioned at the beginning of this paper. He did not refer to it *expressis verbis*, but he used it, as he coined the German term 'Quaternär' (which has obvious linguistic analogy to the French word 'Quaternaire'), something for which we find evidence in handwritten notes (paper inside a letter from Morlot to Simony in 1854, ÖNB, Vienna, 463/28: see Fig. 1) and his publications in 1854 (Morlot 1854*a*, 1859*b*). The fact that as late as 1897 Meyer's renowned *Encyclopedia of Conversation* still gave 'Quartär' as a synonym for 'Quaternär' shows the lasting influence this term enjoyed – which is unusual for the German language – although in 1859 in his review journal, Bronn had already adapted Morlot's term to the German language and *changed it* to 'Quartär' (Bronn 1855, p. 719).

For more than a decade, between 1842 and 1860, Morlot was deeply interested in traces of glaciers in the 'erratic period', as he called it, a fact that is amply documented in his correspondence with Simony, who, in turn, was greatly interested in contemporary glaciology. Simony, initially Warden of the Provincial Museum in Klagenfurt (see Klemun 1992) and after 1849 incumbent of a newly created chair of Geography at the University of Vienna, was particularly interested in movement in the ice cover, which he considered to be a cause of the physical configuration of the Salzkammergut. Morlot's observation that 'diluvial' layers in the Rhone Valley, the classical place for de Charpentier's and Agassiz's researches, are overlain *and* underlain by erratics prompted him, in

1854, to develop the idea of two periods of glaciation. But in the handwritten notes he sent to Simony on the periods of glaciation, Morlot attached no special importance, in his initial thoughts and arguments, to the localities.

Starting tours from Lausanne in 1854, Morlot's first area of analysis was a locality on the Baye de Clarins River on the northern side of Lac Leman, near the village of Tavel, not far from Montreux. There he found terraces of the 'older diluvium' [= alluvial sediment] overlain by 'erratics' [glacial deposits]. He assumed that the alluvial sediment was older than the other glacial deposits found in other localities around Lake Geneva. This observation he presented at a Session of the Société Vaudoise de Science Naturelles (Morlot 1854*b*, p. 39, 1854*c*, pp. 41–45, 1860*b*, pp. 830–838). When Morlot described his observations to the famous Alpine geologist and professor at the newly founded technical university in Zurich, Arnold Escher von Linth (1807–1872), he defended his idea of a two-phase glaciation by asking: if this material were from the Wallis, 'how could it have crossed Lake Geneva without the help of glaciers?' (Wissenschaftshistorische Sammlungen der Eidgenössischen Technischen Hochschule Zürich, Morlot to Escher, No. 1272, 20 February 1854). He followed the bed of the Baye de Clarens River for about 300 feet from the bridge. There he found glacier scratches that were 'very distinct and strong, having a perpendicular position in relationship to the direction they should have might be expected'. And he came to the conclusion that:

here, we have an example of superimposition of the so-called older diluvium onto very characteristic erratics [glacial deposits] caused by magnificent exposure, a finding that is due to the fact that Rock X protected the erratics from being washed away by the turbulent river. Such observations must be very rare, first because they are restricted to the weaker strata of the diluvium, and second because the diluvial rivers wash away and change the erratics almost everywhere. One must be careful not to confuse the actual diluvium with the *alluvion glacière*, which formed numerous terraces in the Wadtland.

(Wissenschaftshistorische Sammlungen der Eidgenössischen Technische Hochschule Zürich, Morlot to Escher, No. 1272, 20 February 1854)

From today's point of view, Morlot had a very important idea and was capable of good observation, but his conclusions and observations on the subject of this locality met with no response; And according to Henri Masson (pers. comm.): 'In fact this locality does not show a superposition of two glacial deposits, but the superposition of an alluvial deposit on the glacial one, a thing which is indeed quite common'.

Nevertheless, Morlot looked further for evidence that would support his idea, as he was not yet fully satisfied with his solution to the problem.

Über die quaternäre Periode in der Schweiz
von A. v. Morlot.

Es wurde bisher angenommen, dass das erratische Gebilde jünger sei als das sogenannte ältere Diluvium, oder schlechtweg das Diluvium; und in der That sieht man, z.B. bei Genf die entsprechende Lagerungsfolge, was auch bei Heidelberg und Wien der Fall ist, da das erratische Lössgebild dort auf dem Diluvium liegt, ebenso in der Lombardei. — Allein eine Schwierigkeit ergibt sich aus dem Umstand, dass das Material der Diluvialterrassen auch an solchen Stellen, gänz alpinisch ist, wo es doch durch gewöhnliche Strömung aus den Alpen nicht hingelangen konnte, z.B. bei Morsee am Genfersee, dann erweist sich der erratische Till des Nordens als älter als von den alten Meeresuferlinien, welche offenbar unserem Diluvium entsprechen, überlagert. Andererseits hat man sowohl in Schottland als in der Schweiz zwei verschiedene Phasen der Gletscherzeit nachgewiesen. Nun ist bei Vivis an einer sehr schön entblössten Stelle die deutliche Auflagerung des Diluviums auf einem ganz charakteristischen erratischen Gebilde zu sehen und wir gelangen daher zu dem Resultat, dass die zwei Phasen der Gletscherzeit durch die Diluvialperiode getrennt waren, eine Periode während welcher die Gletscher nicht nur die Niederungen, sondern auch alle Hauptalpenthäler verlassen hatten, da sich die Diluvialterrassen bis weit in dieselben hinauf verfolgen lassen.

Wir gelangen so zur folgenden Unterabtheilung der quaternären Epoche.

1º. Erste Gletscherperiode, diejenige ihrer grössten Ausdehnung, wo die niedere Schweiz fast ganz mit Eis bedeckt war und der Rhonegletscher z.B. die ihm von H. von Charpentier in seinem klassischen Werk zugewiesene Ausdehnung hatte. Scheint nicht von langer Dauer gewesen zu sein.

2º. Diluvialperiode. Die Gletscher sind ganz verschwunden, auch aus allen Hauptalpenthälern. Der Elephant lebt in der Schweiz. Nach den entsprechenden Wildbachschuttkegeln am Genfersee zu urtheilen mindestens von ebenso langer

463/28 — 23

Fig. 1. Handwritten paper in a letter from Morlot to Friedrich Simony of 1854, ÖNB, Vienna, 463/28, in which he proposed his ideas of a double glaciation. A translation of this paper is given in the Appendix.

Dauer als die moderne Periode, also nach Lyell über 60000 Jahre dauernd.

3⁰. _Zweite Gletscherperiode_, diejenige ihrer mindern Ausdehnung; die niedere Schweiz war zum größten Theil frei von
Eis; der Rhonegletscher z. B. den Jorat nicht überschrei-
tend und den Fuss des Jura nicht erreichend hielt sich
im Genferseebecken, dasselbe einige hundert Fuss hoch
über den gegenwärtigen Spiegel des Sees ausfüllend und
gegen Genf zu ~~sein Ende~~ endigend. Bildungszeit des
Lösses. Der Elephant lebt im Lande. Nach den entsprechenden Ablagerungen zu ~~nicht~~ schliessen von langer Dauer.

4⁰. _Moderne Periode_. ~~Der~~ Continent hebt sich um ein
Geringes und die Flüsse graben sich einen tiefern Lauf
in ihren früheren Diluvialanschwemmungen aus, so dass
diese als Terrassen hervortreten. Der Elephant verschwin-
det und der Mensch erscheint.

Fig. 1. _Continued._

In 1857 he found a second locality situated on the southern side of Lake Geneva, about 6 km SE of the town of Thonon, on the Dranse River, and this was a good place for his observation of the superposition so as to demonstrate a double glaciation (Morlot 1859*c* printed in 1861).

This place near Thonon soon became well known and the botanist Oswald Heer (1809–1883) referred to it in his book *Die Urwelt der Schweiz* [*On the Primeval World of Switzerland*], giving us an idea of what Morlot saw:

The principal evidence for these two glacial periods was found by Professor Morlot in the Dranse Gorge near Thonen. There, at the lowest level, is a twelve-foot layer with scratched stones of Alpine chalk, and above that a 150-foot deposit of layered scree (diluvium); and on this, more scratched and erratic boulders. Between the formation of the first and the third deposit caused by a moraine is the deposit of stratified materials, and these show by their considerable volume that the glaciers had disappeared here for a long time.

(Heer 1865, p. 531)

It should be remarked that there is no source known to me that would indicate that Morlot himself claimed to have proposed the expression 'Quaternär'. But he ensured his fame through his handwritten paper inside a letter from Morlot to Simony in 1854 (ÖNB, Vienna, 463/28) (see Fig. 1) in which he proposed his ideas of a double glaciation. This document was published in various periodicals and was also distributed in manuscript form (Morlot 1854*a, b*, p. 39; Bronn 1855, p. 719).

Morlot differed from other authors, such as Vogt, who postulated a 'parallelism of [terminology for] the Tertiary and Quaternary formations' in his *Lehrbuch der Geologie und Petrefactenkunde* (Vogt 1854, p. 624) and equated Lyell's Pleistocene with Quaternary formations. Morlot restricted his understanding of Quaternary to the post-Pliocene. But those restrictions were compensated for and supported by a differentiation within the categories themselves; and *this* can be seen as Morlot's prime contribution. He linked research on the Ice Age with stratigraphy. Moreover, he differentiated between *two* glacial periods that were separated by a diluvial period and which, together with the modern period, constituted a (*four*fold) 'Quaternary epoch in Switzerland' (i.e. cold–warm–cold–warm [last = Present]). Thus, Morlot's concept of the Quaternary depended on distinct durations or subdivisions of the glaciation processes. It arose from the recognition of a multiple (actually 'double') glaciation in Switzerland, for which he discovered an illustrative locality on the Dranse River near Thonen.

However, Morlot traced this 'new' idea back to *Venetz*. It was he, in fact, who was first convinced of a double glaciation, but his ideas were only published in 1861 (Venetz 1861; Kaiser 1990). (For more about Venetz and his biography, see Truffer (1990, pp. 7–32). For Venetz as an engineer, see Roten & Kalbermatter (1990, pp. 33–46). And for the glacial studies of Venetz see Kaiser (1990, pp. 53–124). All these articles were published in the Swiss local journal *Mitteilungen der Naturforschenden Gesellschaft Oberwallis.*) To Simony, Morlot wrote that no less a man than Venetz – who had given both de Charpentier and Agassiz the idea for the glacial theory with his reflections on glacial movement – already *envisaged a double glaciation*: 'Venetz had known for a long time that there had been two glacial periods separated by the Diluvian, but his friends laughed at him and so he kept silent' (Morlot to Simony, 28 June 1852, ÖNB, Vienna, 463/28). This statement certainly had a real foundation, as the Swiss geologist Bernhard Studer (1794–1887) was also struck by the indications of two ice ages as a result of climatic factors (Studer 1844, p. 310). But he subsumed these phenomena within the period of the Pleistocene.

Also, in a letter to Escher von der Linth in 1853 (Wissenschaftshistorische Sammlungen der Eidgenössischen Technischen Hochschule Zürich, Morlot to Escher, No. 1271, 3 May 1853), Morlot emphasized that Robert Chambers (1802–1871) from Edinburgh had told him personally that he supposed that there had been two different phases of glaciation. As Escher was evidently sceptical of Morlot's theory, Morlot emphasized the difference between 'direct observation and mere deduction'. He considered three alternatives: (1) two glaciers; (2) two diluvia; and finally (3) two glacial periods. He decided in favour of the third alternative, as, in his opinion, it corresponded with all the observed 'facts' (Wissenschaftshistorische Sammlurgen der Eidgenössichen Technischen Hochschule Zürich, Morlot to Escher, No. 1273, 27 May 1856).

Oldroyd (1996, p. 50) has previously mentioned that 'the glacial theory provides a fascinating example of the phenomenon of the theory-ladenness of observations'. When the few geologists – particularly those in Switzerland – had finally 'cognitively adopted and accepted' the theory of glaciation, they saw the landscape with different eyes. Morlot had known of the glacial concept from about 1842 and worked on the collection of evidence in the field, even during his work as geologist/surveyor in Inner Austria. In 1848 he was already stressing to Simony that in a geological representation of the former extent of glacial tracks, traces or marks of the erratic period were indicated on the map: 'I am pleased that you have had at least one opportunity to see typical glacial tracks. When all visible localities in the whole valley have been dealt with, the phenomenon is truly magnificent and impressive' (Morlot to Simony, 8 November 1848, ÖNB, Vienna, 463/28–1).

When reading Albrecht Penck and Eduard Brückner's *Die Alpen im Eiszeitalter* [*The Alps in the Glacial Period*] today, first published in 1909 and written by the most prominent glaciologists of the Eastern Alps, one cannot fail to notice how often Penck referred to and praised Morlot's arguments, almost 50 years after his work (Penck & Brückner 1909). The authors mentioned Morlot in the same breath as John Playfair, Ignatz Venetz, Jean de Charpentier, Louis Agassiz, Friedrich Simony, Oswald Heer and James Geikie (Penck & Brückner 1909 vol. 1, p. 1). According to Penck & Brückner, Quaternary research was justified when three criteria were fulfilled: (1) the expansion of the former glaciers; (2) the periodic recurrence of glaciers; and (3) the influence of those glaciers on the Earth's surface form. Morlot took into consideration – at least in part – all three aspects, as is shown in a handwritten draft document giving an overview of his ideas (paper in a letter from Morlot to Simony in 1854, ÖNB, Vienna, 463/28).

In the first 'period of glaciation' Morlot assumed the greatest expansion of glaciation; in the intermediate period, the glaciers had vanished, and 'the elephant [was] living in Switzerland', as he put it. This period lasted 60 000 years, according to Morlot's estimate. In the second glaciation, a smaller area was covered with glaciers, but 'the continent [wa]s elevated, the elephant disappear[ed] and man arrive[d] on the scene' (paper accompanying a letter from Morlot to Simony of 1854, ÖNB, Vienna, 463/28). Morlot assigned the formation of loess to this second period of glaciation. Then, together with the modern warm epoch, there were the four divisions of his 'Quaternär'. The idea of a double glaciation was soon taken up and discussed by other researchers in respect of questions in the Swiss Alps (e.g. Gras 1860, p. 741).

Why has Morlot's contribution been so largely forgotten?

The term 'Quaternary' did not gain acceptance in the Habsburg countries. Even Morlot's friend, Franz von Hauer (1822–1899), could not bring himself to abandon the old terms 'Diluvial' and 'Alluvial' in his widely read geology textbook (Hauer 1875). In 1854 Mori[t]z Hörnes (1815–1865) had studied the close relation between the Miocene fauna of the Vienna Basin and the Pliocene fauna of Sicily and Cyprus. As the two faunas were difficult to distinguish, he decided to subsume them both under the term 'Neogene'. This idea preoccupied Austrian geologists and diverted them from Morlot's initial ideas. In the first complete mapping of the countries belonging to the Monarchy, co-ordinated by Hauer and conducted by the Geologische Reichsanstalt, Quaternary formations were categorized as Tertiary and were coloured on the maps in the same way as Tertiary formations – something that Penck later pointed out and criticized. The collective resistance towards the Ice Age theory on the part of the leadership of the Imperial Institute in Vienna probably allowed no room in the discussions for innovative reinterpretations of traditional stratigraphy. Such discussions are ultimately documented in the end product – the map – which has to give expression to the widest possible consensus. But the best informed students of the Ice Age in the eastern Alps, Albrecht Penck and Eduard Brückner, commented:

> by virtue of Hauer's summary geological map of the Austrian Monarchy (Sheet 6), these erroneous interpretations of the glacial deposits in the eastern Alps, particularly in the Drau region, as Tertiary have been widely disseminated. If the theoretical standpoint of the older geologists was far removed from those of today, at least they observed with sharp eyes.
>
> (Penck & Brückner 1909, vol. 3, p. 1064).

Yet, attribution of the Quaternary gravel deposits to the Tertiary was maintained even in later editions of the map.

The idea of interglacials that was developed by many protagonists at the same time and moreover the evidence – the most important result of Morlot's research – was, however, adopted by Heer in his aforementioned famous work *Die Urwelt der Schweiz*, with explicit reference to Morlot (Heer 1865, p. 531). Heer was able to support the theory of this intermediary phase biostratigraphcally by investigating the Schieferkohlenflora with more data than were available to Morlot. Thus, Heer is usually associated with 'discovering' the interglacial period, even though he explicitly associated himself with Morlot (and the idea appears, in fact, to go as far back as Venetz).

It was Marc Bloch (1953, p. 28), the founder of the important French historical approach known as *Annales*, who maintained that for historians periodization should not consist of defining epochs, but should involve posing questions and problems that should guide scientists in the categorizing process. In his attempts at periodization, Morlot asked a variety of questions and brought together different approaches and factors for consideration. This was, together with his observations and evidence found in the field, the basis of his creative achievement and the innovative nature of his contribution to the debate on the Quaternary.

Appendix

Handwritten paper in a letter from Morlot to Friedrich Simony of 1854, ÖNB, Vienna, 463/28, in which

Morlot proposed his ideas of a double glaciation (see Fig.1).

On the Quaternary Period in Switzerland
A. von Morlot

It has hitherto been assumed that the erratic formation is younger that the so-called older diluvial period, or simply, the Diluvial Period; and one can, in fact, see in the corresponding sequence of layers around Geneva what is also the case at Heidelberg and Vienna, since the erratic loess formations there lie on the diluvial layer, exactly as in Lombardy. There is just one problem with the fact that the material of the diluvial terraces is quite Alpine in these places too, but it could not have got there by means of a normal flow from the Alps, for example at Morsee on Lake Geneva; and then the erratic Till in the north proves to be overlain by the lines of the old sea-shore, and these clearly correspond to our Diluvial Period. On the other hand, both in Scotland and Switzerland two different phases of the Glacial Period have been demonstrated. And near Vivis, at one well exposed site, the distinct layering of the Diluvial Period may be seen within the very typical erratic formation. We reach the conclusion: that the two phases of glaciation were separated by the Diluvial Period, a period during which the glacier left not only the lower levels but all the principal valleys, since the diluvial terraces may be traced far up these [valleys].

We arrive thus at the following subdivision of the Quaternary Period.

1. First Glacial Period. that of its greatest extent, when Lower Switzerland was almost completely covered with ice and the Rhone Glacier, for example, is of the size attributed to it by J. von Charpentier in his classic work. Does not seem to have lasted long.

2. Diluvial Period. The glaciers have completely disappeared, even in the main valleys. Elephants live in Switzerland. To judge from the corresponding cones in the Wildbach cutting by Lake Geneva, this lasted as long as the Modern Period, and thus, according to Lyell, over 60 000 years.

3. Second Glacial Period. which was of smaller scope; Lower Switzerland was to a large extent free from ice; the Rhone glacier, for example, did not cross the [hills of the] Jorat [district north of Lake Geneva in the Vaud Canton] nor reach the foot of the Jura, but remained within the basin of Lake Geneva, filling this to a few hundred feet higher than modern sea level and ending around Geneva. The age of the formation of loess. Elephants inhabited the country. To judge from the relevant layers: of long duration.

4. Modern Period. The glacier disappears, the continent rises by a small amount and the rivers excavate a deeper course in their former diluvial flood areas so that these take on the appearance of terraces. Elephants disappear and humans arrive.

D. Oldroyd (Sydney) encouraged me to pursue the topic further, with numerous suggestions as to content. I am also indebted to H. Masson (Lausanne) for valued information on the Swiss locations that Morlot visited. I thank them both.

References

AGASSIZ, l. 1841. *Untersuchungen über die Gletscher.* Gent & Gaßmann, Solothurn.

BERGGREN, W. A. 1998. The Cenozoic Era: Lyellian (chrono)stratigraphy and nomenclatural reform at the millennium. *In*: BLUNDELL, D. J. & SCOTT, A. C. (eds) *Lyell: The Past is the Key to the Present.* Geological Society, London, Special Publications, **143**, 111–132.

BERGGREN, W. A. 2005. The Cenozoic Era: Lyellian (chrono)stratigraphy and nomenclatural reform at the millennium. *Stratigraphy*, **2**, 1–12.

BLOCH, M. L. B. 1953. *The Historian's Craft.* Knopf, New York.

BOLLES, E. B. 2003. *Eiszeit. Wie ein Professor, ein Politiker und ein Dichter das ewige Eis entdeckten.* Fischer Taschenbuch, Frankfurt am Main.

BRONN, H. G. 1855. A[dolphe] Morlot. Quartäre[!] Gebilde des Rhone-Gebietes (from: Verhandlungen der allgemeinen Schweizerischen Gesellschaft für Naturwissenschaften bei ihren Versammlungen in St. Gallen). *Neues Jahrbuch für Mineralogie, Geognosie, Geologie und Petrefakten-Kunde*, **25**, 719–721.

CHARPENTIER, J. DE. 1836. Bericht über eines der wichtigsten Resultate der von Hrn. Venetz rücksichtlich des jetzigen und vormaligen Zustandes der Gletscher des Cantons Wallis angestellten Untersuchungen. *Notizen aus dem Gebiete der Natur- und Heilkunde*, **1**, 290–296.

CHARPENTIER, J. DE. 1837. Notice sur la cause probable du transport des blocs erratiques de la Suisse. *Annales des mines*, **7**, 219–236.

CHARPENTIER, J. DE. 1841. *Essai sur les glaciers et sur le terrain erratique de basin du Rhône.* Marc Ducloux, Lausanne.

DESNOYERS, J. 1829. Observations sur un ensemble de dépôts marins plus recents que les terrains tertiaires du bassin de la Seine, et constituant une formation géologique distincte. *Annales des sciences naturelles*, **16**, 171–214, 402–491.

EISZMANN, L. 1974. Die Begründung der Inlandeistheorie für Norddeutschland durch den Schweizer Adolph von Morlot im Jahre 1844. *Abhandlungen und Berichte des Naturkundlichen Museums 'Mauretanium' Altenburg*, **8**, 289–318.

FORBES, E. 1846. On the connexion between the distribution of the existing fauna and flora of the British Isles and the geographical changes which have affected their area, especially during the epoch of the Northern Drift. *Memoirs of the Geological Survey of Great Britain*, **1**, 336–432.

GIBBARD, P. L., SMITH, A. G. ET AL. 2005. What status for the Quaternary? *Boreas*, **34**, 1–6.

GRAS, S. 1860. Über die Nothwendigkeit zwei Gletscher-Perioden im Quartär-Gebirge der Alpen anzunehmen. *Neues Jahrbuch für Mineralogie, Geologie, Geognosie und Petrefakten-Kunde*, **30**, 741–742.

HAUER, F. 1875. *Die Geologie und ihre Anwendung auf die Kenntnis der Bodenbeschaffenheit der österreichisch-ungarischen Monarchie.* Hölder, Vienna.

HEER, O. 1865. *Die Urwelt der Schweiz.* Friedrich Schultheß, Zürich.

HÖRNES, M. 1853. Mittheilungen an Professor Bronn gerichtet. *Neues Jahrbuch für Mineralogie, Geologie, Geognosie und Petrefakten-Kunde*, **23**, 806–810.

KAISER, K. 1990. Ignaz Venetz im Dienste der Eiszeitforschung. *In: Ignaz Venetz, 1788–1859. Ingenieur und Naturforscher. Mitteilungen der Naturforschenden Gesellschaft Oberwallis*, **1**, 53–124.

KAUZ, D. 2000. Wilde und Pfahlbauer. Facetten der Analogisierung. *Preprints zur Kulturgeschichte der Technik*, **11**, 1–25.

KELLER, F. 1865–1866. *Die keltischen Pfahlbauten in den Schweizerseen.* S. Höhr, Zürich.

KLEMUN, M. 1992. Friedrich Simony (1813–1896) – 1. Kustos des Naturhistorischen Museums in Klagenfurt (1848–1850). *Carinthia II*, **182** (= **102**), 375–391.

LEONHARD, C. C. 1844. *Lehrbuch der Geognosie und Geologie.* Winter, Heidelberg.

LYELL, C. 1830–1833. *Principles of Geology, Being an Attempt to Explain the Former Changes of the Earth's Surface by Reference to Causes now in Operation*, Volume 1 (1830), Volume 2 (1832), Volume 3 (1833). John Murray, London.

LYELL, C. 1839. *Nouveaux éléments de géologique.* Pitois-Levranet, Paris.

LYELL, C. 1857. *Geologie oder Entwicklungsgeschichte der Erde und ihrer Bewohner*, transl. by Bernhard Cotta. Duncker & Humblot, Berlin.

MORLOT, A. 1847a. Über Dolomit und seine künstliche Darstellung aus Kalkstein. *Aus den Naturwissenschaftlichen Abhandlungen*, **1**. Braumüller, Vienna.

MORLOT, A. 1847b. *Geologische Übersichtskarte der nordöstlichen Kalkalpen mit Erläuterungen.* Braumüller, Vienna.

MORLOT, A. 1848. Ueber errratisches Diluvium bei Pitten. *Naturwissenschaftliche Abhandlungen*, **4**, 1–20.

MORLOT, A. 1850a. Üeber die Rauchwacke und die Eisenerzlagerstätte bei Pitten. *Aus den Berichten über Mittheilungen von Freunden der Naturwissenschaften in Wien*, **7**, 81–101.

MORLOT, A. 1850b. Einiges über die geologischen Verhältnisse in der nördlichen Steiermark. *Jahrbuch der kaiserlich-königlichen Geologischen Reichsanstalt*, **1**, 99–124.

MORLOT, A. 1850c. Ueber die Spuren eines befestigten römischen Eisenwerkes in der Wochein in Oberkrain. *Jahrbuch der kaiserlich-königlichen Geologischen Reichsanstalt*, **1**, 199–212.

MORLOT, A. 1850d. Ueber die geologischen Verhältnisse von Raibl. *Jahrbuch der kaiserlich-königlichen Geologischen Reichsanstalt*, **1**, 255–267.

MORLOT, A. 1850e. Ueber die geologischen Verhältnisse von Radoboj in Kroatien. *Jahrbuch der kaiserlich-königlichen Geologischen Reichsanstalt*, **1**, 268–279.

MORLOT, A. 1850f. Ueber die geologischen Verhältnisse von Oberkrain. *Jahrbuch der kaiserlich-königlichen Geologischen Reichsanstalt*, **1**, 389–411.

MORLOT, A. 1854a. Ueber die quaternären Gebilde des Rhône-gebiets. *Verhandlungen der allgemeinen schweizerischen Gesellschaft über die gesammten Naturwissenschaften bei ihrer Versammlung in St. Gallen 1854 am 24., 25. und 26. Juli*, **39**, 161–164.

MORLOT, A. 1854b. Observation d'une superposition de diluvium a l'erratique. *Bulletins des Séances de la Société Vaudoise des sciences naturelles*, **4**, 39–40.

MORLOT, A. 1854c. Notice sur le quaternaire en Suisse. *Bulletins des séances de la Société Vaudoise des sciences naturelles*, **4**, 41–45.

MORLOT, A. 1859a. *Allgemeine Bemerkungen über die Altherthumskunde.* Haller, Bern.

MORLOT, A. 1859b. Quartäre Gebilde des Rhône-Gebietes. *Neues Jahrbuch für Mineralogie, Geognosie, Geologie und Petrefacten-Kunde*, **29**, 315–317.

MORLOT, A. 1859c. (printed 1861). Sur le terrain quartaire du bassin du Léman. *Bulletins des Séances de la Société Vaudoise des sciences naturelles*, **6**, 101–108.

MORLOT, A. 1860a (printed 1861). Études géologico-archéologiques en Danemark et en Suisse [Geologisch/archäologische Studien in Dänemark und der Schweitz]. *Bulletin des séances de la Société Vaudoise des science naturelles*, **6**, 263–328.

MORLOT, A. 1860b. Über das Quartär-Gebirge am Genfer-See. *Neues Jahrbuch für Mineralogie, Geognosie, Geologie und Petrefakten-Kunde*, **30**, 830–832.

MORLOT, A. 1865. *Das graue Alterthum. Eine Einleitung in das Studium der Vorzeit.* Bärensprungsche Hofdruckerei, Schwerin. (This work was first published in French in 1861; then in English in 1863; and in Italian in 1863.)

NAUMANN, C. F. 1844. Fels-Schliffe an Porphyr-Hügeln bei Kollmen. *Neues Jahrbuch für Mineralogie, Geologie, Geognosie und Petrefakten-Kunde*, **14**, 557–558, 561–562.

OLDROYD, D. 1996. *Thinking about the Earth: A History of Ideas in Geology.* Athlone Press, London; Harvard University Press, Cambridge, MA.

PENCK, A. & BRÜCKNER, E. 1909. *Die Alpen im Eiszeitalter* (3 vols). C. Hermann Tauchnitz, Leipzig.

POMIAN, K. 1984. *L'ordre du temps.* Gallimard, Paris.

REBOUL, H. 1833. *Géologie de la période quaternaire, et introduction à l'histoire ancienne.* F.-G. Levrault, Paris.

ROTEN, E. & KALBERMATTER, P. 1990. Ignaz Venetz als Ingenieur. *In: Ignaz Venetz, 1788–1859. Ingenieur und Naturforscher. Mitteilungen der Naturforschenden Gesellschaft Oberwallis*, **1**, 33–52.

RUDWICK, M. J. S. 1997. Transposed concepts from the human sciences in the early work of Charles Lyell. *In:* JORDANOVA, L. J. & PORTER, R. (eds) *Images of the Earth: Essays in the History of Environmental Sciences*, 2nd edn. The British Society for the History of Science, Oxford, 77–91 (1st edn published in 1979).

SARJEANT, W. A. S. 1980. *Geologists and the History of Geology.* Arno Press, New York.

SERRES, M. 1830. De la simultaneité des terrains de sédiment supérieurs. *La géographie physique de l'Encyclopédie méthodique*, **5**, 1–125.

STOCKING, G. 1968. *Race, Culture and Evolution*. Free Press, London.

STUDER, B. 1844. *Lehrbuch der physikalischen Geographie und Geologie*. Dalp, Bern, Chur & Leipzig.

TORELL, O. 1875. Protokoll der November-Versammlung über einen Vortrag Torells (über Schliff-Flächen und Schrammen auf der Oberfläche des Muschelkalkes von Rüdersheim). *Zeitschrift der deutschen Geologischen Gesellschaft*, **27**, 961–962.

TORELL, O. 1880. Diskussion über die Gletscher-Erscheinungen bei Rüdersdorf. *Verhandlungen der Berliner Gesellschaft für Anthropologie, Ethnologie und Urgeschichte*, **11**, 152–155.

TRUFFER, B. 1990. Ignatz Venetz (1788–1859). *In: Ignaz Venetz, 1788–1859. Ingenieur und Naturforscher. Mitteilungen der Naturforschenden Gesellschaft Oberwallis*, **1**, 7–32.

VAN RIPER, A. 1993. *Men Among the Mammoths: Victorian Science and the Discovery of Human Prehistory*. Chicago University Press, Chicago, IL.

VAN VEEN, F. R. 2008. Early ideas about erratic boulders and glacial phenomena in The Netherlands. *In*: GRAPES, R. H., OLDROYD, D. R. & GRIGELIS, A. (eds) *History of Geomorphology and Quaternary Geology*. Geological Society, London, Special Publications, **301**, 159–169.

VENETZ, I. 1833. Mémoire sur les variations de la température dans les Alpes de la Suisse. *Denkschriften der allgemeinen Schweitzerischen Gesellschaft für die gesammten Naturwissenschaften*, **1**, 1–38.

VENETZ, I. 1861. Mémoire sur l'extension des anciens glaciers, renfermant quelques explications sur leurs effets remarquables: Ouvrage posthume rédigé en 1857 et 1858. *Neue Denkschrift der allgemeinen Schweitzerischen Gesellschaft für die gesammten Naturwissenschaften*, **18**, 1–33.

VOGT, C. 1854. *Lehrbuch der Geologie und Petrefactenkunde* (2 vols). Friedrich Vieweg und Sohn, Braunschweig.

WAGENBRETH, O. 1999. *Geschichte der Geologie in Deutschland*. Georg Thieme, Stuttgart.

ZITTEL, K. A. 1899. *Geschichte der Geologie und Paläontologie bis Ende des 19. Jahrhunderts*. R. Oldenbourg, Munich.

Manuscript sources

Burgerbibliothek Bern, Switzerland, *Manuscripta Historia helvetica*, **45**: 3–11: Letters from many geologists (de Charpentier, etc.) to Adolphe Morlot 12–22: Papers (Journals, Abstracts, etc.) 23: Lecture notes.

Geologische Bundesanstalt Wien [Geological Survey of Vienna]: letters from Adolphe Morlots to Franz Hauer, AOO209–B104.

ÖNB (Österreichische Nationalbibliothek) [National Library of Austria], Vienna. *Handschriftensammlung*, 463/28: 26 letters from Adolph Morlot to Friedrich Simony (1848–1854).

Wissenschaftshistorische Sammlungen der Eidgenössischen Technischen Hochschule Zürich [Collections of the Swiss Federal Institut of Technology, Zürich]. *Handschriften und Autographen* 142 Handschrift 04, Nos 1264–1275: 12 letters from Adolphe Morlot to Escher von der Linth (1847–1861).

The Spokane Flood debates: historical background and philosophical perspective

VICTOR R. BAKER

*Department of Hydrology and Water Resources, The University of Arizona, Tucson,
AZ 85721–0011, USA (e-mail: baker@hwr.arizona.edu)*

Abstract: The 1920s–1930s debates over the origin of the 'Channeled Scabland' landscape of
eastern Washington, northwestern USA, focused on the cataclysmic flooding hypothesis
of J Harlen Bretz. During the summer of 1922, Bretz began leading field parties of advanced
University of Chicago students into the region. In his first paper, published in the *Bulletin of
the Geological Society of America*, Bretz took special care not to mention cataclysmic origins.
However, in a subsequent paper in the *Journal of Geology*, to the editorial board of which he
had recently been added, Bretz formally described his hypothesis that an immense late Pleistocene
flood, which he named the 'Spokane Flood', had derived from the margins of the nearby
Cordilleran Ice Sheet. This cataclysm neatly accounted for numerous interrelated aspects of the
Channeled Scabland landscape and nearby regions. Nevertheless, the geological community
largely resisted Bretz's hypothesis for decades, despite his enthusiastic and eloquent defence
thereof. Resolution of the controversy came gradually, initially through the recognition by J. T.
Pardee of a plausible source for the flooding: ice-dammed Pleistocene glacial Lake Missoula in
northern Idaho and western Montana. Eventually, by the 1960s, the field evidence for cataclysmic
flooding became overwhelming, and physical processes were found to be completely consistent
with that evidence. The controversy is of philosophical interest in regard to its documentation
of the attitudes of geologists toward hypotheses, which illustrate aspects of geological reasoning
that are distinctive in degree from those of other sciences.

J Harlen Bretz (there can be no period after the
'J' – it was the man's entire first name) was born
in 1882. He initiated one of the great debates in
the history of geomorphology. While not revolutionary in its implications, the debate is, nevertheless, instructive for how it illustrates the workings
of science; where the latter is defined not as a
sterile collection of facts, or even as a logical application of method, but rather as the collective workings and attitudes of human beings passionately
dedicated to the search for truth.

The subject of the debate is an unusual landscape in eastern Washington, northwestern United
States, described by Bretz (1928*a*, pp. 88–89) as
follows:

No one with an eye for land forms can cross eastern Washington
in daylight without encountering and being impressed by the
'scabland.' Like great scars marring the otherwise fair face to
the plateau are these elongated tracts of bare, black rock carved
into mazes of buttes and canyons. Everybody on the plateau
knows scabland . . . . The popular name is a metaphor. The
scablands are wounds only partially healed – great wounds in
the epidermis of soil with which Nature protects the underlying
rock . . . . The region is unique: let the observer take wings of the
morning to the uttermost parts of the earth: he will nowhere
find its likeness.

The controversy that led to the debate arose
because of the interpretation that Bretz made, as
he sought to find a single hypothesis that would
explain the disparate relationships in this landscape.
Bretz (1928*a*, p. 89) argued as follows:

The volume of the invading waters much exceeds the capacity
of the existing streamways. The valleys entered become river
channels, they brim over into neighboring ones, and minor
divides within the system are crossed in hundreds of places . . . .
The topographic features produced during this episode are
wholly river-bottom modifications of the invaded and overswept
drainage network of hills and valleys. Hundreds of cataract
ledges, of basins and canyons eroded into bed rock, of isolated
buttes of the bed rock, of gravel bars piled high above the valley
floors, and of island hills of the weaker overlying formations are
left at the cessation of this episode . . . . Everywhere the record
is of extraordinary vigorous sub-fluvial action. The physiographic
expression of the region is without parallel; it is unique, this channeled scabland of the Columbia Plateau.

Introduction

The unique character of the dry river courses
('coulees') of eastern Washington State was recognized by Reverend Samuel Parker (1838), who provided some of the first scientific observations of the
region. He also offered the first, and most prevalent,
hypothesis for the origin of the Grand Coulee; he
proposed that it was an abandoned former channel
of the Columbia River. Lieutenant T. W. Symons
subsequently made a government survey of Grand
Coulee, describing his travel: 'north through the
coulée, its perpendicular walls forming a vista

From: GRAPES, R. H., OLDROYD, D. & GRIGELIS, A. (eds) *History of Geomorphology and Quaternary Geology.*
Geological Society, London, Special Publications, **301**, 33–50.
DOI: 10.1144/SP301.3   0305-8719/08/$15.00 © The Geological Society of London 2008.

like some grand ruined roofless hall, down which we traveled hour after hour' (Symons 1882). In his report, Symons (1882) proposed the hypothesis that the diversion of the Columbia through the Grand Coulee was caused by a glacial blockage of the river immediately downstream of its junction with the coulee. This view was generally accepted for the next few decades (Russell 1893; Dawson 1898; Salisbury 1901; Calkins 1905).

The Grand Coulee was traversed in 1912 by the American Geographical Society's Transcontinental Excursion. This led to several international papers on the coulee authored by participants on the excursion. Henri Baulig, Université de Rennes, described the loess, coulees, rock basins, plunge pools and dry falls ('cataract desechée de la Columbia') for the area (Baulig 1913). He also noted the immense scale of the erosion (Baulig 1913, p. 159): 'peut-être unique du relief terestre, – unique par ses dimensions, sinon par son origine'. Karl Oestreich, University of Utrecht, considered the Grand Coulee to be 'eines machtigen Flusses Bett . . . ohne jede Spur von Zerfall der frischen Form' (Oestreich 1915). He provided another excellent set of descriptions, including features that he recognized to require a special explanation, including granite hills exhumed from burial by the plateau basalt layers, and almost perpendicular coulee walls locally notched by hanging valleys. Oesterich (1915) proposed that the hanging valleys resulted from glacial erosion of the coulee, but that the deepening of the coulee came about by enhanced fluvial erosion from the glacially enhanced and diverted Columbia River. He also recognized that the coulee was cut across a pre-glacial divide, which he correctly noted to lie just north of Coulee City.

An influential report was written by Oscar E. Meinzer, who later founded many of the hydrological science programmes of the US Geological Survey. Meinzer (1918) proposed that the glacially diverted Columbia River 'cut precipitous gorges several hundred feet deep, developed three cataracts, at least one of which was higher than Niagara . . . and performed an almost incredible amount of work in carrying boulders many miles and gouging out holes as much as two hundred feet deep'. He also recognized that the dipping surface to the basalt on the northern Columbia Plateau was important for generating a steep water-surface gradient to the Columbia River water that was diverted across the plateau by glacial blockage of its pre-glacial valley. The resulting increased velocity of the river water was thus capable of the enhanced erosion indicated by the scabland.

The Channeled Scabland

The cataclysmic flooding hypothesis for the origin of coulees on the Columbia Plateau centred on the work of J Harlen Bretz. Although formally introduced by papers in 1923 (Bretz 1923a, b), the idea of cataclysmic flooding causing the scablands was a certainly a matter of informal discussion before that time. McMacken (1937) credited the hypothesis to A. P. Tooth, of Lewis and Clark High School in Spokane, Washington. Bretz undoubtedly knew of this via his friend and benefactor, Thomas Large, also of Lewis and Clark High School. Nevertheless, Bretz (pers. commun. 1978) recalled that around 1909, while he was teaching high school science in Seattle, Washington, he became intrigued by an unusual landform on the newly published Quincy topographic sheet of the US Geological Survey. The map clearly showed the Potholes Cataract of the western Quincy Basin (Fig. 1). Bretz wondered how such a feature could have been formed without an obvious river course leading to it. His questions to geology faculty members at the University of Washington provided no satisfactory answer.

Bretz conducted extensive studies of the glacial geology of the Puget Sound region near Seattle, and in 1910 he left his high school teaching job to pursue graduate studies in glacial geology at the University of Chicago. Working with Professors R. D. Salisbury and T. C. Chamberlin, Bretz obtained his PhD in 1913, basing his dissertation on the Puget Sound glacial geology (Bretz 1913). After a 1-year appointment at the University of Washington, Bretz returned as a beginning faculty member at the University of Chicago. He began teaching field courses in geology, including an advanced summer course that he taught in the Columbia Gorge, where the Columbia River crosses the Cascade Range between Washington and Oregon.

One of Bretz's early papers derived from the Columbia Gorge fieldwork. It dealt with the widespread erratic boulders that were widely distributed in the region. Thomas Condon, University of Oregon, had earlier described these boulders, and also popularized the idea that they were rafted to their locations by berg ice during a marine submergence (Condon 1902, p. 62). He described the submergence of the Willamette Valley as an ocean sound, as follows:

that fine old Willamette Sound . . . may in the days of the Mammoth and the Broad Faced Ox have welcomed to its scores of sheltered harbors, the ancient hunter who, in his canoe, if he had one, floated one hundred feet or more above the present altitude of the church spires of Portland and Salem.

Bretz (1919) proposed that the inundation of the lower Columbia River (Fig. 2) had been of freshwater, not marine water as advocated by Condon. Bretz reasoned that the immense discharge of the late Pleistocene Columbia River, swollen by glacial melt-water, had maintained the freshwater flow at a time of diastrophic movements that

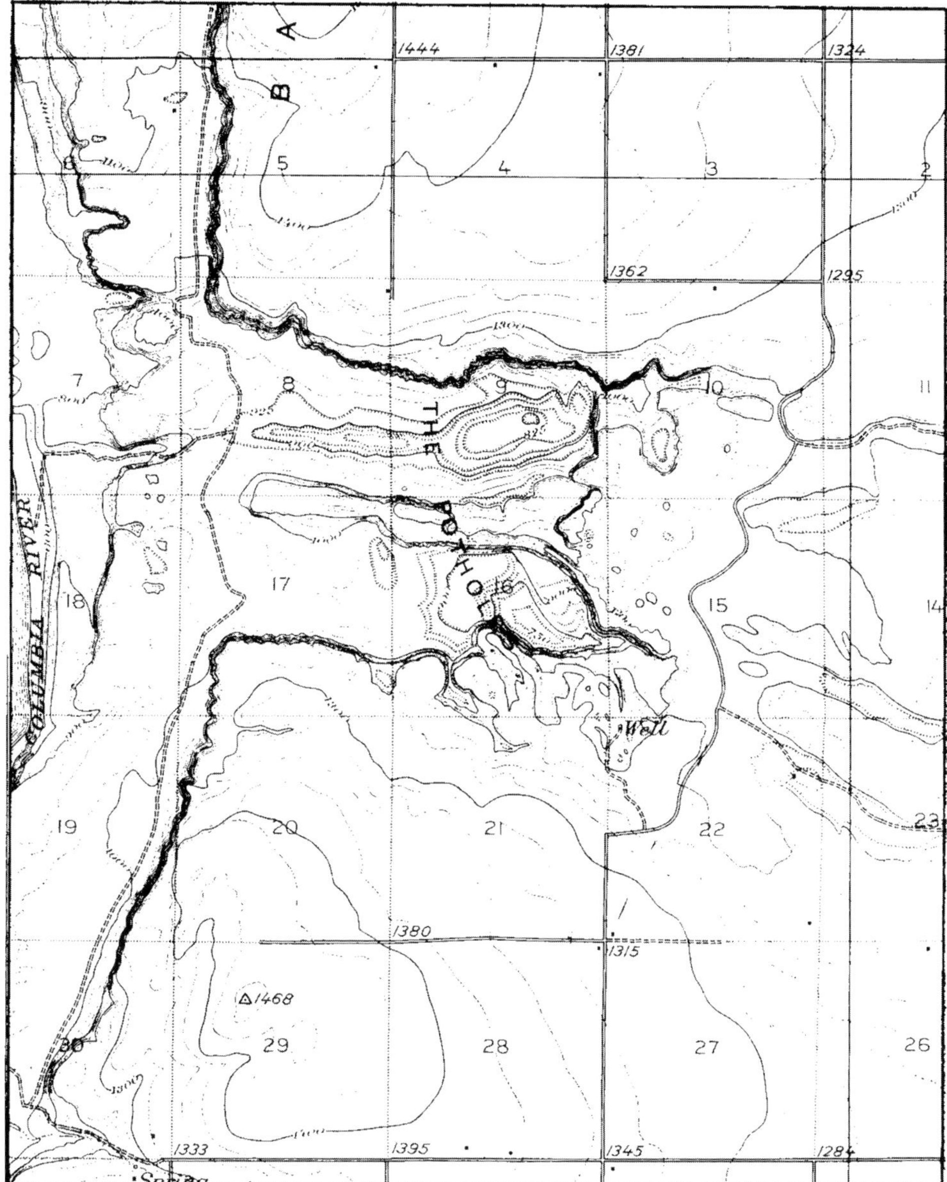

Fig. 1. Topographic map of the Potholes cataract, one of three cataract spillways on the west side of Quincy Basin in the channelled scabland. (From the US Geological Survey Quincy topographic quadrangle map, with 7.6 m-contour interval and 1.6 × 1.6 km numbered section squares).

otherwise would have led to submergence. Nevertheless, he found this explanation problematical in that the duration of ponding had been so brief that it did not create the expected shoreline features or lacustrine plains. The only evidences for the regional extent of ponding were isolated berg deposits and associated silts that Symons (1882) and Russell (1893) had ascribed to a Pleistocene 'Lake Lewis'. Moreover, Bretz could not find evidence of similar inundation or submergence in southwestern Washington State, where he would have expected the hypothetical diastrophism to have also operated. Finally, Bretz was concerned that other known examples of late Pleistocene

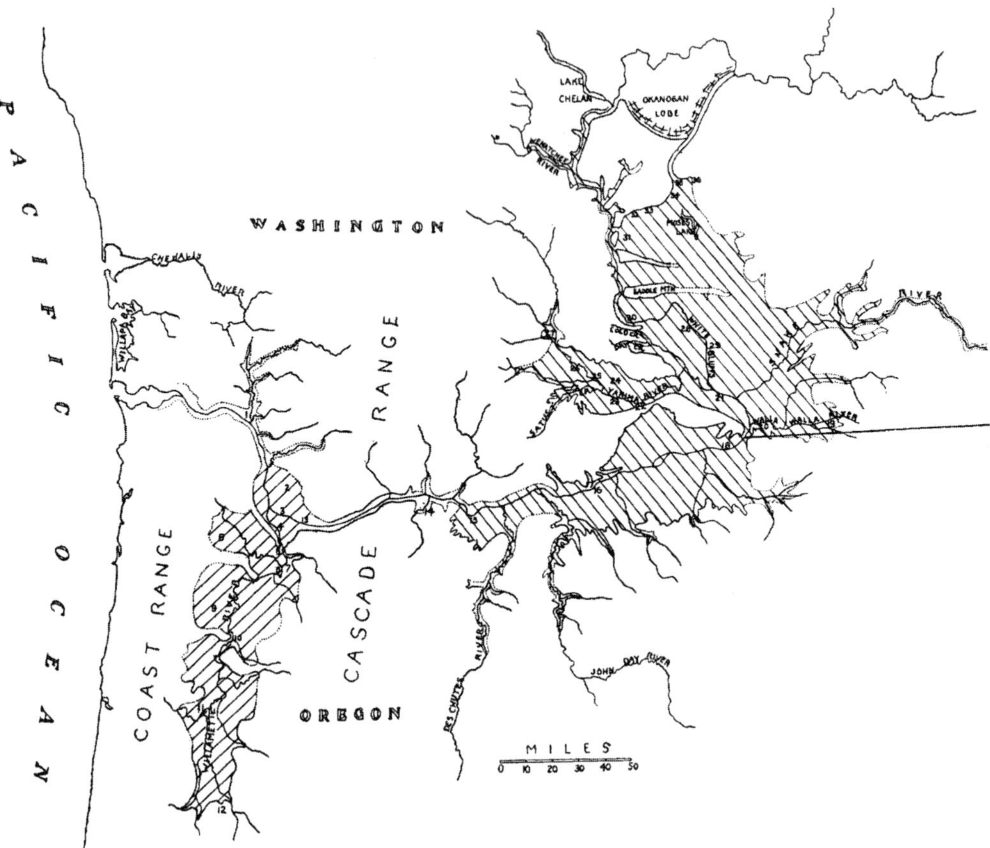

Fig. 2. Hypothesized submergence map of the lower Columbia River system, including the 'Willamette Sound' (NE to SW diagonal ruling) at about 120 m above modern sea level, and the area of 'Lake Lewis' (NW to SE diagonal ruling) at about 380 m above modern sea level (Bretz 1919). Ponded elevations were determined from the heights of ice-rafted erratic boulders.

submergence involved the invasion of marine waters into subsided areas from an immense glacial load had been removed by the rapid retreat of ice sheets. No such Pleistocene glacial load seems to have existed in the Columbia Gorge, yet an immense area (Fig. 2) had clearly been inundated by freshwater during the late Pleistocene.

Beginning in the summer of 1922, Bretz spent eight successive field seasons with parties of advanced geology student in areas of eastern Washington State upstream of the areas described in his 1919 paper. His first season of work led to a paper (Bretz 1923*a*) that was the text of on oral presentation made to the Geological Society of America on 30 December 1922. In that paper Bretz provided detailed descriptions of physiographical relationships in the region, noting the sharp distinction between the Palouse Hills of loess and the eroded tracts of scabland. Although Bretz (1923*a*) did not directly invoke cataclysmic

processes, he did note that the indicated bedrock channel erosion required prodigious quantities of water. Referring to the three outlets at the south end of the Hartline Basin (Dry Coulee, Lenore Canyon and Long Lake Canyon), Bretz (1923*a*, pp. 593–594) stated that 'these are truly distributary canyons. They mark a distributive or braided course of the Spokane glacial flood over a basalt surface which possessed no adequate pre-Spokane valleys'. He also noted that the flows of the glacial melt-water transformed pre-glacial valleys into channels. Referring to Pine Creek valley, south of Spokane, Bretz (1923*a*, p. 584) wrote:

The valley during this episode in its history was but a channel. The glacial stream filled it from side to side for a depth of tens of feet. This is shown a few miles above Malden, where the stream flooded over a low shoulder of basalt, cutting a channel in the rock at least 40 feet deep, though the main valley alongside was a wide open and received gravel deposits.

Bretz also recognized that the extent of removal of the Palouse loess cover, the scarps eroded into the loess and the steep gradients of the channel-ways were all important. He wrote (Bretz 1923*a*, p. 588):

The scablands of the Palouse drainage, with channeled basalt deposits of stratified gravel, and isolated linear groups of Palouse Hills, their marginal slopes steepened notably, bear abundant evidence of a great flood of glacial waters from the north. This flood was born of the Spokane ice-sheet. Its gradient was high, averaging perhaps 25 feet to the mile, and it swept more than 400 square miles of the region clean of the weaker material constituting the Palouse Hills. The hills which have disappeared averaged 200 feet in height, and in some places the glacial torrents eroded 100 to 200 feet into the basalt.

The hypothesis of a truly catastrophic flood appeared in Bretz's second scabland paper (Bretz 1923*b*). In this paper Bretz also introduced the term 'channelled scablands' to refer to areas of loessial soil removal and fluvial scour into the underlying basalt rock. Bretz (1923*b*) provided the first detailed geomorphic map of the entire channelled scabland region, showing its overall pattern of anastomosing channel-ways. The paper also contained the first interpretation of the mounded scabland gravel deposits as subfluvially emplaced gravel bars. Great depths of water were indicated by the need to submerge these bars and by the indicated crossings of divide areas. He concluded (Bretz 1923*b*, p. 649) that '[i]t was a debacle which swept the Columbia Plateau'. The age of this flood, he reasoned, was tied to a glaciation that had occurred prior to the latest Pleistocene glaciation, known as the Wisconsin Glaciation. This older glaciation he named the Spokane Glaciation (Bretz 1923*a*), and its associated debacle was dubbed the Spokane Flood.

It may not be a coincidence that the second paper, with its potentially controversial advocacy of catastrophism, appeared shortly after two important events in Bretz's career: (1) he was promoted to the tenured rank of Associate Professor; and (2) he was named to the editorial board of the *Journal of Geology*, the journal in which the second paper appeared.

In a subsequent paper, Bretz (1924) described the process of subfluvial quarrying that could generate a scabland of anastomosing channels and rock basins cut into the basalt. The process was described for the modern Columbia River at The Dalles, Oregon, but he concluded that only large, vigorous streams could produce these landforms. Bretz (1925*a*, p. 144) then proposed a test of his cataclysmic flood hypothesis:

A crucial test ... is the character of the Snake and Columbia valleys beyond the scablands, for these valleys received all the discharge of the great glacial rivers .... If there is no evidence for a greatly flooded condition of these valleys then the hypothesis is wrong ....

Bretz (1925*b*) traced the Spokane Flood evidence downstream to the Snake and Columbia rivers (Fig. 3). Just below the confluence of those two rivers, where all the floodwaters from various scabland channels had to converge at the Wallula Gateway, he estimated the floodwater depth to be over 200 m above the modern river level. Flood depths of at least 120 m were maintained through the Columbia River gorge all the way to Portland, Oregon, where a huge flood delta was deposited over an area of 500 km^2, with bars over 30 m high on its surface. Bretz (1925*b*) also made the first estimate of the flood discharge, calculated at Wallula Gap using the Chezy Equation. The calculation was actually made by Bretz's colleague, D. F. Higgins, who found a discharge of nearly 2×10^6 m^3 s^{-1} (about 160 km^3 day^{-1}) and flow velocities of 10 m s^{-1}.

These phenomenal quantities of water led Bretz (1925*b*) to think about possible sources. Only two hypotheses seemed reasonable: (1) a very rapid and short-lived climatic amelioration that would have melted the Cordilleran Ice Sheet; or (2) a gigantic outburst flood induced by volcanic activity beneath the ice sheet. The latter phenomena were known to occur in Iceland, where they are known by the term 'jokulhlaup' (which Bretz spelled as 'jokulloup'). Bretz devoted much of the summer field season of 1926 to a search for the source of flooding. In an oral presentation at the Madison meeting of the Geological Society of America, 27 December 1926, Bretz (1927*a*, p. 107) concluded: 'Studies in Washington and British Columbia north of the scabland strongly suggest that basaltic flows which were extruded beneath the Cordilleran ice sheet produced the great flood'.

The Great Debate

The 'Great Scablands Debate' (Gould 1978) took place on 12 January 1927, at the 423rd meeting of the Geological Society of Washington, held at the Cosmos Club in Washington, DC. The entire meeting was devoted to the presentation and subsequent discussion of Bretz's talk, entitled 'Channeled Scabland and the Spokane Flood' (Bretz 1927*b*). In his presentation Bretz provided an overview of his hypothesis and a detailed listing of the numerous, otherwise anomalous phenomena that were explained by it. His talk was then followed by well-prepared criticisms from selected members of the audience, which consisted mainly of scientists from the United States Geological Survey.

W. C. Alden objected to: (1) the immense amounts of water ascribed to rapid melting of the ice sheet; and (2) the requirement that all the

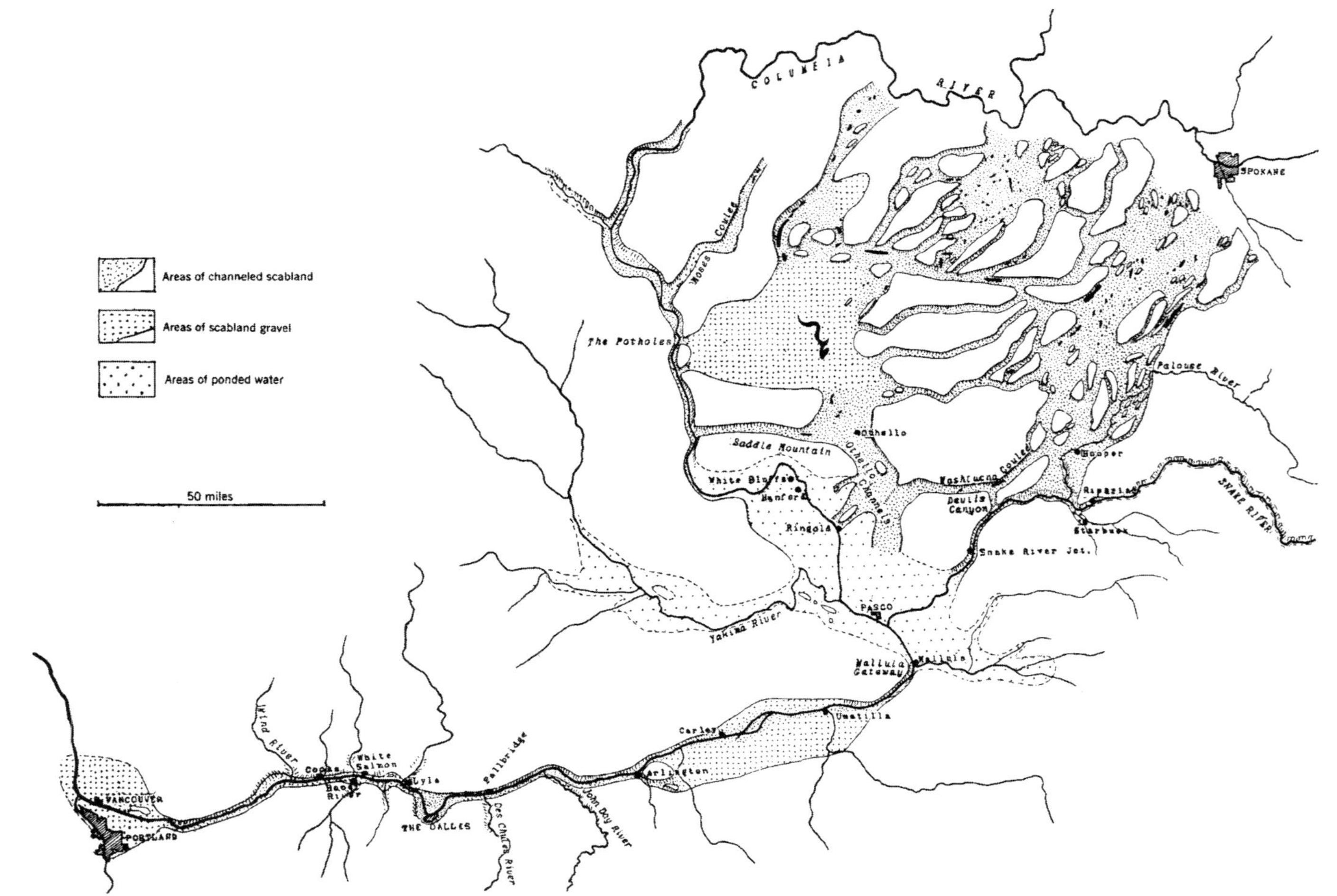

Fig. 3. Sketch map of regions affected by the Spokane Flood extending 500 km from Spokane, Washington, in the NE, to Portland, Oregon, in the SW (Bretz 1925*b*).

scabland channels developed simultaneously over a relatively short period of time. Although he had not visited the field areas, Alden believed these objections to be very serious. James Gilluly expanded on both of Alden's points with quantitative arguments that longer time periods needed to be involved than what Bretz had proposed. Moreover, he calculated that the indicated discharges of water could not be produced either by a climatic amelioration or subglacial volcanism. E. T. McKnight then raised specific objections about field relationships for coulees in the southern part of the scablands, and G. R. Mansfield objected to the rapid erosion of basalt by floodwaters.

Oscar E. Meinzer expanded upon ideas that he had proposed earlier (Meinzer 1918). He stated that the Columbia River, greatly swollen by glacial melt-water, could easily have carved Dry Falls and deposited the great gravel fan of the northern Quincy Basin. He interpreted the gravels of the Quincy Basin as constituting a series of terraces. The high-level crossing of divides could then be explained by successive abandonment as the Columbia progressively cut down to lower levels. Nevertheless, from past work (Schwennessen & Meinzer 1918) he realized that Bretz (1923a) was correct in that the four great spillways out of the Quincy Basin had floors at approximately equal elevation. However, instead of accepting Bretz's interpretation of coincident spillway operation during large-scale ponding of floodwater, Meinzer proposed that the spillways were cut one at a time, but that subsequent Earth movements later brought them to equivalent altitudes. Bretz (pers. commun. 1978) recalled that Meinzer first formulated this unusual hypothesis in response to a question that had Bretz asked him while he was defending his PhD thesis at the University of Chicago.

Bretz answered all these criticisms, both at the meeting itself (Bretz 1927b) and in subsequent papers. However, there was one, relatively minor comment to which Bretz did not reply. This concerned the effects of aspect on rates of talus accumulation, and it was made by J. T. Pardee, who earlier in the same summer as Bretz's first scabland field season had been sent by his supervisor Alden to study the scabland area SW of Spokane. Pardee (1922) reported in print that the area had been glaciated, but he provided few details and never published further results that were anticipated in his preliminary report. Bretz (pers. commun. 1978) suspected, from correspondence with Pardee in the 1920s, that the latter was considering a flood hypothesis for the origin of the scablands. Pardee (1910) had earlier documented the presence of an immense late Pleistocene lake in the western part of Montana. Named

Glacial Lake Missoula, this lake had been impounded by the Cordilleran Ice Sheet, when it cut across the drainage of the modern Clark Fork River, which trends northwestward from western Montana. The lake held as much as 2000 km^3 of water, derived from melt-water inputs from surrounding glaciated areas.

Did Pardee suspect that his Glacial Lake Missoula had released an immense cataclysmic flood? If he did, Bretz surmised, his superior, Alden, had dissuaded him from the idea. Several years after the affair, David White, who was Alden's superior, showed Bretz a memorandum of 22 September 1922, in which Alden had written to White concerning Pardee's scabland studies: 'very significant phenomena were discovered in the region southwest of Spokane .... The results so far ... require caution in their interpretation. The conditions warn against premature publication'. Also in 1922, correspondence between local Spokane residents Thomas Large and Barton W. Everman (Baker 1996a) indicated that Pardee informally told Everman that 'sub-glacial water erosion' involving water under pressure beneath ice was a possible hypothesis to explain the scabland erosion. Moreover, Pardee wrote to Bretz in 1925, suggesting that Lake Missoula be considered as a possible source for the Spokane Flood. In a 1926 letter to J. C. Merriam, Bretz wrote that Pardee proposed Lake Missoula as the Spokane Flood source, and that even Alden, 'our ultra-conservative in Pleistocene geology', agreed that fluvial processes were involved in the origin of the scablands.

Initially, Bretz seems to have resisted Pardee's suggestion of Glacial Lake Missoula as the source of the Spokane Flood. As noted above, he was intrigued by the possibility of subglacial volcanism. He also believed that Cordilleran ice had covered the area between the Lake Missoula ice dam and the heads of the scabland channels south of Spokane, a distance of 100 km. In 1927 he thought that the Spokane Flood was older than the latest Pleistocene glaciation that had formed Lake Missoula, and he also thought the Lake Missoula could not have supplied enough water to have inundated all the scabland channel ways. By 1928, however, Bretz resolved that the source of the Spokane Flood had to be Lake Missoula. Following attendance at Bretz's presentation of this idea at a scientific meeting, Harding (1929) published the first formal paper on this hypothesis without acknowledging either Bretz or Pardee as its source. Bretz made a more extensive presentation of the concept to the Geological Society of America annual meeting in Washington, DC, on 26 December 1929. In the published abstract of this paper, Bretz (1930a, p. 92) stated: '[i]t is suggested that bursting of the ice barrier which confined a large glacial lake among the mountains of western Montana suddenly released a very great quantity of water which escaped across the plateau of eastern Washington and eroded the channeled scablands'. In subsequent publications (e.g. Bretz 1932*a*), Bretz clearly illustrated the relationship of Glacial Lake Missoula to the Channeled Scabland.

After the Washington meeting, Bretz continued to answer in print the various criticisms of his Spokane Flood hypothesis (Bretz 1927*c*, 1928*b*). He added many specifics to hypothesis, including an extensive study of the scabland bars, with detailed descriptions, published in the *Bulletin of the Geological Society of America* (Bretz 1928*c*). An especially eloquent paper was published in *The Geographical Review* (Bretz 1928*a*), the journal of the American Geographical Society. The latter society also supported Bretz by publishing a handsome monograph on the Grand Coulee (Bretz 1932*a*). The results of Bretz's last field seasons, 1928–1930, were presented in papers concerned with valley deposits east and west of the Channeled Scabland (Bretz 1929, 1930*b*). He showed that these tributary valleys contained deposits that indicated emplacement by phenomenally deep floodwater that flowed up the tributaries, away from sources in the scabland channels. Along the Snake River, he traced these deposits to beyond Lewiston, Idaho, about 140 km upstream of the nearest scabland channel (Bretz 1929, p. 509):

Up-valley currents of great depth and vigor are essential . . . . No descending gradient of the valley floor can be held responsible. The gradient must have existed in the surface of that flood. The writer, forced by the field evidence to this hypothesis, though warned times without number that he will not be believed, must call for an unparalleled rapidity in the rise of the scabland rivers.

Alternative hypotheses

The 1933 International Geological Congress field trip provided a kind of transition point in regard to the Spokane Flood controversy. Although Bretz successfully campaigned to include a field excursion to the Channeled Scabland, and even wrote the guidebook for that excursion (Bretz 1932*b*), he had to resign from leadership of the trip in order to participate in the Louise A. Boyd Greenland Expedition of the American Geographical Society. His substitute leader was Ira S. Allison of Oregon State College, who was sympathetic to the Spokane Flood hypothesis. Allison (1932) had even used it to explain landforms in the vicinity of Portland, Oregon. Although he subsequently developed his own hypothesis for the scablands (Allison 1933), he presented Bretz's views quite fairly during the field trip. Allison (pers. commun.

1978) recalled that the international audience of field- trip attendees included several famous geologists, including Alex Du Toit of South Africa and Edward Bailey of Scotland. They were very keen observers, who looked at the field evidence on their own, rather than waiting for the pronouncements of the field-trip leader. This was exactly the kind of scrutiny for his hypothesis that Bretz himself urged.

Allison's (1933) hypothesis was not a denial of the 'Spokane Flood,' but instead a modification of its details. He proposed that the critical factor in the flooding was not the immense volume of water, but instead the effects of floating berg ice. He presented compelling evidence, noted earlier in less detail by others (e.g. Condon 1902; Bretz 1919), that extensive Late Pleistocene ponding of water had occurred all the way from the Columbia River Gorge to Wallula Gap. The ponding was inferred to have arisen from blockage by ice of constricted reaches of the Columbia River in the gorge. The Columbia River, swollen by iceberg-rich melt-water, was impounded by this temporary dam until the inundation reached into eastern Washington. As the water rose to higher levels it spilled across secondary drainage divides, just as Bretz had inferred for the flooding. It thereby created the enigmatic hanging valleys/ channel-ways, high-level gravel deposits and widely distributed erratic boulders. Allison's purpose in proposing this hypothesis was made clear by the concluding sentence in his paper: '[p]erhaps this revision will make the idea of such a flood more generally acceptable' (Allison 1933, p. 722).

Glacial hypotheses for the origin of the Channeled Scabland were proposed by several investigators. Charles Keyes, who published widely on many geological topics (he owned the journal *Pan American Geologist*), made a cursory inspection of the regional geography from topographic maps. His paper (Keyes 1935) concluded that both Moses Coulee and Grand Coulee had formed because of supra-glacial streams that had formed along the axes of the waning ice lobes of the Cordilleran Ice Sheet. These streams were then superimposed on the bedrock surfaces, thereby explaining their crossing of pre-glacial drainage divides. The explanation derived from a presumed analogy to late-glacial Laurentide Ice Sheet lobe in central Idaho, and it did not consider the morphology of bedrock erosion, giant gravel bars and other evidence of cataclysmic flooding.

E. T. Hodge of Oregon State College, Corvallis, was a fairly persistent critic of the Spokane Flood hypothesis, who gave numerous seminars and oral papers attacking it during the 1930s. His alternative explanation (Hodge 1934) was that the seemingly enigmatic features summarized by Bretz could all be explained by complicated alternation of ice advances and drainage changes. He argued that basalt could be quarried by glacial erosion during ice-sheet advances, and that channel complexes could derive their fluvial aspects from the diversion of melt-water streams around blocks of stagnant glacial ice and jams of berg ice that would result from glacial recessions. The published versions of this hypothesis were only presented in outline form as generalizations, and never contained detailed reference to actual field relationships.

Many of Hodge's criticisms were published from 1935 to 1937 as abstracts and notes in the *Geological Newsletter* of the Geological Society of the Oregon Country, which Hodge founded in 1935. The society was apparently founded in part because of a dispute arising from an invitation extended to Bretz to speak on the Spokane Flood at a scientific meeting in Pullman, Washington. Bretz (pers. commun. 1978) recalls that Hodge was so incensed at the invitation that he wrote to the chairman of the meeting demanding that he be allowed to formally debate Bretz after the invited address. When this was refused, Hodge founded his society in order to have it meet concurrently in Pullman. Bretz eventually acceded to an informal debate with Hodge on the day following his formal presentation. After Hodge began the debate with another retelling of his glacial hypothesis, Bretz responded with the suggestion that any interested parties should simply drive out across the Palouse Hills from Pullman to the nearby Cheney–Palouse scabland tract, and look at the field evidence. En route Bretz demonstrated, with an ordinary kitchen colander, how above a certain elevation the loess consisted of silt only, but below that level, in scarps near the scabland tracts, granules and pebbles of basalt and erratic granite indicated that the water flows 30 m deep or more were responsible for their deposition. At the former Palouse–Snake divide, subsequently crossed by floodwater, Bretz pointed out giant gravel bars, prow-pointed hills of loess, abandoned cataracts and spectacular scabland erosion. Thus, the point was made that the field evidence, not Bretz's rhetoric, was the rebuttal to his critics.

Another glacial hypothesis was proposed by the eminent glacial geologist W. H. Hobbs, University of Michigan. Hobbs (1947) stated that he was stimulated in his study by a map of the Channeled Scabland presented during a lecture on evidence for the Spokane Flood. He immediately recognized in the pattern of scabland erosion and the nearby loess accumulation a pattern that he believed to be associated with large ice sheets, like that of Greenland, for which he had made extensive

observation. With grants from the Geological Society of America and the American Philosophical Society, Hobbs spent two field seasons attempting to prove his initial hypothesis. He proposed that a lobe of the Cordilleran Ice Sheet, which he named the 'Scabland Glacial Lobe', had occupied much of the Channeled Scabland (Hobbs 1943). This lobe had eroded much of the scabland topography, while the anticyclonic winds blowing off it had deposited the adjacent loess of the Palouse Hills. Most of the scabland gravel deposits, he proposed, were moraine remnants that had been modified by glacier-border drainage. Although it was presented orally at meetings of the Geological Society of America (Hobbs 1945), the full manuscript exposition this hypothesis was rejected for publication during successive peer reviews by both of the professional societies that had sponsored the research. The manuscript was subsequently printed privately (Hobbs 1947) and distributed by its author.

The most serious challenge to the Spokane Flood hypothesis was posed by Richard Foster Flint of Yale University. Flint had a brief association with Bretz, who served on Flint's PhD examining committee at the University of Chicago in 1925, when Bretz was an associate professor. The two men had very different personalities, which seem to have influenced their approaches to science. Both men worked in the rapidly advancing subject of glacial geology, and Flint eventually became known for writing the most widely used textbooks in the field (Flint 1947, 1957, 1971).

Some of Flint's early research at Yale was focused on the Connecticut River valley ice-stagnation features, associated with the Late Pleistocene ice sheet that had covered New England (Flint 1928, 1932). Interestingly, these same features had been extensively studied by the famous Yale geologist James Dana, who had interpreted them to be evidence of an immense flood that had emanated from the glacial margin (Dana 1882). Flint achieved part of his early reputation by showing that his famous Yale predecessor had been wrong in arguing for a cataclysmic flood origin of the Connecticut valley landforms.

One of Flint's colleagues at Yale, Aaron Waters, recalled (pers. commun. 1978) suggesting to Flint that he would find more interesting examples of ice stagnation on the Waterville Plateau of eastern Washington, where the dry climate meant that landforms were not obscured by vegetation. This is the area where the Okanogan Lobe of the Cordilleran Ice Sheet had advanced across the northwestern Columbia Plateau, blocking the Columbia River valley, thereby diverting its flow and/or any cataclysmic flooding across the plateau in into the Channeled Scabland. During the 1930s Flint published several papers on the glacial features of the region (Flint 1935, 1936, 1937). Waters (pers. commun. 1978), who had been born on the Waterville Plateau, recalled being so incensed by Flint's publications that he initiated his own studies of glacial features in the region. Waters (1933) proposed a cataclysmic flood origin for landforms along the Columbia River downstream of the impoundment created by the Okanogan Lobe, which presumably failed to generate the flood. Flint (1935) disputed Waters' interpretation. Eventually, Waters' growing interest in volcanic geology and his involvement in World War II intervened with his geomorphological research programme so that he never completed his flood-related research.

Flint (1938a, p. 463) described his entry into the Spokane Flood controversy as follows:

I have had the opportunity to examine . . . this system during three field seasons' study . . . . During that time it gradually became apparent that the published explanations of the genesis of these drainage ways did not meet the requirements of the field data as I saw them, and, accordingly, a fourth season (Summer, 1936) was devoted to an intensive examination of one of the principal drainage tracts of the complex.

His paper presented many detailed field descriptions to make the case that the Cheney–Palouse scabland tract, easternmost in the channelled scabland, was the product of 'leisurely streams with normal discharge' (Flint 1938a, p. 472). Flint also enhanced his growing international reputation by the publication of a German-language version of his Cheney–Palouse paper (Flint 1938b).

Flint (1938a) accepted Bretz's (1923b, 1928c) observations that scabland fluvial gravel often: (1) occurred in the lee of island-like areas; (2) had rounded upper surface morphologies; and (3) exhibited a parallelism of surface slopes with the dip of underlying fore-set bedding. Nevertheless, he proposed these 'bar-like' morphologies were merely the expressions of 'non-paired, stream-cut terraces in various states of dissection' (Flint 1938a, p. 475). It was an idea that Bretz (1923a) had briefly entertained, then rejected after further field examination (Bretz 1923b, 1928c). Flint ascribed the dissection to a downstream base-level reduction. Terrace remnants were preferentially preserved in the lee of island-like areas. The low precipitation of the region and the high infiltration capacity of the gravel alluvium prevented gulleying, so that the terrace remnants developed rounded 'bar-like' slopes by dry creep. Flint also found places where the surfaces of the deposits did indeed truncate their internal forest bedding.

Flint (1938a) traced the coarse-grained Cheney–Palouse scabland deposits downstream into the Pasco Basin, where he inferred that the deposits changed in facies from sand and gravel to silt and fine sand containing erratic stones. He named this fine-grained facies the 'Touchet beds'. These deposits had already been described by Bretz (1929, pp. 516–536; 1930b, p. 414), who ascribed them to Spokane Flood water hydraulically ponded upstream of the Wallula Gateway, and by Allison (1933), who ascribed them to normal pro-glacial runoff impounded by ice jams. Flint (1938a) proposed that the damming was provided by a landslide or glacial ice in the Columbia Gorge, as earlier suggested by Russell (1893). Following Symons (1882), Flint named the resulting water body 'Lake Lewis' (see Fig. 1).

At this point Flint had the necessary components for his alternative to the Spokane Flood hypothesis. First, the pro-glacial melt-water streams of normal discharge overran the northern margin of the Cheney–Palouse tract (diagram 1 in Fig. 4). However, as Lake Lewis formed downstream, the 'leisurely' scabland streams aggraded, forming a thick fill (diagram 2 in Fig. 4). This fill blocked pre-glacial tributaries to the scabland tracts, forming marginal lakes that accumulated fine-grained sediments, while the main streams shifted laterally to cut scarps in the loess hills bounding the scabland valleys (diagram 3 in Fig. 4). Finally, when Lake Lewis drained, the streams gradually incised into the fill, thereby forming non-paired terrace remnants, some of which resembled 'bars' (diagram 4 in Fig. 4). Moreover, the enigmatic notched spurs and slot-like hanging valleys crossing divides could all result from superposition of normal streams from the widespread fill, instead of being eroded by the spilling of Spokane Flood water across divides when the pre-glacial valleys could not convey its cataclysmic discharges.

Flint's argument that relatively small streams could cut produce scabland-like erosion in rock was based on examples along the Snake River in Idaho. He correctly noted the presence of scabland landforms near Twin Falls, Idaho, where the Snake flows in a deep, narrow canyon (Flint 1938a, p. 492):

The . . . [basalt] flows yielded to the hydraulic force of the Snake River as similar flows on the Columbia Plateau yielded to the hydraulic force of proglacial streams, yet I am not aware that unusual floods have been held to have affected the upper Snake River.

This line of argument fell into disfavour when Malde (1968) demonstrated that the area was intensely scoured by another cataclysmic flood, the late Pleistocene overflow of Lake Bonneville. The latter was quite similar in its dynamics to Bretz's scabland flooding (O'Connor 1993).

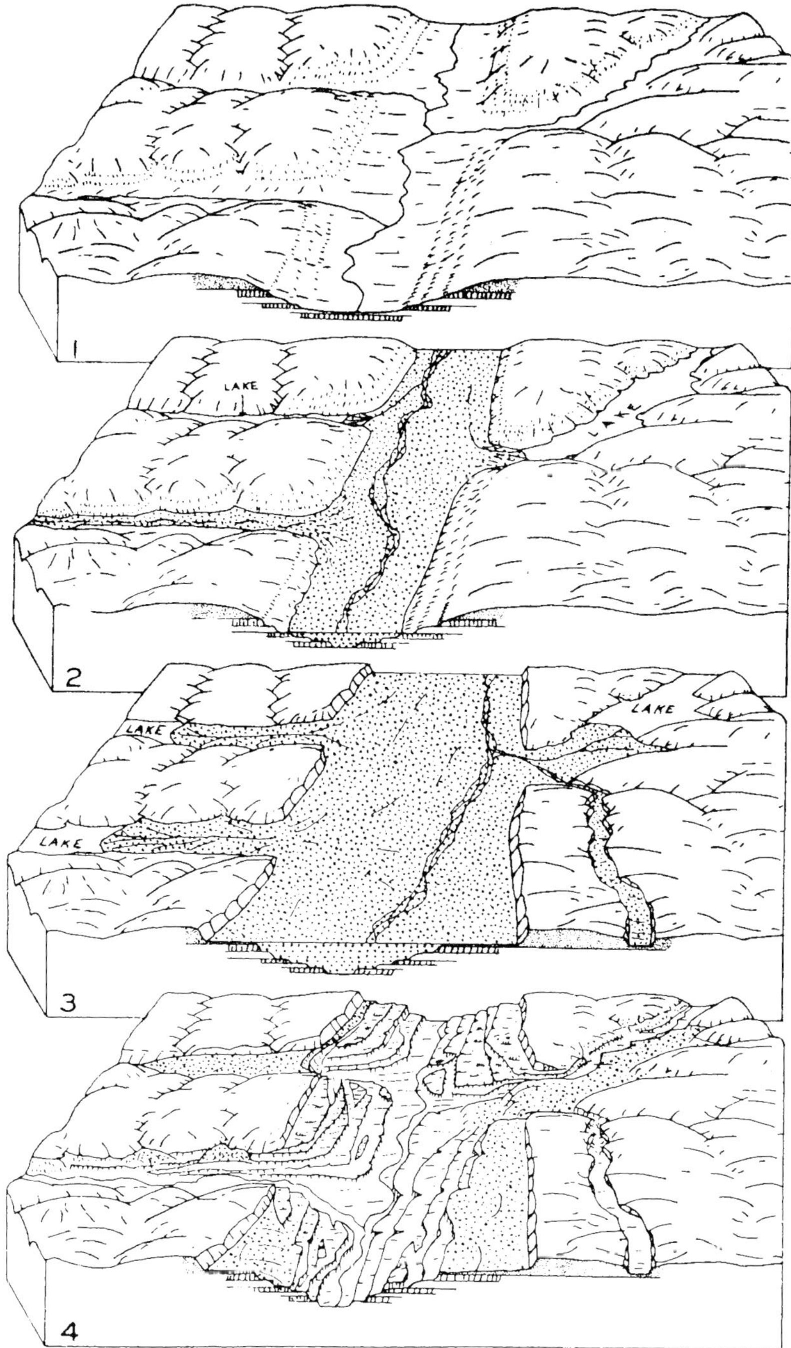

Fig. 4. Diagrams illustrating Flint's fill hypothesis for the origin of a scabland tract (Flint 1938*a*): (1) normal pre-glacial valleys with tributaries eroded into the nearby Palouse loess hills; (2) melt-water stream from nearby glacier front has aggraded the scabland valley floor and begun to backfill tributaries (stippled pattern); (3) many tributaries are completely backfilled, and the main valley stream has migrated to cut scarps on the bordering loess hills; and (4) fluvial dissection has eroded into bedrock and left remnants of the formerly extensive fill, some of which resemble 'bars', and also produced 'divide crossings' by superimposition across spurs.

Resolution

The beginning of change in the mostly skeptical reaction to the Spokane Flood hypothesis came during the course of another scientific meeting. Ironically, however, unlike the 1927 Geological Society of Washington meeting ('Great Scablands Debate'), Bretz was not in attendance. On 17–23 June 1940, the American Association for the Advancement of Science (AAAS) met in Seattle. The association's geological section included a session, 'Quaternary Geology of the Pacific Northwest', held in the afternoon of Tuesday 18 June. The session paper titles suggested that non-cataclysmic origins of the Channeled Scabland would be advocated. A post-meeting field trip to the Channeled Scabland was led by Flint. Although Bretz had been invited to attend, he declined, replying that all his ideas and supporting evidence were in print. If critics would just pay attention to the field evidence, it would speak for itself.

The early papers in the AAAS session merely reiterated the various alternatives for the origin of the Channeled Scabland. Hodge (1940) provided yet another retelling of his generalized scenario involving glacial erosion and complex damming and diversion of melt-water streams. Allison (1940) provided a critique of Flint's fill hypothesis, contrasting it with his own ice-jam theory. In a subsequent, more extended article, Allison (1941) outlined what he believed to be the key shortcomings with regard to Flint's hypothesis. First, Flint's (1938a) 'leisurely streams with normal discharge' were completely inadequate to explain the field evidence of anastomosing scabland channels and rock basins, all eroded deeply into bedrock. Second, the scabland gravels were not contemporaneous with the Touchet Beds of Flint's 'Lake Lewis,' proposing instead that the Touchet lacustrine sequence was younger than the gravels, and therefore unrelated genetically to them. Finally, Allison agreed with Bretz that the mounded gravels and other peculiar shapes of scabland erosional and depositional landforms required shaping by extraordinary fluvial processes. Nevertheless, he still held that such processes could be generated by the action of ice, leading to the complex jamming of various channel ways.

The surprise of the 1940 AAAS meeting came from an unexpected source. A meeting attendee, Howard A. Meyerhoff (pers. commun. 1978), related that no one in attendance expected that any speaker would directly support Bretz's Spokane Flood hypothesis. The eighth speaker of the session was Joseph Thomas Pardee, whose paper was entitled 'Ripple marks (?) in glacial Lake Missoula' (Pardee 1940). Pardee, who apparently was not a dynamic speaker, quietly described Camas Prairie, an intermontane basin in northwestern Montana. On the floor of this basin were ripple marks, composed of coarse gravel and exhibiting a size that was 'extraordinary'. These ripples were up to 15 m high, and were spaced up to 150 m apart (Fig. 5). The ripples had formed in a great ice-dammed lake, Glacial Lake Missoula, which Pardee (1910) had earlier documented to cover an immense area in western Montana. This late Pleistocene lake had held about 2000 km^3 of water, and it was impounded to a maximum depth of about 600 m behind a lobe of the Cordilleran Ice Sheet occupying what is now the modern basin of Lake Pend Oreille that in northern Idaho. In his presentation, and in a subsequent extended paper (Pardee 1942), Pardee presented his new evidence that the ice dam for Glacial Lake Missoula had failed suddenly, with a resulting rapid drainage of the lake. Evidence for the latter included the ripple marks, plus severely eroded constrictions in lake basins and giant gravel bars of current-transported debris. Some of the latter accumulated in high eddy deposits, marginal to the lake basins, and showing that the immensely deep lake waters were rapidly draining westward.

Remarkably, Pardee ended his paper at that point, without saying where the immense flood of 2000 km^3 of water, 600 m deep at its breakout point, would eventually go as it entered the Columbia River system in northeastern Washington. Nevertheless, the obvious connections to the Channeled Scabland and the Spokane Flood were made by members of the audience in the formal discussion that followed Pardee's paper. A group of 34 participants from the meeting, but not including Pardee, then attended an extended field excursion, 1 day of which was devoted to the scablands. Richard Foster Flint, who had not attended the session with Pardee's paper, joined the group to lead the scabland field visit. To his surprise, at each of the stops he had chosen to illustrate his hypothesis for scabland origins, Flint found that many of the trip participants, notably Aaron Waters, countered the Flint story with well-reasoned cataclysmic flood explanations, all of which were firmly tied to a source at Pardee's Lake Missoula.

Pardee retired from the US Geological Survey on his 70th birthday: 30 May 1941. The manuscript of his 1942 paper was submitted to the Geological Society of America several months later. This timing ensured that the paper would not have to pass the required formal review by Alden, Pardee's former supervisor, who was also a long-time critic of the cataclysmic flood hypothesis.

In the summer of 1952, Bretz returned to the Channeled Scabland for an intensive programme of field investigations. The purpose was to use

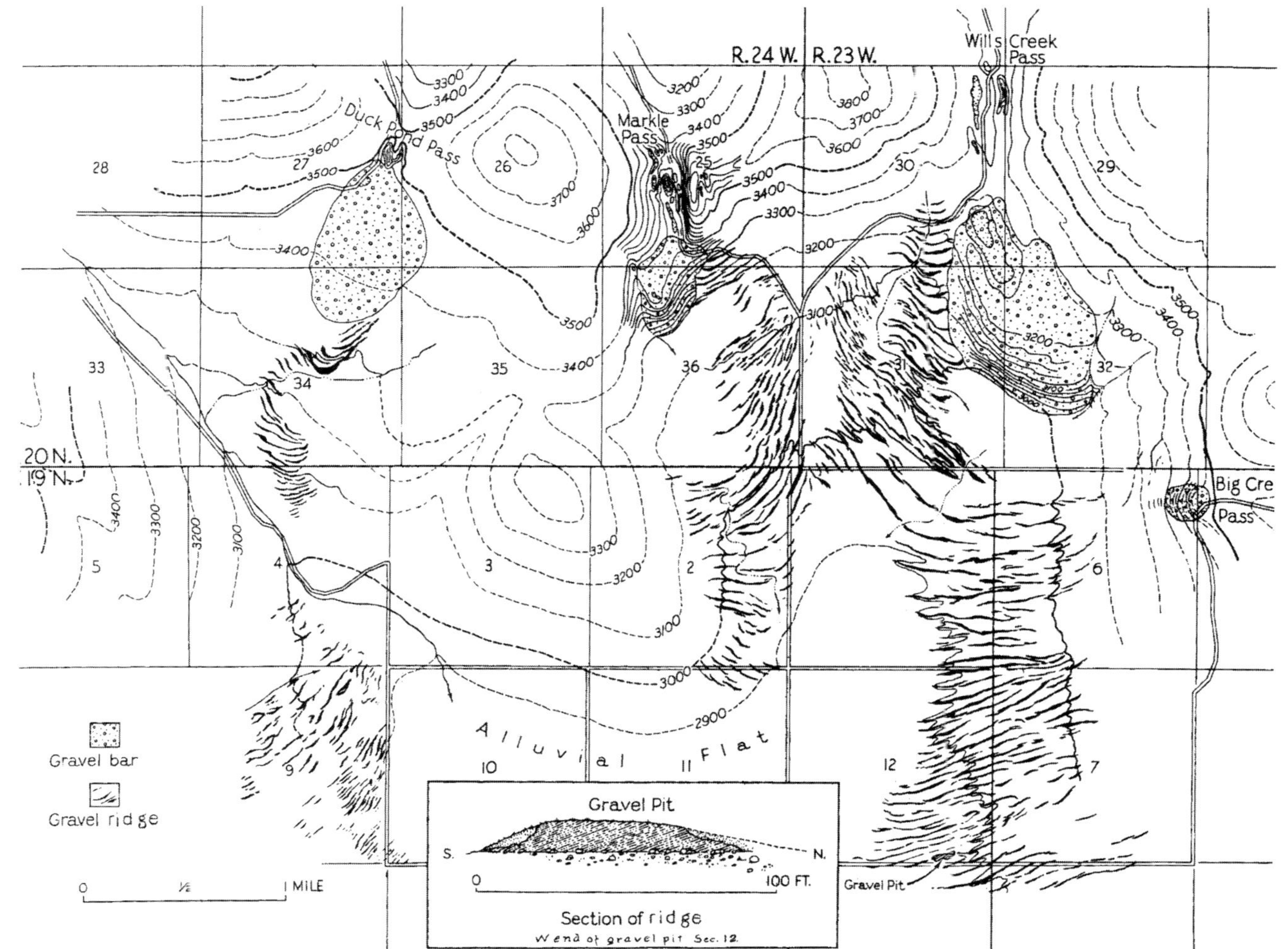

Fig. 5. Map of giant current 'ripple marks' in the northern part of Camus Prairie Basin of Glacial Lake Missoula (Pardee 1942).

new data that had been obtained through surveys for the Columbia Basin Irrigation Project of the United States Bureau of Reclamation. In this new work, Bretz was joined by George E. Neff of the Bureau of Reclamation, and by Professor H. T. U. Smith, University of Massachusetts, who was obliged to act as a 'skeptic for all identifications and interpretations' (Bretz *et al.* 1956, p. 761). In an obvious reference to Chamberlin's (1890) 'multiple working hypotheses' method, Bretz *et al.* (1956, p. 960) stated that: '[t]he field work was carried out with three men working together as a check on the "ruling-theory" tendency'. This new study was supported by a grant from the Geological Society of America. Bretz also received encouragement from several prominent geologists, among them: (1) Waters, who advised, 'the best thing you can do to convince your reader is to beg, cajole, even browbeat him into *really looking* at the region's topographic maps'; and (2) fluvial geomorphologist J. Hoover Mackin, who said: 'to understand the scabland, one must throw away textbook treatments of river work' (Bretz *et al.* 1956, p. 960).

The published results of the new study (Bretz *et al.* 1956) contain an immense amount of data. Excavations from the irrigation project and new topographic maps all showed that the gravel hills interpreted as bars by Bretz (1928c) were indeed subfluvial depositional bed forms of immense scale. Most convincing of all was the presence of giant current ripples on many of these bar surfaces.

At Staircase Rapids Bar, for example, the new maps and photographs showed an extensive set of these ripples on the surface that Flint (1938a, p. 486) had specifically described as a river terrace.

The 1956 paper and subsequent elaborations (Bretz 1959, 1969) successfully countered all opposition to the cataclysmic flood origin of the Channeled Scabland, and the original hypothesis was also extensively modified. It was no longer 'the Spokane Flood', but instead became the Missoula Floods (Bretz 1969). Nevertheless, there was a lingering suspicion that one was dealing with an unusual exception to a general rule. Bretz (1928a, p. 119) had himself claimed: 'The unique assemblage of forms ... described here as the channeled scabland ... records a unique episode in Pleistocene history .... Special causes seem clearly indicated'. Standard textbooks continued to treat the topic as controversial (Thornbury 1969). In the most widely used textbook on glacial and Quaternary phenomena, perhaps the most spectacular ice-age phenomenon was accorded but a single sentence, as follows (Flint 1971, p. 232): 'Similar features, collectively known as channelled scabland, were widely created east of the Grand Coulee by overflow of an ice-margin lake upstream'. Nevertheless, a succession of studies developed Bretz's insights into the cataclysmic flood (e.g. Trimble 1963; Richmond *et al.* 1965; Malde 1968; Baker 1973, 1974, 1981). Bretz outlived most of his critics (Fig. 6) and by the 1980s the controversy shifted from whether cataclysmic flooding had occurred

Fig. 6. Professor J Harlen Bretz (left) at his home with Victor R. Baker (right). Photographed in 1977 by Rhoda Riley.

in the channelled scabland to: (1) how many cataclysms were involved (Baker & Bunker 1985; Waitt 1985); and (2) how the general processes of cataclysmic flooding impacted other regions, both on Mars (Baker 1978, 1982) and on Earth (Baker *et al.* 1993).

Implications

During the intense debates over the Spokane Flood, Bretz held to a particular view of how hypotheses function in the Earth sciences. He wrote (Bretz 1928*b*, p. 701):

Ideas without precedent are generally looked on with disfavor and men are shocked if their conceptions of an orderly world are challenged. A hypothesis earnestly defended begets emotional reaction which may cloud the protagonist's view, but if such hypotheses outrage prevailing modes of thought the view of antagonists may also become fogged.

On the other hand, geology is plagued with extravagant ideas which spring from faulty observation and misinterpretation. They are worse than 'outrageous hypotheses,' for they lead nowhere. The writer's Spokane Flood hypothesis may belong to the latter class, but it cannot be placed there unless errors of observation and direct inference are demonstrated.

The reference to an 'outrageous hypothesis' derives from a 1925 address delivered by William Morris Davis, and subsequently published with the title 'The value of outrageous geological hypotheses,' in which he claimed (Davis 1926, p. 464): '[w]e shall indeed be fortunate if geology is so marvelously enlarged in the next thirty years as physics has been in the last thirty. But to make such progress violence must be done to many of our cherished principles'. Davis was not so much looking to the proposing and testing of hypotheses as he was to having a serious engagement with them (Davis 1926, pp. 467–468):

The idea is set forth simply as an outrage, to do violence to certain generally established views about earth's behavior that perhaps do not deserve to be regarded as established; and it is set forth chiefly as a means of encouraging the contemplation of other possible behaviors; not, however, merely a brief contemplation followed by an off-hand verdict of 'impossible' or 'absurd,' but a contemplation deliberate enough to seed out just what conditions would make the outrage seem permissible and reasonable.

According to the discredited philosophy of logical positivism (which was falsified by its own criteria through the efforts of its own adherents), science (for which the logical positivists took physics as their model) advances by logical propositions and their experimental verification. The latter criterion was modified to falsification, and more complex criteria have been devised by the various analytical philosophical successors to logical positivism. That much of this is of little relevance to classical geology has been argued

elsewhere (Baker 1996*b*, 2000). William Whewell, one of the few philosophers of science to have had any familiarity with geology, suggested from his historical studies of science that the validity of hypotheses (we would now say their productivity or fruitfulness for inquiry) was demonstrated by their ability to bring together *disparate* observations under an overarching explanation and to produce explanatory surprises, such that previously unknown phenomena are also found to fit under that explanation (Whewell 1840) – a procedure that he called 'consilience of inductions' (which involved a kind of coherence theory of truth). Hypotheses are not mere propositions to be tested. They are 'working' elements of inquiry, intimately connected to the phenomena that they explain, and are subject to modification.

The idea that the Spokane Flood hypothesis explains an entire assemblage of features was pervasive throughout Bretz's papers. In an oral meeting presentation on scabland bar morphologies, size, composition, internal structure (fore-set bedding) and distribution on the land surface (including on the summits of prominent pre-glacial divides), Bretz (1928*d*, p. 160) concluded:

The hypothesis of a 'Spokane Flood' alone seems capable of explaining the assemblage. Various alternatives proposed by others or constructed by the author have uniformly failed to explain the field evidence.

Another component of Bretz's reasoning was that a geological hypothesis is not merely tested against the existing data. Rather, the hypothesis is used as a guide to further field study. This is, of course, the 'working' component of Chamberlain's (1890) conception of 'hypothesis'. Bretz (1930*b*, p. 386) made this point as follows:

The flood hypothesis was not to stand or fall on data then in hand ... it could be critically tested by further field study. Certain relationships in unexamined areas must exist if it were incorrect ....

What is discovered in the course of that 'further field study' is to be judged for consistency with the original hypothesis, or its subsequent modification. Moreover, that judgement is not up to the hypothesizer; it is instead a matter for the larger science community that is dedicated to the truth in this matter. Bretz (1930*b*) also refused to separate the act of hypothesizing from the data collection used to test the hypothesis. Instead, he claimed (Bretz 1929, p. 541):

The final accepted explanation for channeled scabland must be built on, limited by, and constructed to include, all the field evidence .... The data in this paper constitute a category of facts not known when the flood hypothesis was first proposed. The relations to that hypothesis are to be decided by the reader. The writer asks only that, if his interpretation be rejected, the date herewith presented be organized to make some coherent explanation.

In all of these attitudes toward his hypothesis, Bretz was following pragmatic practices that had been promoted by his fellow geologists (e.g. Gilbert 1886; Chamberlin 1890; Davis 1926). The philosophical bases for these practices require exposition beyond the limits of this paper, but they can be found in Baker (1996*b*, *c*, 1998, 1999*a*, *b*, 2000).

The differences in attitude between Bretz and his critics or competitors were often subtle. Like Bretz, many of the latter were honestly searching for an explanation for the scabland phenomena. Nevertheless, unlike Bretz, they committed the No. 1 error for pragmatic inquiry: they placed an artificial limitation on that inquiry. 'Do not block the way of inquiry' is absolutely essential to all reasoning (Peirce 1992). The limitation imposed by Bretz's critics was the version of uniformitarianism promoted by Charles Lyell and others, to the detriment of sound geological inquiry (Baker 1998). Bretz was under no such limitation. Shortly before his death, upon receiving the 1979 Penrose Medal, the highest honour of the Geological Society of America, Bretz (1980, p. 1095) wrote, in his acceptance, and last published work: 'Perhaps I can be credited with reviving and demystifying legendary catastrophism and challenging a too rigorous uniformitarianism'.

I thank the late J Harlen Bretz for enthusiastic communications, interviews and insights that greatly aided both my own early studies of cataclysmic flooding and also my exploration of the historical dimensions to the controversy for which his contributions played such a pivotal role.

References

ALLISON, I. S. 1932. Spokane Flood south of Portland, Oregon. *Bulletin of the Geological Society of America Bulletin*, **43**, 133–134.

ALLISON, I. S. 1933. New version of the Spokane Flood. *Bulletin of the Geological Society of America*, **44**, 675–722.

ALLISON, I. S. 1940. Flint's fill hypothesis of origin of scabland. *Bulletin of the Geological Society of America*, **51**, 2016.

ALLISON, I. S. 1941. Flint's fill-hypothesis for channeled scabland. *Journal of Geology*, **49**, 54–73.

BAKER, V. R. 1973. *Paleohydrology and Sedimentology of Lake Missoula Flooding in Eastern Washington*. Geological Society of America, Special Paper, **144**.

BAKER, V. R. 1974. Erosional forms and processes for the catastrophic Pleistocene Missoula floods in eastern Washington. *In*: MORISOWA, M. (ed.) *Fluvial Geomorphology*. State University of New York, Binghamton, New York, 123–148.

BAKER, V. R. 1978. The Spokane Flood controversy and the Martian outflow channels. *Science*, **202**, 1249–1256.

BAKER, V. R. (ed.). 1981. *Catastrophic Flooding: The Origin of the Channeled Scabland*. Dowden, Hutchinson & Ross, Stroudsburg, PA.

BAKER, V. R. 1982. *The Channels of Mars*. University of Texas Press, Austin, TX.

BAKER, V. R. 1996*a*. Joseph Thomas Pardee and the Spokane Flood controversy. *GSA Today*, **5**, 169–173.

BAKER, V. R. 1996*b*. Hypotheses and geomorphological reasoning. *In*: RHOADS, B. L. & THORN, C. E. (eds) *The Scientific Nature of Geomorphology*. Wiley, New York, 57–85.

BAKER, V. R. 1996*c*. The pragmatic roots of American Quaternary geology and geomorphology. *Geomorphology*, **16**, 197–215.

BAKER, V. R. 1998. Catastrophism and uniformitarianism: logical roots and current relevance in geology. *In*: BLUNDELL, D. J. & SCOTT, A. C. (eds) *Lyell: The Past is the Key to the Present*. Geological Society, London, Special Publications, **143**, 171–182.

BAKER, V. R. 1999*a*. The methodological beliefs of geologists. *Earth Sciences History*, **18**, 321–335.

BAKER, V. R. 1999*b*. Geosemiosis. *Bulletin of the Geological Society of America*, **111**, 633–646.

BAKER, V. R. 2000. Conversing with the Earth: the geological approach to understanding. *In*: FRODEMAN, R. (ed.) *Earth Matters: The Earth Sciences, Philosophy, and the Claims of Community*. Prentice-Hall, Englewood Cliffs, NJ, 1–10.

BAKER, V. R. & BUNKER, R. C. 1985. Cataclysmic Late Pleistocene flooding from glacial Lake Missoula: A Review. *Quaternary Science Reviews*, **4**, 1–41.

BAKER, V. R., BENITO, G. & RUDOY, A. N. 1993. Paleohydrology of late Pleistocene superflooding, Altay Mountains, Siberia. *Science*, **259**, 348–350.

BAULIG, H. 1913. Le plateaux del lave du Washington central et la Grand 'Coulee'. *Annales géographie*, **22**, 149–159.

BRETZ, J. H. 1913. *Glaciation of the Puget Sound Region*. Washington Division of Mines and Geology, Bulletin **8**.

BRETZ, J. H. 1919. The late Pleistocene submergence in the Columbia Valley of Oregon and Washington. *Journal of Geology*, **27**, 489–505.

BRETZ, J. H. 1923*a*. Glacial drainage on the Columbia Plateau. *Bulletin of the Geological Society of America*, **34**, 573–608.

BRETZ, J. H. 1923*b*. The Channeled Scabland of the Columbia Plateau. *Journal of Geology*, **31**, 617–649.

BRETZ, J. H. 1924. The Dalles type of river channel. *Journal of Geology*, **32**, 139–149.

BRETZ, J. H. 1925*a*. Spokane Flood beyond the channeled scablands. *Bulletin of the Geological Society of America Bulletin*, **36**, 144.

BRETZ, J. H. 1925*b*. The Spokane Flood beyond the channeled scablands. *Journal of Geology*, **33**, 97–115, 236–259.

BRETZ, J. H. 1927*a*. What caused the Spokane Flood? *Bulletin of the Geological Society of America*, **38**, 107.

BRETZ, J. H. 1927*b*. Channeled Scabland and the Spokane. *Journal of the Washington Academy of Sciences*, **17**, 200–211.

BRETZ, J. H. 1927*c*. The Spokane Flood: a reply. *Journal of Geology*, **35**, 461–468.

BRETZ, J. H. 1928*a*. Channeled Scabland of eastern Washington. *Geographical Review*, **18**, 446–477.

BRETZ, J. H. 1928b. Alternative hypotheses for channeled scabland. *Journal of Geology*, **36**, 193–223, 312–341.

BRETZ, J. H. 1928c. Bars of the Channeled Scabland. *Bulletin of the Geological Society of America*, **39**, 643–702.

BRETZ, J. H. 1928d. Bars of the Channeled Scabland (Abstract). *Bulletin of the Geological Society of America*, **39**, 159–160.

BRETZ, J. H. 1929. Valley deposits immediately east of the channeled scabland of Washington. *Journal of Geology*, **37**, 393–427, 505–541.

BRETZ, J. H. 1930a. Lake Missoula and the Spokane Flood. *Bulletin of the Geological Society of America*, **41**, 92–93.

BRETZ, J. H. 1930b. Valley deposits immediately west of the channeled scabland. *Journal of Geology*, **38**, 385–422.

BRETZ, J. H. 1932a. *The Grand Coulee*. American Geographical Society, Special Publications, **15**.

BRETZ, J. H. 1932b. *The Channeled Scabland*. 16th International Geological Congress, U. S., 1933, Guidebook **22**, Excursion C-2.

BRETZ, J. H. 1959. *Washington's Channeled Scabland*. Washington Division of Mines and Geology, Bulletin, **45**.

BRETZ, J. H. 1969. The Lake Missoula floods and the channeled scabland. *Journal of Geology*, **77**, 505–543.

BRETZ, J. H. 1980. Presentation of the Penrose Medal to J Harlen Bretz: Response. *Bulletin of the Geological Society of America, Part II*, **91**, 1095.

BRETZ, J. H., SMITH, H. T. U. & NEFF, G. E. 1956. Channeled Scabland of Washington: new data and interpretations. *Bulletin of the Geological Society of America*, **67**, 957–1049.

CALKINS, F. C. 1905. *Geology and Water Resources of a Portion of East-Central Washington*. US Geological Survey, Water-Supply Paper, **118**.

CHAMBERLIN, T. C. 1890. The method of multiple working hypotheses. *Science*, **15**, 92–96.

CONDON, T. 1902. *The Two Islands and What Came of Them*. J. K. Gill & Co., Portland, OR.

DANA, J. D. 1882. The flood of the Connecticut River valley from the melting of the Quaternary glacier. *American Journal of Science*, **23**, 87–97, 179–202, 360–373; **24**, 428–433.

DAVIS, W. M. 1926. The value of outrageous geological hypotheses. *Science*, **63**, 463–468.

DAWSON, W. L. 1898. Glacial phenomena in Okanogan County, Washington. *American Geologist*, **22**, 203–217.

FLINT, R. F. 1928. Pleistocene terraces of the lower Connecticut Valley. *Bulletin of the Geological Society of America*, **39**, 955–984.

FLINT, R. F. 1932. Deglaciation of the Connecticut Valley. *American Journal of Science*, **24**, 152–156.

FLINT, R. F. 1935. Glacial features of the southern Okanogan region. *Bulletin of the Geological Society of America*, **46**, 169–194.

FLINT, R. F. 1936. Stratified drift and deglaciation of eastern Washington. *Bulletin of the Geological Society of America*, **47**, 1849–1884.

FLINT, R. F. 1937. Pleistocene drift border in eastern Washington. *Bulletin of the Geological Society of America*, **48**, 203–231.

FLINT, R. F. 1938a. Origin of the Cheney–Palouse scabland tract. *Bulletin of the Geological Society of America*, **48**, 461–524.

FLINT, R. F. 1938b. Scabland auf dem Columbia Plateau im ostlichen Washington. *Journal of Geomorphology*, **1**, 130–139.

FLINT, R. F. 1947. *Glacial Geology and the Pleistocene Epoch*. Wiley, New York.

FLINT, R. F. 1957. *Glacial and Pleistocene Geology*. Wiley, New York.

FLINT, R. F. 1971. *Glacial and Quaternary Geology*. Wiley, New York.

GOULD, S. J. 1978. The great scablands debate. *Natural History*, **87**, 12–18.

GILBERT, G. K. 1886. The inculcation of scientific method by example. *American Journal of Science*, **31**, 284–299.

HARDING, H. T. 1929. Possible supply of water for the channeled scablands. *Science*, **69**, 188–190.

HOBBS, W. H. 1940. Glacial history of southeastern Washington. *Bulletin of the Geological Society of America*, **51**, 2024.

HOBBS, W. H. 1943. Discovery in eastern Washington of a new lobe of the Pleistocene continental glacier. *Science*, **98**, 227–230.

HOBBS, W. H. 1945. Scabland and Okanogan Lobes of the Cordilleran Glaciation, and their lake histories. *Bulletin of the Geological Society of America*, **56**, 1167.

HOBBS, W. H. 1947. *The Glacial History of the Scabland and Okanogan Lobes, Cordilleran Continental Glacier*. J. W. Edwards, Ann Arbor, MI (privately printed).

HODGE, E. T. 1934. Origin of the Washington scabland. *Northwest Science*, **8**, 4–11.

HODGE, E. T. 1940. Glacial history of southeastern Washington. *Bulletin of the Geological Society of America*, **51**, 2024.

KEYES, C. 1935. Glacial origin of the Grand Coulee. *Pan-American Geologist*, **63**, 189–202.

MALDE, H. E. 1968. *The Catastrophic Late Pleistocene Bonneville Flood in the Snake River Plain*. US Geological Survey, Professional Paper, **596**.

MCMACKEN, J. G. 1937. Vicissitudes of Spokane River in late geological times. *Pan American Geologist*, **68**, 111–132.

MEINZER, O. E. 1918. The glacial history of the Columbia River in the Big Bend Region. *Washington Academy of Science Journal*, **8**, 411–412.

O'CONNOR, J. E. 1993. *Hydrology, Hydraulics, and Geomorphology of the Bonneville Flood*. Geological Society of America, Special Paper, **274**.

OESTREICH, K. 1915. Die Grand Coulée. *In: American Geographical Society Memorial Volume of the Transcontinental Excursion of 1912*. American Geographical Society, Special Publications, **1**, 259–273.

PARDEE, J. T. 1910. The glacial Lake Missoula, Montana. *Journal of Geology*, **18**, 376–386.

PARDEE, J. T. 1922. Glaciation in the Cordilleran region. *Science*, **56**, 686–687.

PARDEE, J. T. 1940. Ripple marks (?) in glacial Lake Missoula, Montana. *Bulletin of the Geological Society of America*, **51**, 2028–2029.

PARDEE, J. T. 1942. Unusual currents and glacial Lake Missoula. *Bulletin of the Geological Society of America*, **53**, 1569–1600.

PARKER, S. 1838. *Journal of an Exploring Tour beyond the Rocky Mountains*. Privately published, Albany, NY.

PEIRCE, C. S. 1992. *Reasoning and the Logic of Things: The Cambridge Lectures of 1898*. Harvard University Press, Cambridge, MA.

RICHMOND, G. M., FRYXELL, R., NEFF, G. E. & WEIS, P. L. 1965. The Cordilleran Ice Sheet of the northern Rocky Mountains and related Quaternary history of the Columbia Plateau. *In*: WRIGHT, H. E. & FREY, D. G. (eds) *The Quaternary of the United States*. Princeton University Press, Princeton, NJ, 231–242.

RUSSELL, I. C. 1893. *Geologic Reconnaissance in Central Washington*. US Geological Survey, Bulletin, **108**, 1–108.

SALISBURY, R. D. 1901. Glacial work in the western mountains in 1901. *Journal of Geology*, **9**, 721–724.

SCHWENNESSEN, A. T. & MEINZER, O. E. 1918. Ground water in Quincy Valley, Washington. *US Geological Survey, Water-Supply Paper*, **425**, 131–161.

SYMONS, T. W. 1882. *Report of an Examination of the Upper Columbia River*. US Congress, Senate, 47th Congress, First Session, Senate Document, **186**.

THORNBURY, W. D. 1969. *Principles of Geomorphology*. Wiley, New York.

TRIMBLE, D. E. 1963. *Geology of Portland, Oregon and Adjacent Areas*. US Geological Survey, Bulletin, **1119**.

WAITT, R. B. 1985. The case for periodic colossal jokulhlaups from Pleistocene Glacial Lake Missoula. *Bulletin of the Geological Society of America*, **96**, 1271–1286.

WATERS, A. C. 1933. Terraces and coulees along the Columbia River near Lake Chelan, Washington. *Bulletin of the Geological Society of America*, **44**, 783–820.

WHEWELL, W. 1840. *The Philosophy of the Inductive Sciences Founded Upon Their History* (2 vols). Parker, London.

Pleistocene pluvial lakes of the American West: a short history of research

ANTONY R. ORME

Department of Geography, University of California, Los Angeles, CA 90095, USA
(e-mail: orme@geog.ucla.edu)

Abstract: Scientific investigations of Pleistocene pluvial lakes in the American West occurred in five phases. The pioneer phase prior to 1870 saw former lakes identified by missionary priests, fur trappers, military expeditions and railroad surveyors. The classic phase, between 1870 and 1920, linked initially with independent surveys and, after 1879, with the United States Geological Survey and with irrigation and mining ventures, saw most lakes identified and described by such worthies as Gilbert, Russell, Gale, Waring and Thompson. A consolidation phase from 1920 to 1955 provided synthesis and new data but, in the absence of age controls, saw much speculation about temporal links between pluvial lakes, glacial stages, and climate forcing. The initial dating phase between 1955 and 1980 saw radiocarbon dating applied to late Pleistocene lakes and their Holocene relics and successors. The integrative phase since 1980, supported by enhanced field, remote sensing, laboratory and dating techniques, has seen an array of issues involving pluvial lakes linked to changes in regional ecology and global climate. In the above sequence, progress from one phase to the next reflected changes in the intellectual climate and advances in scientific methods. Today, we reflect on the episodic but cumulative increase in knowledge about late Pleistocene pluvial lakes, especially for Lake Bonneville, Lake Lahontan and the eastern California lake cascade. The record of earlier Pleistocene lakes, in some cases successors to Miocene and Pliocene lakes, is less certain because of deformation and erosion or burial. Continuing challenges involve evaluation of the Pleistocene lake record as a whole in the context of late Cenozoic tectonic and climate change, and of contemporary environmental and water-resource issues.

The modern drylands of western North America include numerous internal drainage basins that contained lakes during the wetter stages of the Pleistocene Epoch. These pluvial lakes have attracted scientific and commercial interest since the mid-nineteenth century, while renewed concerns for climate change and water resources have stimulated much recent research. The Pleistocene lakes have all decreased in volume during the drier Holocene, and those that have desiccated entirely are often termed *dry lakes* or *playas*. These playas may still support shallow lakes after ephemeral floods or more prolonged wet spells, such as those linked with El Niño events.

This paper provides a short history of scientific research on the Pleistocene pluvial lakes of the American West, most of which occur in the intermontane Basin and Range Province. These lakes are pluvial in that they formed in response to more effective precipitation and runoff during Pleistocene cold stages, and have diminished in size and often desiccated under warmer drier Holocene conditions. Despite early assumptions, however, recent research cautions against the easy equation of pluvial and glacial stages. Certainly, many closed basins that have dried out during the Holocene were occupied by lakes during Pleistocene cold stages, but some lake highstands preceded or followed glacial maxima. Nor can synchronous water-level fluctuations between neighbouring lake systems be assumed, essentially because storm tracks, water storage, evaporation and linkages between lakes varied in time and space. The history of pluvial-lake research is thus replete with changing interpretations as more data, better techniques and improved models have served to supplant or enhance earlier views.

This paper identifies five major phases of interest in the Pleistocene pluvial lakes of the region: (1) a pioneer phase before 1870; (2) a classic phase from 1870 to 1920; (3) a consolidation phase from 1920 to 1955; (4) an initial dating phase from 1955 to 1980; and (5) an integrative phase since 1980. The timing of these phases is designed to be helpful rather than rigid. For example, the classic phase could be restricted to the research of Gilbert and Russell, mostly published by 1890, but work of comparable merit over the next 30 years deserves inclusion here. Further, the application of dating methods to questions of lake age, initiated with radiocarbon dating in the 1950s, continues today with a wider variety of techniques. This analysis is selective rather than exhaustive; it attempts to capture the essence of each phase by

From: GRAPES, R. H., OLDROYD, D. & GRIGELIS, A. (eds) *History of Geomorphology and Quaternary Geology.*
Geological Society, London, Special Publications, **301**, 51–78.
DOI: 10.1144/SP301.4   0305-8719/08/$15.00 © The Geological Society of London 2008.

reference to the principal themes and players involved, to scientific methods then in use and to the intellectual climate of the time. Late Pleistocene lakes are emphasized because their record is more visible, but earlier lakes and Holocene events are discussed as appropriate.

Physical setting

The Pleistocene pluvial lakes of the American West lay mostly within the Basin and Range Province that extends diagonally through 15° of latitude and 20° of longitude, from southern Oregon in the NW to Chihuahua, Mexico, in the SE (Fig. 1). Most lakes occurred within the Great Basin of western Utah, Nevada and eastern California. Here lay the large Pleistocene systems of Lake Bonneville, Lake Lahontan and the eastern California lake cascade (Fig. 2). Pluvial lakes also occurred in the Mojave Desert, Mexican Highlands and Sacramento Mountains, and a few existed on the Columbia and Colorado plateaus. The Salton Trough contained a sequence of lakes that formed from channel switching of the Colorado River delta but responded to climate change similarly to pluvial lakes. These pluvial lakes were distinct from the ice-dammed lakes that temporarily existed in the Columbia River basin south of the Cordilleran ice sheet, the glacial and ice-marginal lakes of the Rocky Mountains, and lakes of tectonic, volcanic and glacial origin within the Pacific Mountain System.

Many Pleistocene lakes in the American West were the most recent expressions of lakes that had existed episodically earlier in Cenozoic time (Orme 2002). The stage was set initially by the Laramide Orogeny (80–40 Ma) that expelled Cretaceous seas, raised the Rocky Mountains, reactivated older orogens and formed intermontane basins, such as the Uinta Basin in Utah in which the lacustrine Green River Formation accumulated during Palaeogene time. While relative relief remained low, these lakes experienced fine-clastic and carbonate sedimentation and nutrient-rich ecologies under subhumid climates. Later, as North America's active western margin impinged on the East Pacific Rise, extensional tectonics and block faulting formed many closed basins for Neogene lakes in the evolving Basin and Range Province. In rain shadows afforded by the rising Pacific Mountain System, these basins became more isolated and subject to coarse-clastic and evaporite sedimentation and nutrient-poor ecologies under drier conditions. Although these lake deposits were usually deformed and dissected or buried by later events, their basins were often reoccupied by Pleistocene lakes.

During Pleistocene cold stages, the intermontane basins and plateaus formed by Neogene tectonism lay well south of the Cordilleran and Laurentide ice sheets (Fig. 1). However, the southward deflection of moisture-bearing storms by the strong glacial anticyclone over northern North America brought much rain and snow to the region which, with meltwater from mountain glaciers and reduced evaporation, provided abundant runoff to lake systems. The cooler climates of the glacial stages also favoured downslope movement of alpine, subalpine and forest biota toward the lakes, followed by their upslope retreat during warmer, drier stages. Thus, in addition to physical properties, relict lake deposits contain a biogenic record of climate change. The desert and semi-desert conditions of today's mostly dry lake basins are simply the latest expressions of a sequence of late-Cenozoic environmental changes.

The pioneer phase, before 1870

Archaeological evidence reveals a long association of indigenous peoples with the lake basins of the American West. Prehistoric human immigration to North America, which began in earnest during the terminal Pleistocene, came too late for early migrants to see the pluvial lakes at their optimum, but Holocene peoples seem to have been fully aware of the resources available around surviving lakes. Artifacts, campsites and petroglyphs near many former lakes testify to frequent, if seasonal, human occupance that preceded the first contact with Europeans and persisted well into historic time. For the most part, however, summer heat and drought made these lake basins unattractive for permanent settlement.

The first written records of former lakes occur in the diaries of travellers venturing beyond Spanish colonial centres in New Spain (central Mexico). From early northern outposts, such as Santa Fé (founded in 1609), San Xavier del Bac (1700) and Tubac (1752), missionary priests and military expeditions travelled across the Mexican Highlands, the Sonora and Mojave deserts, and the Colorado Plateau (Fig. 3). Driven mainly by religious zeal and political motives, these were not scientific expeditions in a strict sense, but, armed with quadrant and compass and estimating distance in terms of an hour's plodding on horseback, they did reveal features of interest to later scientists. Their logs and maps refer often to *lagos* and *lagunas* (lakes), *ciénagas* (marshes), *pozos* (waterholes) and *tules* (rushes of swampy bottomlands or *tulares*). A map prepared by Bernardo de Miera y Pacheco (1777) during the Dominguez–Escalante expedition across the Colorado Plateau from Santa

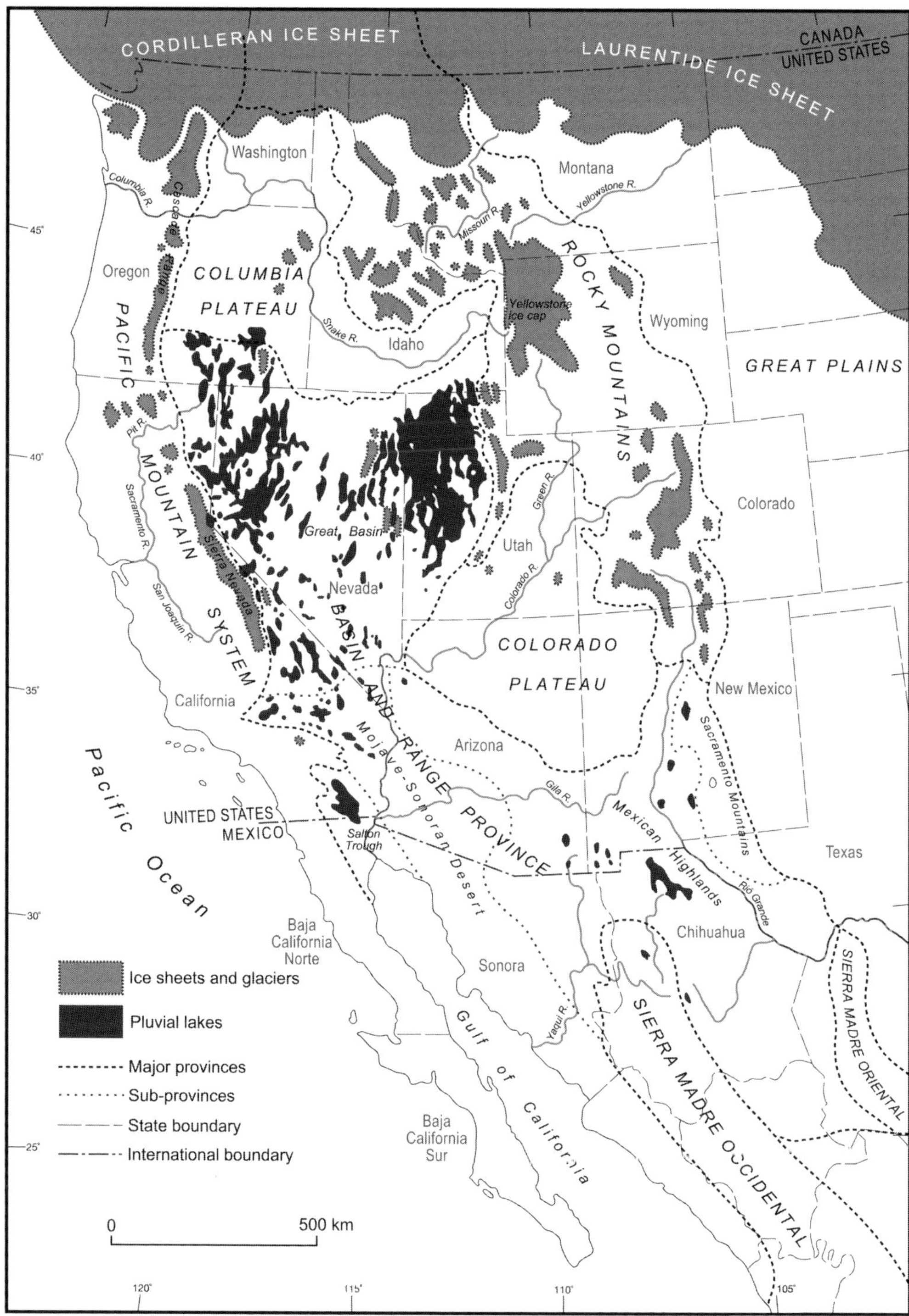

Fig. 1. Late Pleistocene pluvial lakes, ice bodies and geomorphological provinces of the American West (generalized from references in text and author's observations).

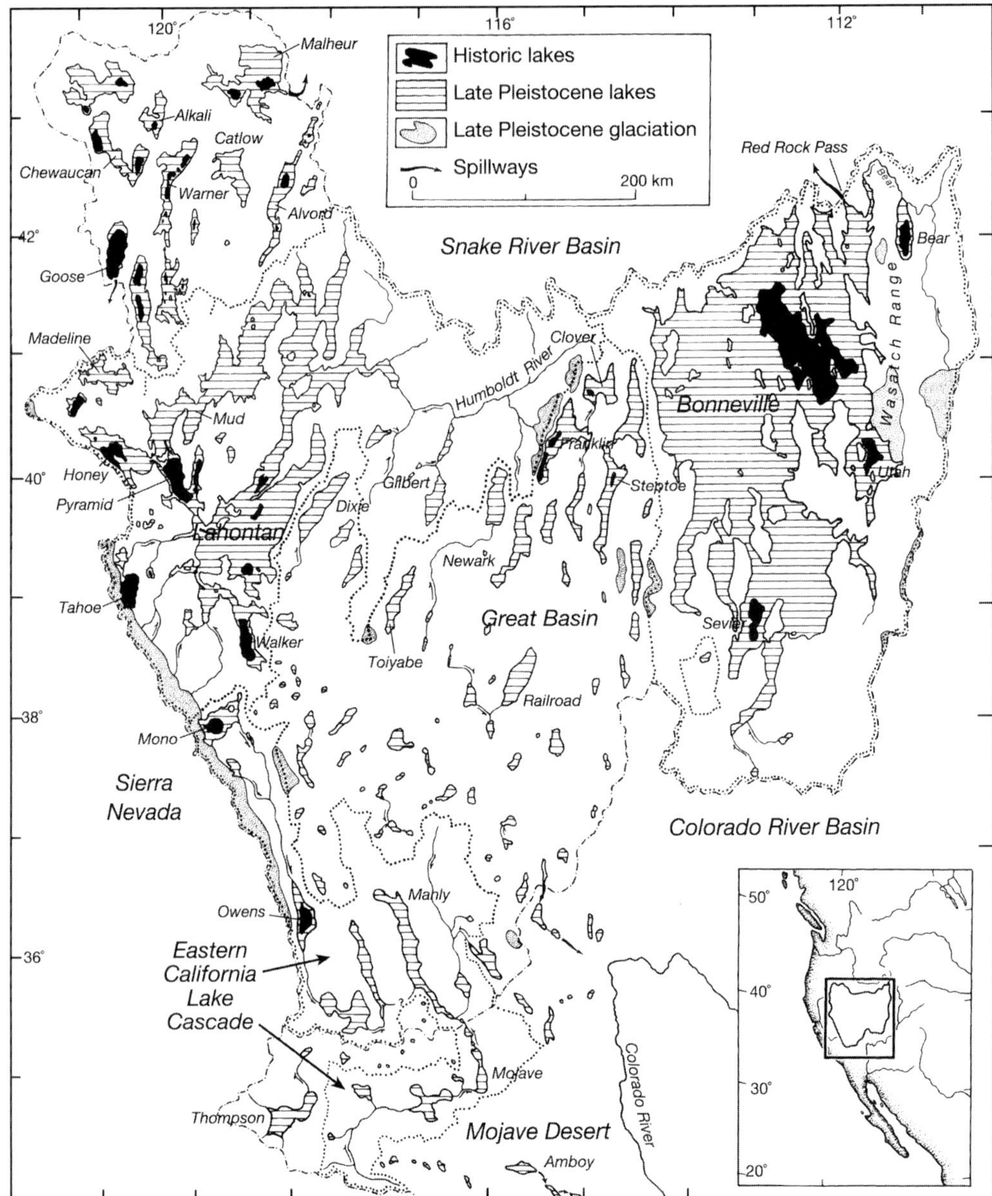

Fig. 2. Late Pleistocene lakes and glaciers of the Great Basin and Mojave Desert (developed from Snyder *et al.* 1964; Morrison 1965; Currey *et al.* 1983; Benson *et al.* 1990; Hershler *et al.* 2002; and the author).

Fé in 1776–1777 shows Laguna de los Timpanogos (Utah Lake), with a hint of drainage northward to the Great Salt Lake, and a Río de San Buenaventura flowing SW into Laguna de Miera (possibly Sevier Lake). The diary of Francisco Garcés, who travelled from Tubac to California and back in 1775–1776, recognizes a former lake in the Mojave Desert, following his eastward descent from the Tehachapi Mountains on 18 May 1776, thus:

96th day. Travelling two and a half leagues in the same direction over an immense plain that clearly was in former times a lake-bed, I found another hole with a little water in it.

(Garcés 1777, p. 59).

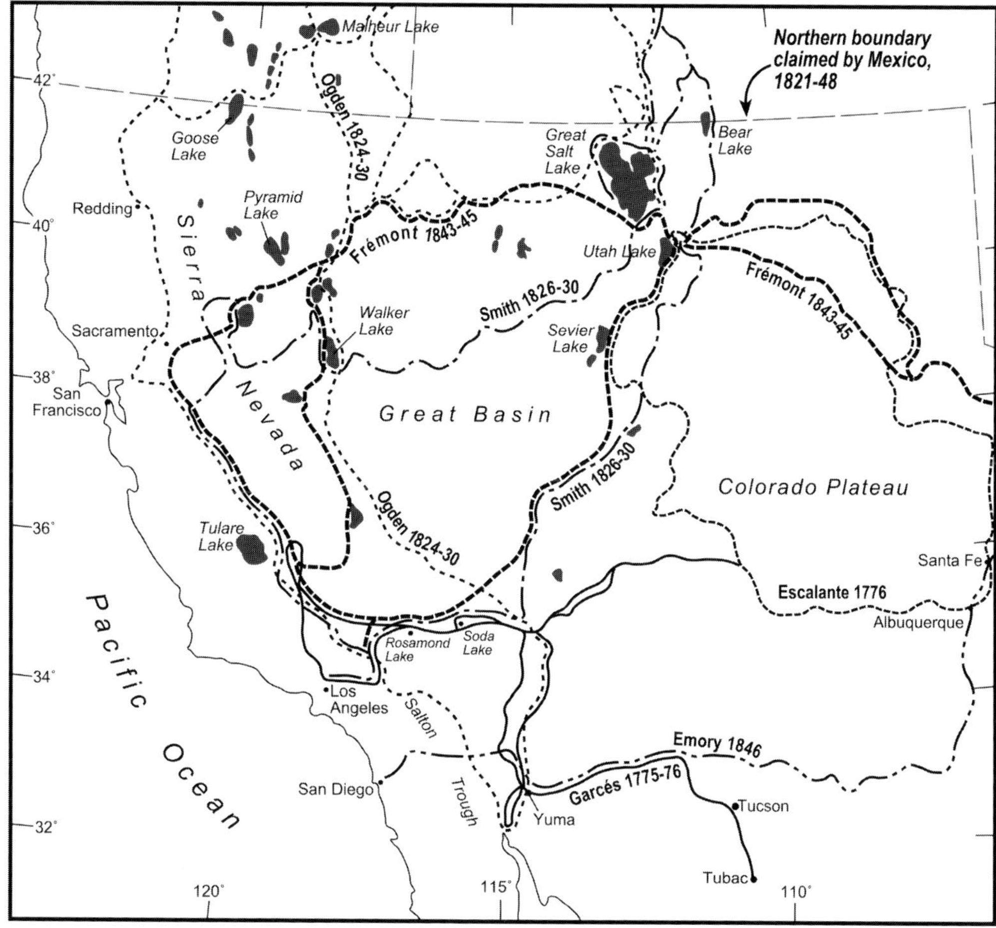

Fig. 3. Routes of some of the pioneer explorers of the American West, 1775–1845 (based on Garcés 1777 (interpreted by Orme 1965); Miera y Pacheco 1777; Frémont 1845, 1846; Emory 1848; and revised from Goetzmann & Williams 1992).

The immense plain was later named the Antelope Valley, and the lake bed was the floor of Pleistocene Lake Thompson.

These Spanish forays notwithstanding, most early nineteenth-century maps of North America left the intermontane region blank (e.g. Arrowsmith 1802), and a map prepared by Alexander von Humboldt in 1811 from Spanish sources did little more than suggest an inland sea, perhaps the mythical *Mer de l'Ouest* of earlier mapmakers (Humboldt 1811). The Louisiana Purchase of 1803, which transferred vast, nominally French, lands to the United States, triggered several expeditions into the intermontane region from the east, just as British and Canadian fur interests were probing from the north. Thus, between 1824 and 1830, Peter Skene Ogden (1794–1854) of the Hudson's

Bay Company surveyed the Humboldt River and former lakes along its course across northern Nevada, and Jedediah Smith (1798–1831) of the Rocky Mountain Fur Company explored the Great Basin (Goetzmann & Williams 1992). Although these were not true scientific expeditions, the information obtained challenged beliefs of earlier travellers and mapmakers in a Mer de l'Ouest and a Río de San Buenaventura that drained to the Pacific Ocean.

Meanwhile, on gaining independence from Spain in 1821, Mexico claimed the 42nd parallel as its northern border – but this was far from Mexico City. American interest in the region thus accelerated, driven by the dubious concept that it was the nation's 'manifest destiny' to control all lands from the Atlantic to the Pacific, and aided by agents such as Benjamin Bonneville (1796–1878) and Joseph Walker (1798–1876), and scouts like James Bridger (1804–1881) and Kit Carson (1809–1868). The author Washington Irving (1783–1859) was so impressed with Bonneville's exploits that, in maps for popular histories, he renamed the Great Salt Lake after him (Irving 1836, 1837). In 1838 the US Army's Corps of Topographical Engineers was formed for the specific purpose of mapping the western frontier – and beyond. Among its personnel, Lieutenant John Charles Frémont (1813–1890) is credited with naming the Great Basin and recognizing that, contrary to earlier belief, it had no outlet to the sea (Frémont 1845).

Territorial rivalries within the region were eventually resolved in favour of the United States by the Oregon Treaty with Britain in 1846, the Treaty of Guadalupe Hidalgo with Mexico in 1848, and the Gadsden Purchase in 1854. The political stage was thus set for more incisive surveys. Of the topographical engineers, Major William Hemsley Emory (1811–1887) mapped dry lakes near the border during the 1846–1848 war with Mexico (Emory 1848), and Captain Howard Stansbury (1806–1863) surveyed 13 former shorelines around the Great Salt Lake in 1849–1850 and observed that 'there must have been here at some former period a vast *inland* sea' (Stansbury 1852, p. 105). By now, with Swiss palaeontologist Louis Agassiz (1807–1873), author of *Etudes sur les Glaciers* (1840), at Harvard University, and with qualified support from Edward Hitchcock (1793–1864) of Amherst College, concepts involving former continental glaciation and related climate change were being debated by American scientists.

The discovery of gold in California in 1848, the anticipation of further mineral wealth in the intermontane West, and the political and commercial desire to link the Atlantic and Pacific coasts of the enlarged nation provided the stimulus for the Pacific railroad surveys, authorized by Congress in 1853. Although these surveys focused primarily on potential railroad routes, the Corps of Topographical Engineers were accompanied by geologists and other scientists charged with identifying and cataloguing resources in the areas traversed. Of the several railroad surveys, three routes directly involved pluvial lakes (Fig. 4). First, a survey party led by Captain John Gunnison reached Sevier Lake in 1853 and was continued across northern Utah Territory in 1854 by Lieutenant Edward Beckwith. The latter surveyed from the Great Salt Lake westward down the Humboldt River near the 41st parallel of latitude, and approached the Sierra Nevada via Mud Lake and Honey Lake. Second, the party led by Lieutenant Amiel Whipple across central New Mexico Territory near the 35th parallel in 1853 and 1854

encountered pluvial lakes in the Mojave Desert. Third, Lieutenant Robert Williamson's survey from central California in 1853 traversed the Mojave Desert en route to the Salton Trough. This southern route was continued eastward in 1854 and 1855 by Lieutenant John Parke, whose parties surveyed from Yuma to the Río Grande along the 32nd parallel and mapped Playa de los Pimas (Wilcox Playa) and other pluvial lakes in southern New Mexico Territory.

The Pacific railroad surveys led to the publication, between 1855 and 1861, of 13 volumes entitled *Reports of Explorations and Surveys, to Ascertain the most Practicable and Economical Route for a Railroad from the Mississippi River to the Pacific Ocean* (US War Department 1855–1861). A sampling of their geology chapters reveals much of interest. Thus, Thomas Phipps Blake (1826–1910) described former lakes encountered by Williamson's 1853 survey and co-ordinated findings from other surveys. From shoreline elevations defined by barometer, and from recognition of tufa, lake clays and the molluscs *Amnicola*, *Anodonta*, *Physa* and *Planorbis*, he identified a former late Holocene freshwater lake in the Salton Trough, which he attributed to beheading of the Gulf of California by a prograding Colorado River delta. On 17 November 1853, he wrote:

These evidences of a former submergence were so vivid and conclusive that it became evident to everyone in the [wagon] train that we were travelling in the dry bed of a former deep lake and extended sheet of water, probably an *Ancient Lake* or an extensive bay.

(Blake 1857, p. 97).

Blake distinguished between horizontal beds of the dry Holocene lake and deformed beds of former Pliocene and Pleistocene lakes, and viewed earthquakes as the cause of their deformation.

Thomas Antisell (1817–1893) of the Parke surveys in 1854 and 1855 found the Río Mimbres terminating in lagoons in the Chihuahuan Desert, noted the pluvial origins of Soda Lake in the Mojave Desert and from the freshwater clam *Anodonta* identified former lake beds (Lake Manix) along the Mojave River farther west (Antisell 1856). He described sediment gradation and salt precipitation at Soda Lake, and summarized the dry lake thus:

The aspect of the playa is remarkably forbidding; a wide expanse, unclad with herbage, bounded by lurid purplish hills without timber, smooth as a bowling green, and glittering in the sun like a snow field, dry brown slopes rising to the margins of the rocks, forms a most dismal picture, and gives an idea of incompleteness and desolation.

(Antisell 1856, p. 102)

Just before the Civil War, Captain James H. Simpson's quest for a good wagon road across

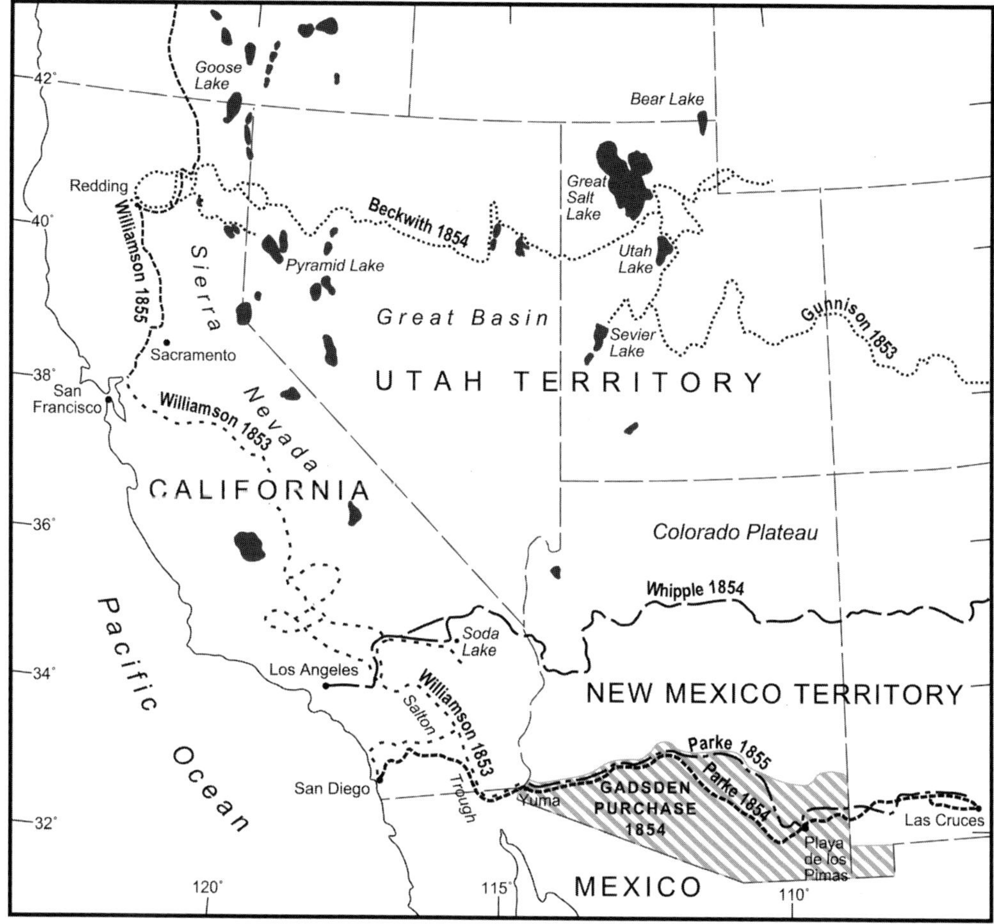

Fig. 4. Railroad survey routes most relevant to identification of pluvial lakes, 1853–1855 (based on Antisell 1856; Blake 1857; US War Department 1855–1861).

the northern Great Basin in 1859, when eventually reported (Simpson 1876), included studies by geologist Henry Engelmann. Above the existing Great Salt Lake, Engelmann (1876) found water marks, shingle benches, calcareous tufa, elevated lake muds containing the freshwater genera *Amnicola*, *Helix*, *Lymnaea* and *Spherium*, and salt beds rich in magnesium sulphate, calcium sulphate and sodium chloride. Such findings paved the way for further exploration after the Civil War (1861–1865). In 1866, the army's Corps of Topographical Engineers was merged with its Corps of Engineers, whose work had been mostly farther east.

The classic phase, 1870–1920

Between 1870 and 1920, scientific understanding of Pleistocene lakes in the American West advanced

significantly. By 1870, growing acceptance of glacial theory had evoked interest in the broader effects of climate change beyond the ice front, stimulated by the findings of European scientists. During the 1860s, for example, Thomas Francis Jamieson (1829–1913) surmised that relict lakes in central Asia had formed originally during glacial stages when evaporation was reduced (Jamieson 1863); Louis Lartet (1840–1899) inferred higher levels for the Dead Sea during the ice age (Lartet 1865); Alfred Tylor (1823–1884) popularized the term *pluvial* to explain river gravels in western Europe which he thought had been deposited under rainier conditions that coincided with glaciers farther north (Tylor 1868); and James Croll (1821–1890) was speculating on Earth's orbital variations as causes of climate change (e.g. Croll 1864). Within the United States, in the aftermath of the Civil War, the eastern scientific community was pondering many of these issues, just as the political establishment and allied commercial interests were again focusing their eyes towards the distant west.

The classic phase of field research in the American West was facilitated by completion of three transcontinental railroads: across northern Utah Territory in 1869, across central New Mexico Territory in 1881, and through the Gadsden Purchase in 1883. Beyond the railheads, however, travel remained challenging and was achieved mostly by saddle horse, pack mule and wagon across difficult terrain in a continental climate of winter cold and snow, and summer heat, thunderstorms and flash floods. Gradually, new railroad links, many directed along the north–south axes of closed basins, improved access. In the haste to tap mineral wealth, however, survey findings were often ignored and railroads built across supposedly dry lake beds. Thus, the Tonopah and Tidewater Railroad (1905–1940), built across Soda and Silver dry lakes in eastern California, was swept away by floods in 1910, 1916 and 1938 (Myrick 1963). Conversely, the transcontinental route of 1869, which ran originally to the north of the Great Salt Lake, was rerouted across the dry lake bed in 1904 and soon suffered from subsidence and flooding. After 1900, motor vehicles began offering greater mobility.

Four major government-sponsored expeditions penetrated the intermontane region after the Civil War: the United States Geological and Geographical Survey of the 40th Parallel (1867–1873) led by civilian geologist Clarence King (1842–1901) under orders from the Corps of Engineers; the United States Geographical Survey West of the One Hundredth Meridian (1867–1879) under Lieutenant George Montague Wheeler (1842–1905), Corps of Engineers; the United States Geographical and Geological Survey of the Rocky Mountain Region (1867–1879) led by John Wesley Powell (1834–1902) and later funded by the Smithsonian Institution; and the United States Geological and Geographical Survey of the Territories (1869–1879), supported initially by the General Land Office and led by veteran geologist Ferdinand Vandeveer Hayden (1829–1887). These were reconnaissance surveys that focused primarily on topographical, geological and resource mapping. In the broader political arena, the surveys were designed to support America's westward expansion, but in reality they were confounded by competition for funds, personnel and equipment. The achievements of their principal scientists have been much reviewed, notably those of Powell along the Colorado River (Chorley *et al.* 1964), of Clarence Edward Dutton (1841–1912) on the Colorado Plateau (Orme 2007) and of Grove Karl Gilbert (1843–1918) on the Henry Mountains and Lake Bonneville (Pyne 1980; Yochelson 1980; Sack 1989; Oviatt 2002). Eventually, in 1879, overlaps between the three remaining surveys were resolved, after much political wrangling, by the formation of a single United Sates Geological Survey (USGS), with Clarence King as its Director. The USGS inherited the scientific work and many of the personnel from the earlier surveys.

Among the most noteworthy studies to emerge from these surveys was Gilbert's research on pluvial Lake Bonneville (Gilbert 1875, 1890). The bathymetry of the Great Salt Lake, which had risen 3 m since Stansbury's survey in 1850, was resurveyed by King in 1869 and Hayden from 1868 to 1871. Gilbert's research there began with the Wheeler survey (1871–1874), expanded with the Powell survey (1875–1879) and concluded with the USGS in 1880–1881. However, owing to administrative duties with the USGS, Gilbert's monograph on Lake Bonneville was delayed until 1890. By then, it was dated and, although numbered Monograph 1, had been preceded by other works including, in 1885, Monograph 11 on Lake Lahontan by I.C. Russell.

Gilbert brought to the West an awareness of lake systems from his earlier work with the Ohio Geological Survey in 1869–1871, for whom he had mapped glacial Lake Maumee (Gilbert 1873). He applied Washington Irving's name Lake Bonneville to the pluvial lake in 1872, defined its former shorelines and maximum extent (51 700 km^2), and measured historic fluctuations of the Great Salt Lake (Gilbert 1875). In brief, Gilbert recognized two major highstands: the Bonneville stage during which the lake spilled northward through Red Rock Pass to the Snake River; and the Provo stage 110 m lower when the lake, now controlled by a bedrock sill in the pass, was invaded by river deltas that invited correlation with the cold wet conditions of a glacial stage. Gilbert provided valuable insights on lakeshore processes (Fig. 5) (Gilbert 1884; Kraft 1980), lake and delta sedimentation during lake 'highstands' (Gilbert 1884, 1890; Hunt 1980), and river and evaporite sedimentation during lowstands. From the domed form of former shorelines across the Bonneville Basin, he also identified isostatic responses to water loading and unloading (Gilbert 1890; Mabey 1980). While lacking absolute chronology, he outlined a temporal framework for the late Pleistocene and Holocene history of Lake Bonneville, thereby contributing to debates involving climate change, and also speculated about earlier lakes in the basin. Gilbert's work heralded a progressive phase of research into pluvial lakes and served as a model for careful reasoning based on field observations and hypothesis testing. Whereas some of his ideas may not

Fig. 5. Shoreline features of Lake Bonneville at the north end of the Oquirrh Mountains, Utah (plate XVI by H. H. Nichols, artist, and G. K. Gilbert, geologist, in Gilbert 1884).

be wholly sustainable today (see Sack 1989), his work at the time was a remarkable achievement and served as the foundation for later research.

Having impressed as an assistant to Gilbert on Lake Bonneville in 1880, Israel Cook Russell (1852–1906) was dispatched to NW Nevada in 1881 and 1882 to study Lake Lahontan, curiously named by King after Baron Louis-Armand de la Hontan (1666–1715), sometime explorer of the distant Mississippi valley. Russell (1885) provided the first clear sense of the complex linkages between the many lakes that, at maximum stage, combined to form this, the second largest lake system (22 300 km^2) in the Great Basin. He inferred from geomorphological, stratigraphic and geochemical principles that the pluvial lake had two Pleistocene highstands (Lower Lacustral Clay and Upper Lacustral Clay, separated by Medial Gravel), but had desiccated before modern Pyramid, Winnemucca and Walker lakes formed in the basin. He also distinguished between lithoid, thinolitic and dendritic tufas within the former lake.

Late in 1882, Russell moved south to study the Mono Basin, completing the fieldwork the following summer. Following Gilbert's belief that glaciers from the Wasatch Range had entered Lake Bonneville during the Bonneville highstand, Russell (1889) now showed how glaciers from the Sierra Nevada had penetrated the high shorelines of Mono Lake (Fig. 6). Certain assumptions would later be challenged, notably his belief that lake highstands coincided with glacier maxima, that the lake never overflowed, and that temperature change was the primary force driving glacier and lake fluctuations. His seminal work at Mono Lake was honoured when Putnam (1949) named its Pleistocene predecessor Lake Russell. After 13 years with the USGS, Russell joined the University of Michigan in 1892, where he wrote a textbook on lakes dedicated to Gilbert (Russell 1894). The map for his 1889 report, while showing lakes Bonneville, Lahontan, Mono and Owens, revealed no pluvial lakes in southern Nevada, the eastern Mojave Desert or Arizona.

Gilbert and Russell paved the way for further research but, with the heady years of military expeditions and scientific discovery now passed, later work was stimulated mainly by surveys of water resources and irrigation potential, and by the economic prospects of evaporite minerals in former lake basins. Pluvial lakes certainly aroused intrinsic scientific interest, but it was the political forces emanating from Washington, DC, and state capitols that drove the search for irrigation water and mineral resources, and therefore much of the agenda of the USGS and state agencies. Irrigation had emerged as an important political issue in the 1870s and, although no panacea for the arid West, soon began driving public policy (Rabbitt 1979). Mormon settlers had been irrigating land near Great Salt Lake since 1847, the Owens and Mojave rivers had been diverted for irrigation in the 1860s, and Oscar Loew (1876) of the Wheeler survey had suggested irrigating dry lakes in the Mojave Desert with Colorado River water. As second director of the USGS from 1881 to 1894, Powell became embroiled in irrigation strategies driven by political whims. He formed an Irrigation Survey in 1888, but congressional discontent led to its closure in 1890. But the die was cast, water issues raged, and in that year the USGS formed a Hydrographic Division with funds for stream gauging and, in 1896, for groundwater studies. By now, the United States Department of Agriculture was also engaged with western irrigation issues.

From the 1870s onward, the groundwater resources of the Mojave Desert were being tapped for irrigation and domestic use. By 1890, more than 100 wells were tapping groundwater beneath the Antelope Valley (Hinton 1891), and by 1921 there were 500 wells (Thompson 1921, 1929). These revealed abundant artesian water trapped beneath thick compact clay of very low permeability, leading Thompson (1929) to speculate on the former existence of a large perennial lake, later named Lake Thompson (Miller 1946). Farther east, Thompson (1929) reported on shells and blue clay 30–60 m beneath Harper dry lake and on dissected lake beds along the middle Mojave and Amargosa rivers. Identifying former highstands, he proposed the name Lake Mohave (now Mojave) for the pluvial lake that once covered Soda and Silver dry lakes and speculated on overflow northward to Death Valley. Thompson (1921) also suggested links between the Mojave, Amargosa and Owens river systems based on relict fish faunas, and between Bristol and Cadiz dry lakes (Lake Amboy), but rejected any connection with Danby dry lake farther SE.

Water-resource investigations also confirmed pluvial lakes elsewhere in the region (Fig. 2). Waring (1908, 1909) mapped lakes Alkali, Alvord, Catlow, Chewaucan, Malheur, Warner and others in SE Oregon; Meinzer (1911) described the beach ridges of pluvial Lake Estancia in New Mexico; Meinzer & Kelton (1913) identified former Lake Cochise around Wilcox Playa in SE Arizona; and Schwennesen (1918) defined pluvial lakes in SW New Mexico, confirming a lake origin for the 20 m-high beach ridge encircling pluvial Lake Cloverdale, earlier viewed as a human earthwork.

Meanwhile, when Los Angeles city engineers began casting their eyes on potential water supply from the Owens River in 1903, they opened a new

Fig. 6. Mono Lake basin and invading glaciers in late Pleistocene time (plate XXIX by W. D. Johnson, topographer, and I. C. Russell, geologist, in Russell 1889). Mono Lake in 1883, as depicted by Johnson and Russell in plate XIX, is outlined for comparison.

chapter on the study of pluvial lakes. Gilbert (1875) and Russell (1885) had found evidence of former pluvial conditions at Owens Lake, the river's downstream terminus; and Lee (1906) now showed that the historic saline lake, which had fallen from 1096 m above sea level in 1872 to 1087 m in 1905 owing to irrigation and reduced precipitation, was the successor to a large freshwater pluvial lake that had spilled southward to Searles Lake. Lee (1912) suggested that this former lake had risen to 1155 m; and Gale (1914), by reference to Lake Bonneville and Lake Lahontan, correlated beach ridges with two Pleistocene glacial stages. Gale (1914) reported a pluvial highstand of 687 m for Searles Lake, almost 200 m above the playa floor, and described its spectacular 30-m high tufa crags, The Pinnacles, around its southern margin. He then invoked overflow from there to a high lake in Panamint Valley, where he found boulder deltas and shells of *Amnicola*, *Carinifex*, *Pompholyx*, *Lymnaea*, *Physa* and *Valvata* embedded in relict tufa. Gale (1914) further speculated about water spillage through the Wingate Pass (603 m) to a former lake in Death Valley that he had identified from shallow probes. Thus, the concept of an integrated late Pleistocene system of lakes and

streams, the Owens River cascade, was formulated (Figs 7 & 8).

Contrary to Russell's emphasis on temperature change, Gale favoured precipitation as the primary control over pluvial-lake fluctuations, and argued that lake maxima would have occurred during or shortly after the initiation of glacier retreat. To him, the lake waters were:

derived partly from the normal precipitation, but chiefly from the release by melting of the immense storage in the fields of glaciation. This affords an explanation for the apparent localization of the larger Quaternary lakes in the Great Basin.

(Gale 1914, p. 320)

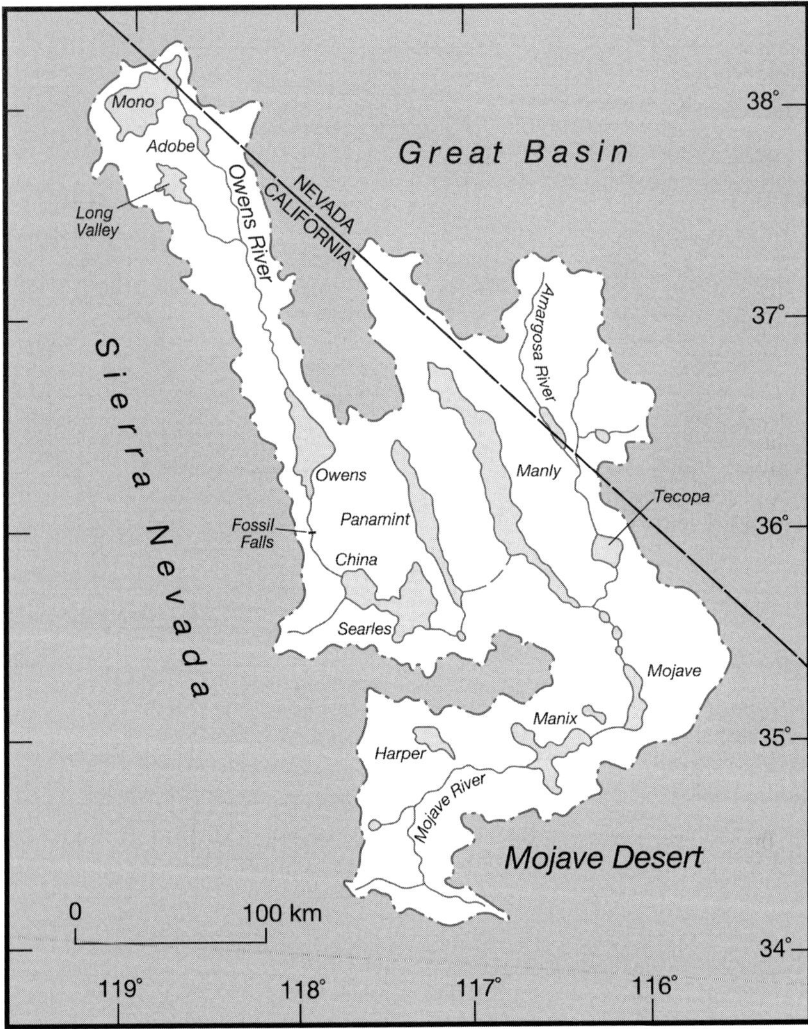

Fig. 7. The Eastern California Lake Cascade at its potential later Pleistocene maximum. Not all lakes or linkages existed simultaneously. Mono Lake became isolated and Lake Manly was deprived of Owens River inflow before the close of Pleistocene time; Lake Tecopa drained after 160 ka; and Lake Manix drained around 20 ka (based on Orme & Orme 2008 and textual references).

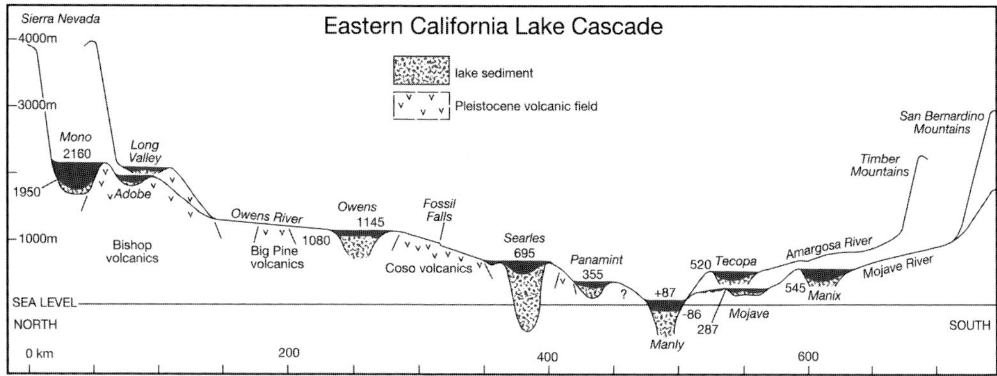

Fig. 8. The Eastern California Lake Cascade, showing in profile the later Pleistocene lakes and their potential linkages (developed from Orme 2002).

Mineral resources had been a focus of earlier surveys in the region and mining interests ensured that they remained so during the early years of the USGS. However, the metallic minerals of most value were in the hard rocks of the mountains or in Cenozoic alluvial deposits of the foothills, rather than in pluvial lakes. Evaporite minerals had been found in Cenozoic lake beds but distance to market and cheaper imports limited mining ventures (Bailey 1902; Campbell 1902). This was to change with the approach of World War I in 1914, which exposed America's dependency on foreign imports of minerals with agricultural, industrial and military value, notably of nitrate from Chile. Thereafter, the saline deposits of pluvial lakes featured prominently in mining activity, which in turn stimulated research on those lakes.

Sodium borate (borax) had been found at Searles Lake in 1862 but shipment by wagon to Los Angeles, or after 1882 to the Mojave railhead, was a tedious affair and, faced with competition from elsewhere, it was not until the Trona Railway was completed in 1914 that these deposits were more thoroughly exploited. Hoyt Stoddard Gale (1876–1949) published the first major evaluation of evaporite minerals beneath Owens, Searles, Panamint and Death Valley dry lakes, and identified the sequence of solute precipitation and evaporation, from carbonates through sulphates, chlorides and borates, to nitrates (Gale 1912*a*, *b*, 1914). From their nature, Gale (1912*a*) questioned Gilbert's 1875 idea that saline minerals were to be found beneath most former lake basins, arguing instead that massive evaporites were to be found only in the deepest and largest lakes.

Meanwhile, Lake Manix was defined by John Peter Buwalda (1886–1954) from dissected late Pleistocene lake beds along the Mojave River east of Barstow. Noted earlier by Antisell (1856), Buwalda (1914) now described upper and lower lake beds of grey-green clay and sand, separated by coarse fluvial sand and gravel, and containing extinct mammals, fish and molluscs. He attributed the sequence to Pleistocene climate changes similar to those at Lake Lahontan, invoked Earth movements to explain lake-bed deformation, and breaching of a downstream ridge by overflow to explain how the lake emptied and its floor became dissected.

Deformed lake beds of much earlier age were also encountered during this classic phase and assigned to earlier Cenozoic epochs beyond the scope of this paper. Nevertheless, speculation developed regarding the significance of certain lake beds that contained both Pliocene and earlier Pleistocene fossils, notably those flanking the Owens Valley and lakes Bonneville and Lahontan.

The consolidation phase, 1920–1955

The history of science is replete with phases of exploration and discovery alternating with phases of synthesis and classification. The study of pluvial lakes is no exception. A consolidation phase began around 1920 which saw scientists synthesize, correlate and augment earlier findings, but generate less new information, mainly because most lakes had already been identified and, apart from increasing use of aerial photography, the techniques needed to spur further advances had yet to develop.

Synthesis was exemplified in two reviews of late Pleistocene lakes that bracket this phase, those by Meinzer (1922) and Snyder *et al.* (1964). Oscar E. Meinzer (1876–1948) showed 68 'ancient lakes' on his map of the Basin and Range Province, most of which had been found by Gilbert and Russell – mainly in the northern Great Basin – and a further 25 mapped farther south by USGS hydrologists. Observing that these lakes decreased southward in number and size, owing to increased evaporation and/or decreased precipitation during the Pleistocene, he stated:

I believe these basins afford a feasible approach to the problem of past climate that has not been exploited. I hope that with further work it will be possible to make a quantitative estimate of Pleistocene climate by comparing the Pleistocene and present hydrology of the basins . . . . The range in humidity (aridity) among the various basins in the Pleistocene epoch was apparently as great as the difference in a given basin between the Pleistocene and the present.

(Meinzer 1922, pp. 547, 549)

Snyder *et al.* (1964) showed 110 pluvial lakes for the Great Basin and western Mojave Desert, and defined their essential dimensions and whether or not they spilled to the outside (shown revised and augmented as Table 1).

Correlation of pluvial lakes with glacial stages was a logical extension of advances in glacial geology wherein, from evidence in mid-continent, several Pleistocene glaciations of North America had been defined. Russell (1889, 1895) had identified three glacier advances at Mono Lake and suggested that the rise and fall of Lake Lahontan might have occurred in step with the advance and retreat of continental glaciers. Atwood (1909) defined two glacial stages in the Wasatch Mountains and endorsed Gilbert's belief that the Bonneville highstand coincided with maximum glaciation. Whereas correlation of pluvial lakes with glacial conditions farther north was defensible, the question of which glacial stage posed problems, as shown when Hay (1927) – based on vertebrate fossil evidence – correlated Lake Bonneville with the Nebraskan stage, the earliest of the then accepted Pleistocene mid-continental glaciations.

Recognition by Elliot Blackwelder (1880–1969), of Stanford University, of four or five glaciations in the Sierra Nevada encouraged their correlation with pluvial episodes in the Great Basin, especially with the more recent Tahoe and Tioga glaciations of presumed early and late Wisconsin glacial age, respectively (Blackwelder 1931). Finding six high shorelines of Mono Lake notched into moraines deposited locally by Tahoe glaciers, Blackwelder assumed that they post-dated the Tahoe stage but could not be more specific. Below the Wasatch Mountains, however, he found deltas of the Bonneville lake stage covering glacial moraines that he considered equivalent in age to the Tioga glaciation of the Sierra Nevada. From this relationship, he concluded that the last glacial maximum predated the Bonneville highstand. Farther from the glaciers, Blackwelder and his student Elmer Ellsworth could do no more than equate highstands at Searles Lake and in the Mojave Desert with the Tahoe and Tioga glaciations (Ellsworth 1932;

Blackwelder & Ellsworth 1936; Blackwelder 1941, 1954). Correlations were a recurrent feature of this phase, with ages of approximately 65 and 25 ka being assigned on geomorphological and astronomical inferences, to the Tahoe and Tioga stages, respectively (e.g. Antevs 1945). The correlative potential of volcanic tephra trapped in lake sediment was also realized (e.g. Allison 1945), but tephra chronologies awaited the development of suitable dating techniques.

Meanwhile, the record of pluvial lakes continued to expand, for example in the upper Owens River basin (Mayo 1934). In Death Valley, where Noble (1926) had noted multiple shorelines on a basalt hill (Shoreline Butte), Blackwelder (1933) concluded that a lake at least 180 m deep had existed during Tahoe time. He named this Lake Manly after William Lewis Manly (1820–1903), who had led an 1849 emigrant party that had strayed into Death Valley to safety in 1850. Further, from alternating evaporites and clays, he surmised that 'Lake Manly was merely the last of a series of Death Valley lakes' (Blackwelder 1933, p. 470). Although he sought vainly for physical evidence, Blackwelder speculated from relict fish evidence that this high lake might have spilled southward to Lake Mojave and then eastward via lakes Amboy and Danby to the Colorado River.

Direct climatic correlations of pluvial lakes were also attempted. Ernst Valdemar Antevs (1888–1974), the Swedish-American scientist, concluded from climatic data that pluvial Lake Estancia in New Mexico post-dated the last glacial maximum and reached its highest stage around 10–12 ka when both temperature and precipitation were higher than today (Antevs 1925, 1935, 1945). In contrast, the USGS hydrologist Luna B. Leopold (1915–2006) used a water-budget approach to argue that lower temperature as well as higher precipitation would have been needed to maintain the lake and that it reached its highest level during the last glacial maximum (Leopold 1951).

Significantly, Antevs, during a symposium on Great Basin natural history held in Salt Lake City in 1942, suggested that the glacial anticyclone overlying the Cordilleran–West Laurentide ice sheets during the late Pleistocene had deflected Pacific storms southward, thereby bringing year-round precipitation, lower temperatures and reduced evaporation to the Great Basin (Antevs 1945, 1948; Oviatt 2002). Then, as anticyclonic conditions over these vast ice sheets began discouraging deep incursions of cyclonic precipitation, ice wastage began to exceed nourishment and the southern ice front retreated northward, even though the Pacific–North American atmospheric circulation system did not change significantly. Based on these assumptions, Antevs theorized that Pacific storms

Table 1. *Selected late Pleistocene pluvial lakes of western North America*

Lake		Basin area maximum (km²)[1]	Lake area maximum (km²)[2]	Water depth estimate (m)[3]	Spillway elevation (m, asl)[4]	Historic surface (m, asl)[5]
Bonneville		140 000	51 700	370	1551	1280
Lahontan		114 700	22 300	280 (Pyramid)	none[6]	1177 (Pyramid)
1	Diamond	8200	760	40		
2	Gilbert	1400	540	80		
3	Madeline	2200	780	40		
4	Tahoe	1400	550			
Eastern California cascade						
5	Mono (Russell)	2100	790	230	2179	
6	Long Valley	1000	230	80		
7	Owens	11 000	530	70 [120]	1146	1081
8	Searles	16 400	910	200	689	493
9	Panamint	19 900	770	280	603	317
10	Harper	1900	220	40		
11	Manix	9400	410	120	571	dissected
12	Mojave	10 500	200	12	287	276
13	Tecopa	10 600	260		eroded	dissected
14	Manly	45 000	1600	180	none	−86
Other California lakes						
15	Cahuilla					
16	Goose (CA/OR)	2900	950	50	?	
17	Pahrump (CA/NV)	2600	630		?	
18	Saline	1700			none	
19	Surprise	3800	1310	200	none	
20	Thompson	5600	950	20 [80]	?	692
Other Oregon lakes						
21	Alkali	2000	550	80	none	
22	Alvord	6100	1270	60	none	
23	Catlow	7700	900	20	none	
24	Chewaucan	3900	1200	90	none	
25	Malheur	14 100	2380	20 [130]	?	
26	Silver-Fossil	5200	1520	30	none	
27	Warner	6900	1250	70	none	
Other Nevada lakes						
28	Clover	2600	890		none	
29	Dixie	6300	1090	70	none	
30	Fish	2800	480		?	
31	Franklin (Ruby)	4900	1220	60	?	
32	Hubbs	1600	530	80		
33	Long (Meinzer)	1900	920	90	?	
34	Newark	3600	930	90	none	
35	Railroad	9200	1360	100	none	
36	Spring	4300	870	80	none	
37	Steptoe	4600	1190	110	?	
38	Toiyabe	3400	650	52	none	
39	Waring	9200	1330	50	none	
Arizona						
40	Cochise		190	150		
41	Red		320	40	none	839
New Mexico and Chihuahua						
42	Animas		390	10	none	1263
43	Cloverdale		30	20	?	1561

(Continued)

Table 1. *Continued*

Lake		Basin area maximum $(km^2)^1$	Lake area maximum $(km^2)^2$	Water depth estimate $(m)^3$	Spillway elevation $(m, asl)^4$	Historic surface $(m, asl)^5$
44	Estancia	4000	1170	50	none	1860
45	Lucero (Otero)		470	20		
46	Playas-Hachita		220		none	1302
47	San Agustín		650	50		
48	King (Salt Basin)		900	20	none	1097
49	Trinity		200	10	none	1425
50	Palomas (CH)	60 000 (El Barreal, Guzman, Santa Mariá, El Fresnel)				

Information is based initially on Snyder *et al.* (1964) for the Great Basin, and then modified and augmented from later studies referenced in this paper. Selected lakes are located in Figures 2–4.

[1] Maximum basin area rounded to nearest 100 km^2.

[2] Maximum lake area rounded to nearest 10 km^2.

[3] Most estimates of water depth, rounded to nearest 10 m, reflect the elevation difference between the maximum highstand of a late Pleistocene lake and the present dry lake bed. This estimate assumes no post-lake erosion, sedimentation or tectonic activity. Where known, the thickness of late Pleistocene lake sediment beneath a dry lakebed is added (in square brackets) to the shoreline level to provide an optimum depth for the late Pleistocene lake.

[4] Spillway elevation is the level to which the upstream lake must fill today before spilling downstream. Owing to subsequent isostatic and tectonic deformation, erosion and alluviation, this is not necessarily the late Pleistocene spillway elevation. 'None' implies that the lake did not spill to an external drainage system.

[5] The historic surface is the average elevation above sea level of either: (a) the historic lake where it exists; or (b) the undissected dry lakebed, which owing to Holocene deflation or sedimentation rarely coincides with the late Pleistocene lakebed.

[6] Lake Lahontan contained internal sills that determined linkages within the larger lake: Mud Lake Slough 1177 m, Emerson Pass 1207 m, Astor Pass 1222 m, Chocolate Pass 1262 m, Darwin Pass 1265 m, Pronto Pass 1292 m, Adrian valley 1308 m.

maintained their southerly track to feed local mountain glaciers and that pluvial lakes 'may have attained their maxima during the first stage of the retreat of the ice sheets', but at about the same time as local mountain glacier maxima (Antevs 1945, p. 1). This was an early attempt to invoke Pacific–North American ocean–atmosphere teleconnections, positive and negative feedbacks and lag effects to explain pluvial-lake responses. However, although favoured by Antevs (1938*a*, *b*), attempts to link lake changes with Earth's orbital variations, following Croll (1864) and now resurrected by Milankovitch (1930), remained a challenge for the future.

As this phase progressed, knowledge about pluvial lakes was augmented by mineral studies, now joined by the search for oil and gas. The quest for nitrate identified a 260 km^2 Pleistocene lake (Lake Tecopa) along the Amargosa River in easternmost California (Noble *et al.* 1922) and improved understanding of pluvial lakes in the Chihuahuan Desert (Noble 1931). There were also retrograde steps, as when Jones (1925) concluded from geochemical evidence that Lake Lahontan had experienced only one lake cycle that had been completed within the past 2000 years.

The ecology of former lakes also attracted interest with respect to fossils preserved within their deposits, notably the rich avian and mammalian faunas in dissected late Pleistocene beds at Lake Manix (Compton 1934) and in deformed Pliocene–Pleistocene lake beds south of Owens Lake (Schultz 1937). Similarities between fish genera (*Cyprinodon*, *Catostomus*, *Sipalates*), now isolated in the region's closed basins, encouraged speculation about former links between drainage systems, and between the Great Basin and the Colorado River. Such links, proposed on physical grounds by railroad geologists of the 1850s, and by Newberry (1871), Jordan (1878) and ichthyologist Snyder (1914), were now revisited by Noble *et al.* (1922) and Blackwelder (1933). Miller (1946) and Hubbs & Miller (1948) then showed convincingly how past and present biogeographies of fish populations could be used to test theories about previous drainage links and subsequent biological isolation, speciation and evolutionary divergence.

The early dating phase, 1955–1980

Until radiometric dating methods began to be applied to pluvial lakes in the 1950s, lake ages had been presumed from correlations involving stratigraphy, palaeontology, sedimentation rates, geomorphology, soils and the like. There were some enterprising approaches, such as the use of chlorine and sodium concentration rates to derive an age of 3500–4200 years for Owens Lake since it last overflowed and became a closed geochemical system (Gale 1914), but these involved assumptions that could not be tested at the time. Antevs (1938*b*), who favoured mid-Holocene drought, suggested that Gale's value could reflect the lapse of time since the later Holocene rebirth of the terminal lake.

Shortly after the discovery of radioactivity in the 1890s, the radioactive decay rates of certain minerals were applied to questions of geological age, for example by Holmes (1911) who used the uranium–lead method to derive an age for Precambrian rocks. Initially, however, such methods lacked the resolution for use with more recent materials. Discovery of the relatively short-lived radioactive isotope ^{14}C in the 1930s and development of radiocarbon dating in the late 1940s (Arnold & Libby 1949) offered a new approach. With a ^{14}C half-life of 5568 $\pm$ 30 years (the Libby half-life, later corrected to 5730 $\pm$ 40 years), and subject to various limitations, radiocarbon dating could be used to constrain the age of the latest Pleistocene lakes and their Holocene relics back to at least 40 000 years BP (before present, where present is AD 1950). The information obtained on these lakes during the previous 100 years certainly provided good opportunities for testing the method. Laboratory analysis also improved as Libby's initial solid carbon approach gave way to increasingly sophisticated gas-counting and liquid-scintillation counting techniques in the 1950s. During this early dating phase, however, and despite a growing awareness of the problem, there was little or no attempt to calibrate radiocarbon years with calendar years based on tree rings and other records. Thus, in this section, ka* is used to designate uncalibrated radiocarbon years before the present (i.e. ^{14}C years BP, rounded to 0.1 ka*).

Searles Lake was among the first pluvial lakes to be treated by radiocarbon dating when Willard Frank Libby (1908–1980), then at the University of Chicago, reported ages of between 24 and 10.5 ka* years for the Parting Mud that lay between salt deposits (Libby 1955). To Flint & Gale (1958), the Parting Mud–Upper Salt couplet indicated a deep freshwater lake, fed by the Owens River, that later evaporated in a closed basin. Stuiver (1964) confirmed these radiocarbon ages, defined three phases of effective wetness based on sedimentation rates, and reflected on sample contamination caused by the incorporation of older and younger carbon. The Bottom Mud–Lower Salt couplet below the Parting Mud was attributed to an earlier lake whose evaporative phase was dated between 32 and 24.5 ka* by comparing ^{14}C and ^{230}Th ages (Peng *et al.* 1978). Scholl (1960) attributed the spectacular pinnacles at Searles Lake to two phases of tufa precipitation coinciding with deep lakes during Tahoe (>32 ka*) and Tioga (23–10 ka*) times.

Radiocarbon-age determinations were soon extended to Lake Bonneville and Lake Lahontan, for which Broecker & Orr (1958) derived 52 radiocarbon ages from freshwater carbonates (shell, marl and tufa). These ages, as reported at the time, revealed low lake levels from before 34 to 25 ka*, a highstand from 25 to 14 ka*, later oscillations with highstands at 11.7 and 10 ka*, and a fall to low Holocene levels after 9 ka*. They suggested that spillage through Red Rock Pass may have occurred as late as 11.7 ka*. This chronology had implications for climate change and, although later revised, provided a basis for debate. The authors thought that a doubling of precipitation and a 5 °C fall in temperature relative to modern annual values would suffice to form these lakes.

Radiocarbon ages were often questioned by scientists whose concepts disagreed with the new findings. Thus, Antevs (1953), familiar with varve chronologies from eastern North America and northern Europe, questioned the validity of certain terminal Pleistocene ages based on samples whose ^{12}C and ^{14}C values he suggested had been contaminated. In general, however, scientists welcomed the opportunities provided by radiocarbon dating and sought to address issues as they arose, including sampling contamination and diagenetic alteration, calibration, and relations of complex carbonates, such as composite tufas, to former lake levels. Thus, Broecker & Kaufman (1965) produced a revised chronology for lakes Bonneville and Lahontan by comparing 80 new radiocarbon ages with ^{230}Th–^{234}U isotope measurements. The latter, one of several U-series disequilibrium methods then being developed to fill the time gap between radiocarbon and potassium–argon (^{40}K–^{40}Ar) dating, was found to be useful for dating calcium carbonate deposited in lakes and for defining ages beyond the temporal range of the radiocarbon method. As reported in 1965, they identified four highstands at 17, 14.5, 12 and 9.5 ka*, with a final downcutting of Red Rock Pass around 12 ka*.

Roger Morrison (1914–2006), however, showed that certain radiocarbon ages based on tufa were reversed relative to chronologies based on soil stratigraphy (Morrison 1964). Despite these ambiguities, Morrison's careful soil studies confirmed most of the earlier work by Russell (1885) and Antevs (1945, 1948) but refuted that of Jones (1925) based on geochemical inferences.

Morrison (1964) concluded that the fluctuations of Lake Bonneville and Lake Lahontan were both similar and synchronous, and that these lakes reached high levels when the Sierra Nevada glaciers were most extensive. Morrison's bold correlations were widely cited during the latter half of this phase but, like many earlier concepts, were challenged as more data and improved age constraints emerged. The need for care in selecting tufa samples was revisited by Benson (1978), whose Lahontan chronology avoided tufas presumed contaminated by younger carbon, recrystallization, and mixing of lake and ground water. Problems posed by small samples and by discrepancies between radiocarbon years and calendar years were also recognized.

Radiocarbon ages for Lake Mojave showed that the main lake event ended around 14.5 ka*, and was followed by shallow oscillations during the 13.75–12, 11–9 and 8.5–7.5 ka* intervals (Ore & Warren 1971). Such evidence suggested a more complex scenario for hydroclimatic forcing of the terminal Pleistocene lake record, the implications of which have since been revisited (e.g. Enzel et al. 2003).

During this phase, advances in isotope geochemistry opened more geochronological avenues, such as dating volcanic ash in lake sediment by using the potassium–argon method that had been formulated around 1950 (Wilcox 1965). In addition, palaeomagnetic signatures of datable changes in Earth's magnetic field began to be identified in lacustrine stratigraphy (Cox et al. 1965).

While the potential of these dating techniques was being pursued, pluvial lakes continued to be explored from surface exposures and cores, and remotely from increasing use of aerial photography and, later in the phase, from satellite-based cameras and sensors. The USGS renewed studies of saline deposits in the Great Basin and Mojave Desert in 1952, publishing many core logs by 1960. Deep cores from the Owens (280 m), China (213 m), Searles (267 m) and Panamint (303 m) basins revealed pluvial events preceding the latest Pleistocene (Smith & Pratt 1957). Although these events were too old for radiocarbon dating, sedimentation rates for Searles Lake suggested that the Bottom Mud was of early Wisconsin age and that the underlying Mixed Layer was of Illinoian age (Smith 1962). In Panamint Valley, Smith (1976) found evidence for five or six deformed shorelines which, extrapolating from a limiting minimum ^{14}C age determination for a shoreline at 648 m, ranged, he suggested, from 111 to 15 ka*. Hunt & Mabey (1966) supported Blackwelder's belief in a Tahoe and Tioga age for Lake Manly, while Hooke (1972) invoked a ^{14}C age of 26–10.5 ka* for the lake's last highstand, and Hooke & Lively (1979) derived U-series ages for tufas showing earlier highstands between 225 and 135 ka. Interfingering muds and evaporites were also mapped in the Bristol, Cadiz and Danby dry lakes (Kupfer & Bassett 1962). To the north, from interbedded lake and tephra deposits around Mono Lake, Lajoie (1968) identified successive deep Pleistocene lakes below 2188 m, the altitude of the spillway to Adobe Valley and the Owens River system (Fig. 7). From sedimentation rates on Paoha Island, he estimated that the lake had existed for at least the last 170 000 years. Lajoie et al. (1982) concluded that deltas had formed when Sierra Nevada glaciers reached their maximum between 26 and 18 ka*, but that Mono Lake rose to its last highstand after glaciers had receded.

Pluvial lakes throughout Nevada were also more thoroughly described during this phase (Mifflin & Wheat 1979), while soil stratigraphy became an important investigative tool for Lake Bonneville and Lake Lahontan (Hunt et al. 1953, Morrison 1964, 1965; Sack 1989). Gilbert's early recognition (1875, 1890) of isostatic deformation at Lake Bonneville was now expanded to explain models of crustal viscosity and lithospheric loading (e.g. Crittenden 1963). In Chihuahua, Mexico, Reeves (1969) suggested that lagunas El Barreal, Guzman, Santa María and El Fresnal had formerly combined to form pluvial Lake Palomas, which as late as early Holocene time covered 5600 km^2 in a in a closed drainage basin covering 60 000 km^2.

During this phase, the retrieval of lake data from much deeper boreholes than had been available earlier also raised questions about the recurrence of pluvial cycles throughout later Cenozoic time. For example, a petroleum test hole in the western Mojave Desert in 1952 passed through 30 m of Lake Thompson deposits to penetrate fluvial and lake deposits of presumed early Pleistocene–Miocene age to a depth of 1700 m (1007 m below sea level) without reaching hard bedrock (Dibblee 1963). Beneath Lake Bonneville, a 307 m-core on the south shore of the Great Salt Lake revealed sequential lacustrine and alluvial events after 3 Ma, including, above the 99 m-deep Matuyama–Brunhes magnetic boundary (c. 790 ka), cyclic alternations involving 28 discrete soils interspersed with lake sediment (Eardley et al. 1973).

The integrative phase, since 1980

Around 1980, the early dating phase, which invited local and regional correlations, passed into an integrative phase as more advanced spatial and temporal methods were applied to questions at continental and global scales over deeper Pleistocene time. This latter phase has undoubtedly been stimulated by increasing concern among scientists, and belatedly in government, for issues of environmental degradation and climate change. Earlier studies of pluvial lakes have been revisited, revised and expanded, and much new work has been initiated. This resurgent interest has yielded a plethora of high-resolution data that has greatly enhanced knowledge of Pleistocene environmental change and posed questions regarding the nature and rate of present and future change. Assessments of recent research have been the focus of many reviews, for example by Smith & Street-Perrott (1983), Sack (1989), Negrini (2002), Reheis *et al.* (2002*b*) Benson (2004) and Phillips (2008). This paper concludes with a brief sample of research in this integrative phase, designed to relate recent work to five principal avenues of enquiry – geochronology, subsurface geology, geomorphology, ecology and climatology – and to the context afforded by earlier studies.

First, advances in geochronology have been promoted by the refinement of radiocarbon dating and the application to pluvial lakes of dating methods often conceived earlier but for which adequate technology had been lacking. Thus, the advent of accelerator-mass spectrometry (AMS) in the late 1970s and further advances in gas- and liquid-scintillation counting methods have permitted high-precision analyses of minute radiocarbon samples from the past 60 000 years (e.g. Thompson *et al.* 1990; Stuiver *et al.* 1998). Within this time frame, however, recognition of the widening discrepancy, with age, between radiocarbon years and calendar years has stressed the need for calibration with tree rings, varves and marine corals (Stuiver *et al.* 1998), and continuing refinement of calibration methods and standards (e.g. Bondevik *et al.* 2006). Increasingly during this phase, published research has specified whether ages are presented in radiocarbon years or calibrated calendar years before present, or both, with minor age discrepancies reflecting different calibration schemes and their continuing refinement. In the present context, and ignoring standard deviation errors, uncalibrated radiocarbon years before present are, as previously mentioned, designated as ka* (^{14}C years BP, rounded to 0.1 ka*), but with calibrated calendar years before present added in parentheses as ka (i.e. cal years BP, rounded to 0.1 ka). Calibrated ages are based on the Cologne Radiocarbon Calibration & Palaeoclimate Research Package (CalPal2007_HULU). Generalized ages based on other methods, again ignoring standard errors, are simply designated as ka and Ma.

Beyond radiocarbon dating, AMS technology has also led to significant advances in the application of other cosmogenic isotopes to lake features. Notable among these has been the use of isotopes with half-lives that fill the gap between the relatively short-lived ^{14}C and the long-lived U-series, ^{40}K–^{40}Ar, and ^{40}Ar–^{39}Ar methods. These include surface-exposure dating using cosmogenic radionuclides, such as ^{41}Ca (half-life 130 000 years), ^{36}Cl (300 000 years), ^{26}Al

(730 000 years) and ^{10}Be (1.5 Ma) (Dorn & Phillips 1991). Thus, based on ^{36}Cl accumulated in shingle, a barrier beach in Death Valley has been dated around 153 ka, or around the Illinoian glacial maximum (Phillips, in Orme & Orme 1991). Saline sediment from Searles Lake has also been dated by ^{36}Cl (Phillips *et al.* 1983). Tephrochronology, aided by electron microprobes, has provided useful age constraints, especially for lakes in the NW Great Basin close to Cascade volcanoes (e.g. Davis 1983; Negrini 2002).

Second, advances in coring technology and laboratory methods have led to a better understanding of pluvial lakes through deeper time. Thus, a 323 m-core (OL-92) from Owens Lake, constrained by mass-accumulation rates, magnetostratigraphy, tephrochronology and radiocarbon dating, penetrated the 790-ka Matuyama–Brunhes magnetic reversal and the 760-ka Bishop volcanic ash near its base, and revealed a shallow lake until 450 ka and then a deeper lake suggestive of tectonic subsidence (Smith & Bischoff 1997). The core showed almost constant sedimentation rates but changes in sediment chemistry (e.g. $CaCO_3$, organic C) and clay minerals (e.g. illite, smectite) revealed lake cycles dominated by 100 000-year periodicities (Bischoff *et al.* 1997). Because oolites occurred only in the topmost deposits, it was assumed that climate since around 5 ka has been drier than at any time over the past 800 000 years. Farther south, a 930 m core (KM-3) from Searles Lake has revealed climate changes since 3.2 Ma, constrained by U-series and K–Ar ages, with muds reflecting deep water fed by Owens River inflow in a wet climate, and evaporites indicating dry conditions (Smith 1984). Climate trends leading to increased regional runoff are recorded later in Searles Lake than in Owens Lake (because the latter had first to overflow), whereas trends leading to decreased regional runoff and desiccation are recorded earlier in Searles Lake than in Owens Lake (Smith & Bischoff 1997). A 186 m core (DV93-1) from Death Valley, dated using the ^{230}Th/^{234}U method, has revealed perennial lakes up to 300 m deep during cold wet intervals from 186 to 135 ka, and again from 35 to 10 ka, whereas mudflats and salt pans characterized the intervening cool dry conditions from 135 to 35 ka, and after 10 ka (Fig. 9) (Lowenstein 2002; Forester *et al.* 2005). Although these three cores are comparable, except for hydrological lags, they raise questions about the timing and magnitude of relative inputs from the Owens, Amargosa and Mojave rivers, and the role of links within and beyond these rivers.

Deep cores well constrained by tephra have also been retrieved elsewhere, for example a 300 000 year-record from Lake Chewaucan in the NW Great Basin (Allison 1982), and a 3 Ma-record

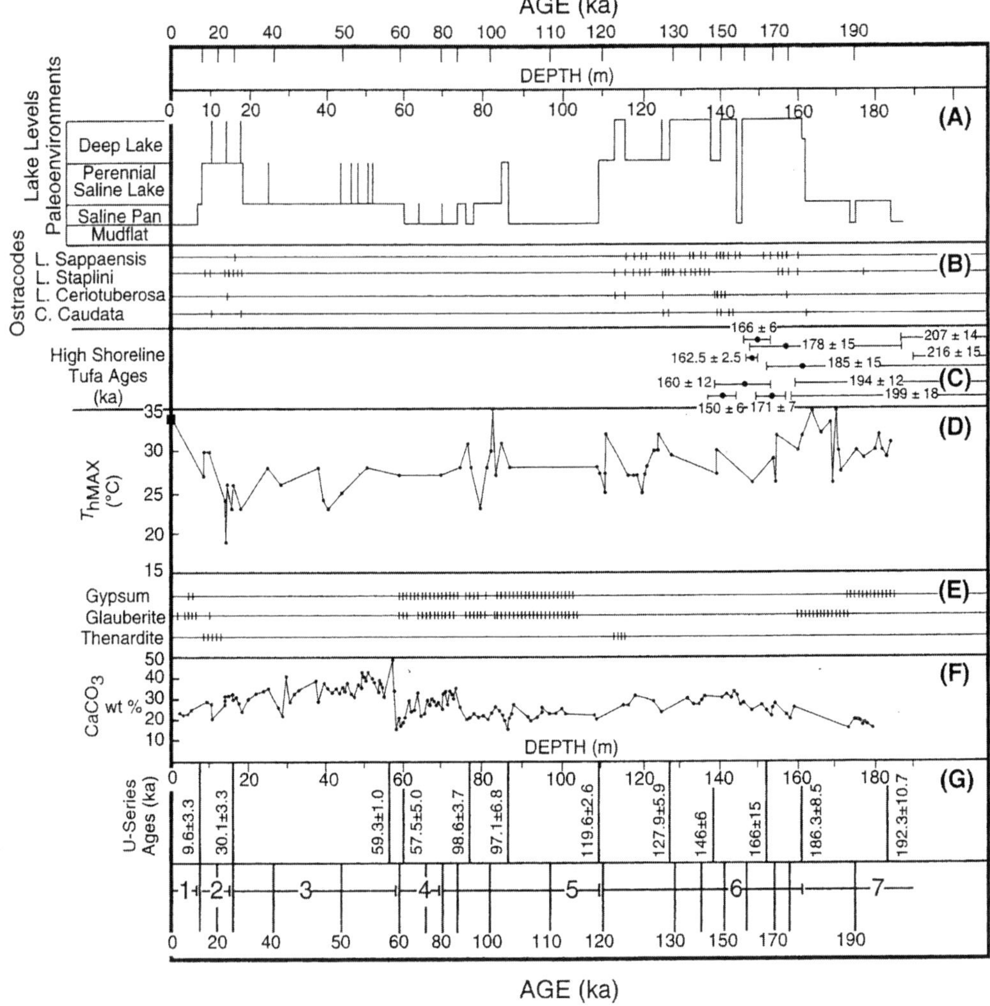

Fig. 9. Palaeoclimate record for Death Valley, California, for the last 200 000 years from core DV-93–1, an example of research during the integrative phase (Lowenstein 2002). (**A**) Lake levels and palaeoenvironments. (**B**) Ostracodes. (**C**) U-series ages of high-shoreline tufa. (**D**) Maximum homogenization temperatures of fluid inclusions in halite. (**E**) Sulphate minerals. (**F**) Weight per cent calcite. (**G**) U-series ages, interpolated ages and marine oxygen isotope stages.

from Tulelake (Lake Modoc) just outside the Great Basin in northern California (Adam *et al.* 1989). Further, older cores, such as that retrieved earlier from the Great Salt Lake (Eardley *et al.* 1973), have been reinterpreted using improved techniques and better understanding of palaeoenvironmental evidence (Oviatt *et al.* 1999).

Third, renewed surface investigations have refined the understanding of late Pleistocene and Holocene lake fluctuations. Whereas cores provide valuable temporal data, such records lack a spatial dimension, other than by inference. Herein lies the significance of datable shoreline features that define the dimensions of former lakes, although unambiguous interpretation is often complicated by later erosion and tectonic activity, and by the preferences of different investigators. Furthermore, earlier ^{14}C-based chronologies have been revised, for example for Lake Bonneville where Scott *et al.* (1983) rejected dates for tufa and marl, and for shells older than 20 ka* or whose ^{230}Th ages were much greater than ^{14}C ages. Even as debate continues, late Pleistocene events at Lake Bonneville are presently thought to involve prolonged closed-basin oscillatory transgressions from before 30 to 15.3 ka* (before 34 to 18.4 ka), followed by the open-basin Bonneville and Provo shoreline phases from 15.3 to 14.2 ka* (18.4–17.4 ka), separated by downcutting of the Red Rock Pass threshold around 14.5 ka* (17.6 ka) and concluding with episodic closed basin regressions after 14.2 ka* (17.4 ka), which lowered the lake to near-modern levels by 12 ka* (14 ka) (Currey *et al.* 1983; Scott *et al.* 1983; Currey 1990; Oviatt *et al.* 1992, 2005; Sack 1999). A minor transgression occurred within the interval 11–10 ka* (12.9–11.5 ka), forming the Gilbert shoreline, before the lake fell towards the late Holocene modern level of the Great Salt Lake.

Farther west, dated features have shown how Lake Lahontan has risen and fallen many times since 40 ka in response to climate forcing and changing intrabasin linkages (Benson *et al.* 1990). In the tradition of Gilbert (1890), who revealed up to 75 m of isostatic deformation at Lake Bonneville, central Lake Lahontan has been found to have risen 22 m from hydroisostatic rebound and faulting since its last highstand at 13 ka* (15.9 ka) (Adams *et al.* 1999). The Pyramid Lake sub-basin of the Lahontan system also saw a terminal Pleistocene highstand at around 10.8 ka* (12.8 ka) and again in late Holocene time (Briggs *et al.* 2005).

At Owens Lake, dated transgressive and regressive sediment facies have defined the timing and extent of lake oscillations from around 31 ka* (35 ka) into Holocene time, with a highstand at around 20 ka* (23.5 ka) during the last glacial maximum, and two prominent terminal Pleistocene highstands at around 12.2 ka* (14.3 ka) and 10.9 ka* (12.8 ka), the latter during the earlier wetter phase of the Younger Dryas cold phase (Orme & Orme 2008). The late Pleistocene shoreline sequence at Owens Lake and the lake's former spillway to the south have since been disrupted by tectonic and magmatic forcing (Orme & Orme 2008).

In Death Valley, Hooke (1999) has correlated shoreline fragments 90 m above sea level with core evidence for a 186–120 ka-deep lake and, allowing for later crustal warping and transpression, has speculated, somewhat controversially, about expansion of that lake southward into the eastern Mojave Desert. Surface evidence has also been invoked to refine Holocene lake changes, notably at Mono Lake (Stine 1990), Lake Mojave (Enzel *et al.* 1989, 2003) and in the Chihuahuan Desert (Castiglia & Fawcett 2006).

Surface investigations of older lake beds, combined with climate and tectonic modelling, have raised questions about the forces shaping Pliocene and early Pleistocene events. Lake Lahontan, for example, may have been much larger during early and middle Pleistocene time, a possibility best explained by invoking wetter climates and later tectonic changes (Reheis *et al.* 2002*a*, *b*). Late Pliocene lake beds in the Waucoba embayment east of Big Pine indicate a westward shift in deposition caused by uplift of the White–Inyo Mountains relative to Owens Valley (Bachman 1978). Further, dissected lake beds interbedded with airfall tephra reveal that a climate much wetter than today sustained Lake Tecopa from 3 Ma until its downstream sill was breached after 160 ka (Hillhouse 1987; Morrison 1991). Such evidence implies that the Sierra Nevada, possibly 1000 m lower around 3 Ma than today, may have presented a less effective barrier to eastbound Pacific storms well into Pleistocene time (Huber 1981; Phillips 2008).

Fourth, although ecological studies of pluvial-lake biota are by no means new, the exchange of information and ideas across disciplinary boundaries has increased significantly. Studies of diatoms, ostracodes, molluscs, insects, fish, pollen, and plant and animal megafossils trapped in lake sediment have refined changes in temperature, salinity, water volume and depth, and basin environments for Pleistocene lakes. Following Hubbs & Miller (1948), recent studies of the depauperate fish fauna of the Great Basin have shown that linkages with neighbouring basins have existed on and off since Miocene time and that small isolated fish populations are prone more to extinction than to speciation (Smith *et al.* 2002). For Lake Bonneville, fossil ostracodes retrieved from the GLAD800 core reveal four major glacial–interglacial sequences after 280 ka (Balch *et al.* 2005), while pollen studies have shown that the last two pluvial optima, during Marine oxygen Isotope Stages (MIS) 6 and 2, saw a regional vegetation mosaic of cold-dry pine woodland and sagebrush steppe (Davis 2002). Extinct mammalian faunas have also received more attention (e.g. Jefferson 1991). Such data feed readily into studies of climate change.

Finally, the search for climatic meaning from increasingly sophisticated data and, conversely, the application of climate models to empirical data, have led to advances in palaeoclimatology, including correlations with marine isotope stages. Better understanding of past climates in turn provides useful information against which to measure present climate trends and to plan accordingly.

Pluvial lakes have climatic meaning over a wide range of scales. At local scales, clastic and evaporite sediment (e.g. Smith 1984; Lowenstein 2002), transgressive–regressive facies (e.g. Orme & Orme 2008), oxygen isotope indicators (e.g. Benson 2004), and biotic proxies such as diatoms, ostracodes and pollen (e.g. Bradbury 1997) reveal much about alternating wet and dry conditions within lake basins. Past wind regimes, inferred from beach properties, lake bathymetry and wave variables, suggest that sustained wind speeds during pluvial maxima were twice those of today (Orme & Orme 1991). Post-pluvial wind erosion is measurable at Lake Thompson where Holocene deflation of the former lake's depocentre has brought moderately deep-water deposits from around 20 ka* (c. 24 ka) to within 4 m of the modern playa surface (Orme 2008).

At the regional scale, the onset and cessation of glaciogenic rock-flour deposition in pluvial lakes constrain the timing of Pleistocene glaciations in nearby mountains (e.g. Bischoff & Cummins 2001). Oxygen isotope data from lake organisms have been compared with a 500 000 year vein-calcite record from Devils Hole, Nevada, to establish regional palaeotemperature trends (Winograd et al. 1992; Smith et al. 1997). Evidence from the smaller Holocene successors to Pleistocene lakes offers valuable insight into regional wet and dry trends lasting from several decades to centuries, notably from the Carson Sink and Pyramid Lake in the Lahontan Basin (Adams 2003; Mensing et al. 2004) and from Mono Lake (Stine 1990). Such information is relevant to water-resource planning in the face of likely prolonged droughts in the future, particularly as many large cities in the American West now rely extensively on water imported from distant drainage basins.

At the continental scale, pluvial lakes have attracted interest as surrogates for Pleistocene climates across North America. Thus, the deep lake in Death Valley has been correlated with the Illinoian glaciation (MIS 6). Later, during the last glacial maximum (MIS 2), most pluvial lakes in the southern Great Basin (30–35°N) rose to high levels between 25 and 18 ka, and then receded from 18 to 15 ka. This implies that maximum continental glaciation and optimum pluviation were responding to comparable climate forcing, with Pacific storm tracks forced far to the south by the strong glacial anticyclone over northern North America (Antevs 1945). In contrast, in the northern Great Basin (c. 40°N), lakes Bonneville, Lahontan and Franklin reached their highest levels at around 18–16 ka, thus lagging behind the last glacial maximum (Thompson et al. 1990). This implies that eastbound storm tracks moved north as continental glaciation waned. Further, following Croll (1864) and Milankovitch (1930), evidence from

lake sediment has revealed cycles that have been linked to Earth's orbital variations (e.g. Smith & Bischoff 1997; Woolfenden 2003).

Terminal Pleistocene lake oscillations also invite speculation about climate change during the transition from MIS 2 to MIS 1. Because glaciogenic inputs to Owens Lake fell rapidly after 15.5 ka (Bischoff & Cummins 2001), transgressions around 15–12 ka were probably triggered not by glacial meltwaters but by continental-scale climate changes involving increased precipitation and runoff and/or reduced evaporation. A later transgression, recorded from Lake Bonneville (Oviatt et al. 2005), Lake Lahontan (Benson & Thompson 1987; Briggs et al. 2005), Mono Lake (Benson et al. 1998) and Owens Lake (Orme & Orme 2008), more certainly correlates with the cold moist phase of the Younger Dryas interlude (13–11.5 ka) now recognized across the northern hemisphere. Thus, by Younger Dryas time, pluvial lakes in and around the Great Basin had approached synchrony. These oscillations reflect climate changes operating at millennial or shorter time-scales that were superimposed onto global warming trends, and invite correlation with high-resolution $\delta^{18}O$ records derived from Greenland ice cores (e.g. Grootes & Stuiver 1997) and marine sediments (e.g. Bond et al. 1993).

Thus, following Meinzer (1922) and Antevs (1945), climate concepts involving latitudinal shifts in the path of the polar jet stream across North America have been resurrected to explain phased pluvial-lake stages across the region on timescales of 10^4–10^5 years (Negrini 2002; Enzel et al. 2003; Benson 2004), while more frequent oscillations have been explained in terms of ocean–atmosphere–cryosphere feedback mechanisms over timescales of 10^3–10^1 years. Even so, spatial heterogeneity and many anomalies still confound such analyses (Mock & Bartlein 1995).

Overview and prospect

Research on the Pleistocene pluvial lakes of the American West developed in five phases. First, the pioneer phase before 1870 identified the surface features of many lakes. Second, the classic research of Gilbert and Russell on lakes Bonneville and Lahontan in the 1870s and 1880s provided broad explanations of late Pleistocene lakes, but raised more questions than could be answered at the time. Third, the consolidation phase between 1920 and 1955 saw much correlation, notably based on Blackwelder's glacial chronology for the Sierra Nevada, and speculation by Antevs concerning the relationship of lake changes to climate forcing. Fourth, radiocarbon dating from the 1950s onwards stimulated geochronologies for late Pleistocene and Holocene lakes, even as cores and deformed surface exposures began revealing more about earlier Pleistocene lakes. Fifth, since 1980, pluvial lakes have been re-examined from broader perspectives of climate and environmental change over deeper time, and the archives contained within them reinterpreted in response to changing scientific concepts and societal needs.

This progress has been cumulative but episodic. The pioneer surveys before 1870 laid the foundation for more discerning studies by Gilbert and Russell, which in turn set the framework for research linked with water and mineral resources before 1920. Lacking age constraints, however, progress then stalled. The resurgence of interest that accompanied the advent of radiocarbon dating has accelerated in recent decades, stimulated by cross-disciplinary studies and a broadening range of investigative techniques. In retrospect, each phase clearly reflected the political, social and economic realities of its time, which in turn shaped the intellectual climate and the resources available. But each phase also provided the foundations upon which subsequent research was built, even as changing realities and new scientific concepts challenged earlier notions and generated the driving force for further enquiry.

Research into past and present lakes of the American West continues apace, driven by changing concepts and needs, not the least of which is the necessity to understand better the temporal and spatial changes that affect the region's water resources.

Despite recent advances, many challenges remain. For example, the Holocene desiccation of Pleistocene lakes and the fluctuations of their successor lakes demand more study as the prelude to understanding continuing climate change; lake changes across the Pleistocene–Holocene boundary need further investigation as surrogates for less accessible environmental changes across earlier glacial–interglacial transitions; pluvial-lake behaviour during the last interglacial–glacial transition needs better understanding in order to reveal more about the onset of terrestrial cold beyond the ice front; and, for the less certain records of pluvial lakes from the earlier Pleistocene, it is important to define the relative roles of climate and tectonism in forcing the changes observed. Historic studies of the pluvial lakes of the American West, as discussed here, provide the essential foundations for future research, even as investigative technologies improve and research agendas change with scientific advances and societal needs.

The author warmly appreciates the thoughtful and constructive reviews of this paper provided by F. M. Phillips and D. Sack, the cartographic expertise of C. Langford, and the editorial input of D. R. Oldroyd.

References

ADAM, D. P., SARNA-WOJCICKI, A. M., RIECK, H. J., BRADBURY, J. P., DEAN, W. E. & FORESTER, R. M. 1989. Tulelake, California: The last 3 million years. *Palaeogeography, Palaeoclimatology, Palaeoecology*, **72**, 89–103.

ADAMS, K. D. 2003. Age and paleoclimatic significance of late Holocene lakes in the Carson Sink, NV, USA. *Quaternary Research*, **60**, 294–306.

ADAMS, K. D., WESNOUSKY, S. G. & BILLS, B. G. 1999. Isostatic rebound, active faulting, and potential geomorphic effects in the Lake Lahontan basin, Nevada and California. *Geological Society of America Bulletin*, **111**, 1739–1756.

AGASSIZ, L. 1840. *Etudes sur les Glaciers*. Jent et Glassmann, Neuchâtel.

ALLISON, I. S. 1945. Pumice beds at Summer Lake, Oregon. *Geological Society of America Bulletin*, **56**, 789–808.

ALLISON, I. S. 1982. *Geology of Pluvial Lake Chewaucan, Lake County, Oregon*. Oregon State Monographs, Studies in Geology, **11**. Oregon State University Press, Corvallis.

ANTEVS, E. V. 1925. On the Pleistocene history of the Great Basin. *In*: JONES, J. C., ANTEVS, E. V. & HUNTINGTON, E. (eds) *Quaternary Climates*, Monograph Series, Carnegie Institution, Washington, DC, **352**, 51–114.

ANTEVS, E. V. 1935. Age of the Clovis Lake clays. *Academy of Natural Sciences of Philadelphia Proceedings*, **87**, 297–312.

ANTEVS, E. V. 1938*a*. Climatic variations during the Last Glaciation in North America. *Bulletin of the American Meteorological Society*, **19**, 12–16.

ANTEVS, E. V. 1938*b*. Post-pluvial climatic variations in the Southwest. *Bulletin of the American Meteorological Society*, **19**, 190–193.

ANTEVS, E. V. 1945. Correlation of Wisconsin glacial maxima. *American Journal of Science*, **243-A**, 1–39.

ANTEVS, E. V. 1948. Climatic changes and pre-white man. *In*: *The Great Basin, With Emphasis on Glacial and Postglacial Times. Bulletin of the University of Utah*, **38**, 167–191.

ANTEVS, E. V. 1953. Geochronology of the Deglacial and Neothermal Ages. *Journal of Geology*, **61**, 195–230.

ANTISELL, T. 1856. Geological Report. *In*: *Routes in California to Connect with the Routes near the Thirty-fifth and Thirty-second Parallels, and Route near the Thirty-second Parallel, between the Rio Grande and Pimas Villages, explored by Lieutenant John G. Parke, Corps Topographical Engineers, in 1854 and 1855*. US War Department, Exploration and Surveys for a Railroad Route from the Mississippi River to the Pacific Ocean, Washington, DC.

ARNOLD, J. R. & LIBBY, W. F. 1949. Age determinations by radiocarbon content: Checks with samples of known age. *Science*, **110**, 678–680.

ARROWSMITH, A. 1802. *A Map Exhibiting All the New Discoveries in the Interior Parts of North America. Inscribed by Permission to the Honorable Governor and Company of Adventurers, Trading into Hudson's Bay, in Testimony of their Liberal Communication . . . to America*. A. Arrowsmith, London (1st edn published in 1795).

ATWOOD, W. W. 1909. *Glaciation of the Uinta and Wasatch Mountains*. United States Geological Survey, Professional Paper, **61**.

BACHMAN, S. 1978. Pliocene–Pleistocene break-up of the Sierra Nevada–White–Inyo Mountains block and formation of the Owens Valley. *Geology*, **6**, 461–463.

BAILEY, G. E. 1902. *Saline Deposits of California*. California State Mining Bureau, Bulletin **24**.

BALCH, D. P., COHEN, A. S. *ET AL*. 2005. Ecosystem and paleohydrological response to Quaternary climate change in the Bonneville Basin, Utah. *Palaeogeography, Palaeoclimatology, Palaeoecology*, **221**, 99–122.

BENSON, L. V. 1978. Fluctuation in the level of pluvial Lake Lahontan during the last 40,000 years. *Quaternary Research*, **9**, 300–318.

BENSON, L. V. 2004. Western lakes. *In*: GILLESPIE, A. R., PORTER, S. C. & ATWATER, B. F. (eds) *The Quaternary Period in the United States*. Elsevier, Amsterdam, 185–204.

BENSON, L. V. & THOMPSON, R. E. 1987. Lake-level variation in the Lahontan Basin for the past 50 000 years. *Quaternary Research*, **28**, 69–85.

BENSON, L. V., CURREY, D. R. *ET AL*. 1990. Chronology of expansion and contraction of four Great Basin lake systems during the past 35 000 years. *Palaeogeography, Palaeoclimatology, Palaeoecology*, **78**, 241–286.

BENSON, L. V., LUND, S. P., BURDETT, J. W., KASHGARIAN, M., ROSE, T. P., SMOOT, J. P. & SCHWARTZ, M. 1998. Correlation of late-Pleistocene lake-level oscillations in Mono Lake, California, with North Atlantic climate events. *Quaternary Research*, **49**, 1–10.

BISCHOFF, J. L. & CUMMINS, K. 2001. Wisconsin glaciation in the Sierra Nevada (79 000–15 000 yr BP) as recorded by rock flour in sediments of Owens Lake, California. *Quaternary Research*, **55**, 14–24.

BISCHOFF, J. L., FITTS, J. P. & FITZPATRICK, J. A. 1997. Responses of sediment geochemistry to climate changes in Owens Lake sediment: an 800-k.y. record of saline/fresh cycles in core OL-92. *In*: SMITH, G. I. & BISCHOFF, J. L. (eds) *An 800 000-year Paleoclimate Record From Owens Lake, California*. Geological Society of America, Special Paper, **317**, 37–48.

BLACKWELDER, E. 1931. Pleistocene glaciation in the Sierra Nevada and basin ranges. *Geological Society of America Bulletin*, **42**, 865–922.

BLACKWELDER, E. 1933. Lake Manly, an extinct lake in Death Valley. *Geographical Review*, **23**, 464–471.

BLACKWELDER, E. 1941. Lakes of two ages in Searles Basin, California. *Geological Society of America Bulletin*, **52**, 1943–1944.

BLACKWELDER, E. 1954. Pleistocene lakes and drainage in the Mojave region, southern California. *In*: JAHNS, R. H. (ed.) *Geology of Southern California*. California Division of Mines and Geology, Bulletin, **170**, 35–40.

BLACKWELDER, E. & ELLSWORTH, E. W. 1936. Pleistocene lakes of the Afton Basin, California. *American Journal of Science*, 5th Series, **31**, 453–463.

BLAKE, W. P. 1857. Geological Report. *In*: *Report of Routes in California, to Connect with the Routes near the Thirty-fifth and Thirty-second Parallels, Explored by Lieut. R. S. Williamson, Corps Topographical Engineers*. US War Department, Exploration and Surveys for a Railroad Route from the Mississippi River to the Pacific Ocean, Washington, DC.

BOND, G., BROECKER, W., JOHNSEN, S., MCMANUS, J., LABEYRIE, L., JOUZEL, J. & BONANI, G. 1993. Correlations between climate records from North Atlantic sediments and Greenland ice. *Nature*, **365**, 143–147.

BONDEVIK, S., MANGERUD, J., BIRKS, H. H., GULLIKSEN, S. & REIMER, P. 2006. Changes in North Atlantic radiocarbon reservoir ages during the Allerød and Younger Dryas. *Science*, **312**, 1514–1517.

BRADBURY, J. P. 1997. A diatom record of climate and hydrology for the past 200 ka from Owens Lake, California, with comparison to other Great Basin records. *Quaternary Science Reviews*, **16**, 203–219.

BRIGGS, R. W., WESNOUSKY, S. G. & ADAMS, K. D. 2005. Late Pleistocene and late Holocene lake high-stands in the Pyramid Lake sub-basin of Lake Lahontan, Nevada, USA. *Quaternary Research*, **64**, 257–263.

BROECKER, W. S. & KAUFMAN, A. 1965. Radiocarbon chronology of Lake Lahontan and Lake Bonneville II, Great Basin. *Geological Society of America Bulletin*, **76**, 537–566.

BROECKER, W. S. & ORR, P. C. 1958 Radiocarbon chronology of Lake Lahontan and Lake Bonneville. *Geological Society of America Bulletin*, **69**, 1009–1032.

BUWALDA, J. P. 1914. Pleistocene beds at Manix in the eastern Mohave Desert region. *University of California Publications, Bulletin of the Department of Geology*, **7**, 443–464.

CAMPBELL, M. R. 1902. *Reconnaissance of Borax Deposits of Death Valley and Mohave Desert*. United States Geological Survey Bulletin, **200**.

CASTIGLIA, P. J. & FAWCETT, P. J. 2006. Large Holocene lakes and climate change in the Chihuahuan Desert. *Geology*, **34**, 113–116.

CHORLEY, R. J., DUNN, A. J. & BECKINSALE, R. P. 1964. *The History of the Study of Landforms or The Development of Geomorphology, Volume 1: Geomorphology Before Davis*. Methuen, London.

COMPTON, L. V. 1934. Fossil bird remains from Manix lake deposits of California. *Condor*, **36**, 166–168.

COX, A., DOELL, R. R. & DALRYMPLE, G. B. 1965. Quaternary paleomagnetic stratigraphy. *In*: WRIGHT, H. E. & FREY, D. G. (eds) *The Quaternary of the United States*. Princeton University Press, Princeton, NJ, 817–830.

CRITTENDEN, M. D. JR. 1963. Effective viscosity of the Earth derived from isostatic loading of Pleistocene Lake Bonneville. *Journal of Geophysical Research*, **68**, 5517–5530.

CROLL, J. 1864. On the physical cause of the change of climate during geological epochs. *Philosophical Magazine*, **28**, 121–137.

CURREY, D. R. 1990. Quaternary palaeolakes in the evolution of semidesert basins, with special emphasis on Lake Bonneville and the great Basin, U.S.A. *Palaeogeography, Palaeoclimatology, Palaeoecology*, **76**, 189–214.

CURREY, D. R., ATWOOD, G. & MABEY, D. R. 1983. *Major Levels of Great Salt Lake and Lake Bonneville*. Utah Geological and Mineral Survey, **Map 73**.

DAVIS, J. O. 1983. Level of Lake Lahontan during deposition of the Trego Hot Springs tephra about 22 400 years ago. *Quaternary Research*, **19**, 312–324.

DAVIS, O. K. 2002. Late Neogene environmental history of the northern Bonneville Basin: a review of palynological studies. *In*: HERSHLER, R., MADSEN, D. B. & CURREY, D. R. (eds) *Great Basin Aquatic Systems History*. Smithsonian Contributions to the Earth Sciences, **33**, 295–308.

DIBBLEE, T. W. 1963. Geology of the Willow Springs and Rosamond quadrangles, California. *United States Geological Survey, Bulletin*, **1089-C**, 141–253.

DORN, R. I. & PHILLIPS, F. M. 1991. Surface-exposure dating: review and critical evaluation. *Physical Geography*, **12**, 303–333.

EARDLEY, A. J., SHUEY, R. T., GVOSDETSKY, V., NASH, W. P., PICARD, M. D., GREY, D. C. & KUKLA, G. J. 1973. Lake cycles in the Bonneville Basin, Utah. *Geological Society of America, Bulletin* **84**, 211–216.

ELLSWORTH, E. W. 1932. *Physiographic history of the Afton Basin, San Bernardino County, California*. PhD dissertation, Stanford University.

EMORY, W. H. 1848. *Notes of a Military Reconnaissance, from Fort Leavenworth, in Missouri, to San Diego, in California, including parts of the Arkansas, del Norte, and Gila Rivers*. United States Congress, 30th session, Senate Executive Document, **7**.

ENGELMANN, H. 1876. Appendix I. Report of the geology of the Country between Fort Leavenworth, K. T., and the Sierra Nevada near Carson Valley. *In*: SIMPSON, J. H. 1876. *Report of Explorations across the Great Basin of the Territory of Utah . . . in 1859*. Engineer Department, United States Army, Washington, DC.

ENZEL, Y., CAYAN, D. R., ANDERSON, R. Y. & WELLS, S. G. 1989. Atmospheric circulation during Holocene lake stands in the Mojave Desert: evidence of regional climate change. *Nature*, **341**, 44–47.

ENZEL, Y., WELLS, S. G. & LANCASTER, N. (eds) 2003. *Paleoenvironments and Paleohydrology of the Mojave and Southern Great Basin Deserts*. Geological Society of America, Special Paper, **368**.

FLINT, R. F. & GALE, W. A. 1958. Stratigraphy and radiocarbon dates at Searles Lake, California. *American Journal of Science*, **256**, 689–714.

FORESTER, R. M., LOWENSTEIN, T. K. & SPENCER, R. J. 2005. An ostracode-based paleolimnologic and paleohydrologic history of Death Valley: 200 to 0 ka. *Geological Society of America Bulletin*, **117**, 1379–1386.

FRÉMONT, J. C. 1845. *Report of the exploring expedition to the Rocky Mountains in the year 1842, and to Oregon and California in the years 1843–44*. United States Congress, 28th session, Senate Executive Document, **174**.

FRÉMONT, J. C. 1846. *Geographical Memoir upon Upper California, in Illustration of his Map of Oregon and California, by John Charles Frémont, Addressed to the Senate of the United States*. Senate, 30th Congress, 1st Session, Miscellaneous Document, **148**.

GALE, H. S. 1912a. Notes on the Quaternary lakes of the Great Basin, with special reference to the deposition of potash and other salines. *United States Geological Survey Bulletin*, **540**, 399–406 (published 1914).

GALE, H. S. 1912b. Prospecting for potash in Death Valley, California. *United States Geological Survey Bulletin*, **540**, 407–415 (published 1914).

GALE, H. S. 1914. *Salines in the Owens, Searles, and Panamint Basins, Southeastern California*. United States Geological Survey Bulletin, **580-L**.

GARCES, F. 1777. *Diario que hà formada el Padre Fr. Francisco Garcés*. Translated by Smithers, A. & Galvin, J. as *A Record of Travels in Arizona and California, 1775–1776, Fr. Francisco Garcés* (*diario* and *plano* interpreted and mapped by A. R. Orme). John Howell Books, San Francisco, CA, 1965.

GILBERT, G. K. 1873. Surface geology of the Maumee Valley. Ohio Geological Survey, **1**, 535–556.

GILBERT, G. K. 1875. Report on the geology of portions of Nevada, Utah, California and Arizona examined in the years 1871 and 1872. *In*: WHEELER, G. M. (ed.) *Report on U.S. Geographical and Geological Surveys West of the 100th Meridian*, Volume 3, *Geology*, Part 1. Government Printing Office, Washington, DC, 17–187.

GILBERT, G. K. 1884. The topographic features of lake shores. *United States Geological Survey, Fifth Annual Report*, 69–123.

GILBERT, G. K. 1890. *Lake Bonneville*. United States Geological Survey, Monograph **1**.

GOETZMANN, W. H. & WILLIAMS, G. 1992. *The Atlas of North American Exploration from the Norse Voyages to the Race to the Pole*. Prentice-Hall, Englewood Cliffs, New York.

GROOTES, P. M. & STUIVER, M. 1997. Oxygen 18/16 variability in Greenland snow and ice with 10^{-3} to 10^{5}-year time resolution. *Journal of Geophysical Research*, **102**, 26455–26470.

HAY, O. P. 1927. *The Pleistocene of the Western Region of North America and its Vertebrate Animals*. Carnegie Institution, Washington, DC, Publication, **322B**.

HERSHLER, R., MADSEN, D. B. & CURREY, D. R. (eds) 2002. *Great Basin Aquatic Systems History*. Smithsonian Contributions to the Earth Sciences, **33**.

HILLHOUSE, J. W. 1987. *Late Tertiary and Quaternary Geology of the Tecopa Basin, Southeastern California*. US Geological Survey Miscellaneous Investigations, **Map I-1728**.

HINTON, R. J. 1891. *Progress Report of Irrigation in the United States*. United States Department of Agriculture, Washington, DC.

HOLMES, A. 1911. The association of lead with uranium in rock minerals, and its application to the measurement of geological time. *Proceedings of the Royal Society of London*, Series A, **85**, 248–256.

HOOKE, R. LE B. 1972. Geomorphic evidence for late-Wisconsin and Holocene tectonic deformation, Death Valley, California. *Geological Society of America Bulletin*, **83**, 2073–2098.

HOOKE, R. LE B. 1999. Lake Manly(?) shorelines in the eastern Mojave Desert, California. *Quaternary Research*, **52**, 328–336.

HOOKE, R. LE B. & LIVELY, R. S. 1979. *Dating of late Quaternary deposits and associated tectonic events by U/Th methods, Death Valley, California*. Final

Report, Grant EAR-7919999, National Science Foundation, Washington, DC.

HUBBS, C. L. & MILLER, R. R. 1948. The zoological evidence: Correlation between fish distribution and hydrographic history in the desert basins of the western United States. *In: The Great Basin, with Emphasis on Glacial and Postglacial Times. Bulletin of the University of Utah*, **38**, 18–166.

HUBER, N. K. 1981. Amount and time of late Cenozoic uplift and tilt of central Sierra Nevada, California – Evidence from the upper San Joaquin River. *United States Geological Survey, Professional Paper*, **1197**, 1–28.

HUMBOLDT A., VON 1811. Carte du Mexique: *In: Atlas géographique et physique du Royaume de la Nouvelle Espagne*. F. Schoell, Paris.

HUNT, C. B. 1980. G. K. Gilbert's Lake Bonneville studies. *In:* YOCHELSON, E. L. (ed.) *The Scientific Ideas of G. K. Gilbert*. Geological Society of America, Special Paper, **183**, 45–59.

HUNT, C. B. & MABEY, D. R. 1966. *Stratigraphy and Structure, Death Valley, California*. United States Geological Survey, Professional Paper, **494-A**.

HUNT, C. B., VARNES, H. D. & THOMAS, H. E. 1953. *Lake Bonneville: Geology of Northern Utah Valley, Utah*. United States Geological Survey, Professional Paper, **257A**.

IRVING, W. 1836. *Astoria, or Anecdotes of an Enterprize beyond the Rocky Mountains*. Carey, Lea & Blanchard, Philadelphia, PA.

IRVING, W. 1837. *The Adventures of Captain Bonneville*. Carey, Lea & Blanchard, Philadelphia, PA.

JAMIESON, T. F. 1863. On the parallel roads of Glen Roy, and their place in the history of the glacial period. *Quarterly Journal of the Geological Society, London*, **19**, 235–259.

JEFFERSON, G. T. 1991. The Camp Cady local fauna: stratigraphy and paleontology of the Lake Manix Basin. *San Bernardino County Museum Association Quarterly*, **38**, 93–99.

JONES, J. C. 1925. Geologic history of Lake Lahontan. *In:* JONES, J. C., ANTEVS, E. V. & HUNTINGTON, E. (eds) *Quaternary Climates*, Monograph Series, Carnegie Institution, Washington, DC, **352**, 3–50.

JORDAN, D. S. 1878. A catalog of the fishes of the fresh waters of North America. *Bulletin of the United States Geological and Geographical Survey*, **4**, 407–442.

KRAFT, J. C. 1980. Grove Karl Gilbert and the origin of barrier shorelines. *In:* YOCHELSON, E. L. (ed.) *The Scientific Ideas of G. K. Gilbert*. Geological Society of America, Special Paper, **183**, 105–113.

KUPFER, D. H. & BASSETT, A. M. 1962. *Geologic Reconnaissance Map of Part of the Southeastern Mojave Desert, California*. United States Geological Survey, Mineral Investigations Field Studies, **Map MF-205**.

LAJOIE, K. R. 1968. *Late Quaternary stratigraphy and geologic history of Mono Basin, eastern California*. PhD dissertation, University of California, Berkeley.

LAJOIE, K. R., ROBINSON, S. W., FORESTER, R. M. & BRADBURY, J. P. 1982. Rapid climatic cycles recorded in closed-basin lakes. *American Quaternary Association Abstracts*, **7**, 53.

LARTET, L. 1865. Sur la formation du basin de la mer morte ou lac asphaltine, et sur les changements survenus dans le niveau de ce lac. *Comptes rendus de l'Académie des Sciences, Paris*, **60**, 796–800.

LEE, C. H. 1912. *An Intensive Study of the Water Resources of a Part of Owens Valley, California*. United States Geological Survey, Water-Supply Paper, **294**.

LEE, W. T. 1906. *Geology and Water Resources of Owens Valley, California*. United States Geological Survey, Water-Supply Paper, **181**.

LEOPOLD, L. B. 1951. Pleistocene climate in New Mexico. *American Journal of Science*, **249**, 152–168.

LIBBY, W. F. 1955. *Radiocarbon Dating*, 2nd edn. University of Chicago Press, Chicago, IL.

LOEW, O. 1876. Report on the alkaline lakes, thermal springs, mineral springs, and brackish waters of southern California and adjacent regions. *In: Annual Report, U.S. Geographical Surveys West of the 100th Meridian*, Appendix **H3**, 188–199.

LOWENSTEIN, T. K. 2002. Pleistocene lakes and paleoclimates (0–200 ka) in Death valley, California. *In:* HERSHLER, R., MADSEN, D. B. & CURREY, D. R. (eds) *Great Basin Aquatic Systems History*. Smithsonian Contributions to the Earth Sciences, **33**, 109–120.

MABEY, D. R. 1980. Pioneering work of G.K. Gilbert on gravity and isostasy. *In:* YOCHELSON, E. L. (ed.) *The Scientific Ideas of G.K. Gilbert*. Geological Society of America, Special Paper, **183**, 61–68.

MAYO, E. B. 1934. The Pleistocene Long Valley lake in eastern California. *Science*, **80**, 95–96.

MEINZER, O. E. 1911. Geology and water resources of Estancia Valley, New Mexico. *United States Geological Survey, Water-Supply Paper*, **275**, 18–23.

MEINZER, O. E. 1922. Map of the Pleistocene lakes of the Basin-and-Range province and its significance. *Geological Society of America Bulletin*, **33**, 541–552.

MEINZER, O. E. & KELTON, F. C. 1913. Geology and water resources of Sulphur Spring Valley, Arizona. *United States Geological Survey, Water-Supply Paper*, **320**, 9–213.

MENSING, S. A., BENSON, L. V., KASHGARIAN, M. & LUND, S. 2004. Holocene pollen record of persistent droughts from Pyramid Lake, Nevada, USA. *Quaternary Research*, **62**, 29–38.

MIERA Y PACHECO, B. 1777. *Plano geografico, de la tierra descubierta y demarcada*. Ministerio de la Guerra, Madrid.

MIFFLIN, M. D. & WHEAT, M. M. 1979. *Pluvial Lakes and Estimated Pluvial Climates of Nevada*. Nevada Bureau of Mines and Geology, Bulletin, **94**.

MILANKOVITCH, M. 1930. Mathematische Klimalehre und astronomische Theorie der Klimaschwankungen. *In:* KÖPPEN, W. & GEIGER, R. (eds) *Handbuch der Klimatologie*. Gerbrüder Borntraeger, Berlin, 1–176.

MILLER, R. R. 1946. Correlation between fish distribution and Pleistocene hydrography in eastern California and southwestern Nevada, with a map of the Pleistocene waters. *Journal of Geology*, **54**, 43–53.

MOCK, C. J. & BARTLEIN, P. J. 1995. Spatial variability of Late-Quaternary paleoclimates in the western United States. *Quaternary Research*, **44**, 425–433.

MORRISON, R. B. 1964. *Lake Lahontan: Geology of the Southern Carson Desert, Nevada*. United States Geological Survey, Professional Paper, **410**.

MORRISON, R. B. 1965. Quaternary geology of the Great Basin. *In:* WRIGHT, H. E. & FRYE, D. G. (eds) *The*

Quaternary of the United States. Princeton University Press, Princeton, NJ, 265–285.

MORRISON, R. B. 1991. Quaternary stratigraphic, hydrologic, and climatic history of the Great Basin, with emphasis on Lakes Lahontan, Bonneville, and Tecopa. *In*: MORRISON, R. B. (ed.) *Quaternary Nonglacial Geology: Conterminous U.S.* Geological Society of America, Boulder, CO, 283–320.

MYRICK, D. F. 1963. *Railroads of Nevada and Eastern California. Volume II: The Southern Roads*. Howell–North Books, Berkeley, CA.

NEGRINI, R. M. 2002. Pluvial lake sizes in the northwestern Great Basin throughout the Quaternary Period. *In*: HERSHLER, R., MADSEN, D. B. & CURREY, D. R. (eds) *Great Basin Aquatic Systems History*. Smithsonian Contributions to the Earth Sciences, **33**, 11–52.

NEWBERRY, J. S. 1871. The ancient lakes of North America, their deposits and drainages. *American Naturalist*, **4**, 641–660.

NOBLE, L. F., MANSFIELD, G. R. *ET AL.* 1922. Nitrate deposits in the Amargosa Region, Southeastern California. *United States Geological Survey, Bulletin*, **724**, 1–99.

NOBLE, L. F. 1926. Note on a colemanite deposit near Shoshone, Calif., with a sketch of the geology of a part of Amargosa Valley. *United States Geological Survey, Bulletin* **785**, 63–73.

NOBLE, L. F. 1931. Nitrate deposits in Southeastern California with notes on deposits in Southeastern Arizona and Southwestern New Mexico. *United States Geological Survey, Bulletin*, **820**, 1–108.

ORE, H. T. & WARREN, C. N. 1971. Late Pleistocene–early Holocene geomorphic history of Lake Mojave, California. *Geological Society of America Bulletin*, **82**, 2553–2562.

ORME, A. J. & ORME, A. R. 1991. Relict barrier beaches as environmental indicators in the California desert. *Physical Geography*, **12**, 334–346.

ORME, A. R. 2002. The Pleistocene legacy: beyond the ice front. *In*: ORME, A. R. (ed.) *The Physical Geography of North America*. Oxford University Press, New York, 55–85.

ORME, A. R. 2007. Clarence Edward Dutton (1841–1912): Soldier, polymath, and aesthete. *In*: WYSE JACKSON, P. N. (ed.) *Four Centuries of Geological Travel*. Geological Society, London, Special Publications, **287**, 271–286.

ORME, A. R. 2008. Lake Thompson, Mojave Desert, California: The late Pleistocene lake system and its Holocene desiccation. *In*: REHEIS, M. C., HERSHLER, R. & MILLER, D. M. (eds) *Late Cenozoic Drainage History of the Southwestern Great Basin and Lower Colorado River Region: Geologic and Biotic Perspectives*. Geological Society of America, Special Paper, **439**, 259–276.

ORME, A. R. & ORME, A. J. 2008. Late Pleistocene shorelines of Owens Lake, California, and their hydroclimatic and tectonic implications. *In*: REHEIS, M. C., HERSHLER, R. & MILLER, D. M. (eds) *Late Cenozoic Drainage History of the Southwestern Great Basin and Lower Colorado River Region: Geologic and Biotic Perspectives*. Geological Society of America, Special Paper, **439**, 207–225.

OVIATT, C. G. 2002. Bonneville Basin lacustrine history: The contributions of G.K. Gilbert and Ernst Antevs. *In*: HERSHLER, R., MADSEN, D. B. & CURREY, D. R. (eds) *Great Basin Aquatic Systems History*. Smithsonian Contributions to the Earth Sciences, **33**, 121–128.

OVIATT, C. G., CURREY, D. R. & SACK, D. 1992. Radiocarbon chronology of Lake Bonneville, eastern Great Basin, USA. *Palaeogeography, Palaeoclimatology, Palaeoecology*, **99**, 225–241.

OVIATT, C. G., MILLER, D. M., McGEEHIN, J. P., ZACHARY, C. & MAHAN, S. 2005. The Younger Dryas phase of Great Salt Lake, Utah, USA. *Palaeogeography, Palaeoclimatology, Palaeoecology*, **219**, 263–284.

OVIATT, C. G., THOMPSON, R. S., KAUFMAN, D. S., BRIGHT, J. & FORESTER, R. M. 1999. Reinterpretation of the Burmester core, Bonneville Basin, Utah. *Quaternary Research*, **52**, 180–184.

PENG, T.-H., GODDARD, J. G. & BROECKER, W. S. 1978. A direct comparison of ^{14}C and ^{230}Th ages at Searles Lake, California. *Quaternary Research*, **9**, 319–329.

PHILLIPS, F. M., SMITH, G. I., BENTLEY, H. W., ELMORE, D. & GOVE, H. E. 1983. Chlorine-36 dating of saline sediments: Preliminary results from Searles Lake, California. *Science*, **222**, 925–928.

PHILLIPS, F. M. 2008. Geological and hydrological history of the paleo-Owens River drainage since the late Miocene. *In*: REHEIS, M. C., HERSHLER, R. & MILLER, D. M. (eds) *Late Cenozoic Drainage History of the Southwestern Great Basin and Lower Colorado River Region: Geologic and Biotic Perspectives*. Geological Society of America, Special Paper, **439**, 115–150.

PUTNAM, W. C. 1949. Quaternary geology of the June Lake district, California. *Geological Society of America Bulletin*, **60**, 1281–1302.

PYNE, S. J. 1980. *Grove Karl Gilbert: A Great Engine of Research*. University of Texas Press, Austin, TX.

RABBITT, M. C. 1979. *Minerals, Lands, and Geology for the Common Defence and General Welfare, Volume 1, Before 1879*. United States Geological Survey, Washington, DC.

REEVES, C. C. 1969. Pluvial Lake Palomas, northwestern Chihuahua, Mexico. *In*: CORDOBA, D. A., WENGERD, S. A. & SHOMAKER, J. W. (eds) *New Mexico Geological Society Guidebook*, **20**, 143–154.

REHEIS, M. C., SARNA-WOJCICKI, A. M., REYNOLDS, R. L., REPENNING, C. A. & MIFFLIN, M. 2002*a*. Pliocene to Middle Pleistocene lakes in the western Great Basin: Ages and connections. *In*: HERSHLER, R., MADSEN, D. B. & CURREY, D. R. (eds) *Great Basin Aquatic Systems History*. Smithsonian Contributions to the Earth Sciences, **33**, 53–108.

REHEIS, M. C., STINE, S. & SARNA-WOJCICKI, A. M. 2002*b*. Drainage reversals in Mono Basin during the late Pliocene and Pleistocene. *Geological Society of America Bulletin*, **114**, 991–1006.

RUSSELL, I. C. 1885. *Geological History of Lake Lahontan: A Quaternary Lake in Northwestern Nevada*. United States Geological Survey, Monograph, **11**.

RUSSELL, I. C. 1889. Quaternary History of Mono Valley, California. *Eighth Annual Report of the United States Geological Survey, 1889*, 267–394.

RUSSELL, I. C. 1894. *Lakes of North America*. Ginn & Co., Boston, MA.

RUSSELL, I. C. 1895. Present and extinct lakes of Nevada. *In*: POWELL, J. W. (ed.) *The Physiography of the United States*. National Geographic Society Monograph 1(4), American Book Company, New York, 101–136.

SACK, D. 1989. Reconstructing the chronology of Lake Bonneville: an historical review. *In*: TINKLER, K. J. (ed.) *History of Geomorphology*, Unwin Hyman, London, 223–256.

SACK, D. 1999. The composite nature of the Provo level of Lake Bonneville, Great Basin, western North America. *Quaternary Research*, **52**, 316–327.

SCHOLL, D. W. 1960. Pleistocene algal pinnacles at Searles Lake, California. *Journal of Sedimentary Petrology*, **30**, 414–431.

SCHULTZ, J. R. 1937. A late Cenozoic vertebrate fauna from the Coso Mountains, Inyo County, California. *In*: *Studies in Cenozoic Vertebrates of Western North America (Contributions to Paleontology)*. Carnegie Institution, Washington, DC, Publication, **487**, 75–109.

SCHWENNESEN, A. T. 1918. *Ground Water in the Animas, Playas, Hachita, and San Luis Basins, New Mexico*. United States Geological Survey, Water-Supply Paper, **422**.

SCOTT, W. E., MCCOY, W. D., SHROBA, R. R. & RUBIN, M. 1983. Reinterpretation of the exposed record of the last two cycles of Lake Bonneville, western United States. *Quaternary Research*, **20**, 261–285.

SIMPSON, J. H. 1876. *Report of Explorations across the Great Basin of the Territory of Utah ... in 1859*. Engineer Department, United States Army, Washington, DC.

SMITH, G. I. 1962. Subsurface stratigraphy of late Quaternary deposits, Searles Lake, California, a summary. *United States Geological Survey, Professional Paper*, **450-C**, 65–69.

SMITH, G. I. 1984. Paleohydrologic regimes in the southwestern Great Basin, 0–3.2 my ago, compared with other long records of 'global' climate. *Quaternary Research*, **22**, 1–17.

SMITH, G. I. & BISCHOFF, J. L. (eds) 1997. *An 800 000-year Paleoclimate Record from Owens Lake, California*. Geological Society of America, Special Paper, **317**.

SMITH, G. I. & PRATT, W. P. 1957. Core logs from Owens, China, Searles, and Panamint basins, California. *US Geological Survey, Bulletin*, **1045-A**, 1–62.

SMITH, G. I. & STREET-PERROTT, F. A. 1983. Pluvial lakes of the western United States. *In*: PORTER, S. C. (ed.) *Late Quaternary of the United States*. University of Minnesota Press, Minneapolis, MN, 190–212.

SMITH, G. I., BISCHOFF, J. L. & BRADBURY, J. P. 1997. Synthesis of the paleoclimate record from Owens Lake core OL-92. *In*: SMITH, G. I. & BISCHOFF, J. L. (eds) *An 800 000-year Paleoclimate Record from Owens Lake, California*. Geological Society of America, Special Paper, **317**, 143–160.

SMITH, G. R., DOWLING, T. E., GOBALET, K. W., LUGASKI, T., SHIOZAWA, D. K. & EVANS, R. P. 2002. Biogeography and timing of evolutionary events among Great Basin fishes. *In*: HERSHLER, R., MADSEN, D. B. & CURREY, D. R. (eds) *Great Basin Aquatic Systems History*. Smithsonian Contributions to the Earth Sciences, **33**, 175–234.

SMITH, R. S. U. 1976. *Late-Quaternary pluvial and tectonic history of Panamint Valley, Inyo and San Bernardino counties, California*. PhD dissertation, California Institute of Technology, Pasadena.

SNYDER, C. T., HARDMAN, G. & ZDENEK, F. F. 1964. *Pleistocene Lakes in the Great Basin*. United States Geological Survey, Miscellaneous Geologic Investigations, **Map I-416**.

SNYDER, J. O. 1914. The fishes of the Lahontan drainage system of Nevada and their relationship to the geology of the region. *Journal of the Washington Academy of Science*, **4**, 299–300.

STANSBURY, H. 1852. *Exploration and Survey of the Valley of the Great Salt Lake of Utah in 1849, including a Reconnaissance of a New Route through the Rocky Mountains*. US Congress, 32nd Session, Senate Executive Document, **3**.

STINE, S. 1990. Late Holocene fluctuations of Mono Lake, eastern California. *Palaeogeography, Palaeoclimatology, Palaeoecology*, **78**, 333–381.

STUIVER, M. 1964. Carbon isotopic distribution and correlated chronology of Searles Lake Sediment. *American Journal of Science*, **262**, 377–392.

STUIVER, M., REIMER, P. J. & BRAZIUNAS, T. F. 1998. High-precision radiocarbon age calibration for terrestrial and marine samples. *Radiocarbon*, **40**, 1127–1151.

THOMPSON, D. G. 1921. Pleistocene lakes along Mohave River, California. *Washington Academy of Sciences Journal*, **11**, 424.

THOMPSON, D. G. 1929. *The Mohave Desert Region, California: A Geographic, Geologic, and Hydrologic Reconnaissance*. United States Geological Survey, Water Supply Paper, **578**.

THOMPSON, R. S., TOOLIN, L. J., FORESTER, R. M. & SPENCER, R. J. 1990. Accelerator-mass spectrometer (AMS) radiocarbon dating of Pleistocene lake sediments in the Great Basin. *Palaeogeography, Palaeolimnology, Palaeoecology*, **78**, 301–313.

TYLOR, A. 1868. On the Amiens gravel. *Quarterly Journal of the Geological Society, London*, **24**, 103–125.

US War Department. 1855–1861. *Reports of Explorations and Surveys, to Ascertain the most Practicable and Economical Route for a Railroad from the Mississippi River to the Pacific Ocean (13 vols)*. US War Department, Washington, DC.

WARING, G. A. 1908. *Geology and Water Resources of a Portion of South–Central Oregon*. United States Geological Survey, Water-Supply Paper, **220**.

WARING, G. A. 1909. *Geology and Water Resources of the Harney Basin Region, Oregon*. United States Geological Survey, Water-Supply Paper, **231**.

WILCOX, R. E. 1965. Volcanic-ash chronology. *In*: WRIGHT, H. E. & FREY, D. G. (eds) *The Quaternary of the United States*. Princeton University Press, Princeton, NJ, 807–816.

WINOGRAD, I. J., COPLEN, T. B. *ET AL*. 1992. Continuous 500 000-year climate record from vein calcite in Devils Hole, Nevada. *Science*, **258**, 255–260.

WOOLFENDEN, W. B. 2003. A 180 000-year pollen record from Owens Lake, CA: terrestrial vegetation change on orbital scales. *Quaternary Research*, **59**, 430–444.

YOCHELSON, E. L. (ed.) 1980. *The Scientific Ideas of G. K. Gilbert*. Geological Society of America, Special Paper, **183**.

Evolution of the theory of continental glaciation in northern and eastern Europe

ANTO RAUKAS

Institute of Geology at Tallinn University of Technology, 5 Ehitajate Road, Tallinn 19086, Estonia (e-mail: anto.raukas@mail.ee)

Abstract: The theory of continental glaciation was worked out independently in different countries, but the idea that glaciers had formerly expanded over much larger areas than today was born in Switzerland (Venetz-Sitten, von Charpentier, Agassiz *et al.*). From the region of 'living glaciers' in the Alps, scientists could make direct comparisons between areas now occupied by ice and those evidently abandoned by ice. Otto Torell in northern Europe and Piotr Kropotkin in Russia are most often named the 'fathers' of the glacial theory. But, in fact, Karl Eduard Eichwald (1795–1876) was the first in the Russian Baltic provinces to consider the possibility of the wide distribution of ice in lowland areas. The glacial theory was strongly supported by the academician Friedrich Schmidt (1832–1908), and features of several glaciations in northern and eastern Europe were first mentioned by Constantin Grewingk in 1879.

The theory of continental glaciation was developed independently in different countries. In northern Europe the starting points are often connected to the well-argued reports by Otto Martin Torell (1828–1900) in the Stockholm Royal Society in 1864 and at the German Geological Society in Berlin in 1875, from which the glacial theory was eventually considered to be scientifically proven.

But long before Torell, the glacial theory had been worked out in mountain areas where scientists could measure the movement of 'living glaciers'. In 1821 Ignaz Venetz-Sitten (1788–1854) argued before the Helvetic Society at Luzern that at some time in the past glaciers had been much more extensive; and in 1829 he stated that not only the Alps but the whole of northern Europe had once been glaciated (Andersen & Borns 1994). In 1824 the Norwegian geologist, Jens Esmark, reached a similar conclusion concerning glaciers in the mountains of Norway. In 1832, the German professor R. Bernhardi (1797–1849) advanced the idea that glacier ice from the far north had once extended across Europe to reach Germany in the south (Kahlke 1984).

In 1834 Johann Georg von Charpentier (1786–1855), a member of the Helvetic Society at Luzern, strongly supported Venetz's views (Andersen & Borns 1994). In 1837 the young zoologist Louis Agassiz (1807–1873) read his epoch-making paper before the Helvetic Society, 'picturing a great ice period' caused by climatic changes and marked by a vast sheet of ice extending from the North Pole to the lowlands. The substance of the address was published in 1840 in his famous work *Études sur les glaciers* (Agassiz 1840). Nowadays, most of glacial geologists and physical geographers acknowledge the pioneering role of Agassiz in the propagation of the glacial theory. He spoke much about glacial ages but paid less attention to the mechanisms of ice motion and the glacial theory itself.

What happened in northern and eastern Europe?

In its early stage of development, the glacial theory was based principally on the occurrence of erratic boulders, which essentially differed from the bedrock on which they were lying. Erratic boulders had long been known but the prevalent theories of their origin were based on the idea of transport by rivers and marine ice. However, it was difficult to attribute huge boulders – for example, Muuga Kabelikivi in Estonia with a diameter of 58 m – to this kind of transportation (Fig. 1).

Estonia is extremely rich in erratic boulders and glacial rafts. Most of the boulders with a circumferences above 30 m (or more than 10 m in length) known in northern Europe occur in Estonia, where their number reaches 62. In addition, there are some 1900 boulders with lengths of more than 3 m (Raukas 1995). Many giant erratic boulders have been found in the sea. For instance, 2–3 km NE of the island of Osmussaar in NW Estonia, there are boulders up to 100 m in diameter and nearly 30 m high. These are supposed to have been carried there by a glacier from the circular ridges of the nearby Neugrund Meteorite Crater, 3–4 km north of the boulder locality (Suuroja 2006).

From: GRAPES, R. H., OLDROYD, D. & GRIGELIS, A. (eds) *History of Geomorphology and Quaternary Geology.* Geological Society, London, Special Publications, **301**, 79–86. DOI: 10.1144/SP301.5   0305-8719/08/$15.00 © The Geological Society of London 2008.

Fig. 1. Muuga Kabelikivi near Tallinn in Estonia. Photograph by the author.

It was no mere chance that Estonia was among the regions where the idea of ancient glaciations was first advanced. This was favoured by the proximity of the territory to the former centre of glaciation and by the similarity of local boulders to the Finnish basement. As early as 1815 the Russian mineralogist, Vassiliy Severgin, advanced the bold idea that boulders in northern Estonia had originated in Finland and had been carried south by ice (Severgin 1815). However, in his later papers, evidently under the influence of Lyell's 'drift theory', he favoured the idea that they had been transported by icebergs and powerful tidal streams.

Karl Eduard Eichwald (1795–1876) (Fig. 2), zoologist and palaeontologist of Tartu University, later professor in Vilnius, Kazan and Sankt-Peterburg, was the first in the Russian Baltic provinces who clearly stated that at least northern Estonia had once been covered by an active glacier, which shaped the topography and carried boulders across its shores. By the mid 1840s, Eichwald (1846) had reached the conclusion that the boulders occurring in diluvial formations had been detached from the Scandinavian mountains by glaciers and transported to Estonia by marine ice. Probably, he was influenced by Roderick Murchison, Edouard de Verneuil and Alexander Keyserling, who a year earlier had vigorously denied the idea of glacial transport of erratic boulders in his study area (Murchison *et al.* 1845). However, Eichwald had collaborated with Louis Agassiz in a study of Devonian placoderms and was well acquainted with Agassiz's idea of a great ice age. Therefore, in this paper he probably tried to combine both concepts (Raukas 1982).

Some years later he (Eichwald 1853) supported the idea that the ancient ice sheet had extended over vast areas and reached at least as far as NW Estonia, leaving behind not only the boulders, but also scratches on the surface of carbonate rocks. Judging by several papers, local geologists were

Fig. 2. Karl Eduard Eichwald (1795–1876). Archive of Tartu University.

well acquainted with Agassiz's ideas. For instance, Wangenheim von Qualen (1852) stressed that Agassiz's theory could not be allied to the territory of Russia, and that the boulders were transported there by water with the help of bottom ice and icebergs. This example shows how in eastern Europe the glacial theory was already a subject of active debate in the 1840s–1850s.

Eichwald also co-operated with Otto Hermann Wilhelm Abich (1806–1866), who worked in Germany in the 1830s and was well acquainted with Agassiz's work in the Alps. In 1842 Abich was recommended for the post of professor at Dorpat (Tartu) University, where he worked until 1847. At the very beginning of his Tartu tenure, he studied the Caucasus where he found irrefutable traces of ancient glaciations and confirmed that the hypothesis Agassiz had proposed for the Alps was also valid for the Caucasus (Raukas 1982).

Although Torell published his land-ice theory in 1872 in the *Proceedings of the Royal Swedish Academy* (Torell 1872), a number of outstanding geologists had already travelled to Sweden to study glacial phenomena in 1864 following Torell's historic speech to the Stockholm Royal Society. Among them was the most prominent Russian geologist of the nineteenth century, the first head of the Russian Geological Committee, academician Gregor Helmersen (1803–1885) (Fig. 3). After a trip to Scandinavia in 1864, he visited Estonia and northern Livonia to study erratic boulders and local geology.

In 1865 Helmersen travelled in Karelia, Finland and Tver Province. The information he obtained from a very large territory served as the basis for his well-known monograph 'Studien über die Wanderblöcke und die Diluvialgebilde Russlands', in which he rejected the drift hypothesis (Helmersen 1869). Helmersen considered the distribution of erratic boulders and the formation of boulder clays and striations to be the result of the combined action of a continental ice sheet, icebergs and erosional processes on land. This conclusion is quite close to modern ideas.

One of the strongest supporters of the glacial theory in Tsarist Russia was the academician Friedrich Schmidt (1832–1908) (Fig. 4). During the summer of 1864 he studied the Quaternary geology of Estonia, and the following year published the results under the title, 'Untersuchungen über die Erscheinung der Glazialformation in Estland und auf Oesel' (Schmidt 1865), in which he demonstrated the undoubted correlation of Estonian and Swedish glacial deposits. He maintained that these deposits had been transported and deposited by an ice sheet spreading from Scandinavia and Finland over the whole of the territory that he had studied. He did not refute the

Fig. 3. Academician Georg Helmersen (1803–1885), the first Head of the Russian Geological Committee. Archive of Tartu University.

Fig. 4. Academician Friedrich Schmidt (1832–1908). Archive of Tartu University.

possibility that glaciers from the mountains of Scandinavia could have descended into the basin of the Baltic Sea where large icebergs broke off and moved southward.

Several years later, Schmidt proved beyond reasonable doubt that during the glacial epoch the Scandinavian ice sheet had extended over the Baltic Sea depression and surrounding territories (Schmidt 1871). As a typical glacial sediment on carbonate bedrock he described the *rihk/rühk* – basal till with angular limestone pebbles and cobbles (Fig. 5).

In the Russian literature, much attention is paid to the studies of Piotr Alekseyevich Kropotkin (1842–1921), son of the General. Being well off, he could travel. In 1869, while investigating erratic boulders on the Island of Suur-Tütarsaar, he concluded that they had been carried there by continental ice (Kropotkin 1869). On 24 October 1873, Kropotkin argued before the Russian Geographical Society that in the past glaciers had covered the territory of Finland. In the same year he published his report of the Olekma–Vitim expedition in Transbaikalia. Here he found polished

Fig. 5. Typical local till (termed '*rihk*' by Friedrich Schmidt) on the carbonaceous bedrock in northern Estonia. Photograph by the author.

and scratched boulders in the river valleys, and concluded that they had been transported and redeposited by a large and thick glacier which had formerly covered the surrounding mountains (Ivanova & Milanovsky 2006). In his lecture before the Russian Geographical Society, on 21 March 1874, he stressed that continental ice covered vast portions of the European part of Russia. In 1876 Kropotkin published his voluminous book *Researches on the Glacial Period*, in which he described in detail different aspects of the morphology, structure and origin of glacial and glacially related deposits and landforms (Kropotkin 1876). As the book was published in Russian, it remained practically unknown in Western countries. However, by that time the glacial theory had already been widely accepted.

Louis Agassiz's pioneering ideas were supported in many classic papers; for example, by T. F. Jamison (1862) and Archibald Geikie (1863) in the UK, and by Timothy Conrad (1839) and Edward Hitchcock (1841) in North America (cited in Flint 1965 and Kahlke 1984). Agassiz himself arrived in the United States in 1842 to become a professor at Harvard where he hastened the wide acceptance of the glacial theory. Publication of a voluminous (575-page) book, *The Great Ice Age and its Relation to the Antiquity of Man* by Geikie in 1874, indicates that in the 1870s the glacial theory had been accepted practically everywhere. From time to time, however, there appeared groups of scientists who interpreted the glacial deposits as being of marine origin based on the presence of marine fauna in tills, and even went so far as to ignore or downplay the glacial evidence. Such ideas were especially strong in the former Soviet Union in the 1960s and 1970s. However, as demonstrated by Dreimanis (1970), marine fossils in the Quaternary deposits are insufficient evidence for marine deposition because fossils can be redeposited.

One or More Glaciations? Further Developments of the Glacial Theory

It was originally assumed that northern and eastern Europe had only been affected by one glaciation, but in 1879 Constantin Caspar Andreas Grewingk (1819–1887) (Fig. 6), Professor of Mineralogy at Tartu University, established that southern Estonia and northern Latvia (Livonia) had, in fact, been glaciated at least twice (Grewingk 1879).

In the Alps, the possibility of more than one glaciation had been known much earlier. By 1822 Venetz had already recorded a layer of lignite between two horizons containing glacial deposits near the shore of Lake Geneva, Switzerland. He suggested that the lignite represented a warm

Fig. 6. Constantin Caspar Andreas Grewingk (1819–1887), who suggested two glaciations in the Baltic area in 1879. Archive of Tartu University.

period that separated two glacial periods (Andersen & Borns 1994). In 1879 Albrecht Penck, in northern Germany, proposed three glaciations. Between 1901–1909 he and Eduard Brückner presented convincing field observations that indicated four glacial advances (Penck & Brückner 1909). Even up until now, the exact number of glaciations is unclear and depends on the definition of glaciation and interglaciation. The Baltic States were covered by ice at least six or seven times.

Before the glacial theory was widely accepted, poorly sorted unconsolidated surficial deposits were thought to be the result of the Biblical flood of Noah. In Germany, the term *diluvium* was widely used. In the British Isles, the same material was called *drift*, a term also used to describe sediments transported along coasts. With the acceptance of the glacial theory to explain the origin of these deposits, glacial drift has come to be the general English term for all sediments transported and deposited by glaciers or associated rivers. Currently, the Scots term *till* is used in the English literature for glacial deposits, while in various languages of eastern Europe the term *moraine* is used for both glacier-depositional landforms and deposits.

During the early studies of glaciers, attention was essentially focused on glacial morphology. Data on sediments were gathered mainly alongside or down-valley from modern Alpine glaciers, where superglacial till was distinguished from basal till.

A basal compact till overlain by a looser and coarser laminated upper till was also distinguished in lowland areas. Earlier, the American T. C. Chamberlin (1883, 1884), one of the founders of modern Quaternary geology, differentiated three classes of drift: subglacial till or 'true till'; upper till (englacial or superglacial till); and subaqueous till (berg till, floe till), and explained their formation. He emphasized that glaciers were merely assumed to be the chief agents; but that secondary and associated agencies were also very important.

Concerning the subglacial transport of debris, two mechanisms were considered in northern and eastern Europe during the nineteenth and the first half of the twentieth century: glacial drag under the ice sole; and transport of debris in the basal part of glacier ice. It was emphasized that till was left behind by passive melting of stagnant ice. Also, two types of marginal landforms were differentiated – push moraines and lateral moraines. Only in the 1970s, was it postulated that till had mostly accumulated below the moving ice and a classification proposed for dynamic facies and subfacies of ground moraines (in the sense of till) (Lavrushin 1970), resulting in several descriptive names of genetically diverse moraines (tills); for example, facies of plastic flow, monolithic moraine, facies of large rafts, facies of imbricate or scaly moraine, and facies of altered moraine.

The Finnish investigator Hans Magnus Hausen (1884–1979), who worked in the NW portion of the East European Plain at the beginning of the twentieth century was the first to attempt a reconstruction of glacial retreat during the last glaciation. In his famous monograph 'Materialen zur Kenntnis der Pleistozänen Bildungen in den russischen Ostseeländern' (Hausen 1913) he presented palaeogeographical schemes that showed successive locations of the active margin of the glacier and the distribution of proglacial lakes. Nowadays, most geologists working in this area support the concept of limited stadial movement of ice cover and the large amount of stagnant ice. However, researchers still disagree as to the extent and thickness of these dead-ice fields. In the course of deglaciation, active, passive and dead-ice conditions probably underwent a gradual change depending on the climate and bedrock topography. Flow tills and collapses of ice cavities were common in dead-ice areas, which is why organic layers with different ages often occur in unusual stratigraphic positions and why they show both horizontal and vertical displacement (Raukas & Karukäpp 1994).

At the beginning of the last century a great contribution to the glacial theory was made by the Swedish geologist Gerard de Geer (1912), who divided the deglaciation history of Scandinavia into three major stages – Daniglacial (20 000–13 000 years ago), Gotiglacial (13 000–10 000 years ago) and Finiglacial (10 000–8000 years ago) – each of which had different palaeoglaciological conditions. de Geer also introduced a new and precise varvological method of investigation, which enabled the establishment of a real geochronology based on counts of annual sediment layers covering the last 12 000 years.

Contemporary studies show (Raukas & Karukäpp 1993) that the Daniglacial was characterized by the development of a thick cover of glacial deposits in the marginal areas of an extensive glacier. Differentiated accumulation of glacial deposits as insular heights on ice divides and accumulative formations between ice streams are typical of the Gotiglacial. The Finiglacial was characterized in erosional glacial formations, radial eskers and drumlins. The Finiglacial ice lobes are bounded by end moraines, marginal eskers and glaciofluvial deltas. In contrast to the Dani- and Finiglacial, the Gotiglacial stage was specified by a high activity of glacial streams and lobes, and a highly variable glacial morphogenesis in different areas.

For more than a century, indicator boulders have been studied in detail in different countries. They show precisely the direction of ice movement, and promote resolution of several stratigraphical and palaegeographical problems. Altogether about 100 different indicator boulders have been described in the Baltic States alone. The results are summarized in the monograph, *Crystalline Indicator Boulders in the Baltic States* (Gudelis 1971). Thanks to common efforts of scientists of different countries, the deglaciation history in northern and eastern Europe has been dated extensively using a variety of methods including ^{14}C, varve chronology, electron spin resonance (ESR), thermoluminescence (TL), optically stimulated luminescence (OSL) and ^{10}Be dating.

In the second half of the century, the great influence of subglacial topography on the movement of ice was pointed out by different authors. The Scandinavian continental glacier had a radial–sectoral structure and formed concentric belts around a glacial centre (Aseyev 1974). Glacial streams with different speeds and energy alternated regularly with ice-divide areas. Glacial accumulation was concentrated into interlobate subglacial uplands, resulting in the formation of a peculiar type of uplands – accumulative insular heights. The formation of such heights included subglacial, englacial, frontal and stagnant ice stages of morphogenesis (Karukäpp *et al.* 1999). These heights have a bedrock core and are composed of several till beds with related intermorainic deposits.

Glacial accumulative insular heights, some thousands of square kilometres in area, form sub-meridional systems comprising from two to four

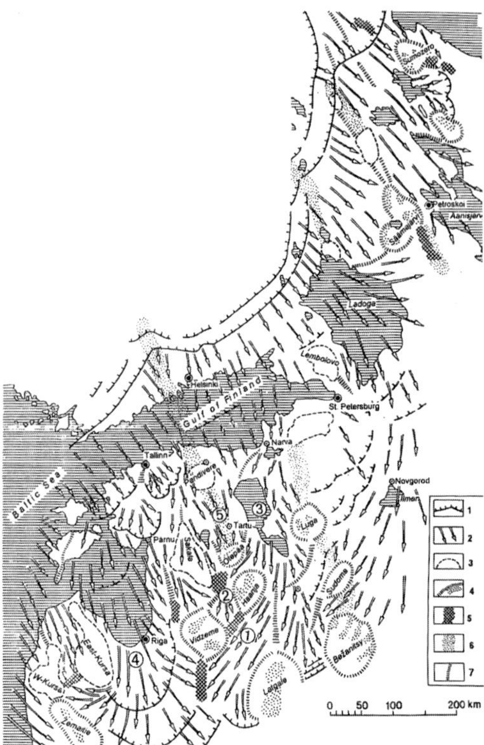

Fig. 7. Time-transgressive reconstruction of the glacial dynamics and morphogenesis of the SE sector of the Scandinavian ice sheet at the end of the Weichselian Glaciation (after Karukäpp 1997): 1, glacier margin; 2, ice flow direction in the final stage of its activity; 3, bedrock upland of ice shed; 4, accumulative insular height; 5, interlobate complex of landforms; 6, ice shed between glacial flows; 7, local ice shed.

units each, possessing a typical hummocky dissected topography and separated from each other by lowlands projecting high above their surroundings. The topmost parts of the heights are represented by high hummocky topography with a complicated structure of glacial deposits, while the landforms gradually diminish in size towards the lower peripheral areas of heights. Establishment of two main types of macrotopography – hummocky island-like uplands; and glacial lobe depressions with almost flat or slightly undulating topography (Fig. 7) – opened the way to the principally new palaeoglaciological and palaeogeographical conclusions in the East European Plain.

Conclusions

Since the middle of the nineteenth century many important studies have been conducted on the glacial geology in northern and eastern Europe. The glaciation theory was first proposed in Switzerland, and embodied the idea that glaciers were formerly more extensive on a huge scale than they are at present. In the Alps, in the area of 'living glaciers', scientists could make direct comparisons between areas now occupied by ice and those evidently abandoned by ice. After the victory of the continental glaciation theory many scientists forgot the different possibilities for boulder transportation and accumulation of boulder clays. To better understand the different possibilities of glacial accumulation, we should look at the modern classifications of tills and related deposits; for example, in the final report of the INQUA Commission on Genesis and Lithology of Glacial Deposits (Goldthwait & Matsch 1989), which presented a consensus of the ideas of over 100 INQUA members from five continents. Different depositional processes and environments are presented in modern classifications, including the deposition of boulder clays in marine environments (marine or basin tills, iceberg tills, etc.). The sediment character is controlled by a number of parameters, especially the place of deposition in relation to the ice and to the water bodies of various types. It means that modern views must be taken into consideration in the evaluation of old theories and old papers, and in the estimation of the role of one or another person in the history of science.

I would like to thank Professor A. Grigelis for inviting me to the INHIGEO conference on the History of Quaternary Geology and Geomorphology in Vilnius. I am much obliged to Dr F. R. van Veen for critical reading of the manuscript and for most helpful comments. Special thanks are due to Professor R. Grapes, who made the final revision of the text. Financial support from the Estonian Ministry of Education and Science (project No. 0320080s07) is gratefully acknowledged.

References

AGASSIZ, L. 1840. *Études sur les glaciers*. Neuchâtel.

ANDERSEN, B. G. & BORNS, H. W., JR. 1994. *The Ice Age World*. Scandinavian University Press. Oslo.

ASEYEV, A. A. 1974. *Ancient Continental Glaciations in Europe*. Nauka, Moscow (in Russian).

CHAMBERLIN, T. C. 1883. Preliminary paper on the terminal moraine of the second glacial epoch. *U. S. Geological Survey 3rd Annual Report*, 291–402.

CHAMBERLIN, T. C. 1884. Proposed genetic classification of Pleistocene glacial formations. *Journal of Geology*, 2, 517–538.

DE GEER, G. 1912. Geochronologie der letzten 12 000 Jahre. *In*: *XI International Geological Congress, Stockholm 1910, Compte Rendu*, 1, 457–471.

DREIMANIS, A. 1970. Are marine fossils in the Quaternary deposits a sufficient evidence for marine deposition? *Baltica*, 4, 313–322.

EICHWALD, E. 1846. Einige vergleichende Bemerkungen zur Geognosie Scandinaviens und der westlichen Provinzen Russlands. *Bulletin dé la Sociéte impériale des Naturalistes de Moscou*. XIX, 1, 3–156.

EICHWALD, E. 1853. *Lethaea Rossica on Paléontologie de la Russie*, Vol. 3. Dernière Période. Stuttgart.

FLINT, R. F. 1965. Introduction: historical perspectives. *In*: WRIGHT, H. E. & FREY, D. G., JR (eds) *The Quaternary of the United States*. Princeton University Press, Princeton, NJ 3–11.

GEIKIE, J. 1874. *The Great Ice Age and its Relation to the Antiquity of Man*. W. Isbister, London.

GOLDTHWHAIT, R. P. & MATSCH, C. L. (eds) 1989. *Genetic Classification of Glacigenic Deposits*. Balkema, Rotterdam.

GREWINGK, C. 1879. Erläuterungen zur zweiten Ausgabe der geognostischen Karte Liv-, Est-und Kurlands. *Archiv für die Naturkunde Liv-, Est- und Kurlands (Dorpat), Ser. 1*, VIII, 343–466.

GUDELIS, V. (ed) 1971. *Crystalline Indicator Boulders in the East Baltic Area*. Mintis, Vilnius (in Russian).

HAUSEN, H. 1913. Materialien zur Kenntnis der Pleistozänen Bildungen in den russischen Ostseeländern. *Fennia*, 34, 2.

HELMERSEN, G. 1869. Studien über die Wanderblöcke und die Diluvialgebilde Russlands. *Mémoires de l'Académie Imperiale des Sciences de St.-Petersbourg, Series 7*, 14, 7.

IVANOVA, T. K. & MILANOVSKY, E. E. 2006. Piotr A. Kropotkin and his *Researches on the Glacial Period*. *In*: GRIGELIS, A. & OLDROYD, D. (eds) *Abstracts of Papers of INHIGEO Conference History of Quaternary Geology and Geomorphology; 28–29 July 2006*, Vilnius, Lithuania. Lithuanian Academy of Sciences, Vilnius, 59–61.

KAHLKE, H. D. 1984. *Das Eiszeitalter*. Urania, Leipzig.

KARUKÄPP, R. 1997. Gotiglacial morphogenesis in the southeastern sector of the Scandinavian continental glacier. *Dissertationes Geologicae Universitatis Tartuensis*, 6. Tartu Ülikooli Kirjastus.

KARUKÄPP, R., RAUKAS, A. & ABOLTINŠ, O. 1999. Glacial accumulative insular heights (Otepääs) – specific topographic features in the Baltic States. *In*: MILLER, U., HACKENS, T., LANG, V., RAUKAS, A. & HICKS, S. (eds) *Environmental and Cultural History of the Eastern Baltic Region. PACT*, Rixensart, 57, 193–205.

KROPOTKIN, P. A. 1869. Some words about the appearance of boulders on the Island of Bolshoi Tyters. *Kronštadtski Vestnik*, 84, 332–333 (in Russian).

KROPOTKIN, P. A. 1876. Studies of the glacial age. *Zapiski Russkogo geographicheskogo obshchestva*, 7, 1 (in Russian).

LAVRUSHIN, Y. A. 1970. Experience of distinguishing dynamic facies and subfacies of ground moraine of continental glaciations. *Litologia i poleznye iskopaemye*, 6, 38–49 (in Russian).

MURCHISON, R. I., VERNEUIL, E. & KEYSERLING, A. 1845. *The Geology of Russia in Europe and the Ural Mountains, 1. Geology*. John Murray, London.

PENCK, A. & BRÜCKNER, E. 1909. *Die Alpen im Eiszeitalter*. Tauchnitz, Leipzig.

RAUKAS, A. 1982. K. E. Eichwala's contribution to the formation of glacial theory. *In*: ILOMETS, T. (ed)

Istoria razvitia, podgotovka kadrov, nautsnye issledovania II. Totsnye i estestvennye nauki. Tartuski gosudarstvennyi universitet, Tartu, 46–54 (in Russian).

RAUKAS, A. 1995. Estonia – a land of big boulders and rafts. *Questiones Geographicae*, (Special Issue), **4**, 247–253.

RAUKAS, A. & KARUKÄPP, R. 1993. Geomorphology in Estonia. *In*: WALKER, H. J. & GRABAU, W. E. (eds) *The Evolution of Geomorphology. A Nation-by-Nation Summary of Development*. Wiley, Chichester, 135–142.

RAUKAS, A. & KARUKÄPP, R. 1994. Stagnant ice features in the eastern Baltic. *Zeitschrift für Geomorphologie, Neue Folge*, **95**, 119–125.

SCHMIDT, F. 1865. Untersuchungen über die Erscheinungen der Glazialformation in Estland und auf Oesel. *Bulletin de l'Académie Imperiale des Sciences de St.-Petersbourg*, **8**, 339–368.

SCHMIDT, F. 1871. Über die Glazialformation in Estland. *Neues Jahrbuch für Mineralogie, Geologie und Palaeontologie*, **7**, 918–921.

SEVERGIN, V. M. 1815. About the potential antiquity of the formation of different Russian mountain ridges. *Umozritelnye issledovanija*. **IV** (in Russian).

SUUROJA, K. 2006. *Baltic Klint in North Estonia as a Symbol of Estonian Nature*. Ministry of Environment, Tallinn.

TORELL, O. 1872. Undersökningar öfver istiden. *Kungliga Vetenskapsakademiens Förhandlingar*, **29**, 10.

WANGENHEIM VON QUALEN, F. 1852. Über eine seculäre langsame Fortbewegung der erratischen Blöcke aus der Tiefe des Meeres aufwärts zur Küste durch Eisschollen und Grundeis. Beobacht an der küste Livlands. *Korrespondenzblatt des Naturforscher-Vereins zu Riga, V Jg.*, **6**, 73–83.

Origin and development of ideas on Pliocene and Quaternary glaciations in northern and eastern Europe, Iceland, Caucasus and Siberia

EVGENY E. MILANOVSKY

Department of Geology, Moscow State University, 119899 Moscow, Russia
(e-mail: milan@geol.msu.ru)

Abstract: A personal account, illustrated by the author's sketches, of ideas about glaciation, neo-tectonics and Quaternary geology, discussing the author's observations made during the course of his work undertaken in eastern Europe and Siberia, the Caucasus and Iceland. The paper is at the same time a contribution to autobiographical literature and to studies in the history of geoscience. The author's discovery of evidence for Pliocene glaciation in the Caucasus is noteworthy.

The present paper presents a short survey of the origin and development of ideas about Pliocene and Quaternary glaciations of northern Europe (including Iceland), in mountainous regions of southern Europe (including the Caucasus Mountains and the Alps), and in regions of western and eastern Siberia. I have been fortunate to have had the opportunity to participate in field studies of Pliocene and Quaternary glacial deposits and geomorphological features of many regions of Eurasia, such as the Caucasus and the Alps, Iceland, the Scandinavian peninsula, the mountain ranges of Central Asia, and the plains and highlands of Europe, Western and Eastern Siberia, etc. The results of personal observations and research, as well as analysis of the enormous amount of data collected and published in the nineteenth and twentieth centuries by many Russian and western European geologists, allow me to offer this short essay in which I characterize some principal features in the development of our knowledge of the history of late Cenozoic glaciations in Eurasia and note problems requiring further investigation.

Glaciation of the Alpine region

I begin by recalling some brief but important comments on the study of the great and high mountainous Alpine region, visited during an INHIGEO excursion in Switzerland in 1998, led by an outstanding authority on Alpine geology, Professor Rudolf Trümpy (Zurich). There one can still see relics of numerous extensive Quaternary glaciers, and can understand how what is now called the Pleistocene glaciation was convincingly established in the 1820s–1840s by Ignaz Venetz, Jean de Charpentier and Louis Agassiz (Hallam 1983). At the end of the nineteenth century, more detailed researches of Quaternary glacial deposits in the Alpine region revealed the existence of five separate Pleistocene glaciations – Donau (Danube), Günz, Mindel, Riss and Würm – named after Alpine rivers.

One of the greatest modern Alpine glaciers (La Mer de Glace), then covering an area of 277 km^2, remained at the highest (4807 m) summit in the Western Alps – the Palaeozoic granite massif of Mont Blanc (Fig. 1). About 130 km north of the Western Alps, on the southern slope of the not particularly high Jura Mountains, near Neuchâtel, there is a huge (more than 1000 m^3) erratic block (nicknamed 'Pierrebot'), composed of Palaeozoic granite (Fig. 2), thought to have been transported from the top of Mont Blanc during the Würm glacial epoch by a very thick alpine glacier and transported over the subalpine foredeep depression to its present deposition site (Fig. 3). This example convincingly demonstrated the reality of the transfer process of the great masses of moraine materials by glaciers during the Quaternary glaciations.

Problems of Quaternary glaciations in eastern Europe and Siberia

On the vast plains of NE Europe and Siberia, the origin of sediments, which we now with confidence identify as glacial moraine deposits, remained unclear until the early 1870s. In the 1820s and 1830s some European geologists had suggested the occurrence of a former glaciation in Norway, Germany and Scotland, but the ideas of Charles Lyell, who suggested the popular 'drift hypothesis' to account for the displacement of erratic boulders by floating ice (Lyell 1863), and of Roderick Murchison (1845), who supported Lyell's ideas

From: GRAPES, R. H., OLDROYD, D. & GRIGELIS, A. (eds) *History of Geomorphology and Quaternary Geology.*
Geological Society, London, Special Publications, **301**, 87–115.
DOI: 10.1144/SP301.6   0305-8719/08/$15.00 © The Geological Society of London 2008.

Fig. 1. Panorama of the Mont Blanc granite massif (4807 m). View from the north (from Chamonix resort).

during his geological expedition to European Russia in 1840–1841, impeded recognition of former glaciations on the East European Plain. By the middle of the nineteenth century some Russian geologists, such as Shchurovsky (1856), G. P. Helmersen (1869) and, especially, F. B. Schmidt (1865) (Fig. 4) were inclined to accept the reality of Quaternary glaciation in the northern part of

Fig. 2. 'Pierrebot': an erratic block of Palaeozoic granite (>1000 m^3) near Neuchâtel, transported during the Würm Glaciation 130 km NNE from Mont Blanc to the southern slopes of the Jura Mountains.

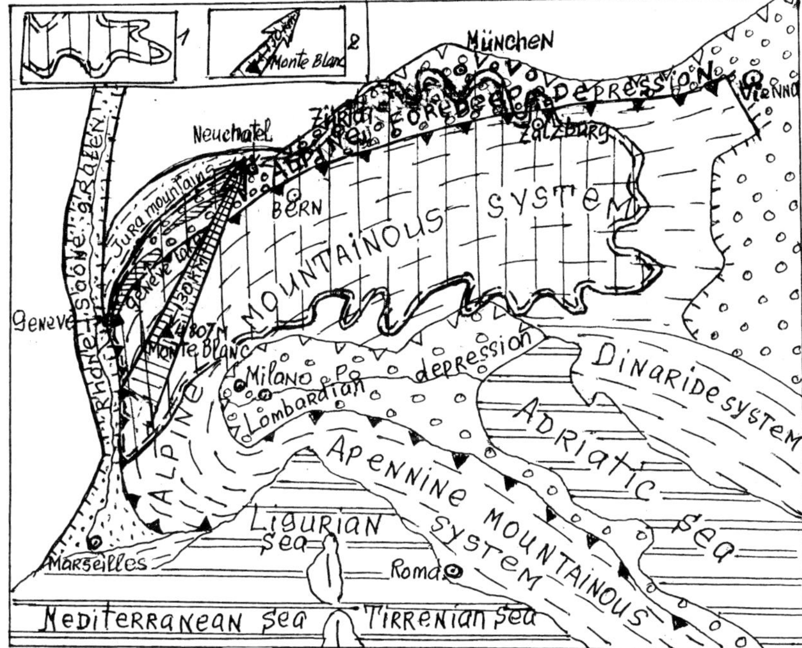

Fig. 3. Alpine mountain system during the Late Pleistocene (Würm) glaciation: 1, high mountainous central part of the Alpine system, covered during the Late Pleistocene (Würm) glacial epoch by a thick glacial sheet; 2, transport direction of erratic blocks of Palaeozoic granite from the summit of Mont Blanc to the southern slopes of the Jura Mountains.

Fig. 4. Academician F. B. Schmidt (1832–1908).

eastern Europe. But it was only at the beginning of the 1870s that the outstanding Swedish geologist and the founder of the Swedish Survey, O. M. Torell (1828–1890), and the talented young Russian geographer and geologist, Prince Piotr Alexeevich Kropotkin (1842–1921), who afterwards became a famous revolutionary and theorist of anarchism, discovered and convincingly traced the extensive ancient glaciation in Eastern Siberia, and later in Finland and Sweden (Fig. 5).

The scientific activity and ideas of Kropotkin, summarized in his classic monograph *Researches on the Glacial Period* (Kropotkin 1876), are described elsewhere in this volume (Ivanova & Markin 2008). Kropotkin's remarkable book initiated an intensive study of Quaternary glacial processes through the enormous territories of Russia. It should be stressed that, according to Kropotkin's thinking, published in the 1870s (Kropotkin 1873, 1874), there had only been *one glacial epoch* on the plains and heights of European Russia and Siberia during the Quaternary.

The next important step in the development history of the study of ancient glaciations in northern Eurasia at the end of nineteenth and beginning of twentieth century was associated with the work of the famous Russian geologist, Academician Alexsey Petrovich Pavlow (1854–1929) (Fig. 6),

Fig. 5. P. A. Kropotkin (1842–1921).

Fig. 6. Academician A. P. Pavlow (1854–1929).

Professor of Moscow University, and later his numerous students who formed the so-called 'Pavlov school of geology'.

During the 40 years preceding the Russian Revolution of 1917, significant progress was made in researches on the Quaternary glacial period in European Russia. The systematic geological mapping of this vast region started in the 1880s. The areas of the Baltic Shield and the Polar Urals were recognized as the main centres of glaciation, and the glaciers' maximal extension to the south was delineated. The study of stratigraphical and lithological successions in various sections of the Quaternary deposits of the Russian plain revealed the presence of alternating glacial (moraine-type) and non-glacial horizons, indicating the existence in this region during the Pleistocene of *several glacial and interglacial epochs*. This phenomenon was for the first time established by Pavlow in 1888 in the western part of the Middle Volga region, and it was confirmed near Moscow in 1890 by his pupil N. I. Krishtafovich. Later, Pavlow recognized on the East European Plain *at least three glacial epochs in the Quaternary* (*Pleistocene*), separated by interglacial epochs (Pavlow 1922). In the later years of his life Pavlow accepted the existence in European Russia of an even earlier, Late Pliocene, glacial epoch. However, some Russian geologists continued to recognize only one Pleistocene glaciation. It is

interesting to mention that even after World War II a well-respected Ukrainian palaeozoologist and geologist, Academician I. A. Pidoplichko, published a substantial monograph in four volumes, *Concerning the Glacial Period* (Pidoplichko 1946–1956), which supported Lyell's explanation of the drift origin of boulders in the Quaternary deposits on the Russian Plain.

Even today, there remain divergent opinions about the origin of the glacial boulder- and pebble-bearing thin-bedded Quaternary sediments in the northern parts of the West Siberian and Pechora plains. One group of geologists considers them to be proper glacial sediments, whereas another group construes them as glacial-marine sediments, including coarse clastic materials, transported by icebergs and floating blocks of ice. A third group of geologists has attempted to interpret these problematic rocks as a combination of the two opinions.

Most Russian and Ukrainian geologists now recognize the evidence of several Quaternary glacial and interglacial epochs in the late Cenozoic of eastern Europe. However, in 1948 some of them, in particular the authoritative Russian researcher, V. I. Gromov (1948), believed that during the greater part of the Pleistocene in the northern parts of eastern Europe and Western Siberia there were permanent ice sheets, which periodically expanded or retreated (but did not disappear). These changes were associated with global climatic changes. Among the Russian polyglacialists there were, however, different opinions about the number of individual Quaternary glaciations. Most of them recognized the existence of three or four glacial epochs, while S. A. Yakovlev (1956) and A. I. Moskvitin (1967) maintained that there were up to six, or even eight, glacial events.

Before the pre-War Soviet Union, an authoritative group of specialists in the study of Quaternary glaciations on the East European and West Siberian regions was gradually formed. Among them (unfortunately, now all are deceased) were such outstanding geologists as G. F. Mirchink (Fig. 7), V. I. Gromov, A. I. Moskvitin, E. V. Shanzer, S. A. Jakovlev, I. P. Gerassimov, K. K. Markov and G. I. Goretsky.

Among the Quaternary moraines and fluvioglacial deposits on the East European Plain the existence of formations belonging to three main glacial epochs was firmly established, and dated as Early, Middle and Late Pleistocene (Fig. 8). The main centres of these glaciations were situated in the central and eastern parts of the Baltic Shield (Fig. 9) and in the northern part of the Urals.

Fig. 7. Professor G. F. Mirchink (1889–1942).

Moraines and fluvioglacial deposits of one of the earliest in the eastern Europe Oka (or Berezina) Glaciation, belonging to the Early Pleistocene, were correlated with glacial deposits of the Early Pleistocene Elster (or Crakow) Glaciation in Central Europe and with the Mindel Glaciation of the Alps; the moraines and fluvioglacial deposits of the Don (or Dnieper) and Moscow glacial epochs in eastern Europe – with deposits of the Saale Glaciation in Central Europe, the Riss Glaciation in the Alps, and the late Pleistocene moraines and fluvioglacial deposits of the Waldai (or Ostashkov) Glaciation in eastern Europe – were correlated with the Vistula Glaciation in Central Europe and the Würm Glaciation in the Alps (Table 1).

After World War II, and especially in recent decades, this correlation scheme was subjected to criticism and modification by some Russian geologists; for example, S. M. Shik (1998), N. V. Makarova and V. I. Makarov (Makarova & Makarov 2004). It was noted that the maximal extension south of the Scandinavian (Baltic) glacial sheet occurred not only in the central part of Europe but also in its eastern part, and had taken place in the *early Pleistocene*. In particular, on the East European Plain the Dnieper and Don 'tongues' of the Scandinavian glacial sheet had their greatest extension to the south in the *early Pleistocene*. Now, most Russian experts in Quaternary stratigraphy, taking into consideration both palaeontological and and palynological data, attribute the Don (or Dnieper) Glaciation to the Early Pleistocene of the Russian Stratigraphic Column for the Quaternary and correlate it with the Elster (Crakow) Glaciation, or even with an earlier glaciation, of the western European chronological scale for the Quaternary (see Fig. 9 and Table 1).

The late Russian geophysicist and Head of the Earth Sciences Museum of Moscow University, Professor S. A. Ushakov, analysing data relating to the Baltic glacial sheet area during its greatest extension, estimated that this gigantic glacier or ice sheet was approximately 3 km thick at its apex during the Early Pleistocene (Ushakov pers. commun.).

Beside the Baltic glacial sheet, similar, but not so extensive, glacial sheets existed during the glacial epochs of the Pleistocene in some other high regions of Eurasia such as the Polar Urals in the NE part of Europe and on the *Putorana* and *Taimyr plateaus* (Fig. 10) in the northern parts of Central Siberia. At times, especially during the Middle Pleistocene, they had merged with each other and united with the great Baltic ice sheet and the smaller glacial sheets on the northern parts of the British Isles and with the other glacial sheets on the archipelagoes of Novaja Zemlya, Severnaja Zemlya, Novaja Sibir, Svalbard, Franz Joseph

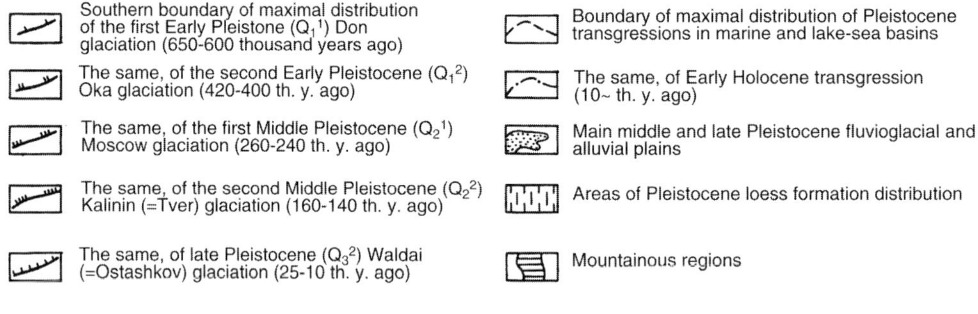

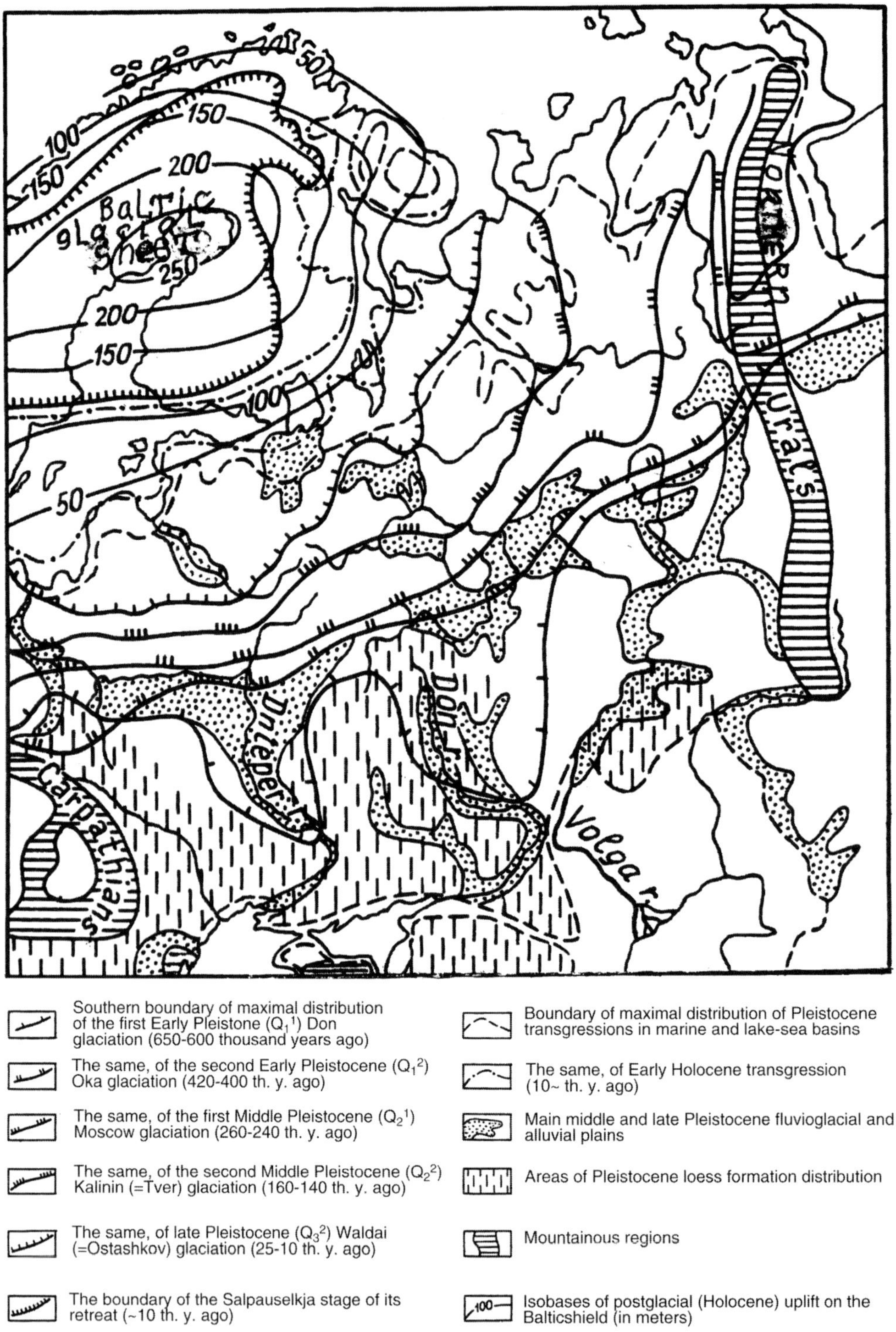

Baltic glacial sheet
50
100
150
200
250
Urals
Dnieper r.
Don r.
Volga r.
Carpathians
Southern boundary of maximal distribution of the first Early Pleistone (Q_1^1) Don glaciation (650-600 thousand years ago)
The same, of the second Early Pleistocene (Q_1^2) Oka glaciation (420-400 th. y. ago)
The same, of the first Middle Pleistocene (Q_2^1) Moscow glaciation (260-240 th. y. ago)
The same, of the second Middle Pleistocene (Q_2^2) Kalinin (=Tver) glaciation (160-140 th. y. ago)
The same, of late Pleistocene (Q_3^2) Waldai (=Ostashkov) glaciation (25-10 th. y. ago)
The boundary of the Salpauselkja stage of its retreat (~10 th. y. ago)
Boundary of maximal distribution of Pleistocene transgressions in marine and lake-sea basins
The same, of Early Holocene transgression (10~ th. y. ago)
Main middle and late Pleistocene fluvioglacial and alluvial plains
Areas of Pleistocene loess formation distribution
Mountainous regions
100
Isobases of postglacial (Holocene) uplift on the Balticshield (in meters)

Fig. 8. Distribution of Quaternary glacial sheets of different ages and some other elements of the Quaternary palaeogeography of the East European plain (compiled by Milanovsky in 1966).

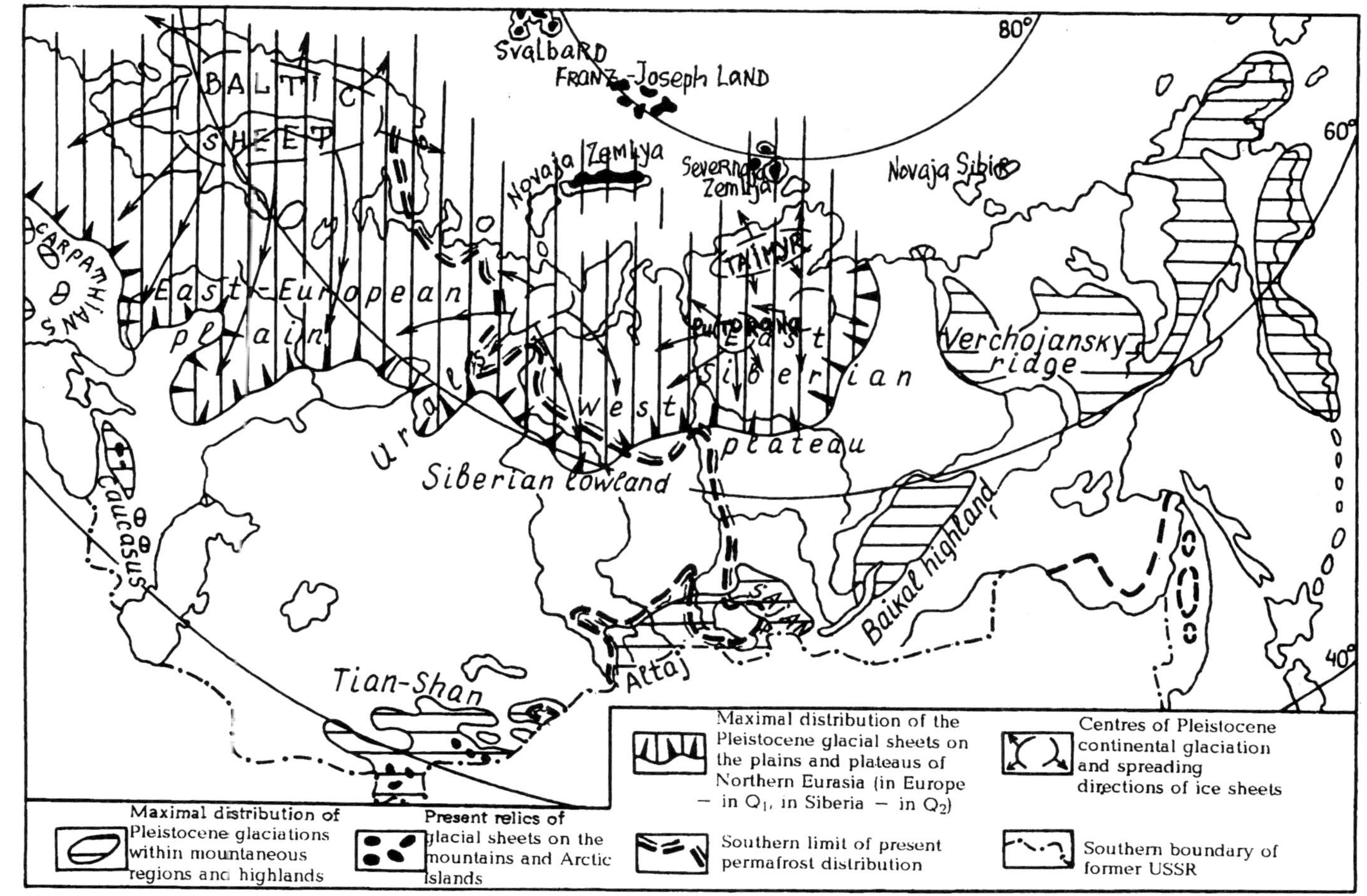

Fig. 9. Maximal distribution of the Quaternary continental glaciations and permafrost regions of northern Eurasia (compiled by E. E. Milanovsky).

Table 1. *Chronological correlation of the Pliocene and Quaternary Glaciations in Eastern Europe, Great Caucasus and Siberia with the same in Central Europe and the Alps, compiled by Milanovsky (1998)*

Abs.age (Th.)	Geological age	West Europe	Russian scale	Alps	Central Europe	Eastern Europe	Great Caucasus	West-Middle Siberia	Altaj Mountains
15	Quaternary = Antropogen Q		Holocene						
40	Quaternary = Antropogen Q	Late Pleistocene		Würm-2 Glaciation	Wistula Glaciation	Late Waldaj (Ostashkov)	Bezengi-2 Glaciation	Zyrian-2 (Sartan) Glaciation	Akkem Glaciation
50 / 75	Quaternary = Antropogen Q	Late Pleistocene		Würm-1 Glaciation	Szczecin cool epoch	Early Waldaj cool epoch	Bezengi-1 Glaciation	Zyrian-1 Glaciation	Chibit Glaciation
110	Quaternary = Antropogen Q	Late Pleistocene		R-W Interglacial	Eem Interglacial				
140	Quaternary = Antropogen Q	Middle Pleistocene	Middle Pleistocene	Riß -2 Glaciation	Saale-2 (Warta) Glaciation	Kalinin Glaciation (Tver)	Terek-2 Glaciation	Taz Glaciation	Chuja Glaciation
180	Quaternary = Antropogen Q	Middle Pleistocene	Middle Pleistocene		Treene Interglacial				
230	Quaternary = Antropogen Q	Middle Pleistocene	Middle Pleistocene	Riß -1 Glaciation	Saale-1 (Drente) Glaciation	Moscow Glaciation	Terek-1 Glaciation	Samarovo Glaciation	Eshtykol Glaciation
370	Quaternary = Antropogen Q	Middle Pleistocene	Middle Pleistocene	M-R Interglacial	Golstein Interglacial	Likhvin Interglacial		Tobol Interglacial	
450	Quaternary = Antropogen Q	Middle Pleistocene	Early Pleistocene	Mindel Glaciation	Elster (Crakow) Glaciation	Oka (Berezina-1) Glaciation	Eltübü-2 Glaciation	Shaitan Glaciation	Late Katun Glaciation
650	Quaternary = Antropogen Q	Middle Pleistocene	Early Pleistocene	G-M Interglacial	Kromer Interglacial	Belovezha Interglacial			
650	Quaternary = Antropogen Q	Middle Pleistocene	Early Pleistocene	Haslach Glaciation?	?	Don (Dzuki) Glaciation	Eltübü-1 Glaciation	?	Early Katun Glaciation
750 / 770	Quaternary = Antropogen Q	Middle Pleistocene	Early Pleistocene						
850	Quaternary = Antropogen Q	Early Pleistocene	Eopleistocene	Gunz Glaciation	Menapian cool epoch	Olonets Glaciation (?)	2. Eopleistocene Glaciation ?		Bashkatau Glaciation
1300	Quaternary = Antropogen Q	Early Pleistocene	Eopleistocene	D-G Interglacial	Vaal warm epoch	Scyphian warm epoch			
1550	Quaternary = Antropogen Q	Early Pleistocene	Eopleistocene	Danub Glaciation	Eburon cool epoch	Domashkino cool epoch	1. Eopleistocene Glaciation ?		
1750	Quaternary = Antropogen Q	Early Pleistocene	Eopleistocene	B-D Interglacial	Tegelen warm epoch				
2100 / 2500	Pliocene N2	Late Pliocene	Akchagyl stage	Biber Glaciation	Pre-tegelen cool epoch		Chegem Glaciation		
3500	Pliocene N2	Late Pliocene	Akchagyl stage				Elbrus Glaciation		
5200	Pliocene N2	Early Pliocene	Kimmerian (Balakhan) stage						
	Late Miocene								

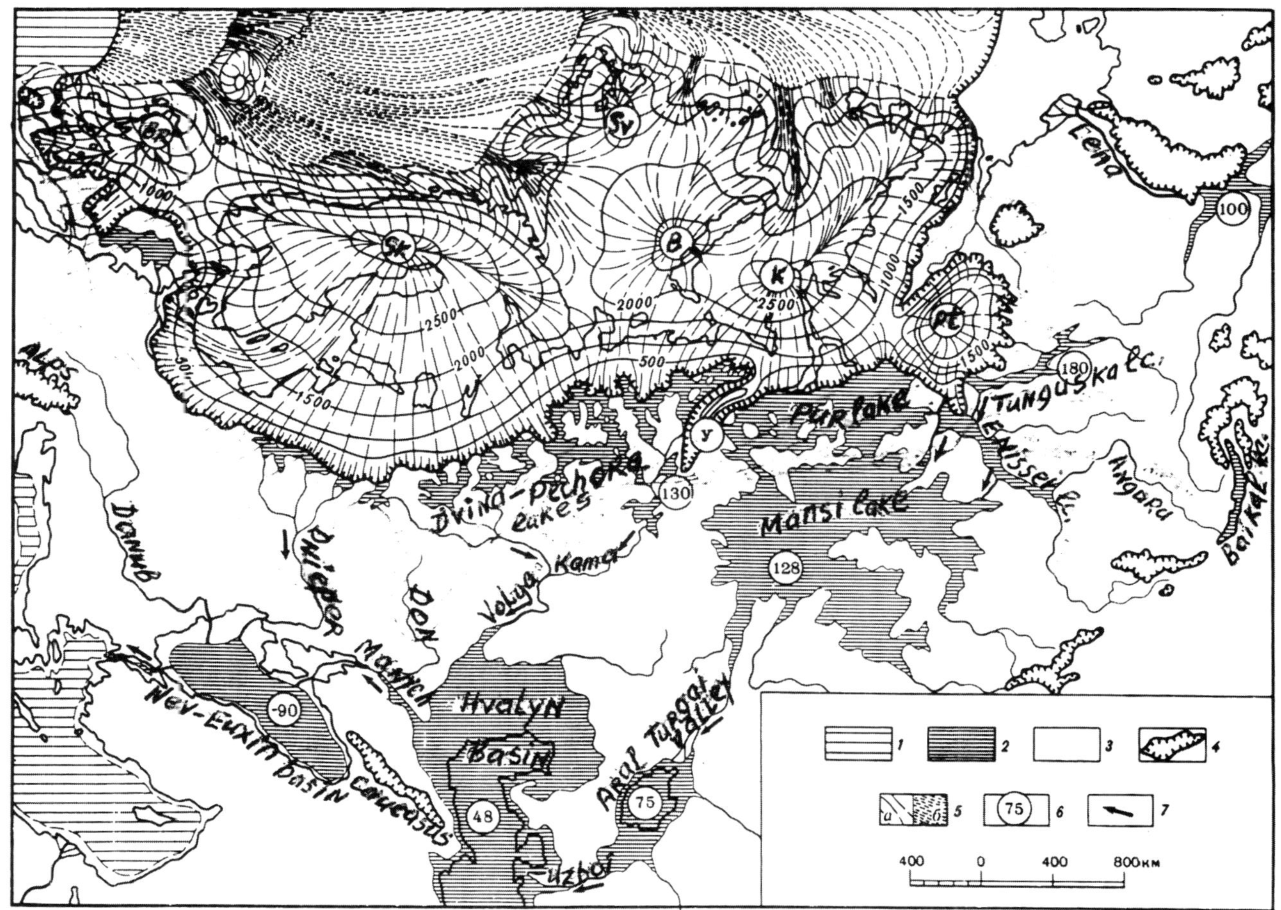

Fig. 10. Reconstruction of the glacial sheets of northern Eurasia in the Late Pleistocene (Würm) epoch (based on Grosswald 1983). Key: 1, ocean free of glaciers; 2, freshwater basins; 3, dry land, free of ice; 4, glacier boundaries; 5, lines of movements of glaciers: along the Earth and floating glaciers; 6, levels of intracontinental basins (relatively modern ocean); 7, currents of near-glacial waters.

Land, etc., forming one great North Eurasian glacial sheet (Fig. 10). It is plausible to assume that on some of these islands and archipelagoes, and in some parts of the Baltic, sheet glaciers first appeared not at the beginning of the Pleistocene but, in fact, had already occurred in the late Pliocene.

With regard to the palaeogeographic conditions that existed in the Pleistocene at localities of relatively shallow sea basins, such as those of Barents, Kara, Laptev, East Siberian and some other seas of the Arctic Ocean, the Russian glaciologist M. G. Grosswald (1983) expressed the interesting, although somewhat questionable, idea that during the Pleistocene glacial epochs these basins were covered with thick icy 'cuirasses', similar to the contemporary shelf glaciers that cover the southern parts of the Ross and Wedell seas around Antarctica, as if 'soldered' to the glacial sheet. According to Grosswald's hypothesis, such glacial sheets or 'cuirasses' covered the shelves of the Barents and Kara seas and other polar North Eurasian glacial sheets during the last Pleistocene glaciation (Fig. 11) and during the stages of its degradation (Fig. 12). During the glacial epochs in the Pleistocene and probably already in the late Pliocene, these Arctic seas, as well as the islands within them, preserve evidence of glacio-eustatic sea-level falls (Grosswald 1983).

Interglacial epochs of the Pleistocene and Holocene (the last 10 ka) are correlated with significant glacio-eustatic rises of sea level and, consequently, with transgressions and regressions of marginal seas of the Arctic Ocean, the Black Sea and the Baltic Sea in the Holocene. They also correlate with the marine ingressions in the valleys of the Severnaya Dvina, Pechora, Ob' and Yenisei rivers (which flow northwards into the Arctic Ocean). By contrast, transgressions and ingressions in the isolated Caspian Lake/Sea coincided with the end of the glacial epochs and with the beginnings of the interglacial epochs.

On the West Siberian Plain, the first (Shaitan) glaciation only began at the end of the Early Pleistocene (Table 1), i.e. later than the first glaciation on the East European Plain. The most significant Samarovo Glaciation in the middle part of the Middle Pleistocene spread southwards to 60°N–62°N. The less important Taz Glaciation took place at the end of Middle Pleistocene. The relatively limited Zyrian (or early Zyrian) and Sartan (or late Zyrian) glaciations in Siberia occurred in the Late Pleistocene (Table 1). The glacial sheets in Western Siberia spread SE from the Polar Urals and SW from the Putorana Highland in the NW part of the East Siberian Plateau. They completely (or almost completely) joined in the central part of the West Siberian Plain. These glaciers formed temporary glacial and moraine barriers, which dammed the northward flow of the main West Siberian rivers (Tobol, Irtysh, Ob' and Yenisei). This process led to the creation of a series of extensive lakes and the huge West Siberian Lake situated to the south of this glacial barrier. During such epochs the waters partly escaped to the SW into the basins of the Aral and Caspian lakes/seas. Some geologists, however, believe that a considerable part of the quasi 'glacial' sediments in the northern part of Western Siberia in fact represent the rocks of marine-glacial origin or may even be of marine origin.

In Eastern Siberia, where climatic conditions in the Quaternary were colder and drier than in Western Siberia and eastern Europe, the epochs of global cooling intensified the process of freezing and formation of a permafrost zone (Fig. 10), whose thickness reached up to several hundred metres and even 11.5 km. However, glacial sheets of relatively limited dimensions had periodically appeared on the Eastern Siberian highlands of Putorana, Anabara, Taimyr, Transbaikalia and Anadyr, and also on the high mountainous ridges of Altai, Western and Eastern Sayan, Stanovoj, Verkhoyanskij, Chersky and Koryak, and on the ridges of the Kamchatka Peninsula.

Some peculiarities of glaciations in Iceland

The history of glaciations in the Neogene and Anthropogene is unusual and particularly interesting in the islands of Greenland and Iceland. In Greenland, and especially in the northern part, the world's largest 'non-continent' island, between 60° and 83°N, glaciation probably began in the Miocene. On the smaller island of Iceland, situated between 64° and 67°N, glaciation began in the Pliocene, and the rather numerous oldest glacial sediments–moraine horizons, lying between date-able basaltic lava flows, are about 3 Ma, i.e. Middle Pliocene (Fig. 13). Research in Iceland has been carried out since 1971 by the Iceland Geodynamic Expedition of the Soviet Academy of Sciences under the leadership of its corresponding member V. V. Beloussov, and with the author's participation for three summer seasons. The results are described in a series of monographs entitled *Iceland and the Mid-Oceanic Ridge* published by Nauka between 1978 and 1980. In particular, the problems of late Cenozoic and contemporary glaciations in Iceland are considered in detail in the volume *Stratigraphy: Lithology* by Achmetjev *et al.* (1978), and in the volume *Geomorphology: Tectonics* (1979) by Milanovsky *et al.* (1979). The scheme for the geological structure of Iceland that was compiled is shown in Figure 14.

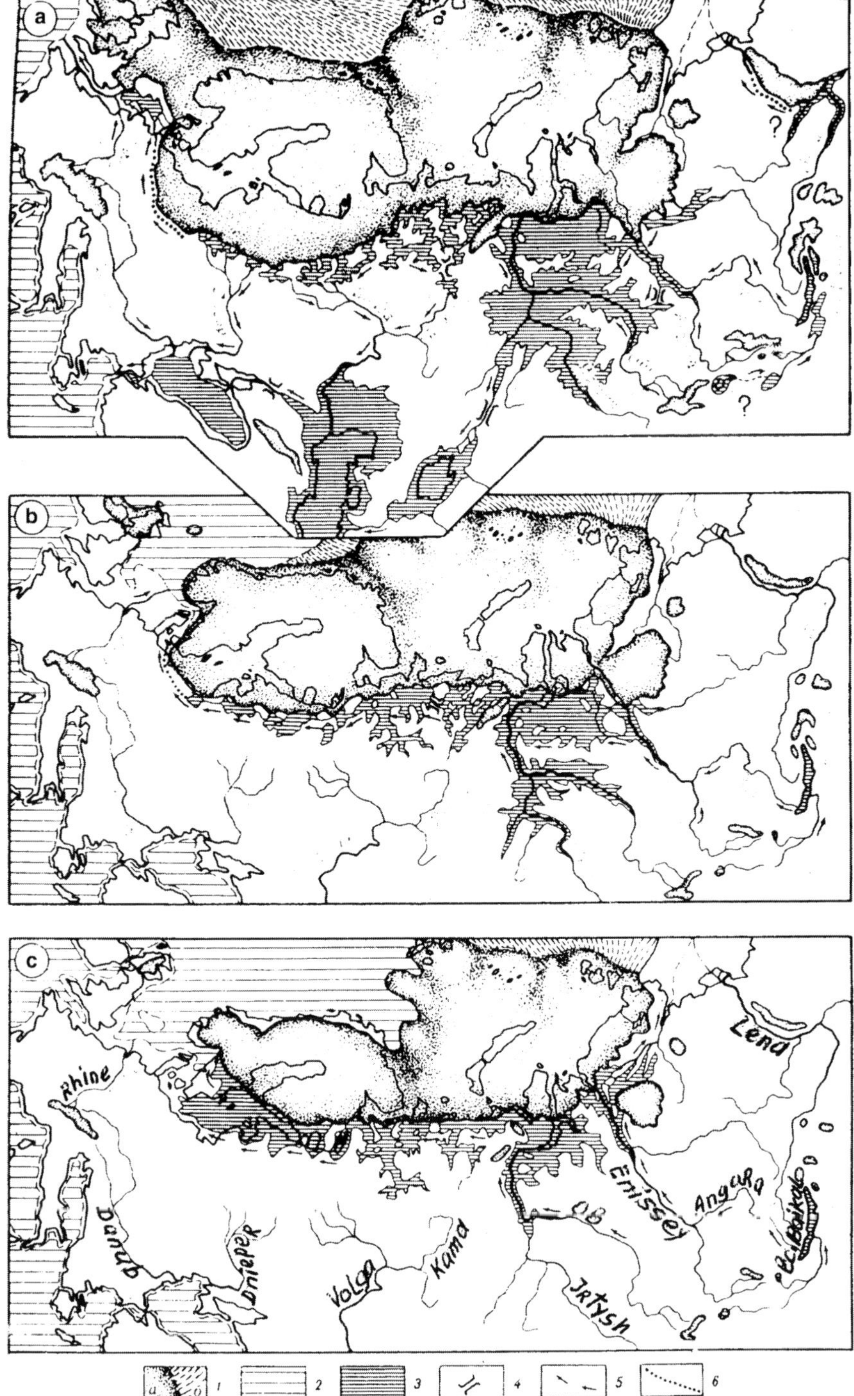

Fig. 11. Three stages of Late Pleistocene glaciations in northern Eurasia (according to Grosswald 1983). Key: 1, (**a**) overlying earth and (**b**) floating parts of a glacial cover; 2, ocean free of glaciers; 3, lakes; 4, major spillways; 5, directions of water flow; 6, pre-existing valleys.

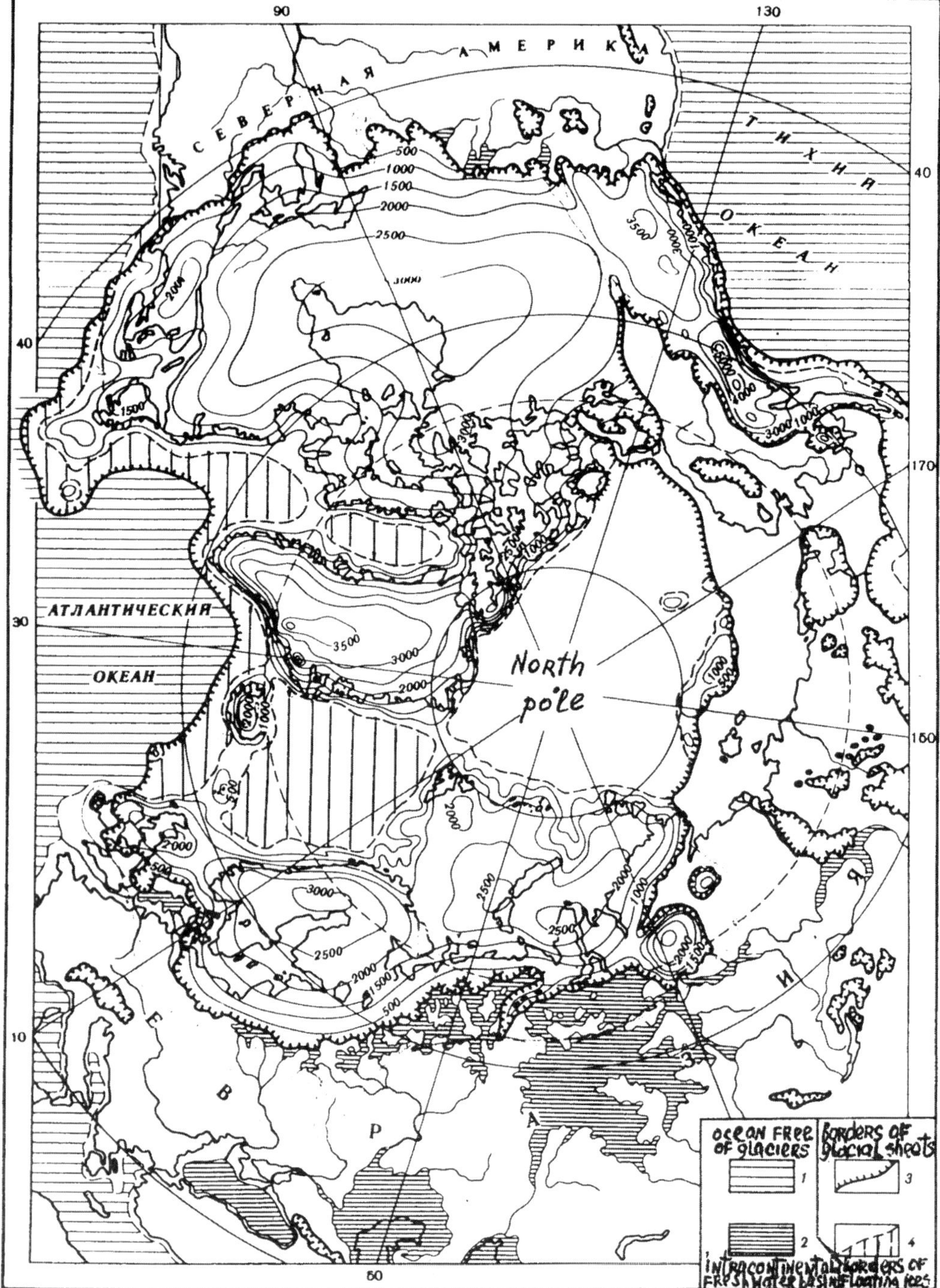

Fig. 12. Surface relief of the last Pan-Arctic glacial sheet. Key: 1, ocean free of glaciers; 2, intracontinental freshwater basin; 3, borders of glacial sheets; 4, borders of floating ice.

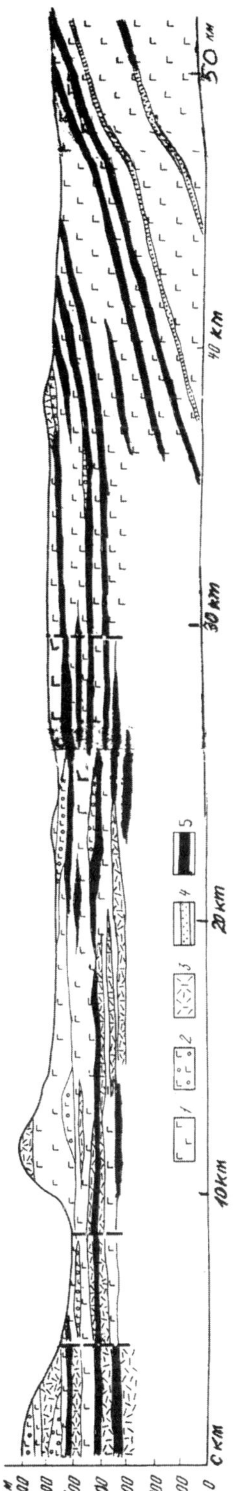

Fig. 13. Geological profile along the western flank of neotectonic zone of Iceland, showing alternation of Pliocene basaltic lavas and numerous horizons of tillites (moraines) of Middle Pliocene age – approximately 3 Ma (Wensink 1964). Key: 1, subaeral basalts; 2, pillow lavas of basalts; 3, volcanoclastics; 4, volcano-sedimentary deposits; 5, moraines (tillites).

As mentioned earlier, the oldest horizons of glacial sediments (moraines), interlayered with basaltic lavas in the NW and eastern parts of Iceland, were formed between 3 and 2 Ma (Fig. 13). Lava sheets in the NW part of Iceland are gently inclined to the SE; and in the eastern part to the west or NW. They dip toward the island's axial rift zone. During several glacial epochs, beginning from the Middle Pliocene, and especially in the Late Pliocene and Pleistocene, when the most of Iceland was buried beneath a thick (>1 km) ice cover, volcanic activity, especially in the main eastern rift zone, continued, and the thickest glacial 'cover' (up to 1.2 km) was formed above this wide zone of the eastern rift depression (Fig. 14).

The numerous active volcanic vents of central and linear types in the eastern rift zone, extending through the whole island from SW to NE, led to melting of the overlying glaciers; and under the thick glacial cover the melting of ice led to the formation of small cavities, filled with thermal waters, which had the form of 'caverns' and 'passages', hidden beneath the icy 'armour'.

Eruptions from the volcanic vents of central and linear type beneath these water reservoirs often led to the opening of their 'roofs' and to sudden catastrophic expulsion of powerful hot-water fountains and mud flows or debris streams – so called *jökulshlaup* or *jökulklaup* – which flowed down or under the inclined surfaces of the glaciers to the coast.

In the course of such volcano-glacial processes occurring within the glacial sheets, peculiar volcanic structures were formed – peculiar with respect to the hyaloclastic *móberg* or *lakagigar* ridges produced by subglacial eruptions. The so-called volcanic 'table mountains' (*stapi*) are widely distributed and typical of several of the central regions of Iceland (Fig. 15a, b). They were formed within the thick glacial sheets above volcanic orifices. Inside the glacial sheets were formed cylindrical reservoirs filled with water, into which basaltic lava was intruded from below, which, at the contact with the water, formed 'pillows' to produce thick complexes of basaltic pillow lavas (Fig. 15). Only the thin uppermost part of the volcanic edifice formed higher than the surface of the water, filling the cylindrical water reservoir. This consists of normal basaltic lava.

During the main glacial epochs of the Quaternary – in the Early, Middle and Late Pleistocene – almost all of Iceland was covered with a thick glacial sheet. Its thickness ranged from 1 to several kilometres (at present it is no more than 1 km thick), and its area considerably exceeded that of modern Iceland (1030 km^2). Thick tongues of the Icelandic glacial sheet extended radially to the ocean to produce icebergs – a process that

E. E. MILANOVSKY

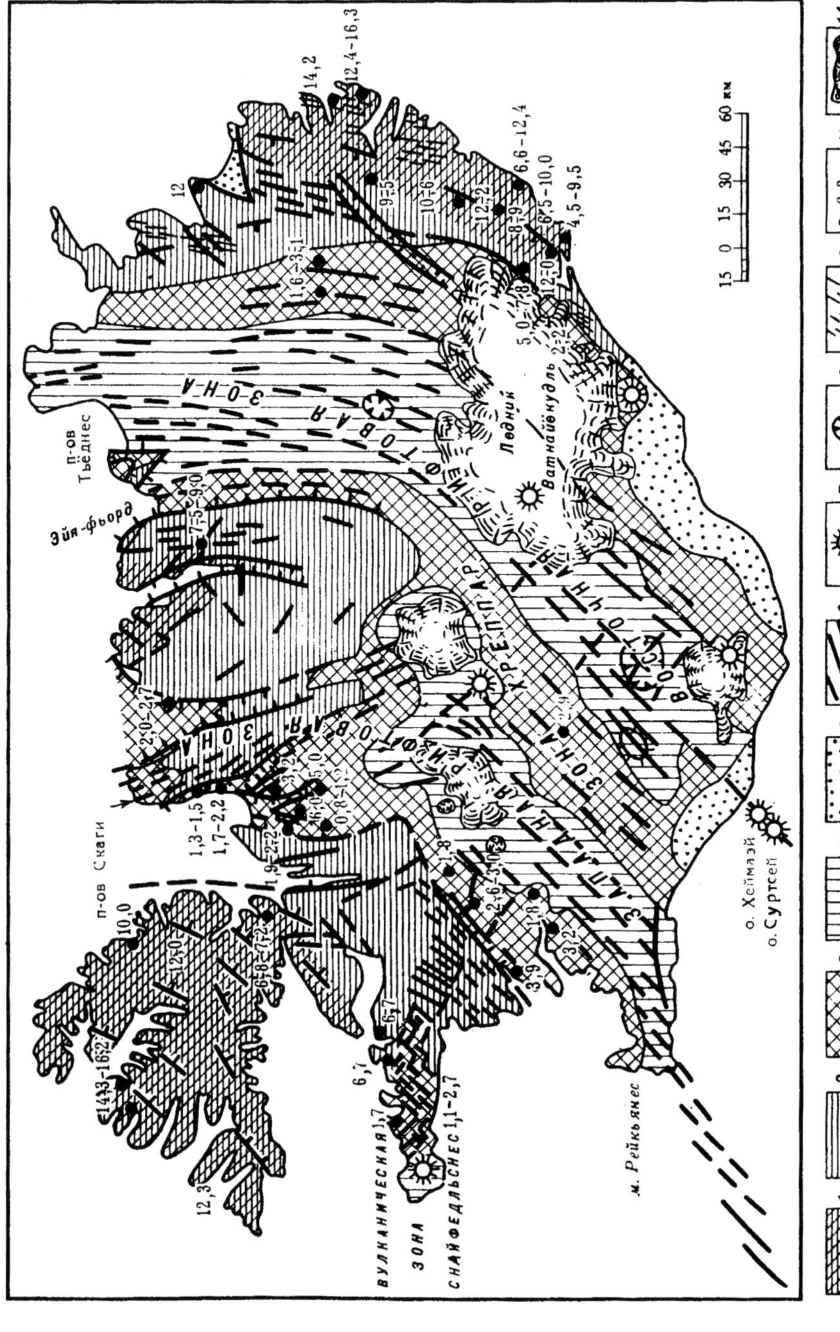

Fig. 14. Scheme of the geological structure of Iceland, compiled by Milanovsky in 1978. Key: 1, complex of Miocene plateau-basalts, lower part (9–16 Ma); 2, the same, upper part (5–9 Ma); 3, volcanic–sedimentary Pliocene–Eopleistocene complex (1–5 Ma); 4, Pleistocene sedimentary complex (younger than 7 Ma); 5, Holocene sedimentary complex (younger than 10 ka); 6, faults and fissures of extension; 7, largest Late Quaternary volcanoes; 8, largest Late Quaternary calderas; 9, swarms of Neogene calderas; 10, absolute age (in millions of years); 11, Contemporary glaciers.

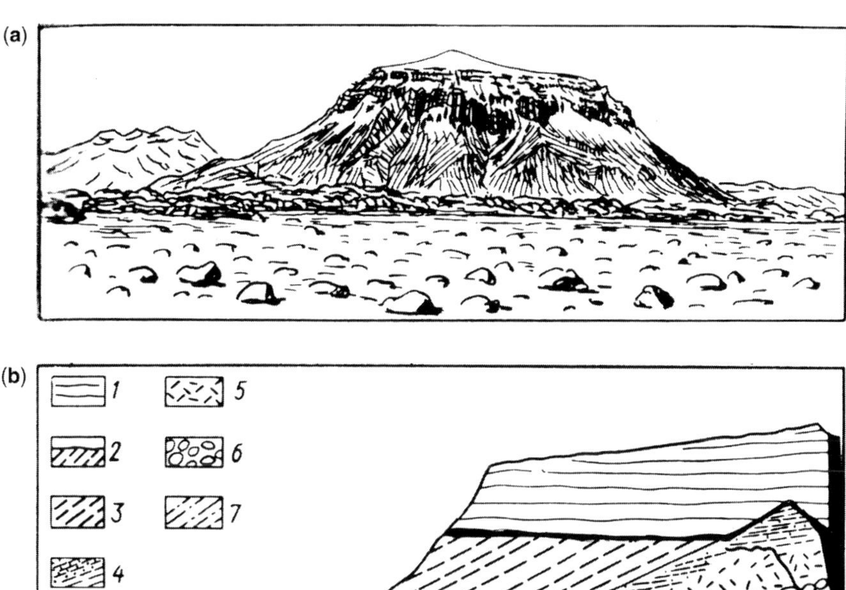

Fig. 15. Typical structures of Late Pleistocene volcanic table mountains. (**a**) Herðubreið, Late Pleistocene volcanic table mountain (1682 m), NE Iceland (drawing by E. E. Milanovsky in 1972). (**b**) Schematic cross-section of the Late Pleistocene volcanic table mountain in Iceland (based on Sæmundsson 1967). Key: 1, near-top basaltic lavas; 2, line of water level; 3, breccias of the subwater slope of the volcano; 4, stratified hyaloclastites; 5, pillow breccias; 6, pillow lavas; 7, diluvium.

continues to this day at the SE margin of the largest Icelandic glacier, Vatnajökull. Under the thickest central part of this glacier, having the form of a cone with a height of 1725 m above sea level, was the summit of Iceland's highest volcano, Grímsvötn. This edifice is now dissected by several canyons that channel melt water to the numerous fjords on the northern and NE coasts of Iceland.

The central part of Grímsvötn now has the form of a glacial dome, which has a maximum height of about 2 km at Kverktjöll. Glaciers extend from the dome to the SE and southern coasts of Iceland, and are the source of numerous small icebergs (Figs 16–18).

To the north, the glacial sheet of Vatnajökull has been subjected to intensive melting and erosion. Numerous rivers, abounding in waterfalls, flow northward along submeridional gorges cut into the basalt to the coasts of the Greenland and Norwegian seas (Fig. 19).

The so-called 'Primary Dettifoss' surface and many others surfaces of the Pleistocene glacial sheet in the NE part of Iceland can be reconstructed from the summit heights of the numerous volcanic table mountains, taking into consideration the absolute heights of the summits of the numerous Pleistocene volcanic table-mountains (Fig. 20).

The well-known Hekla (Fig. 21), Iceland's most active volcano, is a stratovolcano, but also forms part of a volcanic ridge.

Ideas on former glaciation of the Caucasus

Geological and glaciological research in the Caucasian mountains began after the area was acquired by Russia in the 1850s–1860s, i.e. later than in the Western Alps. From the very beginning Russian scientists believed in the existence of a significant former glaciation in the Greater Caucasus, for the simple reason that these mountains are much

Fig. 16. Glacial 'tongue', Hofsjökull, descending from the great glacial sheet of Vatnajökull to the southeast to Hornafjörður (Horna Fiord) of the Norwegian Sea (drawing by E. E. Milanovsky in 1972).

Fig. 17. Near Lake Jocudlgarlon, a glacial lake with numerous small icebergs at the edge of the Breidamerkürjökull (southeastern tongue of the Vatnajökull Glacier) (drawing by E. E. Milanovsky in 1972).

Fig. 18. Southern border of the Vatnajökull glacial sheet and the dust storm on the *sandur* [=outwash] plain; Skeidara sandur to the south (drawing by E. E. Milanovsky in 1972).

higher than the Alps. The size of glaciers in this region is similar to that in the Alps, where extensive Pleistocene glaciation had already been convincingly established in the first half of nineteenth century.

Between 1844 and 1850, Professor Hermann Wilhem Abich (1806–1886), a pioneer of Caucasian geology, was invited to Russia from Germany – initially as a professor of Dorpat (Tartu) University and later for work in the Caucasus, where he collected and described the first scientific data on the present and former glaciations of the Greater Caucasus and the Armenian Highlands (Fig. 22), producing several publications on the glaciation of the region (Abich 1853, 1870 [1871], 1871, 1878, 1878–1887). In particular, he was the first scientist to visit the highest volcanic mountains of Elbrus (5642 m) (Fig. 23) and Ararat

Fig. 19. Dettifoss Waterfall at the canyon-like cataract gorge near the Jökulsá á Fjöllum River, flowing northwards from the Vatnajökull glacial sheet in southeastern Iceland into the wide Öxarfjörður (Axarfjord) of the Greenland Sea of the Arctic Ocean (drawing by E. E. in Milanovsky in July 1973).

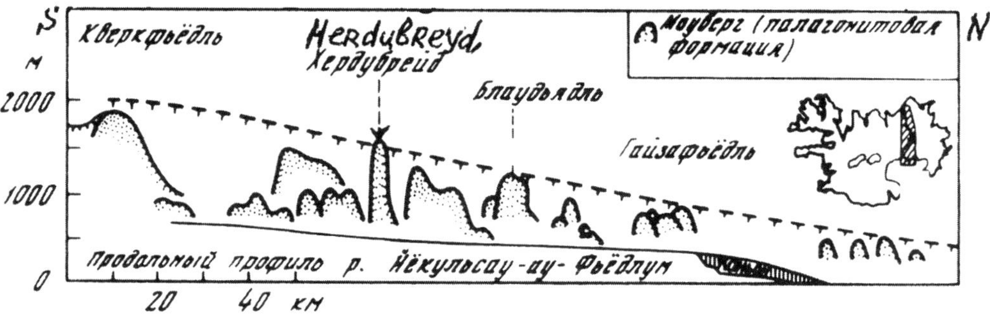

Fig. 20. Reconstruction of the Iceland Pleistocene glacial sheet surface, based on the heights of the volcanic table mountains in northeastern Iceland: surface shown by the dashed line of summit heights.

Fig. 21. The largest active volcano in Iceland, Hekla (1491 m), situated at the southern part of island. Near the summit it is covered by small glaciers and with basaltic lava flows and volcanic cones on its slopes resulting from the last eruption in 1970 (seen from the north) (drawing by E. E. Milanovsky in August 1973).

Fig. 22. The German geologist: Wilhelm Abich (1806–1886).

Fig. 23. The Mount Elbrus Volcano, from the SW. In the foreground are rocks of the pre-Mesozoic basement overlain by andesite lavas of Q_{2-3}; and above them lavas, Q_4, erupted from the eastern volcanic cone of Elbrus.

(5165 m) (Fig. 24) in the Armenian Highlands, as well as Kazbek volcano (5033 m) in the central part of the Great Caucasus (Fig. 25). At the southern foot of Elbrus, near the source of the Baksan River, he found and described indications of a very recent advance of glaciers that had descended from this largest Caucasian volcano. One 'tongue' of the Elbrus glacier 'invaded' a pine forest at the bottom of the Azau River basin. Now, 150 years after Abich's observations, the termination of the Azau Glacier has retreated several kilometres; this is typical of many Caucasian glaciers, most of which are now retreating.

In the second half of the nineteenth century, glaciologists discovered terminal moraines in some U-shaped valleys in the central part of the

Fig. 24. View of Little and Great Ararat volcanoes from the Ararat Valley, drained by the Arax River.

Fig. 25. Kazbek Volcano: view from the east (from a camp in the valley of Gudoshauri Aragvi).

Greater Caucasus. The moraines correspond to different stages of the still-continuing Late Pleistocene–Holocene glaciation. Remnants of older U-shaped troughs are preserved in the slopes of these valleys, which were produced by valley-type glaciers formed during earlier glacial epochs, but these glacial features remain poorly studied.

In the first half and middle of the twentieth century, many Russian geologists and geomorphologists worked in the different parts of Caucasus. A. P. Gerassimov (1911), A. L. Reinhard (1936, 1947), V. P. Rengarten (1932), Gerassimov and K. K. Markov (1939), L. A. Vardanyants (1932, 1948), N. V. Dumitrashko (1949), P. V. Kovalev (1960), S. L. Kushev (1964), G. K. Tushinsky (1949, 1958), L. I. Maruashvili (1971), I. N. Safronov (1960), E. E. Milanovsky (1960, 1964, 1966), A. V. Kozhenikov and Milanovsky (1984), Kozhevnikov, Milanovsky and Yu. V. Sayadyan (1977), and many others proposed a detailed time scale for the Quaternary Caucasian glaciations, similar to that proposed for the Alps between 1901 and 1909 by Penck & Brückner (and included the Danube–Günz–Mindel–Riss–Würm glacial epochs). However, detailed correlation between the Alpine and Caucasian timescales remained uncertain. Initially, it was suggested, by analogy with the Alps, that during their maximum advance the ancient glaciers on the northern slope of the Central Caucasus extended to the Cis-Caucasian fore-deep plain, because moraine-like accumulations were found in some places consisting of large boulders and blocks transported from the inner part of the Caucasus. In the 1950s–1960s, it was established that these deposits of coarse clastic material were transported and deposited mainly by periodic catastrophic mud flows from the mountains into the fore-deep plains. In fact, in contrast to the Alps, even during the coldest stages of the Pleistocene the glaciers on both slopes of the Greater Caucasus did not reach the foothill plains, but ended in the trough-like valleys within the Caucasian mountains. Both present and former glaciations reached their maximum intensity in the highest, central part of the Greater Caucasus, where many summits are as high as 4000–5000 m and even exceed 5600 m on Mount Elbrus (Fig. 23). The mountains in the eastern part of the Greater Caucasus are also high (up to 4000 m and more), but the climate in this part of Caucasus is dry. Correspondingly, indications of the present and former glaciations on the eastern Caucasus are much more limited than in the central area. In the Lesser Caucasus and in the Armenian volcanic highlands the mountains do not exceed 3000–4000 m (with the exception of the Mount Ararat), and the local climate is even more arid. Ancient glaciation was insignificant in this region, and now only small glaciers exist on Great Ararat (5000 m) and on Lesser Ararat (4000 m).

In most Eurasian mountainous systems the *in situ* pre-Würm and pre-Riss moraines have been eroded. However, in many places of the central high mountainous part of the Greater Caucasus, and especially on its northern slopes, moraines have been preserved under Late Pliocene–Holocene lavas, ignimbrites and tuffs that were erupted during several phases in the Elbrus (Fig. 23) and Kazbek (Fig. 25) areas. The stratigraphic and geomorphological correlation of glacial deposits and young volcanic rocks provides a unique opportunity to date glacial deposits 'protected' by their volcanic cover.

In the 1950s and 1960s the author and his colleagues from the Moscow University (A. V. Kozhevnikov and N. W. Koronovsky) analysed the relationships between late Cenozoic glacial and volcanic complexes in the central part of the Greater Caucasus. We proposed a chronostratigraphic scale for the late Pliocene–Holocene glaciations in the Caucasian region, based on the universally adopted stratigraphic scheme for the Caucasus. The late Pliocene rhyolite–dacite ignimbrites are particularly important for age determination of late Pliocene and Eopleistocene glacial sediments. The volcanic rocks, underlying and overlying the oldest moraines, are widely distributed on the northern slopes of the Central Caucasus and in the Elbrus volcanic area. They extend from Elbrus volcano in the west to the Verkhniy Chegem Highland in the SE, and the Nizhiniy Chegem Highland to the NE (Fig. 26). The isotopic age of the ignimbrites is 2.8 Ma (Lipman *et al.* 1993).

On the eastern slope of Elbrus volcano, Milanovsky & Koronovsky (1961) found the oldest moraine deposits in the Caucasian region (belonging to the Elbrus Glaciation) (Fig. 27). This 20 m-thick moraine occurs at an altitude of approximately 3700 m and overlies weathered Palaeozoic basement, while layered ignimbrites overlie it unconformably. These ignimbrites are, in turn, unconformably overlain by Holocene moraine of the modern Elbrus glacial sheet. The moraine deposits of the older Elbrus Glaciation are clearly older than 2.8 Ma and, taking into account the unconformable contact of the overlying ignimbrites, the moraine could be as old as 3.0–3.5 Ma. It would thus correspond to the beginning of the Late Pliocene and would coincide with the first glaciation in Iceland. This means that the axial zone in the central segment of Greater Caucasus, situated at 42°–43°N, experienced uplift to attain an altitude of at least 2000–3000 m and was thus capable of producing glaciation. It is necessary to take into account the fact that the Elbrus volcano was active and formed a sizable caldera during the Late Pliocene and Quaternary. The moraine deposits of the Elbrus Glaciation occur inside the eastern slope of this caldera and were involved in its relative subsidence.

In the Verkhniy Chegem Volcanic Highland, at an altitude of 3500 m, rhyolite–dacite ignimbrites, 2000–3000 m thick, are conformably overlain (but with a hiatus) by 50 m of typical moraine deposits of the Chegem Glaciation (Fig. 28). This moraine includes large boulders of Palaeozoic granites and metamorphic rocks, and is covered, with a hiatus but without unconformity, by andesitic flows. These lavas were erupted from the volcanic centres on the Verkhniy Chegem Plateau and have preserved the moraine deposits from erosion. The eruption of andesite represented the last, post-caldera phase of the eruptive cycle in the Verkhniy Chegem volcanic area that began with deposition of voluminous ignimbrites. The age of the Chegem Glaciation, which occurred between these two volcanic phases, was evidently not very different from the age of the underlying rhyodacite ignimbrites and that of the overlying andesite lavas. Although the andesites have not been dated, the Chegem Glaciation may correspond to the interval of 2.8–2.5 Ma; but in any case, it occurred more than 2 Ma ago. It is obvious that the moraines, as well as the andesite lavas overlying them, are located at at an altitude of 3.5 km on the surface of the volcanic plateau that was formed before intensive uplift of the Chegem volcanic highland and erosion to form the gorge of the Chegem River, whose bed is at a height of 1500 m, or some 2000 m below the Chegem Glaciation moraine.

Other moraine remnants of the Chegem Glaciation occur further east, on the watershed between the left tributaries of the Terek River (Chegem, Nalchik, Cherek Bezengiysky and Cherek Balkarsky) at an altitude of about 2500–3500 m, or more than 1000 m above the present beds of these rivers. The fluvioglacial sediments of the Elbrus and Chegem glaciations belong to a sheet- or half-sheet-type glaciation and accumulated when the internal parts of the rapidly rising mountain ridge of the Greater Caucasus were not yet deeply dissected by rivers.

In contrast, the later Quaternary glacial epochs of the Greater and Lesser Caucasus belong to the mountain-valley type. The glaciers were confined to cirques and U-shaped valleys. The valleys were successively deepened by excavation during several glacial and interglacial periods, which took place during an arch-like uplift of the Greater Caucasus, the amplitude of which was more than 1000 m. Detailed geological and geomorphological studies of successive Quaternary glaciations on the northern slopes of the Central Caucasus, in the U-shaped valleys of the major Caucasian rivers (Kuban, Terek, and their tributaries Malka, Baksan, Chegem, Cherek and Urukh) provide a basis for the chronostratigraphic correlation of the Quaternary glaciations in the Caucasus.

This chronostratigraphic scale consists of three main glacial and interglacial epochs, subdivided into phases and stages, corresponding to the Early Pleistocene, the Middle Pleistocene and the Late

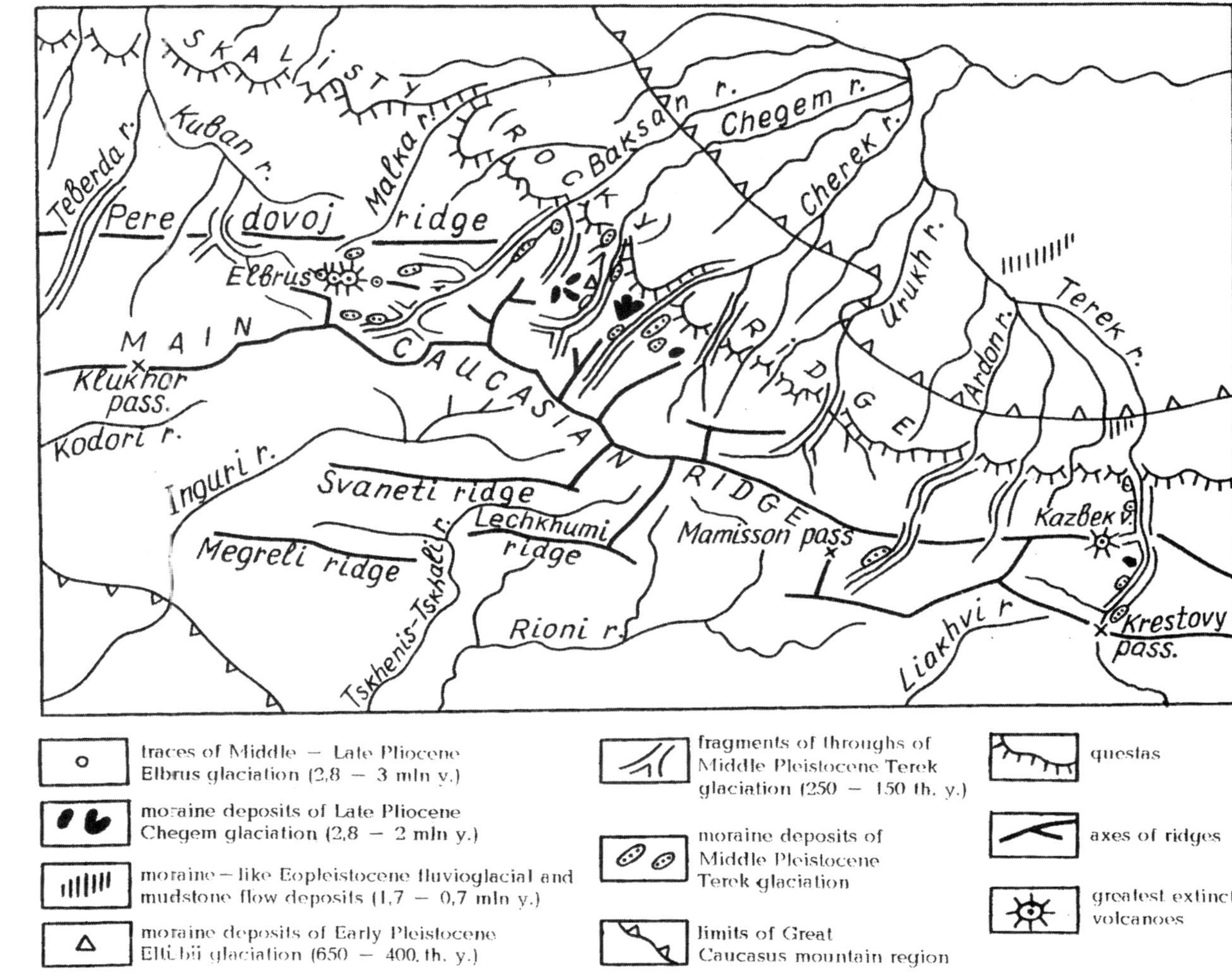

Fig. 26. Map of Pliocene, Eopleistocene and Middle Pleistocene glaciations in the central part of Great Caucasus (compiled by E. E. Milanovsky).

E. E. MILANOVSKY

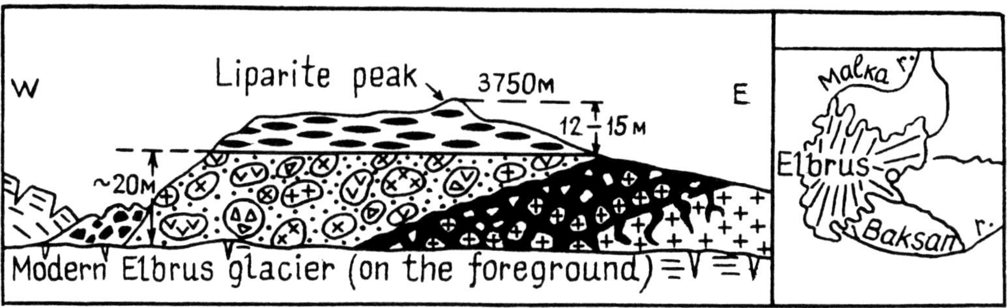

Present glacier of Elbrus (on the line of crossection)

Its moraine of 19-20 century

Upper Pliocene (2,8 mln y.) striped rhyolite and rhyo-dacite ignimbrites similar to the same of Upper Chegem volcanic highland

Coarse clastic non-sratified moraine-like deposits consisting of big boulders and angular fragments mainly of Palaeozoic granites and gneisses, Jurassic argillites and massive rhyo-dacite lavas in the sandi-gravel cement

Early-Middle Pliocene weathering crust consisting of products of mechanical and chemical weathering of Upper Palaeozoic grey granites and migmatites

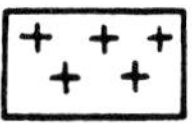
Non-weathered Upper Palaeozoic grey granites and migmatites

Fig. 27. Map showing traces of Middle (?)–Late Pliocene Elbrus Glaciation on the eastern slope of Elbrus Volcano under the 'liparite peak', between the Jrik–Chat and Jika–Uchen–Kez glaciers. The geographic position is shown on the right (compiled by E. E. Milanovsky and N. V. Koronovsky).

Pleistocene–Holocene. The Quaternary glacial epochs, named by the author as the Eltübü, Terek and Bezingi glaciations (Fig. 29) are chronological equivalents of the Mindel, Riss and Würm–Bühl epochs in the Alps (Milanovsky 1966).

During the Quaternary, rivers gradually deepened and narrowed their valleys in response to nearly 1000 m of uplift of the Greater Caucasus. During the course of intensive glacial erosion, the valleys became U-shaped troughs, whereas during interglacial epochs river erosion transformed them into V-shaped gorges or canyons (Fig. 30a–c). Fragments of the bottoms and 'shoulders' of the relatively broad troughs of the Early Pleistocene Eltübü Glaciation are preserved in the upper parts of the slopes, between 400 and 700 m above the present river bed. In some places, especially in the Chegem Valley near Eltübü Village and in the upper part of the Baksan Valley, remnants of the Quaternary moraine are preserved on the valley shoulders. The relatively narrow and deep troughs of the Middle Pleistocene Terek Glaciation occur on the inside of the Early Pleistocene Eltübü troughs. As a rule, fragments of their bases can be traced at 180–350 m above the present valley bottoms. Even narrower and deeper troughs of the Late Pleistocene Bezengi Glaciation occur inside the troughs of the Terek Glaciation. In some places there are moraine deposits on the bottom of the Middle Pleistocene troughs. In the upper part of the Baksan and Terek valleys this moraine is covered by Middle Pleistocene lavas, confirming that it belongs to the Terek Glaciation. The bottom of the troughs, formed by the Late Pleistocene–Holocene Bezengi Glaciation, are usually near the elevation of the present rivers beds. Most belong to the late Bezengi Phase, which began about 25 000 ka, reaching its maximum 15 000–20 000 ka, and is still continuing (Fig. 30). In some valleys the bottom of the Late Pleistocene troughs is below the level of presen river bed – for example, in the upper part of the Baksan Valley it is 100 m lower (Fig. 30a), in the upper part of the Kuban Valley it is 200 m lower and in

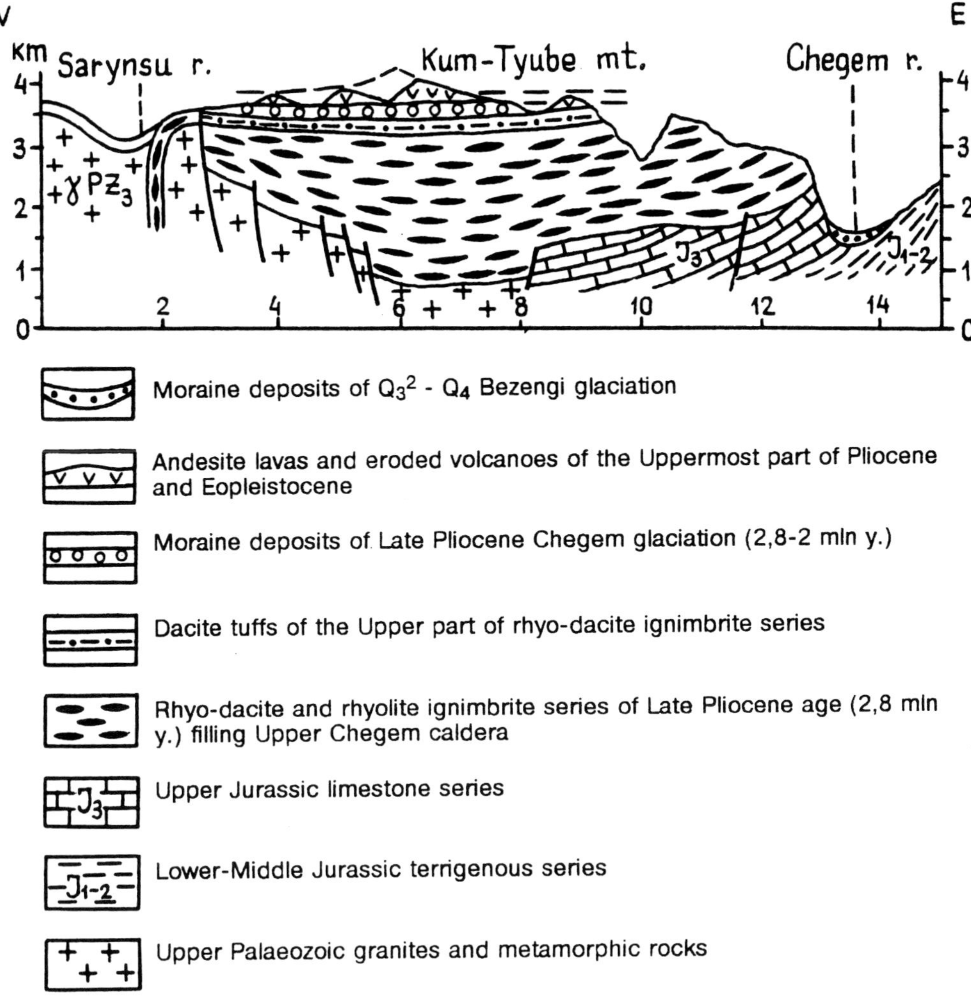

Fig. 28. Cross-section of the Upper Chegem volcanic highlands, showing the stratigraphic and geomorphological position of the Upper Pliocene moraine deposits of the Chegem Glaciation (compiled by Milanovsky 1960, 1964).

the upper part of the Terek Valley as much as 300–450 m lower (Fig. 30c). In these areas drilling has revealed the presence of thick glacial, fluvioglacial and glacio-lacustrine sediments. The glacial deposits of the Bezengi Glaciation, formed during its second phase and during several stages of retreat, and the topographic features of glacial deposition and erosion are well preserved and exposed. In several U-shaped valleys, lacustrine sediments accumulated in glacial lakes during the different phases and stages of the Bezengi Glaciation. These temporary lakes formed between the termination of the valley glaciers and dams formed by moraines or landslides, or they have an avalanche origin.

Conclusions

A chronological scheme of the Pliocene–Quaternary glaciations in different regions of northern Eurasia, resulting from 150 years of research, has been given in Table 1. The distribution and development of glaciations depend on geographic position, climatic conditions, topography and neotectonic movements in the regions where the glaciations occurred.

(1)   The correlation scheme suggests that between 3.0 and 3.5 Ma there was repeated alternation of cold and relatively warmer climates, particularly in northern Eurasia, but probably

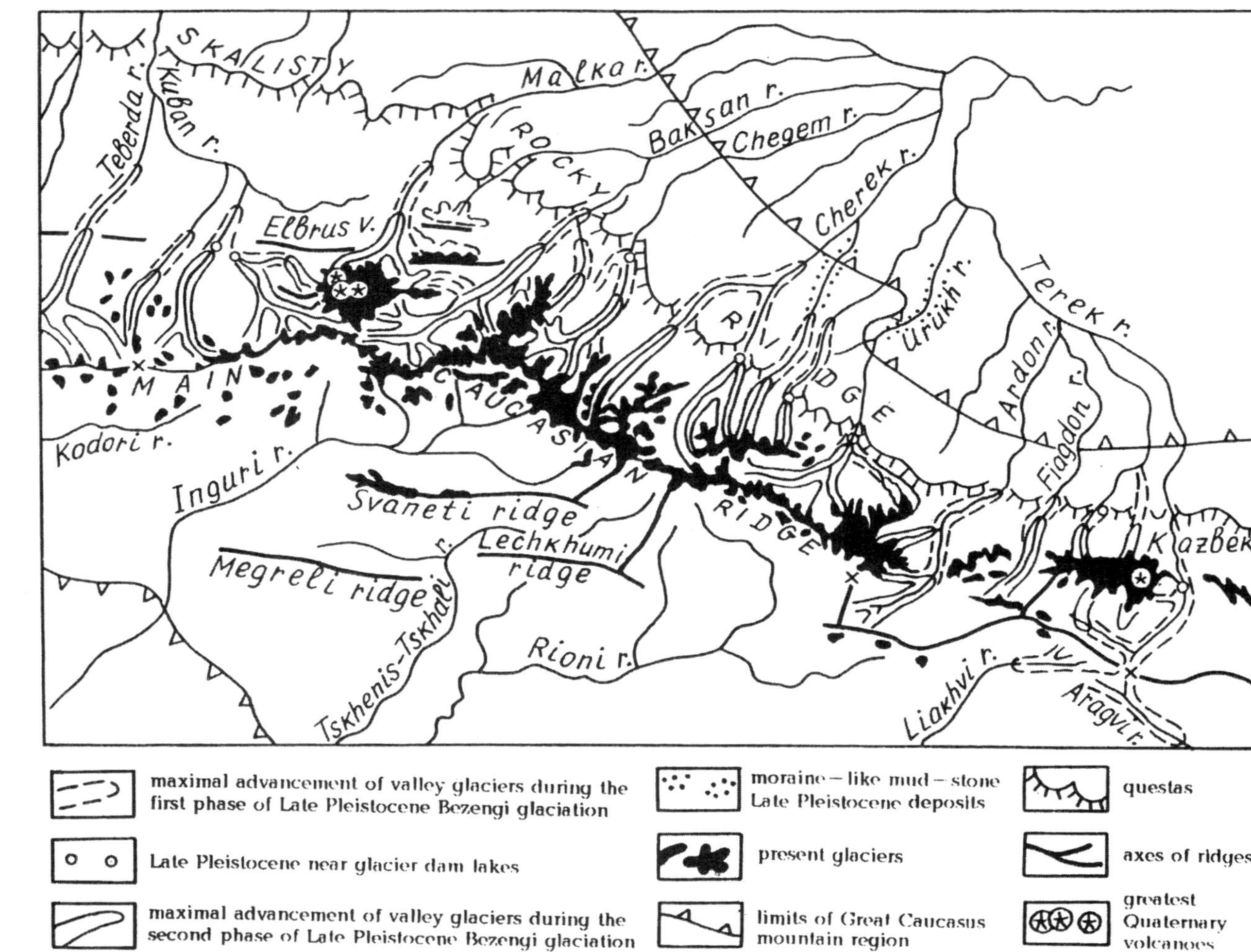

Fig. 29. Map showing the distribution of the Late Pleistocene Bezengi Glaciation and modern glaciers on the northern slope of the central part of Great Caucasus (compiled by E. E. Milanovsky).

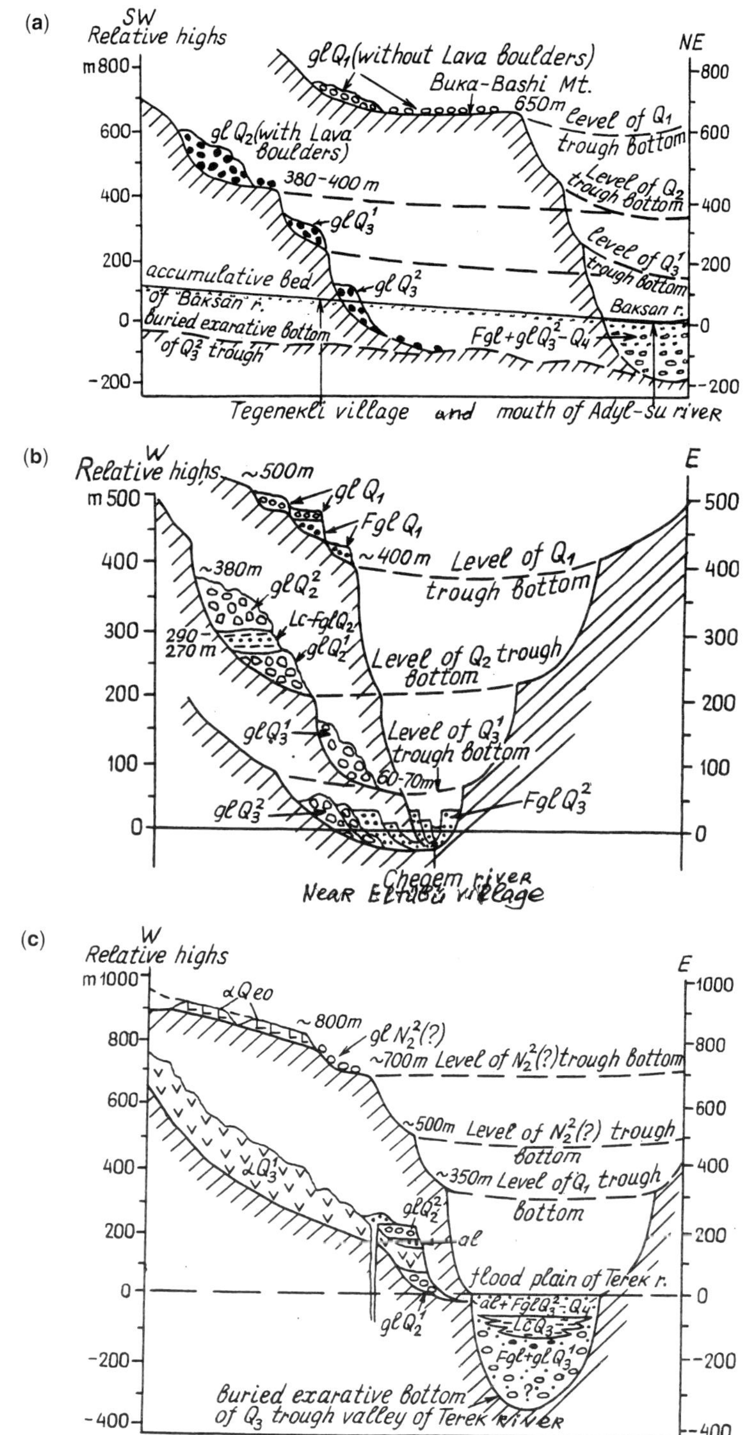

Fig. 30. Conditions of occurrence and interrelations of glacial and other types of Eopleistocene and Pleistocene continental deposits and forms of relief in the U-shaped valleys on the northern slope of Central Caucasus (compiled by E. E. Milanovsky 1966). (**a**) Tegenekli Village and mouth of the Adyl-su River. (**b**) Chegem River near Eltübü village. (**c**) Buried exavated bottom of the Q_3 U-shaped valley of the Terek River.

over the whole of the Earth's surface. Glaciation periodically occurred on vast plains, plateaus and in mountainous regions of Eurasia, whereas during the warmer periods glacial sheets considerably diminished or completely disappeared. The first evidence of the Cenozoic global-scale cooling is recorded in Antarctica, were glaciation began in the Oligocene (Dingle & Lavelle 1998). In parts of northern Eurasia and North America the first glaciation occurred in the Pliocene, but in Greenland it may have started in the Late Miocene. In general, glaciation in the western part of northern Eurasia reached its maximum between 0.7 and 0.4 Ma, i.e. in the Early Pleistocene according to the Russian Chronostratigraphic Scale, whereas in the western part of northern Eurasia it occurred about 0.2 Ma ago, i.e. by the end of the Middle Pleistocene.

(2) The regional extent of glaciations, and the maximum thickness and volume of glacial sheets in different parts of northern Eurasia, depended on climatic factors such as average annual temperature, relative humidity and the amount of precipitation in the form of snow. During the Pliocene and Quaternary, both the average annual temperature and precipitation decreased considerably towards the east of northern Eurasia. This eastward increase in aridity manifested itself in a decrease of glaciation intensity, and in the gradual delay of the beginning of the first and of the maximal glaciation. For instance, in Scandinavia and on the plains of Central and eastern Europe glaciation started in the Eopleistocene, reaching its maximum in the Early Pleistocene, whereas on the West Siberian Plain the first glaciation took place at the end of the Early Pleistocene and reached its maximum in the Middle Pleistocene. Its extent was considerably smaller than the Early Pleistocene glaciation in Central and eastern Europe. In East Siberia the first glaciation took place at the end of the Early Pleistocene, with its maximum in the Middle Pleistocene, but it was even more restricted than in West Siberia, occurring only on the plateaus, highlands and mountains, where precipitation was more abundant.

At the same time the dominant arid climate in the east led to an extensive development of permafrost. In the eastern parts of Europe (on the Kola Peninsula and Pechora Lowlands) permafrost is limited to north of the Arctic Circle (67°N), to 60°N in West Siberia. In contrast, in the northern part of the West Siberian Lowlands and to east of the Yenisei River, permafrost exists practically everywhere at latitudes higher than 48°–50°N, near the southern border of Russia, in North Mongolia and North China. In many places in East Siberia the depth of permafrost attains 1000 m or even 1500 m.

(3) The orographic factor also played an important role. Decrease in the average annual temperature is in direct correlation with altitude, so that precipitation is in the form of snow during the winter. In the high mountains between 40°N and 52°N, such as the Alps, the Greater Caucasus and the Altai, the first glaciation started in the Late Pliocene (Alps, Caucasus) or in the Eopleistocene (Altai), whereas farther north, on the plains of Eurasia, glaciation began only at the end of the Eopleistocene in Europe or in the Early Pleistocene in Siberia.

Although the elevations of the Greater Caucasus are higher than those of the Alps, ancient and present glaciations are much more important in the Alps. During the Würm glacial period almost the whole of the Alps, except for small peaks on the watersheds, were covered by a thick (up to 1000–2000 m) ice sheet, which spread out onto the northern piedmont plains in some places. However, in the Greater Caucasus the glaciers of the Late Pleistocene (Bezengi Glaciation) remained confined to cirques and U-shaped valleys as 50–60 km-long ice-'tongues', covering only about 12% of the whole area. Their thickness did not exceed a few hundred metres. These differences between the Alps, the Tien-Shan, mountainous regions of Siberia and the Greater Caucasus reflect a general eastward increase in aridity in northern Eurasia.

(4) Neotectonic factors, such as the intensity, extent and timing of uplift, also played a very important role in the development of the late Cenozoic glaciation. The first significant glaciation in the centre of the Great Caucasus and in the Alps took place in the first half of the Late Pliocene (2.5–3.0 Ma), i.e. much earlier than the first glaciation on the plains of eastern and Central Europe. It was almost synchronous with the first glaciation in Iceland, near the Arctic Circle. This shows that the axial zones of the alpine orogenic systems were already uplifted to the altitudes that were necessary to produce glacial sheets. Markov *et al.* (1965) estimated that the mountains were 2000–3000 m or even 3000–4000 m high. The present altitudes are of the order of 4000–5000 m in the Great Caucasus (5600 m on Elbrus volcano) and

3000–4800 m in the Alps. The amplitude of neotectonic uplift after the first glaciation, i.e. during the last 2.5 Ma, was at least 1000–2000 m in the axial zones of the central Greater Caucasus and 500–1500 m in the Alps. Erosion produced deep river valleys on the slopes of the Greater Caucasus. In the process of intensive uplift, the Late Pliocene and Eopleistocene sheet-type glaciers were gradually replaced by mountain-valley- or trough-type glaciers during the Pleistocene.

(5) In contrast to the glaciated mountainous regions of Eurasia, the central part of the Greater Caucasus reveals the close interaction of glacial and volcanic processes with volcano-tectonic deformation, in particular caldera formation. Alternation of glacial epochs and phases of volcanic activity led to the formation of a relatively hard volcanic 'armour' that protected moraines from total denudation in the Caucasus. For instance, in the *Verkhniy* (=Upper) Chegem Highlands and on the Elbrus volcanic massif, Late Pliocene glacial deposits were preserved in calderas. Glaciation there may have been synchronous with subglacial volcanic eruptions, such as are represented as a widespread phenomenon in Iceland.

The author of this paper would like to express his gratitude to the Russian Science Programme, Grant NSCH-5280.2006.5, for financial support for this work.

All geological and other maps and geological cross-sections (with the exception of illustrations whose authors are indicated in the explanations to such pictures) were prepared by the author. He has added several portraits of scientists who studied the geological structure and development of European and Asiatic regions, and samples of his landscape sketches (drawn in ink) of various places in Europe, Eurasia and Iceland that experienced glaciation during the Late Cenozoic. D. Oldroyd and R. Grapes are thanked for their help with editing the original manuscript.

References

ABICH, G. (H.) VON. 1853. Explanation of the geological sequences on the northern slopes of the Caucasus from Elbrus to Beshtau. *Kavkazskiy Kalendar*, Part 4, 440–471 (in Russian).

ABICH, G. (H.) VON. 1870. *Études sur les glaciers actuels et anciens du Caucase. Première partie par H. Abich.* L'Imprimerie de l'Administration Civile au Caucase, Tiflis (Russian translation in 1871 by F. G. von Koshkul. *In: Collection of Data on the Caucasus*, Volume 1, 85–126, Tiflis).

ABICH, G. (H.) VON. 1871. Bemerkungen über Geröll und Trümmerablagerungen aus des Gletscherzeit im Kaucasus. *Bulletin de l'Académie Impériale des Sciences de St.-Pétersbourg*, **16**, 245–265.

ABICH, G. (H.) VON. 1878. Über die Lage die Schneegrenze und die Gletscher die Gegenwart im Kaukasus. *Bulletin de l'Académie Impériale des Sciences de St.-Pétersbourg*, **24**, 258–282.

ABICH, G. (H.) VON. 1878–1887. *Geologische Forschungen in den kaukasischen Ländern* (3 vols). A. Hölder, Vienna.

ACHMETJEV, M. A., GEPTNER, A. R., GLADENKOV, YU. B., MILANOVSKY, E. E. & TRIFONOV, V. G. 1978. *Iceland and the Mid-Oceanic Ridge. Stratigraphy: Lithology.* Nauka, Moscow, 1–204 (in Russian).

DINGLE, R. V. & LAVELLE, M. 1998. Antarctic Peninsular cryosphere: Early Oligocene (*c*. Ma) initiation and a revised glacial chronology. *Journal of the Geological Society*, **155**, 433–437.

DUMITRASHKO, N. V. 1949. About the ancient glaciation of the Lesser Caucasus. *Trudy Instituta Geographii Akademii Nauk SSSR [Proceedings of the Geographical Institute of the Academy of Sciences of The USSR]*, **43** (in Russian).

GERASSIMOV, A. P. 1911. The northeastern feet of Elbrus. *Trudy Geologicheskogo Komiteta [Proceedings of the Geological Committee]*, **30**, 77–152 (in Russian).

GERASSIMOV, I. P. & MARKOV, K. K. 1939. *Glacial Period on the Territory of the USSR.* Akademii Nauk SSSR, Moscow (in Russian).

GROMOV, V. I. 1948. The palaeontological and archaeological basis of the stratigraphy of the continental deposits of the Quaternary Period in the Territory of the USSR. *Trudy Geologicheskogo Instituta Akademii Nauk SSSR [Proceedings of the Geological Institute of the Academy of Sciences of the USSR]*, **43**, 33–52 (in Russian).

GROSSWALD, M. G. 1983. *Ice Sheets of the Continental Shelves.* Nauka, Moscow (in Russian).

HALLAM, A. 1983. *Great Geological Controversies.* Oxford University Press, Oxford.

HELMERSEN [OR GELMERSEN], G. P. 1869. Studien über die Wanderblöcke und die Diluvial gebilde Russlands. *Mémoires de l'Académie Impériale des Sciences de St.-Petersburg*, Series 7, **14**, No. 7.

IVANOVA, T. K. & MARKIN, V. A. 2008. Piotr Alekseevich Kropotkin and his monograph *Researches on the Glacial Period* (1876). *In*: GRAPES, R. H., OLDROYD, D. R. & GRIGELIS, A. (eds) *History of Geomorphology and Quaternary Geology.* Geological Society, London, Special Publications, **301**, 117–128.

KOVALEV, P. V. 1960. Traces of ancient glaciation in the Kabarda-Balkarian ASSR. *Materials of the Caucasian Expedition*, **1**, 119–207. Kharkov University, Kharkov (in Russian).

KROPOTKIN, P. A. 1873. Report of the Olekma–Vitim Expedition's investigations of the Cattle route from Nerchinsk to Olekma Districts. *Reports of the Russian Geographical Society*, **3**, Part 13 (in Russian).

KROPOTKIN, P. A. 1874. On the study of glacial deposits of Finland. *Reports of the Russian Geographical Society*, **10**, Part 6, Section 1, 323–332 (in Russian).

KROPOTKIN, P. A. 1876. Researches on the Glacial Period. *Reports of the Russian Geographical Society*, **7**, Part 1 (with a separate booklet of maps, cross-sections and drawings) (in Russian).

KUSHNEV, S. L. 1964. The Bezengi Glaciation of the Central Caucasus. *In: Proceedings of the International Conference 'Oledenenie Kavkaza'*, **10**, 184–202 (in Russian).

LIPMAN, P. W., BOGATIKOV, O. A. *ET AL.* 1993. 2.8 Ma ash flow caldera at the Chegem River in the Northern Caucasus Mountains (Russia): contemporaneous granites and associated ore deposits. *Journal of Volcanological and Geothermal Research*, **37**, 85–124.

LYELL, C. 1863. *The Geological Evidences of the Antiquity of Man with Remarks on Theories of the Origin of Species by Variation.* John Murray, London.

MAKAROVA, N. V. & MAKAROV, V. I. 2004. On the stratigraphy of the Quaternary sediments in the centre of Russian Plain (controversial questions). *Vestnik, Moscow University, Series 4, Geology*, **6**, 19–25 (in Russian).

MARKOV, K. K., LASUKOV, G. I. & NIKOLAEV, V. A. 1965. *The Quaternary Period: The Glacial and Anthropogene (I & II).* Moscow University Press, Moscow (in Russian).

MARUASHVILI, L. I. 1956. *Provisional Revision of the Present Concepts Concerning the Palaeogeographic Conditions of the Glacial Period in the Caucasus.* Akademiya Nauk Gruzinskoy SSR [Academy of Science of the Georgian SSR], Tbilisi (in Russian).

MARUASHVILI, L. I. 1971. Glacial and ancient glacial forms of the Quaternary period (Post-Kimmerian). *In:* MARUASHVILI, L. I. (ed.) *Geomorphology of Georgia.* Metsniereba, Tiblisi (in Russian).

MILANOVSKY, E. E. 1960. Concerning the traces of the Late Pliocene glaciation in the high mountainous parts of the Central Caucasus. *Doklady Akademii Nauk SSSR [Reports of the Academy of Sciences of the USSR]*, **130**, 158–161 (in Russian).

MILANOVSKY, E. E. 1964. Concerning the Upper Pliocene glaciation of the Central Caucasus. *In: Selected Works International Geophysical Year*, Volume 10. Moscow University Press, Moscow, 9–43 (in Russian).

MILANOVSKY, E. E. 1966. Principal problems of the ancient glaciation history of the Central Caucasus. *In:* MILANOVSKY, E. E. (ed.) *Problems of Geology and Palaeogeography of the Anthropogene.* Moscow University Press, Moscow, 5–49 (in Russian).

MILANOVSKY, E. E. 2000. The Plio-Pleistocene glaciation in Eastern Europe, Siberia and Caucasus: evolution of thoughts. *Eclogae Geologicae Helvetiae*, **93**, 379–394.

MILANOVSKY, E. E. & KORONOVSKY, N. V. 1960. Geological structure and formation history of the Elbrus Volcano. *In: Materials Geology and Metallogeny of Central Caucasus. Transactions of the All-Union Aerogeological Trust*, **6**, 92–127 (in Russian).

MILANOVSKY, E. E. & KORONOVSKY, N. V. 1961. New data on the most ancient development stages of the Elbrus Volcano. *Doklady Akademii Nauk SSSR [Reports of the Academy of Sciences of the USSR]*, **14**, 433–436 (in Russian).

MILANOVSKY, E. E., TRIFONOV, V. G., GORJACHEV, A. V. & LOMIZE, M. G. 1979. *Iceland and the Mid-Oceanic Ridge. Geomorphology: Tectonics.* Nauka, Moscow, 7–211 (in Russian).

MOSKVITIN, A. I. 1967. Stratigraphy of the Pleistocene of the European part of the USSR. *Trudy Geologicheskogo Instituta Akademii Nauk SSSR [Proceedings of the Geological Institute of the Academy of Sciences of the USSR]*, **156** (in Russian).

MURCHISON, R. I. 1845. *The Geology of Russia in Europe and the Ural Mountains*, Volume 1. John Murray, London.

PAVLOW, A. P. 1922. Époques glaciers et interglaciares de l'Europe et leur rapport à l'histoire de l'homme fossile. *Bulletin of the Moscow Society of Naturalists*, **31**, 23–76.

PENCK, A. & BRUCKNER, E. 1901–1909. *Die Alpen im Eiszeitalter* (3 vols). Tauchnitz, Leipzig.

PIDOPLICHKO, I. G. 1946–1956. *Concerning the Glacial Period* (4 vols). Academy of Sciences of the Ukranian SSR, Kiev (in Russian).

REINHARD, A. L. 1936. Stratigraphy of the Glacial Period of the Alps according to Bruckner and Penck and the glaciation of the Caucasus. *In: Materials on the Study of the Quaternary Period in the USSR*, Volume 1. Akademiya Nauk SSSR [Academy of Sciences of the USSR], Moscow, 33–45 (in Russian).

REINHARD, A. L. 1947. The Quaternary System and the Geomorphology of the Northern Caucasus. *In: Geology of the USSR, Volume 9: Northern Caucasus.* Gosgeolizdat, Moscow, 291–502 (in Russian).

RENGARTEN, V. P. 1932. A geological essay concerning the region of the Georgian Military Road. *Trudy Vsesoyuznogo Geologorazvedochnogo Obyedineniya [Proceedings of All-Union Prospecting Association]*, **148**, 3–67 (in Russian).

SÈMUNDSSON, K. 1967. Vulkanismus und Tektonik des Hengill-Gebietes in Südwest Island. *Acta naturalia Islandica*, **2** (in German with English summary).

SAFRONOV, I. N. 1960. On the ancient glaciation of the North-West Caucasus. *Trudy Stavropolskogo Pedagogicheskogo Instituta, Estestvenno-Geographicheskiy phakul'tet [Proceedings of the Stavropol Pedagogical Institute]*, **18**, 3–16 (in Russian).

SCHMIDT, F. B. 1865. Untersuchungen über die Erscheinungen der Glacialformation in Estland und auf Oesel. *Bulletin de l'Académie Impériale des Sciences de St.-Petersbourg*, Series 1, **8**, 339–368.

SHANZER, E. V. 1976. The role of P. A. Kropotkin in the formulation of glacial theory. *Bulletin of the Moscow Society of Naturalists. Geological Section*, **51**, 64–76 (in Russian).

SHCHUROVSKY, G. E. 1856. Erratic phenomena. *Russky Vestnik [The Russian Bulletin]*, **18**, No. 5 (in Russian).

SHIK, S. M. 1998. Climatic rhythmicity in the Pleistocene of the East European platform. *Stratigraphy and Geological Correlation*, **4**, 105–109 (in Russian).

TUSHINSKY, G. K. 1949. Modern and ancient glaciation of the Teberda Region. *In:* TUSHINSKY, G. K. (ed.) *The Conquered Summits, Annals of Soviet Alpinism.* Geographgiz, Moscow, 263–316 (in Russian).

TUSHINSKY, G. K. 1958. Post-lava glaciation of Elbrus and its dynamics. *In:* TUSHINSKY, G. K. (ed.) *Collected Information from Works on the International*

Geophysical Year. 2nd Elbrusskaya Expedition of the Moscow State University and the Institute of Applied Physics of the Academy of Sciences of the USSR. Moscow University Press, Moscow, 117–166 (in Russian).

VARDANYANTS, L. A. 1932. The Glacial Epoch in Northern Ossetia. *Doklady Russkogo Geographicheskogo obshchestva [Reports of the Russian Geographical Society]*, **64**, 499–537 (in Russian).

VARDANYANTS, L. A. 1948. *Post-Pliocene History of the Caucasus–Black Sea–Caspian Region.* Akademiya Nauk Armenian SSR [The Academy of Sciences of Armenian SSR], Yerevan (in Russian).

WENSINK, H., 1964. Paleomagnetic stratigraphy of younger. basalts and intercalated Plio-Pleistocene tillites in Iceland. *Geologische Rundschau*, **54**, 364–384.

YAKOVLEV, S. A. 1956. Principles of the geology of the Quaternary deposits on the Russian Plain. *Trudy Vsesoyuznogo Geologicheskogo Instituta [Proceedings of All-Union Geological Institute]*, New Series, **17** (in Russian).

Piotr Alekseevich Kropotkin and his monograph *Researches on the Glacial Period* (1876)

TATIANA K. IVANOVA[1] & VYACHESLAV A. MARKIN[2]

[1]*Earth Science Museum, Moscow State University, 119899 Moscow, Russia (e-mail: tkivan@mes.msu.ru)*

[2]*Institute of the History of Natural Sciences and Technology, Russian Academy of Sciences, 1/5 Steropansky Per., 109012 Moscow, Russia (e-mail: zevs@naukaran.ru)*

Abstract: The notable Russian scientist Piotr Alekseevich Kropotkin (1842–1921) is well known in Western countries for his writings on anarchist philosophy and various historical and political themes, but his geological and geographical work is less familiar, and his great treatise on Quaternary geology is virtually unknown in the Western world. The present paper provides a summary account of Kropotkin's Quaternary studies and his travels in the glaciated regions of Siberia and Scandinavia. He was an exponent of the 'land-ice' theory and traced the movements of glaciers in Scandinavia, paying particular attention to the form and structure of eskers.

Kropotkin's career

Kropotkin's treatise, *Researches on the Glacial Period* (Kropotkin 1876), had a profound impact in Russia and convinced the majority of Russian scientists of the validity of the glacial theory. His work showed a development of this theory and of the methods of glacier research employed in Western Europe and America. *Researches* is a fundamental monograph in which the author discussed different aspects of the morphology, structure and origin of glacial (and glacially related) sediments, and the forms of glacial landscapes.

Kropotkin (see Fig. 1) was born, in Moscow on 9 December 1842, into an ancient aristocratic family, descended from the legendary Rurik, a founder of the Russian state more than 1000 years ago. Rurik or Riurik (Russian: Рюрик, Old East Norse: *Rørik*; *c.* 830–*c.* 879) was a Viking chief who captured Ladoga in 862, built the settlement of Holmgard near Novgorod and founded the Rurik Dynasty, which ruled Russia until the seventeenth century.

The future famous scientist and political philosopher graduated from the exclusive St Petersburg Page Corps in 1862, with excellent marks. The Page Corps was founded in St Petersburg in 1759 as a school for training pages and chamber pages. To meet the need for trained officers for the Imperial Guard it was reorganized in 1802 into an educational establishment similar to a cadet school, accepting pages from the royal court only. Thus, it became a privileged military establishment in Imperial Russia that prepared aristocratic children for military service. Kropotkin's great opportunity for scientific exploration occurred when he was posted to Siberia shortly thereafter.

Kropotkin's creative career can be divided into three quite distinct periods as follows (Markin 1985).

1862–1867

Kropotkin worked as aide-de-camp to the military chief-of-staff of the Transbaikal region, and participated in several East Siberian scientific expeditions. In 1865 he became a member of the East Siberian Branch of the Russian Geographical Society and helped organize five scientific expeditions into unexplored regions: Lake Baikal, the Amur river, Mountains Sayan, Manchuria, and the Patom Highlands between the Vitim and Olekma rivers. He described parts of these territories and compiled maps of the mountain relief of Eastern Siberia (Postnikov 1993, 2002). Kropotkin's researches were of a high quality and he was awarded a gold medal by the Royal Geographical Society for his work. In the course of his travels in the different regions of Siberia he found numerous indications of a former glacial period. In 1868 he presented a report at the First Congress of Russian Naturalists. Before then, in 1867, he wrote the manuscript of an article, 'On traces of the Glacial Period in Siberia', but this was only published for the first time in 1998 (Kropotkin 1998*b*).

1867–1876

Returning to St Petersburg, Kropotkin worked in collaboration with the Russian Geographical Society, where he was elected a Secretary of its Physical Geography Section (1868). He worked on the completion of his Siberian materials and

From: GRAPES, R. H., OLDROYD, D. & GRIGELIS, A. (eds) *History of Geomorphology and Quaternary Geology.* Geological Society, London, Special Publications, **301**, 117–128. DOI: 10.1144/SP301.7   0305-8719/08/$15.00 © The Geological Society of London 2008.

Fig. 1. Photoportrait of Piotr A. Kropotkin (1886). State Archives of the Russian Federation, Fund 1129.

published them as 'Report of the Olyokma–Vitim Expedition' (1873) and 'Introductory outline of East Siberian orography' (1875). He also proposed some other important projects, including a great plan for the exploration of the Arctic seas. In 1871, the Society dispatched Kropotkin on an official journey to Finland and Sweden to study the evidence there for former glaciation. This expedition was highly successful and on his return he began to write his principal scientific book: *Researches on the Glacial Period*.

Unfortunately, however, Kropotkin was not able to complete this task as he was arrested on account of his political views and activities, and spent 2 years (1874–1876) in the prison of St Petersburg's Peter-and-Paul Fortress. However, he continued his scientific work while in prison and completed the first volume of his monograph on glacial theory, supplemented by a short summary of the contents of the second volume. An unfinished manuscript of the second volume has been preserved and was published in 1998, but it did not appear during his lifetime.

1876–1917

With the assistance of friends and supporters, Kropotkin made an ingenious dramatic (and famous) escape from prison, following which he travelled to Western Europe via Finland. As a political refugee, he lived mostly in Britain, near London. But he was again arrested for his political activities and spent 2 years in prison in France. In England he worked as a scientific publicist and contributor to scientific magazines and newspapers (e.g. *Nature, Nineteenth Century, Geographical Journal* and *Proceedings of the Royal Geographical Society*). He also collaborated with the Royal Geographical Society. In 1893 he was elected to the British Association for the Advancement of Science. From 1888 to 1891 Kropotkin was head of a section on 'Recent Science' for the magazine *Nineteenth Century*.

Kropotkin's main scientific works published in the period when he was in the West were: 'La plasticité de la glace' (1884), 'The glaciation of Asia' (1897*a*), *Anarchism: Its Philosophy and Ideal* (1897*b*), 'Some of the resources of Canada' (1898), *Fields, Factories and Workshops* (1899), *Modern Science and Anarchism* (1903), *Mutual Aid: A Factor of Evolution* (1902) and 'Desiccation of Eur-Asia' (1904). He published about 200 articles in the 9th, 10th and 11th editions of the *Encyclopaedia Britannica*, mostly on topics concerned with Russia. He was also the author of the *Great French Revolution* (1909). These works were published in Russia after 1917.

Old age: 1917–1921

Returning to Russia in 1917, after the Revolution, Kropotkin lived in Petrograd, then in Moscow and, finally, in the small town of Dmitrov, about 100 km north of the centre of the capital, where he established the Dmitrov Museum of Regional Studies. He preserved an interest in glaciation and glacial theory throughout his life. One of his last projects was a plan of a book entitled *Glacial and Lacustrine Periods* (1919), but it was only published in 1998 (Kropotkin 1998*a*). Kropotkin died on 8 February 1921, and was buried with honour in Novodevichy Cemetery, Moscow.

Early ideas on glacial theory

Before considering Kropotkin' glaciological researches, we should briefly recall some aspects of the early history of glaciation investigations in Western Europe.

The first steps in the study of what came to be called Pleistocene glaciation were made by Ignaz Venetz, Jean de Charpentier and Louis Agassiz in the Alps in the 1820s–1840s. In addition, some geoscientists suggested the existence of former glaciation not only in the Alps but also in the uplands and plains of Norway (Jan Esmark), Germany

(Reinhard Bernhardi, Karl Schimper) and Scotland (Louis Agassiz, William Buckland and Thomas Jamieson).

In 1833 Charles Lyell proposed the so-called 'drift hypothesis' to account for the dispersal of erratic boulders by floating ice, which was supported by the majority of geologists at the time. Among them was Roderick Murchison, who travelled through European Russia in 1840–1841. Some Russian geologists (Grigorii Shchurovsky [1856], Friedrich Schmidt [1865] and Grigorii Helmersen [1869]) also generally supported the drift hypothesis.

The originally Estonian stratigrapher, Friedrich Bogdanovich Schmidt (1832–1908), described the Quaternary boulder-bearing sediments of the Baltic States. He also indicated that clastic materials were transported there from north to south (i.e. from Finland and Sweden), based on the analysis of the direction of the scratches or striations observed below the boulder-bearing deposits. At the same time, Schmidt in Russia and some Scandinavian geoscientists interpreted åsar (or eskers) as marine coastal banks or strand-lines, formed under the influence of ice blocks floating in the cold seas of the Glacial Period. These geologists thus attempted to combine the elements of the drift and glacial hypotheses. Later, following Kropotkin, Schmidt became a strong supporter of the land-ice theory. The most complete and convincing justifications in favour of the concept of the great Pleistocene glaciation in northern Eurasia were given in the 1870s in the classic works by Otto M. Torell (1828–1900) (see van Veen 2008 in this volume), who was the founder and first Director of the Geological Survey of Sweden, and by Kropotkin.

The present paper deals mainly with the period when Kropotkin was collecting scientific information on the evidence of glaciation and was creating a version of the theory of a glacial period. He was the first to discover traces of former glaciers in Siberia and, subsequently, he recognized similar indications in Finland and Sweden.

The Siberian investigations

In 1864–1866 Kroptkin was a young officer in the East Siberian Administration, working as a military geographer/cartographer. During this period he organized and carried out expeditions to Eastern Sajan across the Great Hingan Range in Manchuria, later to East Sajan Ridge (1865) and the Patom Highlands (1866). Figure 2 shows the main routes of Kropotkin's expeditions in the Baikalian region. In this territory scientists found the first time traces of former glaciation.

During his expeditions in Eastern Siberia, Kropotkin found and examined polished and scratched

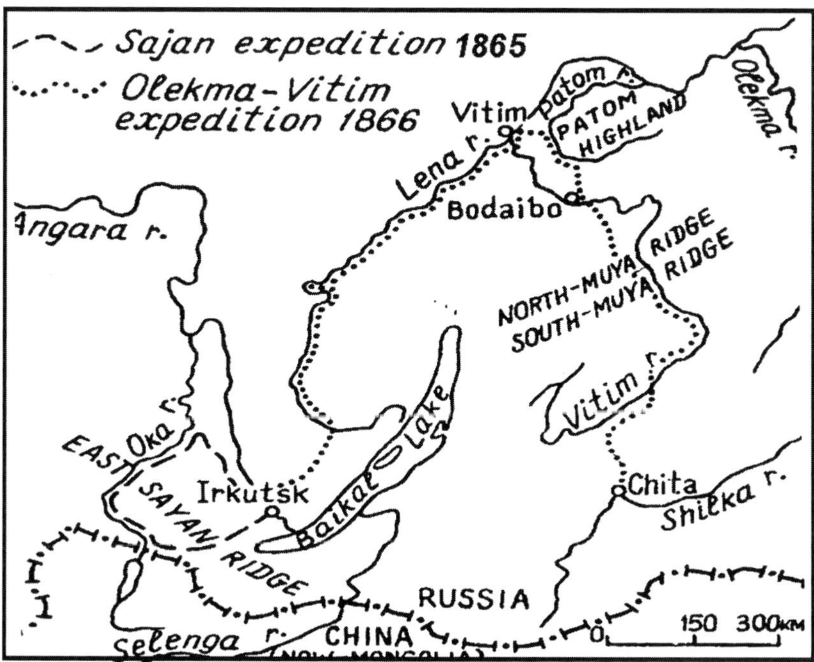

Fig. 2. The main routes of Kropotkin's expeditions in Eastern Siberia, 1865–1866 (Markin 2002, p. 192).

Fig. 3. Landscape of the Muya River valley. Modern photograph, from Markin (2002, p. 96).

boulders in the river valleys (see Figs 3 & 4). He also discovered granite boulders, evidently not *in situ* and undoubtedly transported from adjacent regions. Moreover, he noted the indications of glacial smoothing and glacial furrows on hard rock surfaces beneath the cover of boulder-clay (till). He concluded that the boulders could only have been transported and redeposited by a large and thick glacier, which formerly covered the surrounding mountains and extended down to an altitude of 700 m. (The present glaciers in Eastern Siberia do not come lower than 3000 metres: Kropotkin [1873].)

Kropotkin explained these unusual structures by invoking the idea of the action of a moving glacier mass that ground up and smoothed every form of relief.

The Scandinavian investigations

In 1871 the Geographical Society sent Kropotkin to Finland and Sweden to study the evidence of former glaciations in that part of the world (Kropotkin 1874). During this expedition he collected valuable field data and studied various collections in Swedish

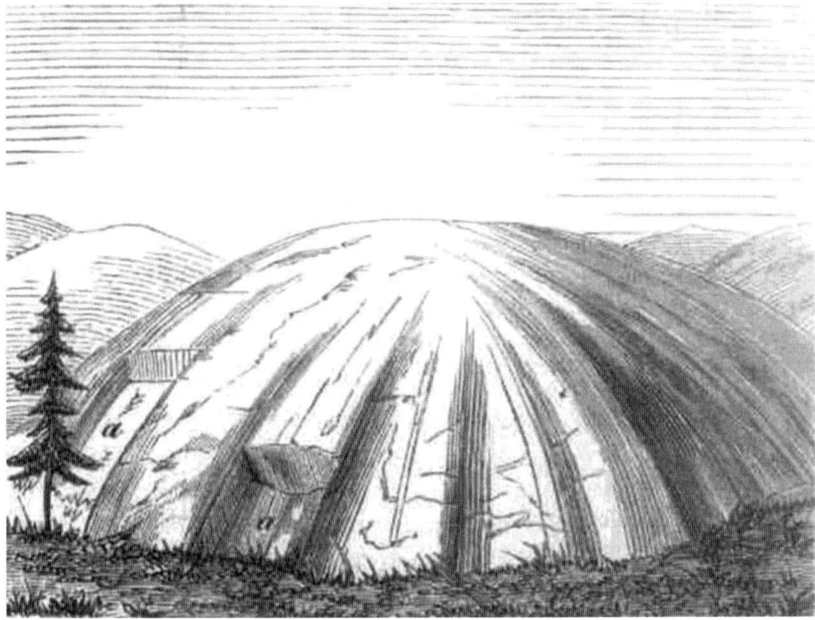

Fig. 4. Dome-like form of ridges rounded by a former glacier in the Oka Valley, Eastern Sayan Mountains (Supplement to *Researches on the Glacial Period*, 1876, p. 43).

museums. His data and his critical analysis of the international literature convinced him that the former glaciation had had a very wide extension in the northern hemisphere. His observations in Scandinavia and Siberia, and his outline of the principal problems and hypotheses concerning the glacial period, were collated and published in *Researches on the Glacial Period.*

Researches on the Glacial Period was written – much of it in prison – during the years from 1871 to 1876. Fascinated by the problem of a former continental glaciation in Siberia, Kropotkin studied the available European and American literature on the topic, and in his book he referred to the publications of William Hopkins, Charles Lyell, John Tyndall, Edward Forbes, James Croll, James Dana, Adolf Nordenskiold, Theodor Kjerulf, Axel Erdmann and Hampus von Post, among others. He also utilized the researches of Russian geologists and geographers such as Alexander Voeikov, Grigorii Helmersen, Friedrich Schmidt and Stepan Kutorga.

The Russian Geographical Society, together with Kropotkin's brother Alexander, helped Kropotkin prepare his manuscript for publication in the *Memoirs of the Russian Geographical Society.* The volume appeared in 1876, almost simultaneously with Kropotkin's escape from the prison hospital. Only the first volume was published, although it was supplemented by a short summary of the contents of some chapters from the second volume (see Fig. 5). Some fragments from the second volume were subsequently found in archives and published in Moscow in 1998 (Kropotkin 1998*b*).

The left-hand page in Figure 5 reads:

Memoirs of the Imperial Russian Geographic Society
On General Geography (Sections of Mathematical and Physical Geography).
Volume Seven.
Edited by
A. Kropotkin and Iv. Poliakov.
With maps, section and pictures in a separate brochure.
S. Petersburg.
1876.

The right-hand page in Figure 5 reads:

Researches on the Glacial Period.
P. Kropotkin.
I. On the glacial deposits of Finland (Report on the journey into Finland and Sweden in 1871 on behalf of the Imperial Russian Geographic Society).
II. On the foundation of hypotheses concerning the glacial period.
With maps, section and pictures in a separate brochure.
S. Petersburg.
M. Stasjulevich typography.
Vassilijevsky Island, 2 L, 7.
1876.

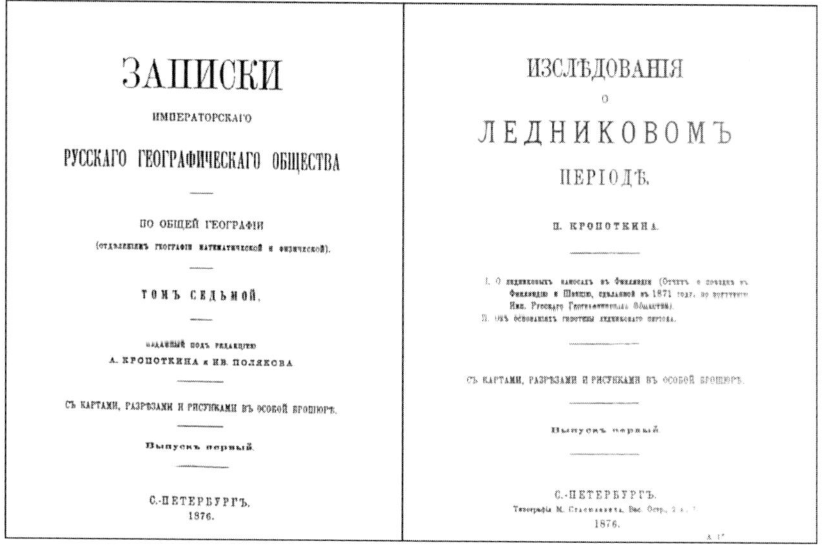

Fig. 5. The title pages of Kropotkin's *Researches on the Glacial Period* (with maps, section and illustrations in a separate brochure).

The first volume of Kropotkin's treatise consisted of 828 pages (16 chapters), and a separate fascicle with supplementary illustrations, including 100 maps, cross-sections and original drawings. In 13 chapters Kropotkin gave a description of his routes and data of his observations of alluvium, the traces of different glaciers, and the forms of relief. In addition, information was given on the investigations of Pungaharju Mountain and the lakes and ridges of South Finland. In Chapter 14 ('On the basis of the glacial hypothesis') and in Chapter 15 ('The furrowing of rocks') he considered the processes involved in the formation of glacial covers, the plasticity and brittleness of ice, and different explanations of the movement of glaciers. Chapter 16 was devoted to the forms of mountains and rocks. All the illustrations were drawn by the author himself.

In the second (posthumous) volume, Kropotkin included chapters on 'Boulders', 'On the classification of post-Pleistocene alluvium' and 'Moraines and *oses*' (Kropotkin 1998*b*).

Kropotkin's Scandinavian expedition started out from the Russian town of Vyborg, close to the Finnish border (formerly the Finnish Viipuri), and then continued into Finland, visiting such places

Fig. 6. Photoportrait of Academician Fedor Bogdanovich Schmidt (1832–1908). State Archives of the Russian Federation, Fund 1129.

Fig. 7. Photoportrait of Academician Grigorii Petrovich Helmersen (1803–1885): the first Director of the Geological Committee of Russia. Archives of the Department for the History of Geology, Vernadsky State Geological Museum, Moscow.

Fig. 8. Photograph of Kropotkin (right) and his colleague Michail P. Rebinder (left). State Archives of the Russian Federation, Fund 1129.

as Pungaharju Ridge, the Imatra Waterfall, Savonlinna, Yoensu, Scansland, Turku (Åbo), Kuopio, Tampere, Kajana and other localities. The party then moved on to Sweden, observing the Uppsala and Enköping eskers, and visiting the towns of Göteborg, Nyköping and Norrköping. During the last part of the Scandinavian journey (from Tavatshus to Helsinki) Kropotkin passed along the line of a new railway, thus enabling hum to observe numerous fresh sections.

Friedrich (Fedor) Schmidt (see Fig. 6), and the geologists Gregor Helmersen (or Gelmerson) (Fig. 7) and Michail Rebinder (dates not known) (see Fig. 8), all specialists in the geology of Finland, participated in the first part of the Kropotkin's Scandinavian expedition and they discussed together the peculiarities of the landscape near the town of Vyborg.

Helmersen interpreted the scratched boulders and the polished, rounded surfaces of the low granite hills as the result of water action. He had been under the influence of Charles Lyell and supported the old hypothesis of 'drift' (transport of boulders, etc., by floating icebergs). But Kropotkin and Schmidt considered that the granite boulders could only have been moved by large masses of ice, moving over land and leaving marks on the rocks as they slowly passed over them.

In both Finland and Sweden Kropotkin observed interesting forms of relief (*åsar* or eskers) stretching as linear ridges. *Ås* is the Swedish word for esker (plural *åsar*). The Russian term is *os* (or *oz*, if the Cyrillic script is transliterated). Kropotkin drew the *åsar* that he investigated on a hypsometric map prepared by Edvard Erdmann (1840–1923) for the Swedish Survey, and also indicated the

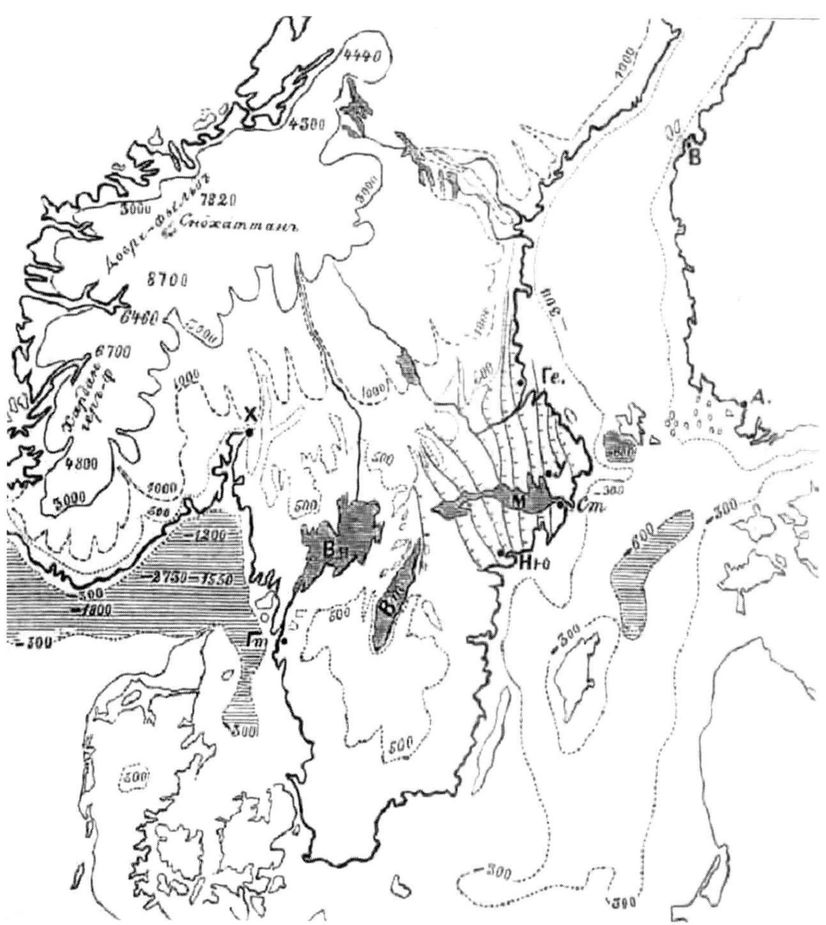

Fig. 9. Hypsometrical map of the southern part of the Scandinavian Peninsula (by Edvard Erdmann): scale 1: 9 000 000 (Supplement Researches on the Glacial Period, p. 15). Lines with small dashes on their right sides, *åsar;* Ге, Gävle; CM, Stockholm; J, Uppsala; Hю, Norrköping; X, Oslo.

Fig. 10. Photoportrait of N.A.E. Nordenskiold. State Archives of Russian Federation, Fund 1129.

names of the various towns and lakes in Russian (see Fig. 9). In Stockholm, Kropotkin met the famous explorer and geographer Nils Adolf Erik Nordenskiold (1832–1901) (see Fig. 10), with whom he established a friendship.

The Swedish geologists (Erdmann, amongst others) had supposed that *åsar* were of marine origin. But Kropotkin noticed that their directions usually corresponded to the lines of glacial movement, running down from the Scandinavian mountains. The boulders and the glacial marks (striations) on rocks clearly demonstrated the direction of movement and it was evident to Kropotkin that a huge glacial mass had moved southward without regard to the relief. It had crossed the Scandinavian Melar lowland (*Melar* or *Melur* = type of vegetation found in Iceland; here the name refers to a plain in Sweden), and then moved *up*hill – the mountains could not stop it! Such independence of relief is typical for large glaciers. Indeed, a glacier in part creates the relief. *Åsar* are the most interesting forms of post-glacial relief.

Kropotkin first observed *åsar* in Finland in the 'Pig Mountains' – or *Pungaharju*. In Sweden, he studied in detail the large and famous Uppsala *ås*, located near the ancient university town of Uppsala. Kropotkin drew a cross-section of this *ås* and different parts of the outcrop (see Fig. 11).

Figure 12 shows various profiles of an *ås* near Hvittis (or Huittinen). Figure 13 shows the ridge of the Uveskul *ås* (near Alvaervi), and a section of

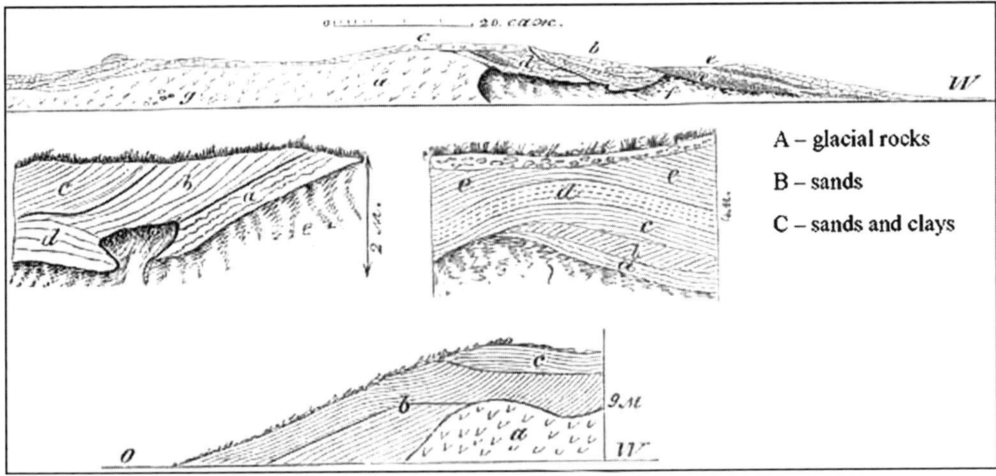

Fig. 11. Cross-sections of the Uppsala *ås* (Kropotkin 1876, Supplement, figs 17–20, pp. 17 and 18). *Top*: general profile: E–W. *a*, glacial detritus; *b*, sands and gravel; *c*, pebbles; *d*, sands, gravel and silt; *e*, layered clay; *f*, talus; *g*, elliptical boulders. *Centre left*: cross layers in the upper parts of the cover. a, grey sand with ripple marks; b, greenish sand; c, muddy-brown sand; d, grey sand; e, talus. *Centre right*: eastern part, viewed from the north. a, grey sand with ripple marks; b, c, e, fine layered sands; d, coarse sands; f, fine pebbles. *Bottom*: eastern part, viewed from the north. a, glacial detritus; b, sands and gravel; c, sands and clay.

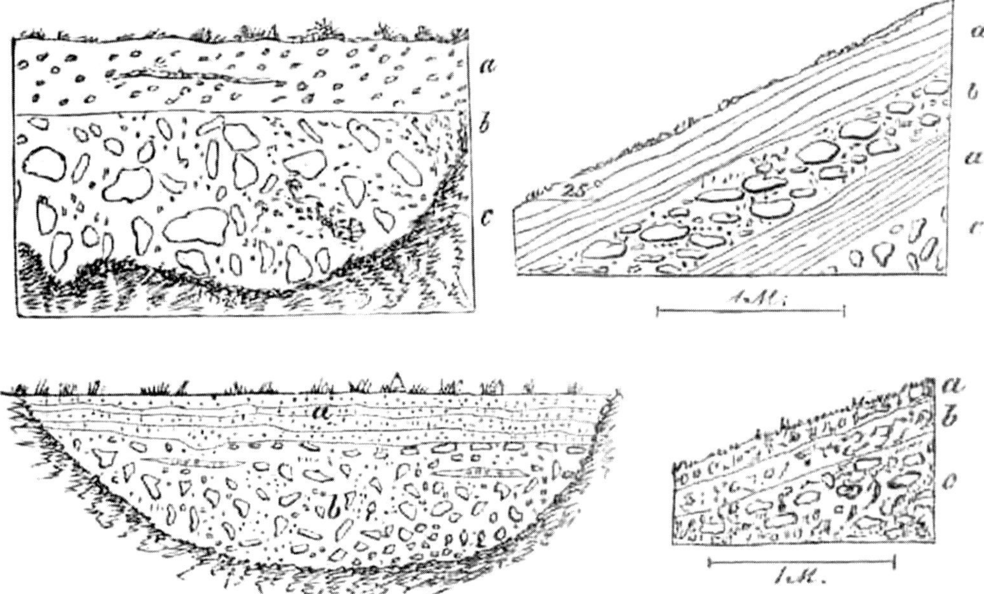

Fig. 12. *Ås* near Hvittis: different sections of the outcrop (Kropotkin 1876, Supplement, p. 17). *Top left*: *a*, pebbles; *b*, glacial detritus; *c*, talus. *Top right*: *a*, fine yellow glacial detritus; *b*, the same with white flour; *c*, the same but washed. *Bottom left*: *a*, stratum of sand with small pebbles; *b*, typical yellow glacial detritus. *Bottom right*: inclined layers above talus. *a*, sand; *b*, pebbles with sand; *c*, glacial detritus.

Fig. 13. Ridge of the Uveskul *ås* with glacial detritus (Kropotkin 1876, Supplement, p. 20).

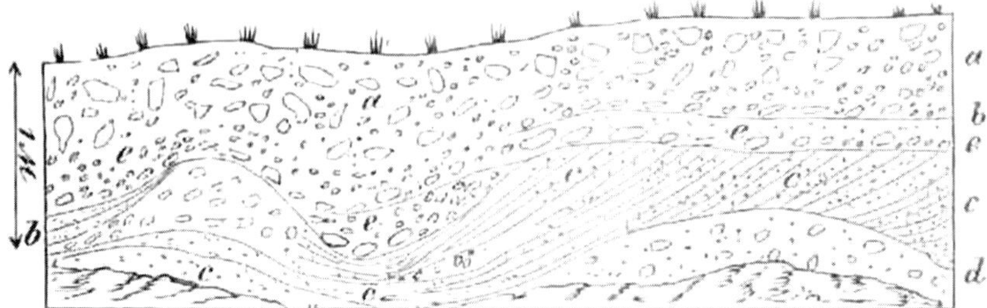

Fig. 14. Section of *ås* near Kuopio (Kropotkin 1876, Supplement, 20). *a*, glacial gravel wash, with angular, glacial detritus; *b*, the same, but with small and slightly rounded clasts; *c*, fine layer gray sand; *d*, the same with pebbles; *e*, yellow sand with pebbles in the lower part of the outcrop.

Fig. 15. Moraine near Lohikoski (Kropotkin 1876, Supplement, p. 18).

Fig. 16. Moraine near Korpilaksa (Kropotkin 1876, Supplement, p. 18).

another one near Kuopio is shown in Figure 14. Figures 15 and 16 represent moraines at Lohikoski and Korpilaksa, respectively. All these were *åsar* and moraines investigated by Kropotkin in Finland.

Also in Finland, Kropotkin noticed that the lakes have a preferred orientation. He compared the direction of lake banks, ridge moraines, and marks on boulders and rocks, and concluded that all three phenomena were produced by a single cause − a large glacier.

As is well known, *åsar* are linearly elongated, narrow ridges with steep slopes of material of glaciofluvial origin, composed of cross-bedded sands, gravels and pebbles. They have the appearance of embankments, their heights may reach 100 m and they may extend to hundreds of kilometres in length. The smallest *åsar* have a height of about 2−3 m and are only a few hundred metres long. All *åsar* are situated at the bottom of depressions and/or valleys. In plan, they are seen to curve or meander. The formation of *åsar* is thought to be related to the filling of subglacial tunnels and fissures by moraine-type materials. Most of the classic eskers are situated in Sweden and other Baltic countries, but they also occur elsewhere, such as in Ireland for example. They are formed as a result of deltaic accumulation as ice melts and retreats. But their main mass is formed within or beneath the glaciers. The direction of ice movement (NE−SW in the Swedish cases examined by Kropotkin) is indicated by the orientation of the *åsar*.

Åsar are the most interesting forms of postglacial relief. But Kropotkin found the main evidence of their glacial origin in their internal composition. The material is mostly unsorted, and typically has glacial detritus in the form of a rounded, fine dust (flour). The preservation of structures in such material is impossible in the presence of moving water. In Kropotkin's opinion, therefore, the Scandinavian *åsar* were formed from material accumulated in tunnels within (or beneath) glaciers. The Swedish geologists had interpreted *åsar* as marine deposits, but Kropotkin essentially achieved the modern understanding.

Perhaps the most significant part of Kropotkin's work, which preserves its value up to the present, was his original concept of the dynamics of a continental ice sheet as a result of its slow plastic flow, induced by its weight. The flow was believed to start from the highest 'internal' area of the ice sheet, where the ice was thickest and might be several kilometres thick. However, some aspects of Kropotkin's glacial theory do not correspond to modern opinions. For instance, he only recognized the occurrence of a single glaciation.

In his monograph, Kropotkin discussed different aspects of glacial geomorphology in detail, as well as the structure and origin of glacial (and glacially related) sediments, and the forms of glacial landscapes. His main conclusion was that the existence of a specific complex of indications of glacial activity in former times confirmed the reality of a great continental glaciation of the northern hemisphere during the Pleistocene. His ideas have subsequently been confirmed by modern Russian geologists working on the widely distributed evidence of glaciation in Siberia.

The tradition at Lomonosov Moscow State University is to preserve the memory of outstanding Russian and foreign scientists. The Earth Science Museum continues this fine tradition. The Glacial Period exhibition is located in the Earth Science Museum in the hall devoted to the Cenozoic Era. The results of Kropotkin's scientific investigations are outlined in posters and a bronze bust of him is located there.

The Moscow side-street where Kropotkin was born is named after him, as is the local Metro station. A memorial museum located at his last place of residence in Dmitrov was opened in 2007.

References

KROPOTKIN, P. A. 1873. Report of the Olyokma–Vitim Expedition on investigations of the cattle route from Nerchinsk to Olekma Districts. *Russian Geographical Society*, **3** (in Russian).

KROPOTKIN, P. A. 1874. On the study of glacial deposits of Finland. *Russian Geographical Society*, **10**, Section 1, 323–332 (in Russian).

KROPOTKIN, P. A. 1875. Obshchy ocherk orografii Vostochnoi Sibiri [Introductory outline of East Siberian orography]. *Proceedings of the Russian Geographical Society on General Geography*, **9** (in Russian).

KROPOTKIN, P. A. 1876. *Researches on the Glacial Period. Russian Geographical Society*, **7** (in Russian).

KROPOTKIN, P. A. 1884. La plasticité de la glace. *Revue scientifique*, **33**, 37–48.

KROPOTKIN, P. A. 1897a. The glaciation of Asia. *Report of the 67th Meeting of the British Association for the Advancement of Science. Section C, Geology, London*, **19**, 774–775.

KROPOTKIN, P. A. 1897b. *Anarchism: Its Philosophy and Ideal*. T. H. Kneell, London.

KROPOTKIN, P. A. 1898. Some of the resources of Canada. *Nineteenth Century*, **43**, 494–511.

KROPOTKIN, P. A. 1899. *Fields, Factories, and Workshops*. Hutchinson & Co., London.

KROPOTKIN, P. A. 1902. *Mutual Aid. A Factor of Evolution*. Heinemann, London.

KROPOTKIN, P. A. 1903. *Modern Science and Anarchism . . . Translated from the Original Russian by D. A. Modell*. Social Sciences Club, Philadelphia, PA.

KROPOTKIN, P. A. 1904. The desiccation of Eur-Asia. *Geographical Journal*, **23**, 722–734.

KROPOTKIN, P. A. 1909. *The Great French Revolution, 1789–1793 . . . Translated from the French by N. F. Dryhurst*. Heinemann, London; G. P. Putnam's Sons, New York.

KROPOTKIN, P. A. 1998*a*. *Lednikovyi i ozernyi periody (Glacial and Lacustrine Periods). In: Natural Science Works.* Science, Moscow (in Russian).

KROPOTKIN, P. A. 1998*b*. *Natural Science Works: Science Heritage.* Science, Moscow (in Russian).

MARKIN, V. A. 1985. *Peter Alekseevich Kropotkin.* Science, Moscow (in Russian).

MARKIN, V. A. 2002. *The Unknown Kropotkin.* Olma Press, Moscow (in Russian).

POSTNIKOV, A. V. 1993. Kartografia v tvorchestve P. A. Kropotkina [Cartography in the work of P. A. Kropotkin]. *In: P. A. Kropotkin i Sovremennost.* Moscow Branch of the Russian Geographical Society, Moscow, 80–93.

POSTNIKOV, A. V. 2002. Kartografia v tvorchestve P. A. Kropotkina [Cartography in the work of P. A. Kropotkin]. *Trudy Mezhdunarodnoy Nauchnoy Konferentsii, Posvyashchennoy 150-letiyu so Dnya Rozhdeniya P. A. Kropotkin,* Issue 4. Institut Ekonomiki RAN, Kommissiya po Tvorcheskomu Naslediyu P. A Kropotkin, Moscow, 72–88.

VAN VEEN, F. R. 2008. Early ideas about erratic boulders and glacial phenomena in The Netherlands. *In:* GRAPES, R. H., OLDROYD, D. R. & GRIGELIS, A. (eds) *History of Geomorphology and Quaternary Geology.* Geological Society, London, Special Publications, **301**, 159–169.

Quaternary research in the Baltic countries

ALGIRDAS GAIGALAS

Department of Geology and Mineralogy, Vilnius University, Vilnius, Lithuania
(e-mail: Algirdas.Gaigalas@gf.vu.lt)

Abstract: The development of Quaternary research in the Baltic countries was determined by geopolitical position as well as by economical and social conditions, the formation of science centres, the progress of geological thought, the natural environment and the specific geological development of this area of Eastern Europe. One of the peculiarities of Baltic Quaternary research was that it was undertaken initially by French, German, Russian, English, Finnish and Polish scholars, and later by Lithuanian, Latvian and Estonian researchers. The different nationalities infused Quaternary research with a variety of ideas and methods of study. Thus, within the course of investigations of both Quaternary and bedrock geology, specific periods related to progress in science and historical events in the Baltic States can be recognized: (1) scholastic period, drift hypothesis, Ice Age, glacialism, polyglacialism (Russian Empire administration until 1914); (2) geomorphological investigations (independent republics interwar years 1918–1939); (3) detailed investigations and mapping (Soviet administration 1940–1990); and (4) modern Quaternary studies (restoration of independence of Baltic States from 1990).

Investigations of Quaternary geology in the Baltic region were, and are, an important link between the East and West in northern Europe, which was covered with ice during the Pleistocene derived from Scandinavian glaciation centres (Fig. 1). Quaternary glacial deposits from 10 glaciations or major stadials from the Fennoscandian ice centre are extensively distributed in the Baltic States. The first glaciations covered the Baltic area before 800 000 years ago and the maximum thickness of glacial deposits was more than 300 m. The Baltic countries were probably freed from the continental ice sheet in Dani- and Gotiglacial times, i.e. approximately 18 000–11 000 years ago.

Quaternary research has undergone a long and complicated development in the Baltic countries, determined by their geopolitical position and dependence on Russia in particular, as well as economic and social conditions and the formation of science centres. The foundation of universities in Vilnius (in 1579) and Tartu (in 1632) can be regarded as starting points affecting the development of natural sciences and its progress. Geology began to be taught as a part of natural philosophy in Vilnius University, and the development of Quaternary geology was related to the founding of the Principal School of the Great Duchy of Lithuanian (in 1773) and the establishment of a Natural History Department at the Vilnius University in 1781. However, the earliest investigations in Quaternary geology in the Baltic countries began a year earlier in 1780, when Jean-Emmanuel Gilibert (1780) published observations on surface sands during a trip from Grodna (in present Belarus) to Vilnius in Lithuania. This was followed by a note on the occurrence of Quaternary sediments in Latvia (Fischer 1784).

One of the peculiarities of Baltic Quaternary research is that it was initially conducted by French, English, German, Russian, Finnish and Polish scholars, and only later by Lithuanian, Latvian and Estonian researchers. The national traits of all these researchers affected the ideas and methods applied to their Quaternary studies. The dependence of the Baltic countries on other states and the resolution of their people to gain independence was another important factor influencing scientific research. Thus, investigations related to both Quaternary and general bedrock geology can be subdivided into specific periods related to historical events affecting the Baltic countries as detailed below.

Periods of Quaternary research

The history of investigating Quaternary geology of the Baltic countries may be divided into several periods, each reflecting a stage in the development of scientific thought and progress of investigation.

Scholastic period, drift hypothesis, Ice Age, glacialism, polyglacialism (Russian Empire Administration until 1914)

During this period, the Baltic countries were part of the Northwest Administrative Region of Russia, and geological research was directed from Russian institutions and carried out by researchers from Russia, Poland, Finland and local German

From: GRAPES, R. H., OLDROYD, D. & GRIGELIS, A. (eds) *History of Geomorphology and Quaternary Geology.* Geological Society, London, Special Publications, **301**, 129–140. DOI: 10.1144/SP301.8   0305-8719/08/$15.00 © The Geological Society of London 2008.

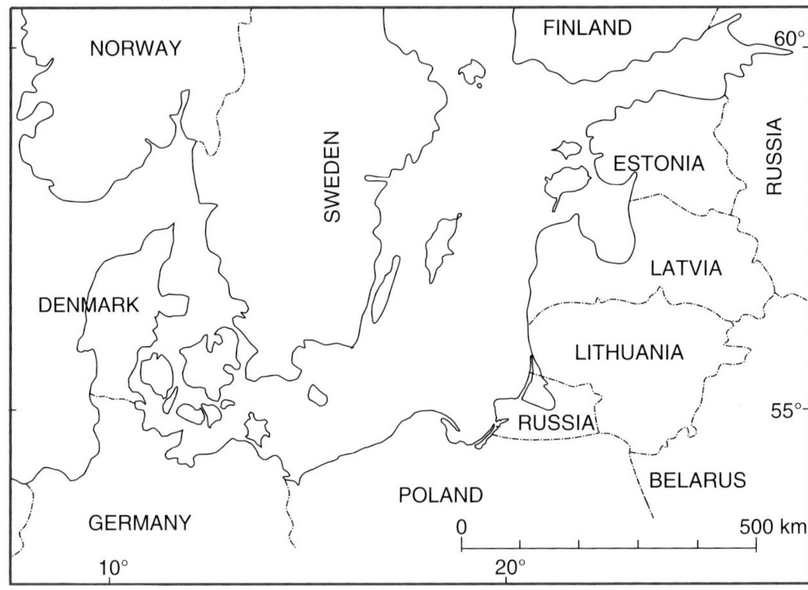

Fig. 1. Map showing location of the Baltic States in the Peribaltic region.

colonists. Only in Lithuania Minor (Klaipėda Region), which was under the Administration of Eastern Prussia, were studies undertaken by German scientists, mainly from Königsberg University.

After the closure of Vilnius University (established by Stephan Batory, King of Poland and Lithuania in 1579) in 1832 by the Tsarist government, followed by the Vilnius Medical–Surgical Academy in 1837, the Baltic region came under the control of geological research institutions of the Russian Empire (the Geological Committee of Russia established in 1882 and the Mining Department of Russia, which had been established since 1805). At the same time, scholars from Tartu (Dorpat) University, as well as Königsberg and Warsaw universities, showed considerable interest in the Quaternary geology of the region.

Erratic boulders had been noted by the earliest natural historians of the Baltic countries, and the first hypotheses of their origin appeared in the seventeenth century when it was considered that the boulders originated from volcanic eruptions on the Moon, or were the products of exploded planets (Raukas 1971). From the beginning until the middle of the nineteenth century the erratics were considered either to have had a catastrophic flood (diluvial) origin or had been derived from icebergs (drift origin). The first published note on the erratic boulders was by the academician Vasiliy M. Severgin (1803), who observed them during a journey from Vilnius to Grodna in 1802. He

compared the Lithuanian boulders, with those in St Petersburg, noting that their surface appearance was different, and concluded that they were derived from different places and were of different rock types. Erratic boulders that are common in all the Baltic countries were subsequently described by others such as Jan von Ulman (1827), Frederik Dubois de Montpereaux (1830), Eduard Carol Eichwald (1830), Ignacy Jakowicki (1831) and Roderick Murchison (1845), to name the best known. They recognized boulder rock types, and on this basis attempted to locate their provenance and in doing so determine their transport direction. For example, some researchers (Rasoumowsky 1816) considered that the majority of boulders were derived from Fennoscandia – and identified rapakivi granite from Vyborg, quartzite from Shoksha and sandstone from the Onega Lake beaches. However, at that time there was no systematic study of the erratic boulders in the Baltic countries. This only came later, when the theory of continental glaciation became generally accepted, thus greatly expanding the possibilities for the study of Quaternary geology. In the beginning, most investigators adopted the idea of a single glaciation, mainly because the drift theory of the origin of glacial deposits dominated the thinking for much of the nineteenth-century.

The most important contributions to the study of Quaternary geology of the Baltic region are summarized below. The famous geologist and professor at the Freiberg Academy of Mining in Germany,

Fig. 2. Abraham Gottlob Werner (1749–1817).

Abraham Gottlob Werner (1749–1817) (Fig. 2), was extremely influential in the development of geological thought at Vilnius University at the end of the eighteenth century. Werner's Neptunistic ideas were adopted and clarified in a thesis by Roman Symonowicz published in Vilnius (Symonowicz 1806). At Vilnius University, Werner's Neptunism survived the longest – for example, as propounded by Symonowicz (1806), Felix Drzewiński (1816) and Ignacy Jakowicki (1825, 1827) – not only because of Werner's eminence, but also because of the geology of Lithuania, which favoured interpretation in terms of the Neptunist hypothesis. The drift hypothesis prevailed and no Lithuanian observers immediately adopted Jean Louis Rodolphe Agassiz's idea of an Ice Age to account for the occurrence of the erratic boulders in the region, although the idea that icebergs were responsible for the peculiarities of relief of Vilnius region had been known for some time.

Eduard Carol Eichwald (1795–1876) (Fig. 3), palaeontologist at Tartu University, professor at Vilnius, Kazan' and St Petersburg universities, paid particular attention to the regional distribution of sands and regarded them as the result of shallow marine sedimentation. In his earlier work (Eichwald 1830) he was a supporter of the drift theory, and, although drawing attention to the widespread distribution of erratic boulders, he did not attempt to

Fig. 3. Eduard Carol Eichwald (1795–1876).

determine their origin at this time. Later, in 1853, he was the first in the Baltic provinces to consider the idea of continental glaciation, and the Baltic States were among the first places where this theory was developed. Eichwald (1853) thought that all the erratic boulders had been transported by, and were deposited from, an ice sheet. He also noted that erratic boulders from the bed of the Danė River near Klaipėda were similar to rock outcrops in Finland and Sweden.

Carl Friedrich Schmidt (1832–1908) (Fig. 4), a full member of the St Petersburg Academy, can be regarded as the founder of Quaternary geology in Estonia, and was one of the creators of the theory of continental glaciation (Schmidt 1865). In Russia, Piotr A. Kropotkin's work concerning continental glaciation was published in 1876. Schmidt was the first to discover remains of the freshwater snail *Ancylus fluviatilis* on Saaremaa Island, and from this he established the freshwater pre-Littorina stage in the Quaternary history of the Baltic. He also recognized many unique landforms, namely the meteor craters at Kaali and Ilumetsa, the astrobleme at Paluküla, coastal dunes on Cùrónian spit (Kuršių Nerija), natural monuments (the largest crystalline boulders in northern Europe); and he drew attention to the extensive insular heights in the East European Plain, examples of which were practically unknown elsewhere in the world. Later in his career, he contemplated the idea of several glaciation events. As early as 1879, Carl Andreas Constantin Grewingk (1819–1887) (Fig. 5) had maintained that the Baltic States were glaciated at least twice, and that the glaciers had formed the

Fig. 4. Carl Friedrich Schmidt (1832–1908).

topography and left tills of different colour and lithology (Grewingk 1861, 1879). As professor at Tartu University, he was the author of the first stratigraphic scheme of Quaternary deposits of the Baltic countries.

After examining erratic boulders, together with field observations in extensive areas of the islands in the Riga Gulf and environs of St Petersburg, Józef Siemiradzki of Dorpat (Tartu) University concluded that boulder deposits in the eastern part of the Baltic region (granite, porphyry, red quartz sandstone) had been transported mainly from Finland and environs of the Ladoga and Onega lakes, whilst those found in the western part were mainly derived from Sweden, including sedimentary rocks containing Silurian fossils (Siemiradzki 1882). The Upper Silurian fauna of the erratic boulders studied by Czesław Chmelewski in 1900 confirmed their provenance.

In Lithuania, the initiator of detailed studies of Quaternary geology was Antanas Giedraitis (Anton Giedroyć) (1848–1909) (Fig. 6), who was a proponent of the idea of several glaciation events (polyglacialism) (Giedroyć 1886). After describing many Quaternary exposures, he was able to identify two well-developed till horizons in the Augustów and Kaunas provinces, and a third horizon in the Vilnius region. Following on from Giedroyć's work, one of the most significant contributions to Quaternary geology in the late nineteenth–early twentieth centuries was made by

Fig. 5. Caspar Andreas Constantin Grewingk (1819–1887).

Fig. 6. Anton Giedroyć (1848–1909).

the Russian palaeobotanist, Nikolai Krischtapo-witch (1910), who examined all the main Quaternary exposures described by Giedroyć and was able to identify interglacial deposits.

Finally, the Finnish researcher, Hans Magnus Hausen (1884–1979) (Fig. 7), is known for several outstanding contributions to Baltic Quaternary geology. He was the first to attempt a reconstruction of the glacial retreat during the Last Glaciation in the Eastern Baltic using palaeogeographical maps showing successive ice-marginal positions and ice-dammed lakes (Hausen 1912, 1913).

In summary, although some observations and publications concerning Ice Age deposits in the Baltic contries were made during the latter part of the eighteenth century, this early period of Quaternary studies, ending in 1914, was one characterized by the accumulation of descriptive data. The drift theory of the origin of glacial deposits was dominant in the earlier part of this period with ideas changing towards the notion of continental glaciation in the latter part and with a polyglacial origin for the Quaternary deposits taking hold in the early twentieth century.

Fig. 7. Hans Magnus Hausen (1884–1979).

Geomorphological investigations (World War I and independent republics interwar years 1918–1939)

World War I (1914–1917) changed the political and economic situation of the Baltic countries. Only short-term investigations by German geologists – for example Hans Mortensen, Ernst Karl Kraus and others such as H. Philipp, Fr. Wahnschaffe and F. Schucht – were undertaken at that time, with the former compiling a morphological map of Lithuania from aerial observation (Mortensen 1924).

After World War I, the Baltic countries gained their independence, and national researchers of all three Baltic States – Lithuania, Latvia and Estonia – took over geological investigations. During this time, national research centres and schools at the universities of Kaunas, Riga and Tartu were established. In the Vilnius area, which was under Polish occupation, investigations continued to be carried out by Polish geologists (Halicki 1934).

Geomorphological investigations are a characteristic feature of this period (Vaitiekūnas 1960, 1969). Baltic scientists took an active part in gathering material to produce the Quaternary map of Europe, but this was not published immediately owing to the start of World War II. In Lithuania, important investigations were carried out by Mykolas Kaveckis, Juozas Dalinkevičius and Česlovas Pakuckas; in Latvia by Ernst Karl Kraus, Aleksis Dreimanis, and others such as Z. Lancmanis, I. Sleinis and V. Zans; and in Estonia by August Tammekann, Karl Orviku, Johannes Gabriel Granö, Paul William Thomson and A. Laasi. Valuable glaciomorphological conclusions were made on the basis of these investigations, such as the determination of the limits of the greatest extent of the last glaciations, recognition of three stages of the Last Glaciation and the oscillatory nature of moraines.

Together with geomorphological investigations, stratigraphic surveys were also carried out at this time, which led to the discovery of interglacial and interstadial intertill peat and gyttja horizons in Lithuania (Pakuckas 1935), Latvia (Zans 1936) and Estonia (Orviku 1939) As well as valuable palynological data obtained from these sections, the petrography of tills in the Vilnius region of East Lithuania were described by Jaroszewicz-Kłyszińska (1938) and in Latvia by Dreimanis (1939). The first data were also obtained on the bedded character and thickness of glacial deposits coinciding with ideas on Pleistocene tectonics enunciated by Kraus (1924, 1928).

The more detailed and systematic studies of this 1914–1940 period revealed both the common distribution of Quaternary sediments in the Baltic countries and provided information on important mineral resources. It also heralded the beginning of stratigraphic studies of glacial deposits.

Detailed investigations and mapping (Soviet administration 1940–1990)

In 1940 the Baltic States lost their independence and at first were incorporated into the Soviet Union, with Quaternary research co-ordinated by central Soviet institutions and only localized investigations being made; for exmaple, glaciolacustrine varved clay prospecting. In 1941, the institutions of the Geological Survey were founded.

During the German occupation between 1941 and 1944, soil and agricultural mapping for colonization purposes was instigated, including research related to mineral resources of Quaternary deposits. In Lithuania, the study of the Quaternary deposits was undertaken by Bronislaw Halicki and his assistants. Halicki compiled geological maps of several localities in the valleys of the Nemunas and Neris rivers, and proposed a stratigraphic scheme for the Lithuanian Pleistocene. The results were published in 1950 (Halicka & Halicki 1950). Also, during this time, Quaternary deposits in Lithuania were correlated with four glaciations by Juozas Dalinkevičius (1944), based on an analysis of the results of previous research.

The second Soviet occupation of the Baltic countries lasted from 1944 to 1990, and during these years a policy of industrialization and collectivization resulted in the acceleration of geological and, in particular, Quaternary investigations. Extensive geological and geomorphological mapping, at scales of 1:200 000 and 1:50 000, of all the Baltic Republics was carried out, with the rapid development of geological work associated with the training of highly skilled specialists. During this time Quaternary geology was established as an independent branch of geology in the Baltic countries, and a large amount of data relating to Quaternary stratigraphy, geomorphology, palaeogeography, palynology, lithology, erratic boulders and isotopic characteristics were accumulated from geological surveys and deep boreholes (e.g. Viiding *et al.* 1971; Danilans 1973; Gudelis 1973; Raukas 1978). Quaternary and geomorphological maps of the Baltic Republics, on a scale of 1:500 000 with explanatory notes and compiled on the basis of stratigraphic–genetic principals, were compiled in 1978 based on data provided by the geological boards of Estonia, Latvia, Lithuania and the northwestern region of Russia (Grigelis 1980*a,b*). These maps showed the distribution of the deposits of the last and penultimate advances of the Quaternary glaciers, as well as post-glacial sediments,

delineated rock composition, marginal glacial deposits, major topographic forms (drumlins, kames, eskers), the ancient coastline of the Baltic Sea, the thickness of the deposits and the deglaciation boundaries of the Last Glaciation, with an inset map illustrating the 'Degradation scheme of the last Baltic glaciation' on the scale of 1:1 700 000. The relief of the sub-Quaternary surface was shown to have a major effect on the distribution and thickness of glacial sediments and buried valleys in sub-Quaternary surface (Melešytė 1976).

Pleistocene glaciosedimentation cycles were distinguished on the basis of correlation of the ages of glacial deposits (Gaigalas 1979), and the lithogenetic, lithostratigraphic and palaeogeographic importance of glacial deposits was emphasized. Sources of fragmental material and the directions of ice movement during different times during the Pleistocene were determined, and questions of ground-moraine formation solved. A genetic classification of glaciofluvial deposits was also developed based on their different textural, compositional and structural properties (Jurgaitis 1984), and criteria for the recognition of different types of glaciofluvial deposits were proposed. In 1955 a new stage in the elaboration of a regional stratigraphical scheme of the Baltic Quaternary was initiated at a regional conference in Vilnius, resulting in stratigraphic charts being proposed and modified (Grigelis 1978; Krasnov & Zarina 1986). Because fossil-bearing interglacial sediments are rare, only generalized regional stratigraphic schemes of Quaternary sediments could be made with help from the results of previous research on the structure and composition of tills.

Modern Quaternary studies (restoration of independence of the Baltic States from 1990)

A new stage in Quaternary research began after the collapse of the USSR when the Baltic Republics once more regained their independence. Research became centred on geochronology (radiometric and dosimetric), palaeobotany, stratigraphy and lithology, and the interaction between climate, man and environment. These investigations were, and remain, an important link between east and west in the northern European region that had been ice-covered during the Pleistocene. The geology of the Baltic States was shown on 1:200 000-scale maps (including maps of Quaternary deposits, geomorphology and bedrock topography), with large areas also mapped on a scale of 1:50 000. The Baltic area also became the principal research region for the well-known Saale Complex (Middle Pleistocene Interglacial–Glacial

Table 1. *Correlation of the Saale Complex (Gaigalas 2004)*

West Europe (Bowen *et al.* 1986)	Lithuania (Gaigalas 2003)			Baltic Countries (Gaigalas 2003)		East-European Plain (Bolikhovskaya 1995)	Oxygen isotope stages
EEMIAN	Merkinė Interglacial III mr	−122 −132	5	Merkinė Interglacial		MIKULINO Interglacial III mk	
WARTHE — Saale 3, Rugen, Saale 2	Medininkai Glacial II md		6	UPPER UGANDI (Medininkai) Stadial	UGANDI	DNIEPER GLACIAL II dn	6
Drenthe-Warthe	Snaigupėlė Interglacial II sn	198 −252	7	MIDDLE UGANDI (Snaigupėlė) interglacial		CHEREPET Interglacial II chr	7
SAALE (*s. s.*) (Drenthe)	Žemaitija Glacial II žm		8	LOWER UGANDI (Žemaitija) Stadial		ZHIZDRA Glacial II zh	8
HOLSTEIN — DOMNITZ (Wacken)		−302	9			CHEKALIN Interglacial II ch	9
FUHINE (Mechlbeck)	BUTĒNAI Interglacial II bt	−338 −352	10	BUTĒNAI Interglacial		KALUGA Glacial II kl	10
Holstein (*s. s.*)			11			LIKHVIN Interglacial (*s. s.*) II	11

transitions) of the Quaternary in Europe (Table 1). Absolute dating of sediments helped to solve the problem of the existence and timing of the Interglacial period within marine oxygen isotopic Stage (MIS) 7 (Gaigalas 2004; Gaigalas *et al.* 2007). Insular (interlobe, island) highlands in the Baltic countries are now recognized as unique depositional forms of glacial relief that are rare in the other regions of ancient glaciation (Fig. 8). According to recent data (Raukas 1993; Raukas *et al.* 1995), these depositional features developed between active ice flows (lobes) during the course of subglacial, englacial, marginal and stagnant ice stages. As the eastern Baltic area was probably free of continental ice in the Dani- and Gotiglacial times, between about 18 000 and 11 000 years ago, it is a key region of the Quaternary stratigraphic record. The most important type sites are shown in Figure 9. The catalogue of stratotypes for the Quaternary of the Baltic Region was compiled for the first time (Kondratienė 1993), and the stratotypes make up the basis for the stratigraphical division of the Quaternary deposits of the Baltic States (Table 2). Taking into account the specificity of palaeogeographic conditions during the Quaternary, stratigraphic units are recognized on the basis of many criteria, including geomorphological, so that a particular stratotype is not based on one section, but on a certain area (Baltrūnas 1995). As a result, the areal stratotype was established. According to the Regional Stratigraphic Scheme of the Quaternary Deposits, up to five interglacial layers and up to nine independent till beds can be differentiated. The Merkinė (Eemian,

Mikulinan) and Butėnai (Holsteinian, Likhvinan) interglacial deposits are considered as key horizons (Sidaravičienė Lithuanian stratigraphic units 1999). In the Baltic States both are represented by interglacial marine and continental lacustrine-bog deposits, and are clearly differentiated by palynological data; the possibility of comparing the Baltic Quaternary scheme with those of neighbouring countries was suggested by Kondratienė 1996).

Activities of the International Geological Correlation Programme Projects have proved successful in many fields of Quaternary geology: such as stratigraphy, geomorphology, palaeogeography, lithology and isotopic investigations, and, today, specific features of Quaternary sediments continue to be studied. These include peculiarities of Quaternary sediments, surfaces of pre-Quaternary rocks, Quaternary sediments resulting from endogenous and exogenous processes, sedimentation and palaeogeography of glacial deposits, sedimentation of inter-till clast deposits, specific features of changes of vegetation and climate during interglacial periods, and the genesis of periglacial deposits.

With respect to resources, Quaternary clay, sand and gravel in the building material industry, and peat in agriculture, are mostly used. Groundwater from Quaternary sediments is the main source of potable and curative water in the Baltic countries. Amber reserves rewashed from Palaeogene deposits are found in the south of the Baltic region. Large accumulations of erratic boulders of crystalline origin occur on the surface. Some regions have rich peat deposits. Placers of heavy minerals

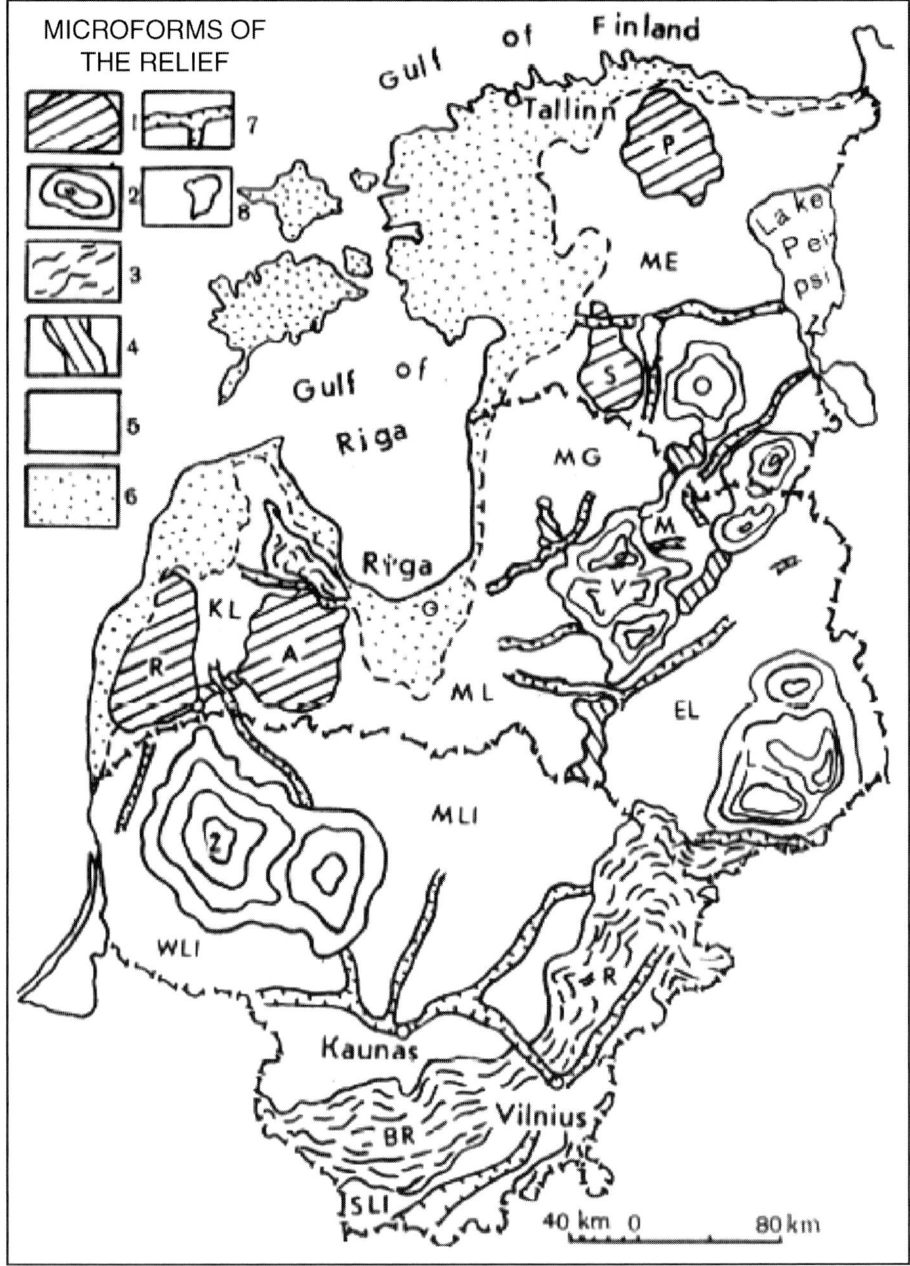

Fig. 8. Schematic geomorphological map of the Baltic States (Āboltinš 1993). **Highlands**: 1, insular exaration-accumulative (depositional) (plinth-type heights); 2, insular glacio-structure-accumulative (depositional); 3, ice marginal glacio-structural-accumulative (depositional); 4, interlobate high. **Lowlands**: 5, glacio-depressional lowlands (till plains, drumlin fields, ribbed moraines, uval moraines (moraine ridges), flutings, eskers and local ice-dammed lake plains); 6, abrasion-accumulation (depositional) plains of the Baltic Ice Lake and younger stages of the Baltic Sea; 7, largest spillway valleys; 8, largest lakes. **Major lowlands**: ME, Middle-Estonian Lowland; MG, Middle-Gauja Lowland; ML, Middle-Latvian Lowland; EL, East-Latvian Lowland; MLL, Middle-Lithuanian Lowland; WLl, West-Lithuanian Lowland. **Major higlands**: R, Rietumkursa; A, Austrumkursa; S, Sakala; P, Pandivere; O, Otepaa; H, Haanja (together with the Aluksne Highland); V, Vidzeme; L, Latgale; Z, Žemaitija; BR, Baltic Ridge (together with the Augšzeme Highland); ZK, Ziemelkurzeme.

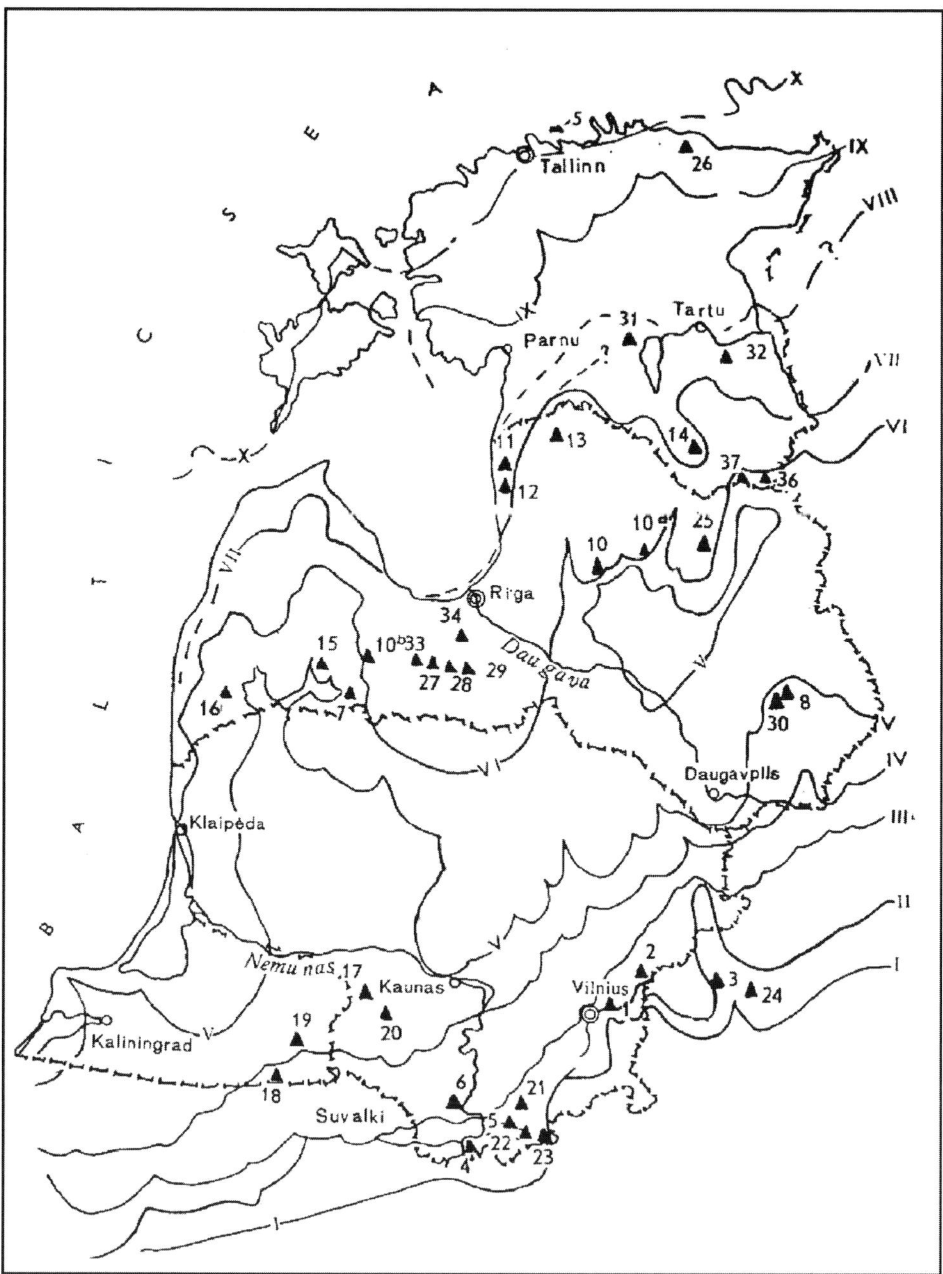

Fig. 9. The most important ice-marginal zones, interstadial and interphasial sections in the Baltic States (INQUA 1995). Compiled by O. Aboltinš, A. Gaigalas, M. Melešytė & A. Raukas. Stages and stades: I, Brandenburgian (Grūda); II, Frankfurthian (Žiogeliai); III, Pomeranian (Aukštaičiai); IV, South-Lithuanian; V, Middle-Lithuanian; VI, Luga (North-Lithuanian, Linkuva, Haanja); VII, Otepää (Pliens); VIII, Sakala (Valdemarpils); IX, Pandivere (Neva); X, Palivere. **Sections**: 1, Antaviliai; 2, Buivydžiai; 3, Kamariškės; 4, Druskininkai; 5, Mančiagirė; 6, Krikštonys; 7, Lieclautse; 8, Burzava; 9, Kanini; 10, Rauna; 11, Bridagi; 12, Upitē s; 13, Aloja; 14, Kurenurme; 15, Saldus; 16, Priekulė; 17, Nopaitis; 18, Manturiai; 19, Gumbinė (Gusev); 20, Gabiauriškis; 21, Pamerkės; 22, Zervynos; 23, Rudnia; 24, Studenets; 25, Dulupite; 26, Kunda; 27, 'Progress'; 28, 'Sarkanais Mals'; 29, Tetele; 30, Rezekne; 31, Viljandi; 32, Tartu; 33, Kalnčiems; 34, 'Purmali'; 35, Prangli; 36, Viitka; 37, Petruse.

Table 2. *Regional stratigraphic scheme for the Pleistocene of the East Baltic area (Raukas 1993)*

Main subdivisions	Super stage	Stage		Substage
		Holocene		
Pleistocene	Upper	Nemunas	Latvia	Baltija, stadial
				Grūda, stadial
			Lejasčiems, interstadial	
			Varduva, stadial	
		Merkinė interglacial		
	Middle	Ugandi	Upper Ugandi (Medininkai), stadial	
			Middle Ugandi (Snaigupėlė), interstadial (?)	
			Lower Ugandi (Žemaitija), stadial	
		Butėnai, interglacial		
		Lithuania	Dainava	Upper, stadial
				Middle, interstadial
				Lower, stadial
			Turgeliai, interglacial (?)	
			Dzūkija	Upper, stadial
				Middle, interstadial
				Lower, stadial
			Katleriai	Židini, interglacial
				Lower, glacial
	Lower	Daumantai	Giliai (cold)	
			Šilelis (mild)	
			Jundzikis (cold)	
			Šlavė (mild)	
			Šlavėnai (cold)	
Pliocene		Upper Pliocene (Anykščiai stage)		

(ilmenite, rutile, zircon) are associated with the underwater slope of the Baltic Sea. Sapropel accumulations and bog (limonitic) ores are also found in Holocene deposits.

Concluding remarks

Quaternary geological research in the Baltic States of Lithuania, Latvia and Estonia since the middle of the nineteenth century has provided evidence confirming continental glaciation in northern Europe, and support for the idea of multiple glaciations and has contributed to the advancement of glaciomorphological studies. Between World War I and II Quaternary investigations in the Baltic Republics provided crucial material for the compilation of the 1:1 500 000 International European Quaternary map, the publication of which was delayed until after World War II. Also, during this time, the first steps were made towards climatostratigraphic correlation that provided the basis for the establishment of glaciomorphological zones. The most recent Quaternary geological research in the Baltic Republics is concerned with medium- and large-scale mapping that has allowed the establishment of climatostratigraphic divisions and the creation of regional stratigraphical schemes.

Nowadays, the territories of all three Baltic countries have Quaternary geological, bedrock topography and geomorphological maps on the scale of 1:200 000, with large areas mapped on scales of 1:25 000 or 1:50 000. All known gravel and construction sand deposits are correlated with Quaternary glacial and aquaglacial, as well as aeolian, deposits. Large volumes of fresh underground water are also utilized.

The authors are very grateful to Professors R.H. Grapes and D.R. Oldroyd, and an anonymous reviewer, for constructive comments and suggestions, as well as improvements to the English and revision of the manuscript.

References

ĀBOLTINŠ, O. (compiles). 1993. Schematic geomorphological map of the Baltic States. *In: Pleistocene Stratigraphy, Ice Marginal Formations and Deglaciation of Baltic States. Excursion Guide.* 7–19 June 1993, IGCP, **253**, Talinn.

BALTRŪNAS, V. 1995. *Stratigraphy and Correlation of Pleistocene.* Academia, Vilnius (in Lithuanian).

BOWEN, D. Q., RICHMOND, G. M., FULLERTON, D. S., SIBRAVA, V., FULTON, R. J. & VELICHKO, A. A. 1986. Correlation of Quaternary glaciations in the Northern Hemisphere. *Quaternary Sciences Review*, **5**, 509–510.

BOLIKHOVSKAYA, N. S. 1995. *Evolution of the Loess Palaeosoil Formation of the Northern Eurasia.* Moscow University Press, Moscow (in Russian).

DALINKEVIČIUS, J. 1944. Stratigraphical outlines of Lithuanian Pleistocene. *Technika*, **4–5**, 10–14 (in Lithuanian).

DANILANS, I. 1973. *Pleistocene Deposits of Latvia.* Zinātne, Riga (in Russian).

DREIMANIS, A. 1939. Eine neue Methode der quantitativen Geschiebeforschung. *Zeitschrift für Geschiebeforschung und Flachlandsgeologie*, **15**.

DRZEWIŃSKI, F. 1816. Początki mineralogii podług Wernera. Wilno, 615 s.Dubois F., 1830. *Geognostische Bemerkungen über Lithauen – Archiv Mineral. Geogn.*, **B.2**, 135–158.

DUBOIS DE MONTPEREAUX, F. 1830. Geognostische Bemerkungen über Lithauen. *Archiv Für Mineralogie, Geognosie, Bergbau und Hüttenkunde, Bd. II*, 135–156. [Zusatz von Leupold von Duch, 156 158.]

EICHWALD, E. 1830. *Naturhistorische skizze von Lithauen, Volynien und Podolien in geognostisch – mineralogischer, botanischer und zoologischer Hinsicht.* Gedrukt bei Joseph Zawadzki, Wilna.

EICHWALD, E. 1853. *Lethaea Rossica on Paléontologie de la Russie, 3, Dernière Période.* Stuttgart.

FISCHER, J. 1784. *Zusatze zu einem Versuche einer Naturgeschichte von Livland, nebst einigen Anmerkungen zur physischen Erdbeschreibung von Kurland, entworfen von J. J. Ferber.* Riga.

GAIGALAS, A. 1979. *Glaciosedimentation Cycles of the Lithuanian Pleistocene.* Mokslas, Vilnius (in Russian).

GAIGALAS, A. 2004. Succession, chronostratigraphic position and palaeoenvironmental evolution during the formation of the Saale Complex (Middle Pleistocene Interglacial/Glacial transitions) in Baltic countries and East Europe (with special attention to an Intra-Saale warm period). *Praehistoria Thuringica*, **10**, S. 67–71.

GAIGALAS, A., ARSLANOV, KH. A., MAKSIMOV, F. E., KUZNETSOV, VL. YU. & CHERNOV, S. B. 2007. Uranium–thorium isochron dating results of penultimate (Late Mid-Pleistocene) Interglacial in Lithuania from Mardasavas site. *Geologija*, **57**, 21–29.

GIEDROYĆ, A. 1886. *Sprawozdanie z poszukiwań geologicznych, dokonanych w gub. Grodzieńskiej i przyleglych jej powiatach Królewstwa Polskiego i Litwy.* "Pamiętnik fizjograficany", *c*. VI, dzial II. Str. 3–16, Warszawa.

GILIBERT, J.-E. 1780. *Pètrification remarquable. Acta Academia Sciences Petropolitanae, II, Petropoli*, **45**.

GREWINGK, C. 1861. Geologie von Liv- und Kurland mit Inbegriff einiger angrenzenden Gebiete. *Archiv Für die Naturkunde Liv-, Est-, und Kurlands* (Dorpat), Ser. 1, Bd. 2, 479–776.

GREWINGK, C. 1879. Erläuterungen zur zweiten Ausgabe der geognostischer Karte Liv- und Kurlands. *Archiv Für die Naturkunde Liv-, Est-, und Kurlands* (Dorpat), Ser. 1, Bd. 8, 343–465.

GRIGELIS, A. A. (ed.) 1978. Reszeniya mezhvedom stvenovo regional'novo stratigraficzeskovo sovesczaniya po razrabotke unificiro vannich stratigraficzeskich schem pribaltiki (Resolution of regional stratigraphical meeting for compiling unification stratigraphical schemes of East Baltic region.) LitNIGRI, Leningrad, 1–86, s tablicami, 1–14.

GRIGELIS, A. (editor-in-chief). 1980a. *Geological Map of the Quaternary Deposits of the Soviet Baltic Republics with Explanatory Note. Scale: 1:500 000.* Editor-in-chief GRIGELIS, A.

GRIGELIS, A. (editor-in-chief). 1980b. *Geomorphological Map of the Soviet Baltic Republics with Explanatory Note. Scale: 1:500 000.*

GUDELIS, V. 1973. *Relief and Quaternary of the East Baltic Region.* Mintis, Vilnius (in Russian).

HALICKI, B. 1934. O zasięgu zlodowacenia w północnowschodniej Polsce. *Posiedziena naukowe Państwowego instytutu geologicznego*, **41**, 4–6.

HALICKA, A. & HALICKI, B. 1950. La stratigraphie du Quaternaire dans le basin du Niemen. *International Geological Congress, Report of the 18th Session, Great Britain, 1948*, p. IX, *Proceedings of Section H, London*, 47–53.

HAUSEN, H. 1912. Studier öfver der Sydfinska ledblockens spridning i Russland, jämte en öfversikt af is-recessionens forlöpp i Ostbaltikum. *Bulletin de la Commission géologique de Finlande*, **32**.

HAUSEN, H. 1913. Materialien zur Kenntnis der pleistozänen Bildungen in der Russischen Ostseeländern. *Fennia*, **34**, 181.

INQUA. 1995. *Quaternary Field Trips in Central Europe.* SCHIRMER, W. (ed.) *3. Baltic Traverse.* Verlag Dr. Friedrich Pfeil, München.

JAKOWICKI, J. 1825. *Krótki wykład oryktognozyi i geognozyi. Podlóg ostatniemu układu Wernera.* Nakladem A. Marcinowskiego, Wilno.

JAKOWICKI, J. 1827. *Wykład oryktognozyi i poczatków geognozyi.* W Drukarni Manesa i Zymela, na ulicy Zomkowéy pod N. 185, Wilno.

JAKOWICKI, J. 1831. *Obserwacje geognostyczne w guberniach zachodnich i południowych Panstwa Rossyjskiego*. Nakladem A. Marinowskiego, Wilno.

JAROSZEWICZ-KŁYSZYŃSKA, A. 1938. O utworach morenowych Łysej Góry pod Wilnem. *Starunia*, **15**, 129–136.

JURGAITIS, A. 1984. *Lithogenesis of Glaciofluvial Deposits of the Region of Last Continental Glaciation*. NEDRA, Maskva. (in Russian).

KONDRATIENĖ, O. (complier). 1993. *Catalogue of Quaternary Stratotypes of the Baltic Region*. Baltic Stratigraphic Association, Lithuanian Geological Institute, Vilnius (in Russian).

KONDRATIENĖ, O. 1996. *The Quaternary Stratigraphy and Paleogeography of Lithuania Based on Paleobotanic Studies*. Academia, Vilnius (in Russian).

KRASNOV, I. I. & ZARINA, E. P. (eds) 1986. Reszeniya 2-vo mezhvedomstvenovo sovesczaniya po czetverfiozney sistene Vostoczno–Evropeyskoy platformi (Leningrad–Poltava–Moskva, 1983 g.) s regional' nimi stratigraficzeskimi schemami. (Resolution of second interdepartment stratigraphical meeting for Quaternary system of East European platform.) VSEGEI, Leningrad, 1–156.

KRAUS, E. 1924. Der Abschmelzungsmechanismus des jungdiluvialen Eises im Gebiet des ostpreussischen Mauersees. *Jahrbuch der Preussischen Geologischen Landesanstalt zu Berlin*, **XLIV**.

KRAUS, E. 1928. Tertiär und Quartär des Ostbaltikums (Ostbaltikum, II. Teil.). *Die Kriegschauplätze 1914–1918*, **10**, I, Berlin, 142 (in German).

KRISCHTAFOWITCH, N.-J. 1910. Sur la dernière période glaciaire en Europe et dans l'Amérique du Nord en rapport avec la question de la cause des périodes glaciaires en général. (Traduit du russe par Mr. W. P.). *Extrait du Bulletin de la Société Belge de Géologie de Paléontologie et D'hydrologie (Bruxelles)* **XXIV**, 1–112.

KROPOTKIN, P. A. 1876. *Researches on the Glacial Period*. Zapisky Rossiyskogo Geograficheskogo Obchestva, [Russian Geographical Society], **7**(1) (with a separate booklet of maps, cross-section and drawings) (in Russian).

MELEŠYTĖ, M. 1976. *Expansion and Genesis of Paleoincisions. The Buried Paleo-incisions of Sub-Quaternary Rock Surface of the South-east Baltic Region*. Vilnius, 17–37.

MORTENSEN, H. 1924. Beiträge zur Entwicklung der glacialen Morphologie Litauens. *Geologisches Archiv*. **3**, 1–93.

MURCHISON, R. I. 1845. *The Geology of Russia in Europe and the Ural Mountains*, Volume I. *Geology*. London.

ORVIKU, K. 1939. Rôngu interglatsiaal-esimene interglasiaalse vanusega organogeensete setete leid Eestist. *Eesti Loodusa VII*, **1**.

PAKUCKAS, Č. 1935. Lietuvos žemės praeitis. *Žemėtvarka ir melioracija*, No. 2, Kaunas, 25–66.

RASOUMOWSKY, G. 1816. *Coup d'ocil geognostique sur de Nord de l'Europe in general et particulicrement de la Russie*. St Petersbourg.

RAUKAS, A. 1971. *Outline of the Indicator Crystalline Boulder Investigations in Northern Europe. Crystalline Indicator Boulders in the East Baltic Area*. Mintis, Vilnius, 9–21 (in Russian).

RAUKAS, A. 1978. Pleistocene Deposits of the Estonian SSR. Valgus, Tallinn (in Russian).

RAUKAS, A. (compiler and ed.). 1993. *Pleistocene Stratigraphy, Ice Marginal Formations and Deglaciation of the Baltic States. Excursion Guide, 14–19 June 1993. RIGCP*, **253**.

RAUKAS, A., ĀBOLTINŠ, O. & GAIGALAS, A. 1995. *The Baltic States. Quaternary Field Trips in Central Europe. International Union for Quaternary Research XIV International Congress August 3–10, 1995, Berlin, Germany. Volume 1. Regional Field Trips*. Dr. Friedrich Pfeil, München, 146–151.

Reszenija mezhvedomstvenovo regionalnovo stratigraficzeskovo soveczianaya po razrabotke unificirovanych stratigraficzeskich schem Pribaltiki 1976, s unificirovanimi stratigraficzeskimi koreliacionimi tablicami. 1978. Leningrad.

Reszenije 2-vo mezhvedomstvenovo stratigraficzeskovo soveczianiya po czetverticznoi sisteme Vostocznoeuropeyskoy platformi. (LENINGRAD–POLTAVA–MASKVA, 1983) s regionalnimi stratigraficzeskimi schemami. 1986. Leningrad.

SCHMIDT, F. 1865. Untersuchungen über die Erscheinungen der Glacialformation in Estland und auf Oesel. *Bulletin de l'Academie imperiale des Sciences de St.-Petersbourg*, **8**, 207–248.

SEVERGIN, V. 1803. *Zapiski puteshestviya po zapadnym provintsiyam Rossiyskogo gosudarstva, ili mineralogicheskiye, chozyaistvennye i drugiye primechaniya, uchinennye vo vremya proyezda cherez onye v 1802 godu*. Pri Imperatoskoy akademii nauk, St Petersburg.

SIDARAVIČIENE, N. (compiles). 1999. *Lithuanian Stratigraphic Units*. Lietuvos geologijos tarnyba, Vilnius.

SIEMIRADZKI, J. 1882. Nasze glazy narzutowe. *Pamiętnik Fizjograficzny*, **2**, 87–122.

SYMONOWICZ, R. 1806. *O stanie dzisiejszym mineralogii*. Zawadzki, Wilno.

VAITIEKŪNAS, P. 1960. *The History of Quaternary Geology in Lithuania. Collected Papers for the XXI Session of the International Geological Congress*. Academy of Sciences of the Lithuanian SSR, Geological and Geographical Institute, Vilnius, 159–173.

VAITIEKŪNAS, P. 1969. Zur Geschichte der Anhäufung und Verallgemeinerung des Materials über die glazigenen bildungen in den altischen Ländern. Inlandvereisung und Glazigene Morphogenese. *In: Zum VIII INQUA Kongress Frankreich, 1969*. Litauisches Ministerium für Hoch- und Mittel-Speziellebldïng; Staatliche V. Kapsukas Universität Vilnius, Sowjetische Sektion INQUA, Vilnius. 11–64 (in Russian).

VIIDING, H., GAIGALAS, A., GUDELIS, V., RAUKAS, A. & TARVYDAS, R. 1971. *Crystalline Indicator Boulders in the East Baltic Area*. Mintis, Vilnius (in Russian).

VON ULMAN, J. 1827. *Obezrjenie geognostyczne gubernij Wileńskiej, Grodzieńskiej i t. d. Dziennik Wileński*, II, *Umiejętności i sztuki*, 246–265.

ZANS, V. 1936. *Leduslaikmets un pēcleduslaikmets Latvijā. Gramatā: 'Latvijas zeme, daba un tauta'*, I sēj. Riga.

Česlovas Pakuckas (or Czesław Pachucki): pioneer of modern glaciomorphology in Lithuania and Poland

ALGIRDAS GAIGALAS[1], MAREK GRANICZNY[2], JONAS SATKŪNAS[3] & HALINA URBAN[2]

[1]*Department of Geology and Mineralogy, Vilnius University, Vilnius, Lithuania*
(e-mail: Algirdas.Gaigalas@gf.vu.lt)

[2]*Polish Geological Institute, Warsaw, Poland*

[3]*Geological Survey of Lithuania, Vilnius, Lithuania*

Abstract: The contributions of the Lithuanian geologist Česlovas Pakuckas (1898–1956) (in Poland – Czesław Pachucki) to Pleistocene geology are of major significance, and he is regarded as the pioneer of modern glaciomorphological investigations of the Baltic marginal highlands in Lithuania and Poland. Pakuckas published his first paper on the glacial morphology of south Lithuania in 1934, followed in 1936 with one on the orientation marginal moraines in the east Lithuanian highlands and their origin, and in 1938 with another on the glacial morphology of south Lithuania. His ideas on glacial morphology were presented at the First Conference of Lithuanian and Latvian geologists held in Kaunas in 1940. Working in Poland after World War II, Pakuckas continued his glaciomorphological research and compiled data on the correlation of end moraines of NE Poland and south Lithuania with those in west Belarus as well as the Peribalticum. He concluded that during the Last Glaciation, the continental glacier was not a single ice sheet but consisted of a number of flows, each dependent on topography and each with its specific glacial centre. In Lithuania, Pakuckas defined one large field of lateral moraines, the so-called Baltic Highland consisting of the Sūduva, Dzūkija and Aukštaitija highlands. He traced this formation SW as the Mazurian arc of lateral moraines (in NE Poland) and found that the Švenčionys–Naroch or north Belarus field of lateral moraines extend to the NE of the Baltic Highland.

The pioneer of modern glaciomorphological investigations of the Baltic margin highlands in Lithuania (pre-World War II) and Poland (post-World War II) was the Lithuanian geologist, Professor Česlovas Pakuckas (or Czesław Pachucki, as he is named in Poland) (1898–1956) (Fig. 1).

In 1922 Pakuckas finished his secondary school education in Kaunas, Lithuania. Higher education at the universities in Münster and Vienna followed, and he obtained his doctorate in Vienna in 1927 with the thesis 'Die Nachträge zur mitteleren und oberen Trias–Fauna von der Insel Timor' (Indonesia) ('Additions to the Middle and Upper Triassic fauna of the island of Timor'). His first position was as a teacher in a 'gymnasium' in Klaipėda, Lithuania and from 1930 he became an assistant and *Privatdozent* in the Lithuanian University in Kaunas. In 1934 he passed the 'Habilitation' examination at Vytautas, the Great University in Kaunas. Two publications in the *Transactions of the Faculty of Natural Science and Mathematics* at the Lithuanian University followed; 'Papilės oksfordo ir kelovėjo amonitų fauna' ['The Oxfordian and Kelloway ammonite fauna of Papilė'] in 1932, and 'Papilės juros stratigrafinė apžvalga remiantis amonitų fauna' ['A stratigraphic overview of the Jurassic deposits of Papilė on the basis of their ammonite fauna'] in 1933.

In 1933, after a visit to Sweden and attending lecture courses of the prominent Quaternary geologists Lennart von Post and Gerhard de Geer, Pakuckas began his studies of modern glaciomorphology in Lithuania. With Kaunas University professors Mykolas Kaveckis and Juozas Dalinkevičius, Pakuckas took an active part in gathering material in Lithuania for the Quaternary map of Europe, involving mapping in southern Lithuania near Poland. A geomorphological map of Lithuania on a scale of 1:750 000 was compiled in 1936 (Fig. 2), on which glaciomorphological information relating to eastern Lithuania was included (Pakuckas 1936, 1938). This map was first published in a manual for high schools (Pakuckas & Viliamas 1938) (Fig. 3).

Pakuckas published his first paper on the glacial morphology of southern Lithuania in 1934, which dealt with the Lithuanian territory south of the 55th parallel (Pakuckas 1934). From the glaciomorphological point of view this territory can be divided into two different landscapes: (1) a plain of ground moraines or gently rolling country; and (2) a hilly zone in the south. The

From: GRAPES, R. H., OLDROYD, D. & GRIGELIS, A. (eds) *History of Geomorphology and Quaternary Geology.*
Geological Society, London, Special Publications, **301**, 141–147.
DOI: 10.1144/SP301.9   0305-8719/08/\$15.00 © The Geological Society of London 2008.

Fig. 1. Česlovas Pakuckas (Czeslaw Pachucki) as Professor at Marie Curie-Sklodowska University, Lubin, Poland (about 1960).

geological boundary between these two landscapes followed approximately the geomorphological changes. Starting from the German frontier near Pilupėnai, the boundary extends east through Pajievonys, Bartninkai, Kalvarija, Liudvinavas, 8 km NW from Prienai, Darsūniškis, Kruonis and Žiežmariai (the locality between Kruonis and Vievis). This line marks the northern limit of the Suvalkai Highlands, which is an extension of the 'Baltic Highlands'. To the north of this boundary is the plain of ground moraines, whereas to the south there is a distinctly hilly landscape. In the hilly zone, there is a continuous chain of terminal moraines, characterized by varied relief, with sands, gravels and numerous boulder erratics. North of the chain of terminal moraines there follows a pronounced hilly landscape, consisting mostly of clay. To the south, spread numerous sandurs (Icelandic word for outwash plains), although these are not to be found near every part of the chain. In the extreme western and eastern limits (near Vištytis and Vievis) the terminal moraines follow a southeasterly direction, passing linearly into the Lithuania (Fig. 4).

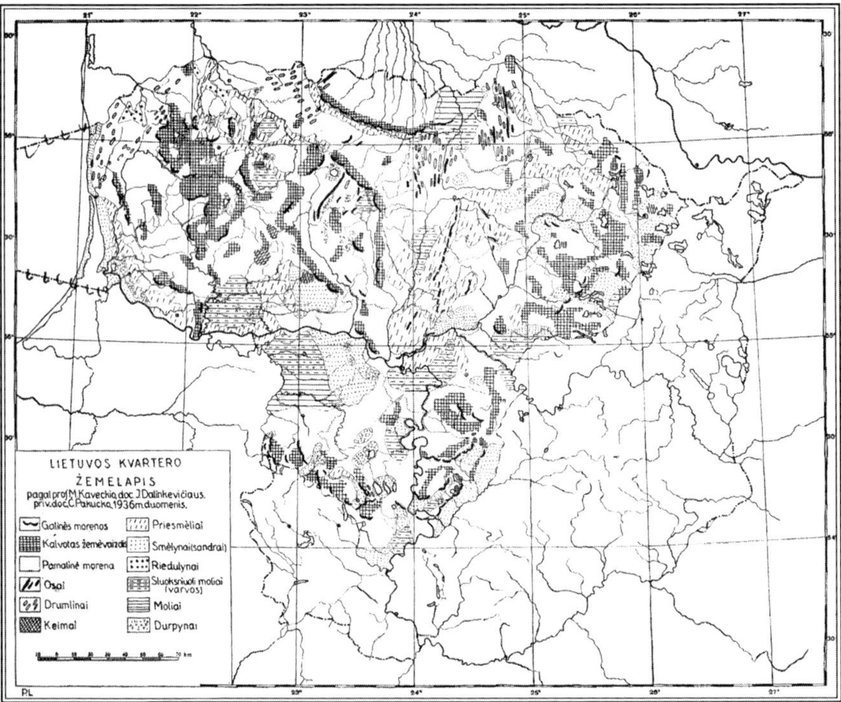

Fig. 2. Quaternary map of Lithuania compiled by M. Kaveckis, J. Dalinkevičius and Pakuckas in 1936 (Pakuckas & Viliamas 1938). Galinės morenos, End moraines; Kalvotas žemėvaizdis, hilly landscape; Pagrindinė morena, ground moraine; Ozai, eskers; Drumlinai, drumlins; Keimai, kames; Priesmėliai, sandy loam; Smėlynai (sandrai), outwash plain; Riedulynai, boulder fields; Sluoksniuoti moliai (varvos), varved clay; Moliai, clay; Durpynai, peat-bogs.

Dr. Č. PAKUCKAS ir Dr. VI. VILIAMAS

GEOLOGIJA

IR

FIZINĖ GEOGRAFIJA

AUKŠTESNIOSIOMS GIMNAZIJŲ KLASĖMS

Fig. 3. Title page of the manual *Geology and Physical Geography* (in Lithuanian and published in 1938 in Kaunas) by Pakuckas & Viliamas with a dedication from Pakuckas to his son, Aloyzas, dated 13th January 1953.

In the area from Kalvarija to Aukšdvaris, the terminal moraines have a curved form with their convex sides facing south. The most southern arc is to be found in the Kučiūnai–Veisiejai section. Another arc begins about 10 km north of the first and follows the line: Lazdijai–Seirijai–Merkinė–Nedzingė–Onuškis. This is the longest chain of terminal moraines in southern Lithuania, where it stretches continuously for about 120 km. In front of this chain there is a level area of sandur.

A third morainic arc (Nemunaitis–Daugai) is somewhat shorter, but it is clearly pronounced. It is followed by sandur on its southern or frontal side. A fourth arc extends between Alytus and Aukštadvaris. Between Aukštadvaris and Vievis terminal moraines follow a NW–SE direction as almost straight belts. However, one of these, east from Semeliškės, trends in a north–south direction. Apart from these two straight belts, sandur are scarce and are only to be found in small areas. The western part of this region has only a few lakes, while in the eastern part there are numerous lakes, and many of them are of rinne type (tunnel valley). Sizable streams flowing beneath the ice and not loaded with coarse sediment cut shallow trenches in till and other loose material at the surface. The rinne valleys of the Baltic region are thought to have originated in this way.

In 1936 Pakuckas published a paper dealing with orientation of marginal moraines of the eastern Lithuanian highlands and their origin (Pakuckas 1936). In 1938, he published another article on the glacial morphology of southern Lithuania (Pakuckas 1938). His generalized glacial–morpological ideas and conclusions were presented in 1940 at the First Conference of Lithuanian and Latvian geologists held in Kaunas. These conclusions were:

- that during the advance of the Last Glaciation, the ice sheet consisted of a number of flows;
- each one was dependent on the topography;
- each had its own specific glacial centre;
- the Lithuanian glacier (from the Žemaičiai highland to Zarasai) flowed from SW Finland, the Gulf of Bothnia and the area of the Åland Archipelago.
- The western slope of the Žemaičiai Highland and the maritime plain were covered by a glacial flow from Sweden across the depression of the Baltic Sea. The eastern side of the region of Lithuania was covered by a third flow from South Karelia along the depressions of the Narva River, which drains Lake Peipsi (divided between Estonia and Russia) into the Gulf of Finland, the Velikaya River, which today flows northwards into Lake Peipsi, and the lowland around Polock in northern Belarus. This was named the Belarus Glacier.

From 1940 to 1944 Pakuckas was the first Director of the Lithuanian Geological Survey, and between 1942 and 1944 was Head of the Department of Geology at Vilnius University. He helped Polish geologists in Lithuania during the war years and at the end of the war he emigrated, residing initially in Germany and Austria. In 1946 Pakuckas started to work as an adjunct at Wrocław University, Poland, in the Department of Stratigraphical Geology. From 1954 to 1965 he was the Head of the Department of Geology at Maria Curie–Skłodowska University in Lublin as an Associate Professor, and was promoted to Professor in 1960.

In Poland, Pakuckas (or Pachucki as he was called) continued his glacial–morphological research. He compiled a geological map of the Suwalki region (NE Poland) at a scale of 1:300 000

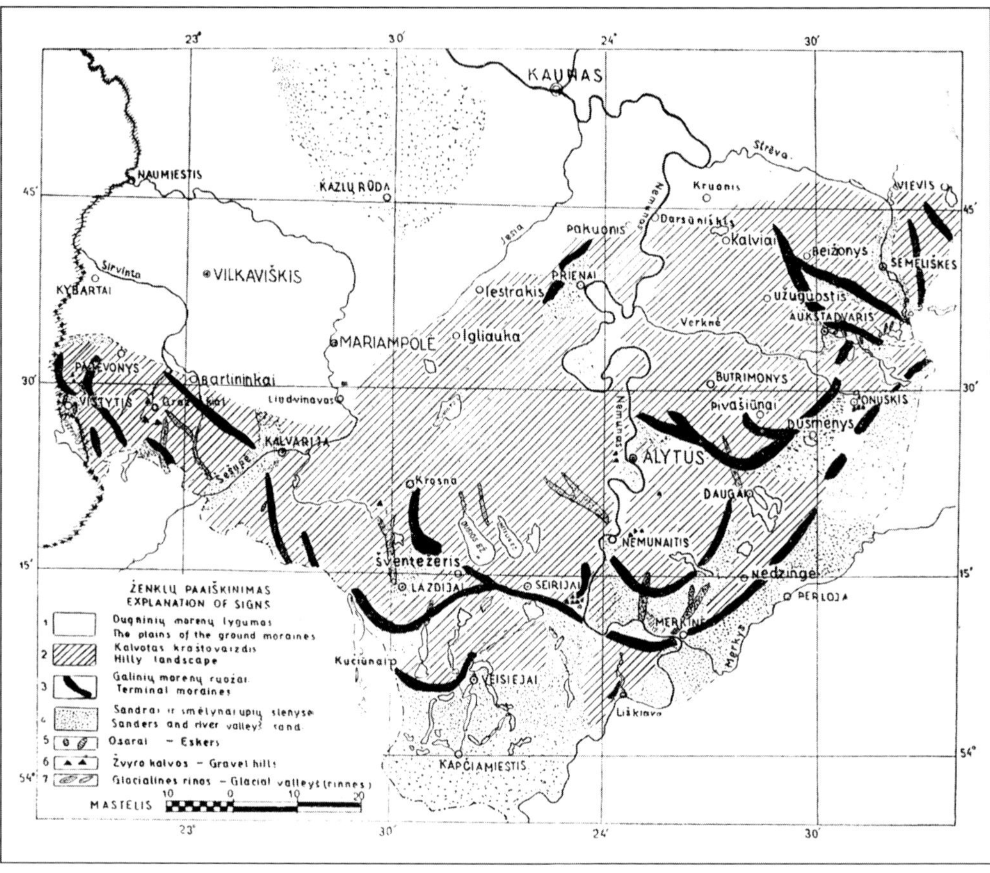

Fig. 4. Map showing glacial features of southern Lithuania (Pakuckas 1934).

(as part of the General Geological Map of Poland) in 1948, and between 1952 and 1961 published significant work on the correlation of terminal moraines in NE Poland and southern Lithuania with western Belarus as well as the 'Peribalticum'. Pakuckas was of the opinion that the ridges of the Lithuanian glacial tongue terminal moraines are contemporaneous with those of the Lake Narocz area and in the west with the glacial tongue of Gdańsk. Within Lithuania, Pakuckas had delineated a large field of marginal moraines – the so-called Baltic Highland, extending across the Sūduva district (SW Lithuania), Dzūkija (SE Lithuania) and Aukštaitija (in NE Lithuania) Highlands (Fig. 5). In the SW this formation extends as the Mazurian arc of marginal moraines (NE Poland), and the Švenčionys–Naroch, or North Belarus, field of marginal moraines extends to the NE part of the Baltic Highlands. These glaciomorphological units correspond to separate glacial flows: Western, Middle and Eastern, each consisting of smaller marginal moraine units (lobes) (Fig. 6). All the mentioned terminal moraine ridges of the Baltic marginal highlands are associated with three glacial

stages: the Brandenburgian, Poznanian and Pomeranian, or belong to the Daniglacial (Fig. 6). Further to the north, beyond the extent of the Baltic Hills, Pakuckas correlated moraines with the younger Gotiglacial stage and later included the Brandenburgian end moraines as a phase of the Frankfurtian stage (Pachucki 1961*a*, *b*). The last glaciation, in the Peribalticum area, was up to now divided into three stages: Brandenburgian, Frankfurtian and Pomeranian. Pakuckas recognized only two stages: the Frankfurtian and the Pomeranian. He considered that the Brandenburgian was a phase of the Frankfurtian stage, because no interstage organogenic sediments were found between the Brandenburgian and Frankfurtian moraines.

In the interval between the greatest range of the Frankfurtian stage and that of the Pomeranian stage, the courses of the three phases were delineated by Pachucki (1961*b*):

- Phase 1 Chozież–Lubaniec, runs from Chodzież to the NE, across the Wągrowiec, Janowiec, Mogilno, Radzejów and Lubaniec;

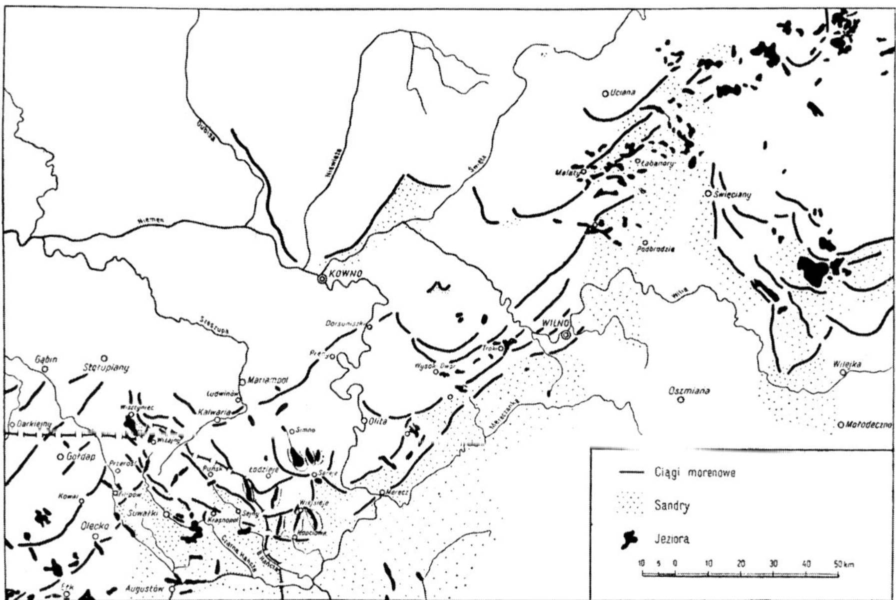

Fig. 5. Map showing the distribution of terminal moraines of the Last Glaciation in south Lithuania, NE Poland and neighbouring countries (Pachucki 1952). Ciągi morenowe, end moraines ridges; Sandry, outwash plains; Jeziora, lakes.

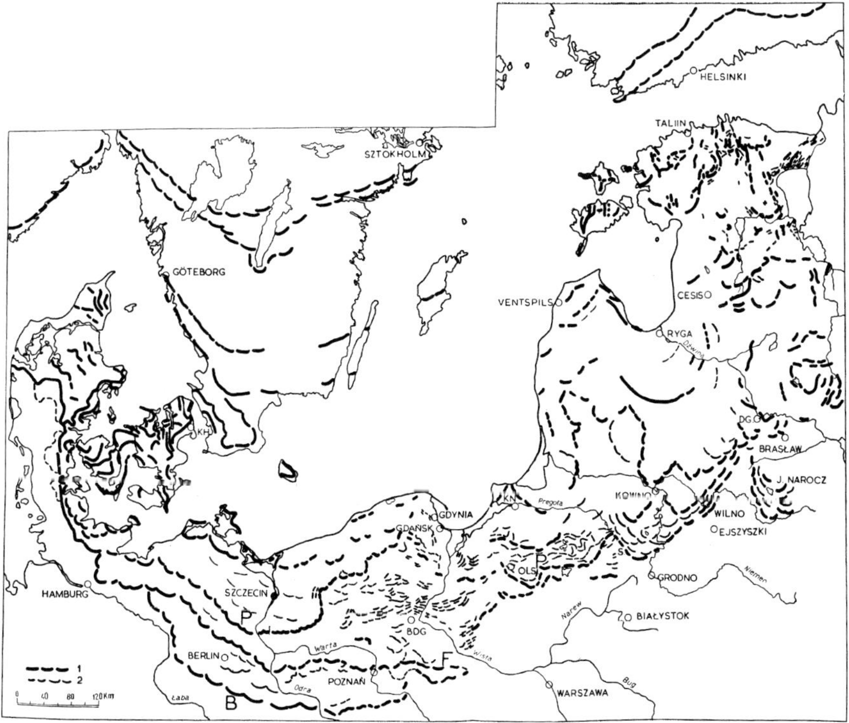

Fig. 6. Map showing distribution of end moraines from the Jutlandian peninsula through North Germany and Poland, Baltic countries to Finland (Pachucki 1961*a*). 1, end moraines of stages and main phases; 2, end moraines of recessions and oscillations; P, Pomeranian stage; F, Frankfurtian stage; B, Brandenburgian stage. BDG, Bydgoszcz; Dg, Dyneburg (Davgapils); S, Suwałki; Ols, Olsztyn; KN, Kaliningrad; KH, Kopenhaga.

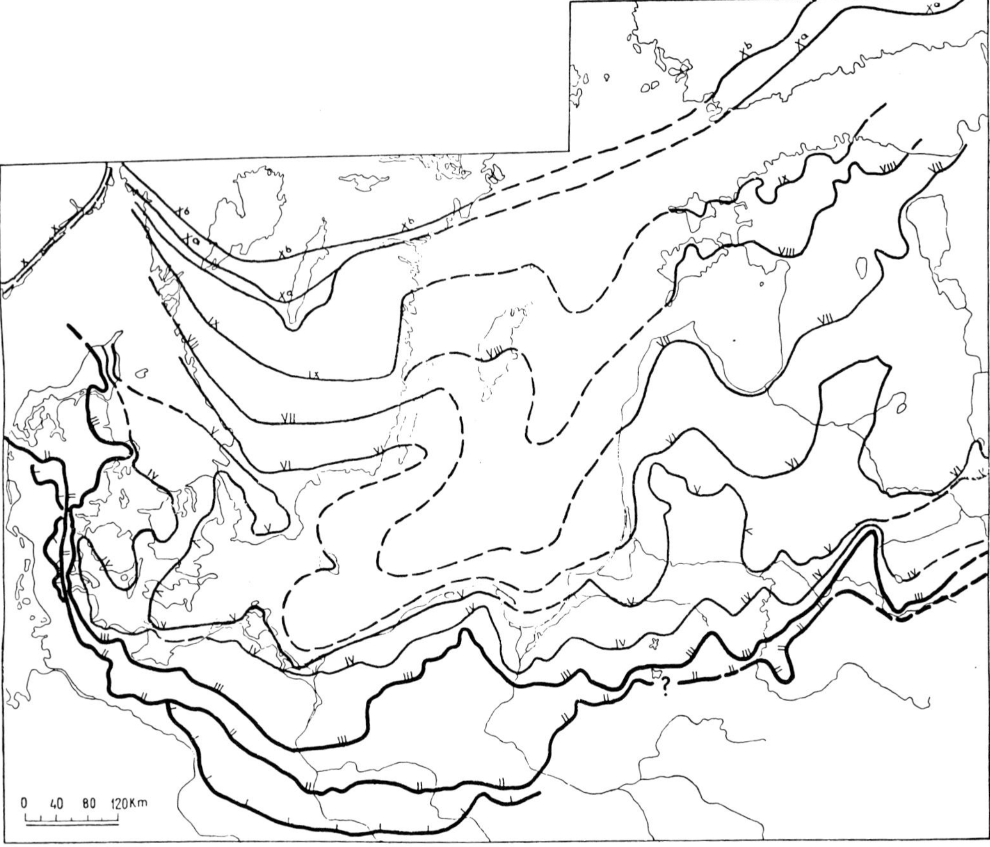

Fig. 7. Stages (bold lines I–III) and phases (IV–X^b) of the Last Glaciation in the Peribalticum area (Pachucki 1961*a*).

- Phase 2 Więcbork–Wąbrzeźno, runs almost parallel from Jastrów across Więcbork, and on the eastern side of the Vistula River is marked in the Wąbrzeźno–Brodnica region;
- Phase 3 Człuchów–Chojnice–Tuchola–Grudziądz.
- The ranges mentioned prove that the Frankfurt stage had the greatest expansion and persisted for a long time.

Pomeranian frontal moraines indicated a separate stage with regard to their dominant transgression characteristics. The inland ice of the Pomeranian stage retreated with certain stages, stopping on the way and leaving the series of frontal moraines, classifed as recessive phases (Fig. 7, stops IV–X^b). The maximal extent of the Pomeranian inland ice-formed series is represented by south-directed 'lobes' of the Odra, Vistula and Łyna rivers, the Mazurian lakes, the Nemunas River and Lake Narocz. The area beyond the Pomeranian stage frontal moraines is one of extensive sander,

indicating the maximum advance of the youngest glaciation of this stage (Pachucki 1962).

Together with geomorphological investigations, Pakuckas also worked on the Quaternary stratigraphy of Lithuania. He presented a stratigraphic overview of the Pleistocene deposits in the region of Vilnius, detailing evidence for the presence of three glaciations, moraines of oscillatory character and the limit of Last Glaciation through Vilnius (Pakuckas 1940*a*, *b*). He also discovered and investigated buried interglacial organogenic deposits (peat and gyttja) on the banks of the Nemunas River in the area of Merkinė, SE Lithuania (Pachucki 1952). This section is today recognized as the stratotype of the Merkinė (Eemian) Interglacial in Lithuania, and has a thickness ranging from 100–250 m that contains four separate boulder clay horizons with intervening interglacial or interstadial deposits.

Pakuckas (or Pachucki) can be regarded as the leading pioneer of modern glacial–morphological investigations of the Baltic marginal highlands of

Lithuania and Poland. Probably, his most significant contribution was the conclusion that during the Last Glaciation, the continental glacier was not a single ice sheet but consisted of a number of flows, each being dependent on topography and each with its separate glacial centre.

The authors are very grateful to Professors R. H. Grapes and D. R. Oldroyd for constructive comments and suggestions, as well as improvements to the English and revision of the manuscript.

References

PAKUCKAS, Č. 1934. Pietinės Lietuvos reljefo glacialiniai elementai [The glacial elements in the relief of Southern Lithuania]. *Kosmos*, **15**, 185–200.

PAKUCKAS, Č. 1936. Galinių morenų kryptis rytinės Lietuvos aukštumose ir tų aukštumų kilmė [Courses of the end moraines in the highlands of East Lithuania and the origin of these highlands]. *Kosmos*, **17**, 323–334.

PAKUCKAS, Č. 1938. Pietinės Lietuvos glacialimorfologiniai bruožai [Glacialmorphological sketch of the Soutern Lithuania]. *Kosmos*, **19**, 321–333.

PAKUCKAS, Č. 1940a. Galinių morenų kryptis ir paskutiniojo ledlaikio ledynų fazės Lietuvoje [Courses of end moraines and phases of last glaciation in Lithuania]. *In*: *Pirmojo lietuvių-latvių geologų suvažiavimo darbai* [Transactions of the First Conference of Lithuanian and Latrian Geologists (complied this article by B. Tijūnaitytė). *Gamta*, 3–4.

PAKUCKAS, Č. 1940b. Vilniaus krašto pleistoceno klausimais [On the questions of Pleistocene of Vilnius region]. *Kosmos*, **21**, 98–106.

PAKUCKAS, Č. & VILIAMAS, VL. 1938. *Geologija ir Fizinė Geografija Geology and Physical Geography.* Sakalas, Kaunas.

PACHUCKI, CZ. 1952. O przebiegu moren czołowych ostatniego zlodowacienia północno-wschodniej Polski i terenów sąsiednich [Direction of the course of the terminal moraines of the last glaciation in the northeastern Poland and the neighbouring countries]. *Z badán czwartorzędy w Polsce*, 1. Państwowy Instytut Geologiczny (Warszawa), Bulletin, **65**, 599–625.

PACHUCKI, CZ. 1952. *Badania geologiczne na arkuszach 1 : 100 000 Trzebnica i Syców.* Państwowy Instytut Geologiczny (Warszawa). Bulletin, **66**.

PACHUCKI, CZ. 1961a. Moreny czolowe ostatniego zlodowacenia na obszare Peribalticum [End moraines of the last glaciation in the Peribalticum]. *Rocznik Polskiego Towarzistwa geologicznego (Kraków)*, **31**, 303–318.

PACHUCKI, CZ. 1961b. The stages and phases of the latest glaciation in the Peribalticum's area. *In*: *Abstract of Papers. INQUA VIth Congress, Wydawnictwa Geologiczne Warszawa*, 163–164.

PACHUCKI, CZ. 1962. Uwagi o geomorfologii doliny Czarnej Hańczy. [Comment on the geomorphology of the Czarna Hańcza Valley]. *Kwartalnik geologiczny*, **6**, 176–186.

Valerija Čepulytė (1904–1987) and her studies of the Quaternary formations in Lithuania

ONA KONDRATIENĖ & MIGLE STANČIKAITĖ

Department of Quaternary Research, Institute of Geology and Geography, Ševčenkos Street 13, LT03223, Vilnius, Lithuania (e-mail: o.kondratiene@gmail.com; stancikaite@geo.lt)

Abstract: Valerija Čepulytė (1904–1987) devoted her life to the development and promotion of geographical/geological research in Lithuania. The different aspects of Quaternary stratigraphy, palaeogeography, extent of the Last Glacial (Weichselian) and deglaciation of the territory, together with the development of the geomorphological terminology and methodology of the geomorphological mapping, were of primary interest to her during her long scientific career. Graduating from Vilnius University in 1939, Čepulytė became the first woman in Lithuania to take a doctoral degree in geographical science (1948) and in 1968 she was awarded the Doctor Habilitus degree. Her scientific innovations and ideas were important for generations of research workers dealing with the Quaternary system both in Lithuania and abroad.

Valerija Čepulytė was born in St Petersburg (Russia) into a working-class family. Her parents – small farmers from Tverečius district in the NE part of the present Lithuania – had moved to St Petersburg seeking employment, which was quite common at that time when Lithuania was a part of the Russian Empire. Her father worked at a bakery and later at a police station. Her mother was a housewife and produced a large family of eight children, Valerija being the eldest. Tragically, all her brothers and sisters died young except for her youngest brother, Julius, who perished at the beginning of World War II. Her mother died at the early age of 42. Thus, Valerija was the only child who survived. As a result of such a difficult life she became a rather uncommunicative person.

Čepulytė obtained her elementary education in St Petersburg, at a private elementary school, but also attended a Lithuanian Kotryna Gymnasium. In 1919, aged 15, she returned to Lazinkai village, Tverečius district, along with her parents. At that time Vilnius and the Vilnius Region were occupied by the Polish Army and had been incorporated into Poland. Difficult living conditions forced her to take a job as an elementary school teacher, and from 1924 to 1927 she studied at the Vilnius teacher seminary. Having finished there, she returned to her village to work as a teacher, where she became an active member of the social life and participated in the different Lithuanian public organizations. Čepulytė was a member of the Teachers Union Council, which included teachers in the Vilnius Region. She also published a few humoristic pieces of literature that were later performed as plays by local groups. She was familiar with the traditions and customs of the local inhabitants, and used different aspects of their lives in her writing.

At that time a teacher's seminary certificate was insufficient for admission to Polish universities. However, in 1934 she successfully passed all the necessary matriculation examinations at the Vytautas Magnus Gymnasium in Vilnius, and in the autumn of that year she entered Stepan Bator University in Vilnius as a student of geography in the Faculty of Mathematics and Natural Sciences.

From her reminiscences kept in the archives of the Lithuanian Academy of Sciences, Department of Manuscripts (F 290–310), it is clear that Čepulytė would have liked to have chosen geology, had it been available in the university curriculum. Be that as it may, she graduated from the University in 1939, and was awarded an MPhil degree for scientific research work, her dissertation being titled 'The North East of Gardinas Land: Geographical Features'. The Gardinas area lies in the western part of the present Belarus Republic. After graduating, she worked for a short time as a teacher at Vilnius Gymnasium.

Čepulytė was unemployed for some time after the outbreak of World War II when the German Army occupied Vilnius. However, Professor Juozas Dalinkevičius (the then Principal) invited her to undertake scientific research work at the Institute of Geology and Geography (Fig. 1), and on 1 November 1941 she embarked on a new career, one that she had always dreamed of. She remained with the Institute, although its name and affiliation changed many times during her lifetime. She worked in different positions: Assistant, Junior Researcher, Senior Researcher, Head of Subdivision, Head of Division and, finally, Scientific Adviser. In 1944–1945 she was also Head of the Department of Archaeology in the Lithuanian Institute of History.

From: GRAPES, R. H., OLDROYD, D. & GRIGELIS, A. (eds) *History of Geomorphology and Quaternary Geology.* Geological Society, London, Special Publications, **301**, 149–158. DOI: 10.1144/SP301.10   0305-8719/08/$15.00 © The Geological Society of London 2008.

Fig. 1. Valerija Čepulytė, photographed about 1975, at the Institute of Geology, Vilnius. Photograph by O. Kondratienė.

Dalinkevičius suggested that Čepulytė should investigate the Lithuanian Quaternary sediments as this field of research was most closely related to her geographical education. The suggestion was accepted and Čepulytė set to work. Having studied the relevant literature, she quickly came to the conclusion that the Quaternary stratigraphy was anything but clearly established. The available stratigraphic schemes were based merely on 'logic' (rather than field evidence) and schemes from neighbouring territories. They lacked scientific substance and there were no criteria for the satisfactory classification of Quaternary sediments in Lithuania (Čepulytė 1946), and, without a sound stratigraphical basis, Quaternary studies made no sense. With this in mind, Čepulytė sought to devise viable stratigraphic criteria.

Work under the wartime conditions was not easy. There was no laboratory and, what was more important, there were too few good-quality boreholes penetrating through the Quaternary deposits. Only sparse core material collected at the Institute of Geology and Geography and drilled by the 'Biske' concern was available. The remainder of the information had been collected by the Čepulytė herself from visiting more than 150 outcrops and numerous quarries (Čepulytė 1946). On the basis of all the data Čepulytė managed to determine the lithological characteristics of the Quaternary tills, which could be taken as stratigraphical criteria. The tills in the *lower part* of the Quaternary are grey, bluish grey or greenish grey; they were rich in small grey boulders of Silurian limestone, and in places contain faunal remains. She named these tills the 'Grey complex'. The *overlying tills* are characterized by a dark brown or greyish-brown colour, occasionally with a greenish tint. They are rich in yellowish–brownish and greyish–brownish clastic material of Devonian dolomite, and contain a greater proportion of sand than the tills of the underlying 'Grey complex'. In addition, the amount of clastic material of Silurian limestone in the till has decreased to 5–10%. These tills were classified as the 'Brown complex'. The *upper tills* are pinkish, and can be distinguished from the other tills by their friability, variable petrographic composition and the large quantity of small weathered crystalline boulders. These tills were classified as the 'Pinkish complex' or the 'Variegated complex'.

In 1946 Čepulytė described these and other characteristic features of the lithological composition of Quaternary tills in a study entitled, 'Stratigraphic types of tills in the territory of the Lithuanian SSR'. For this work, she was awarded a 'Candidate of Geographical Sciences' degree (equivalent to a doctorate) in 1948. She associated the three complexes with three of the four then-accepted ice ages – Mindel, Riss and Würm. Although the Alpine glaciation terminology was in widespread use at the time, she did not directly associate the glacial cover in Lithuania with the Alpine glaciations, and wrote that the 'Alpine stratigraphical scheme is hardly acceptable or even inadaptable in the lowland areas'

(Čepulytė 1946, p. 26). Later, the Lithuanian till complexes were referred to as Eo-, Meso- and Neo-Pleistocene or Early, Middle and Later Pleistocene complexes, and, by some, as the Likhvin, Dnieper and Valdai complexes.

Subsequently, Čepulytė specified and detailed the petrographic criteria for her three till units and they became the main basis for classification of Quaternary sections in Lithuania. Yet, it did not take her long to realize that her criteria were applicable only within certain restricted areas and that 'Silurian and Devonian carbonate correlatives are important for the southeast Lithuania and the Gardinas region, western Belarus' (Čepulytė 1967, p. 20). But in western Lithuania and east of the Gardinas region these correlatives lost their importance. The percentage of ultrabasic rocks was, in her opinion, a more reliable criterion for identification of tills belonging to Mindel complex in western Lithuania. As to the lower boundary of the Quaternary, Čepulytė adhered to the common attitude accepted in the then Soviet Union, i.e. it embraced only the supposed glacial Quaternary sediments.

Čepulytė's proposed principles for the classification of the Quaternary in Lithuania should be regarded as the first empirical, stratigraphy-based work undertaken. It should be noted that sections later identified on the basis of biostratigraphic data in most cases correlated well with the tripartite classification that she established for the Lithuanian Quaternary. But Čepulytė saw no possibility of any detailed classification of Quaternary sediments on the basis of lithological data. In her opinion, the available data were insufficient. Thus, in 1967 (p. 32) she wrote:

The sediment age in concrete sections cannot be identified more objectively and in greater detail than by dividing the sediments into three parts. The lithological differences are too small. Quantitative variations within the tills of the same age are greater than between tills of different age.

Čepulytė's further research was mainly related to the structure of Quaternary deposits in the distinct orographic regions of Lithuania. She was interested in the possible influence of the pre-Quaternary surface and the palaeosurfaces of the particular Quaternary stages on the present-day relief of Lithuania. She made an attempt to find methods for the evaluation of the relief's heredity in the context of the whole Quaternary System (Fig. 2), and in 1958–1968 she published several articles relating to these issues. It is interesting to note that during her career Čepulytė preferred to work and publish her data alone. There were only two co-authored publications.

After World War II intensive geological investigations began in the territory of Lithuania. In 1957 the Board of Geology was founded, which was responsible for all geological investigations. Complex geological mapping with numerous cores provided geologists with much new data to describe the Quaternary sediments. Nevertheless, together with colleagues from the Institute of Geology and Geography and students of Vilnius University, Čepulytė organized field expeditions and collected a lot of new data herself. Talking about her scientific career we should not forget her civic activities. Čepulytė was a councillor in the Vilnius City Council between 1955 and 1958, was elected to the Highest Council of the Lithuania SSR, and was a member of numerous scientific and social boards and commissions. For her civic activities she received many medal awards, and in 1965 she was bestowed with an Honorary Rank for cultural achievement.

In 1968 Čepulytė was awarded a Doctor of Geology–Mineralogy degree (equivalent to a Doctor Habilitus degree in physical sciences) for her published work (Čepulytė 1968). It should be pointed out that she never used the term 'palaeogeomorphology' but mostly used the term 'palaeosurfaces'. She maintained that it was impossible to reconstruct in detail the Pleistocene palaeorelief because it had been destroyed in different degrees by subsequent processes. In her opinion, only the general features of a palaeosurface could be reconstructed.

The issues of the structure and 'heredity' of the Lithuanian Quaternary complex could not be solved without a reconstruction of the pre-Quaternary surface and without knowledge of morphogenesis of recent relief. However, knowledge of the pre-Quaternary surface was very scanty at that time and Čepulytė found some information in the work of Lewiński & Samsonowicz (1918). Analysing the pre-Quaternary surface of Poland, these researchers had included part of Lithuania. However, according to Čepulytė, this study did not provide a clear overview of the pre-Quaternary surface in Lithuania. In 1930 Dalinkevičius presented a more complete picture of the pre-Quaternary surface, including data obtained from exposures of pre-Quaternary strata, uplifts and depressions, plotted on a map (Dalinkevičius 1930). Using data from numerous boreholes, Čepulytė produced a detailed map of the pre-Quaternary surface of certain regions in Lithuania that provided a basis for an evaluation of the connections between the pre-Quaternary surface and present relief (Čepulytė 1959, 1971).

Čepulytė (1955, 1956) came to the conclusion that the general features of the pre-Quaternary surface were largely reflected in the structure of recent orographic units, whereas such elements as eskers and kames are unrelated to the pre-Quaternary surface (Fig. 3).

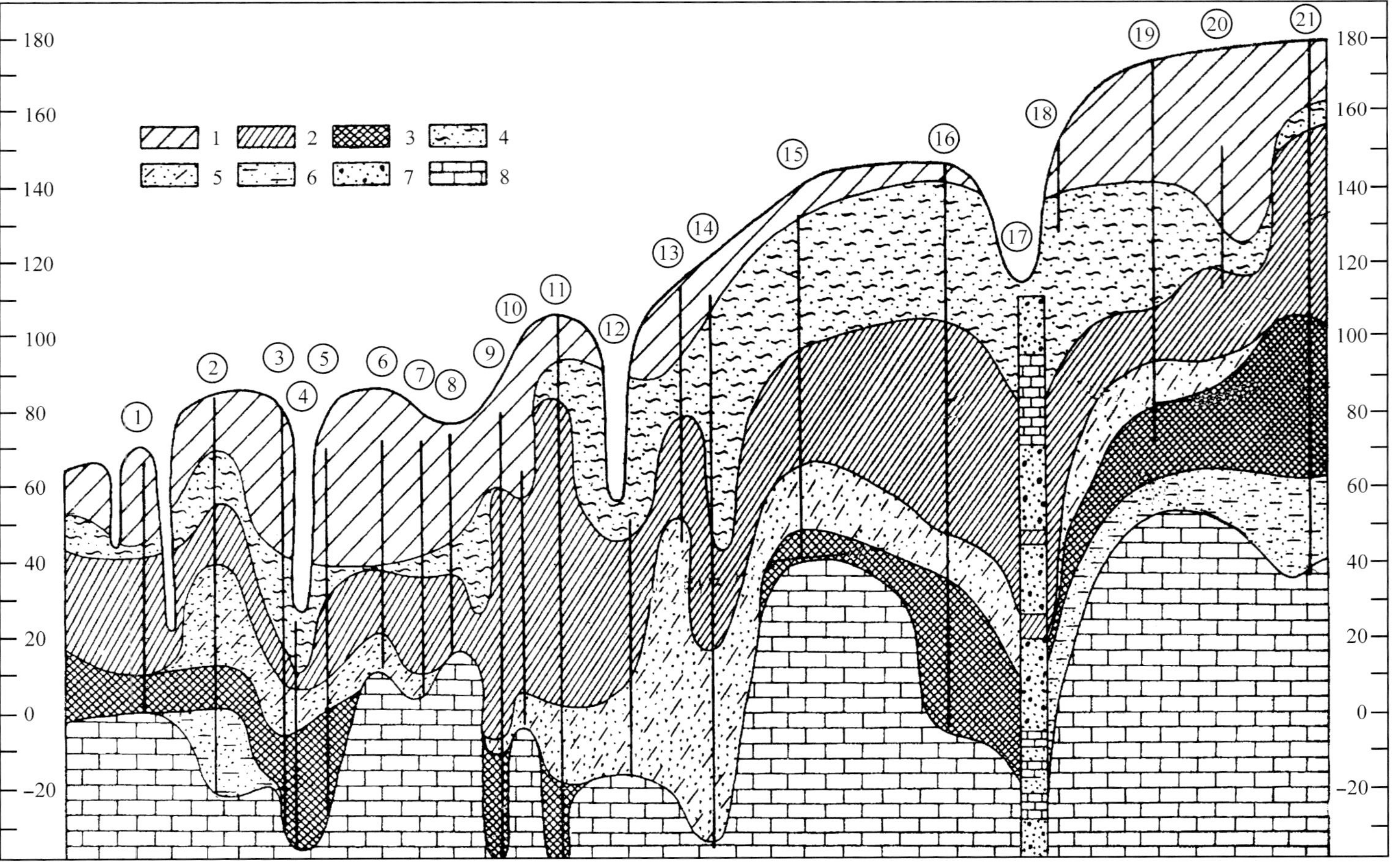

Fig. 2. Cross-section of the Quaternary sediments in SE Lithuania according to Čepulytė (1965). 1. Basal till of the Late Pleistocene (Würm) age; 2. basal till of the Middle Pleistocene (Riss) age; 3. basal till of the Early Pleistocene (Mindel) age; 4. sandy and clayey deposits of the Eemian Interglacial; 5. sandy and clayey deposits of the Holsteinian Interglacial; 6. sandy and clayey deposits of Neogene and pre-glacial age; 7. beds of sand with gravel in Pamerkės section; 8. deposits of Cretaceous age. *Cores*: 1. Purviškiai; 2. Vilkija; 3. Miškiniai; 4. Kulautuva; 5. Raudondvaris; 6. Kaunas (Aleksotas); 7. Jonučiai; 8. Garliava; 9. Išlaužas; 10. Linksmakalnis; 11. Ošminta; 12. Birštonas; 13. Jieznas; 14. Vincentava; 15. Žaliai; 16. Puodžiai; 17. Pamerkės; 18. Matuizos; 19. Kalesnikai; 20. Eišiškės; 21. Račkūnai.

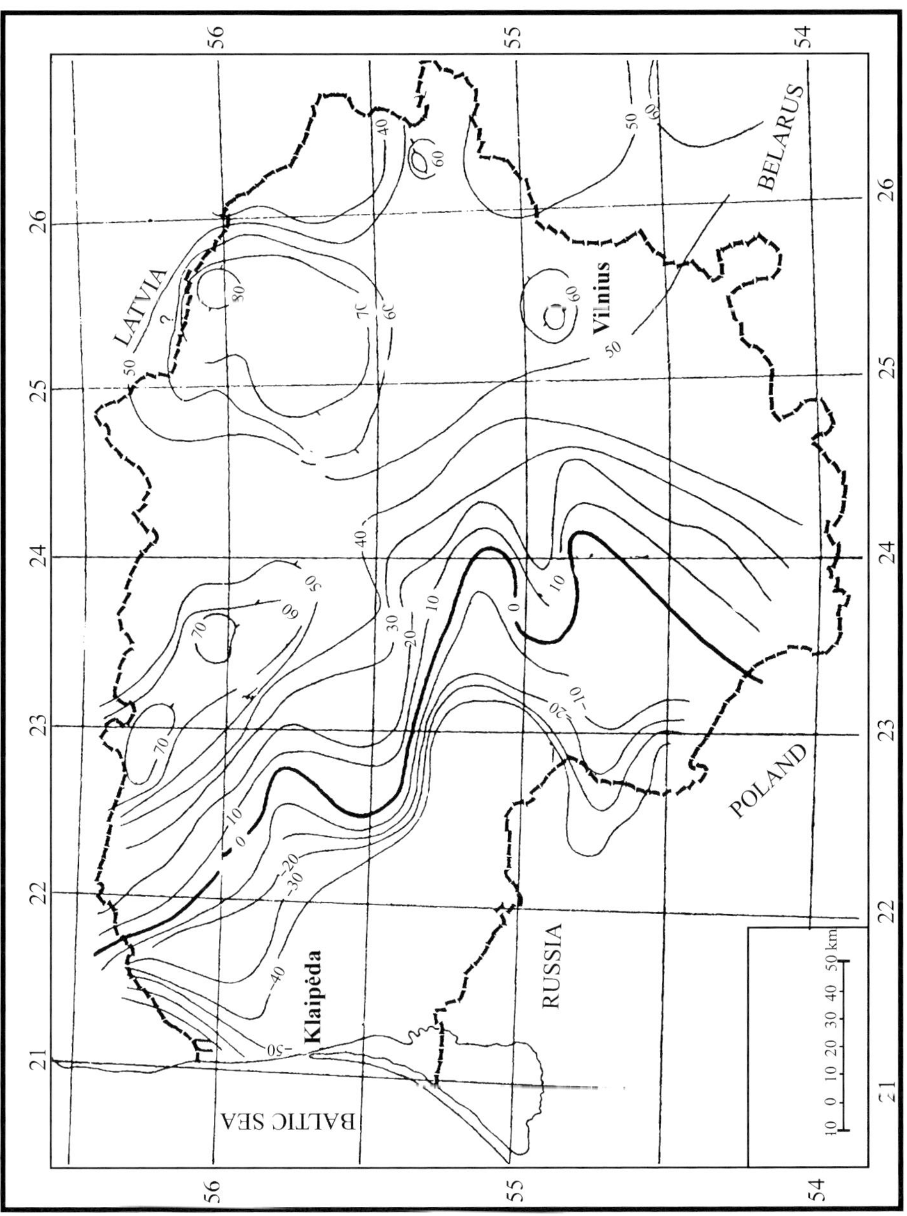

Fig. 3. Contours (m) of pre-Quaternary relief according to Čepulytė (1959).

She determined that the formation and distribution of recent morphogenetic types of relief were pre-determined by the palaeosurface of the 'Brown complex', which she had identified as being Middle Pleistocene. Gradually filling the map with data from new boreholes, she suggested the possibility of determining the location of pre-Quaternary valleys and evaluating the histories of some of the recent river valleys and their separate sectors. Later, in 1965, she identified a large pre-Quaternary drainage artery, trending from Kaunas in central Lithuania across Druskininkai towards Poland, and later indicated many smaller palaeo-incisions orientated from the Žemaičiai Upland towards the Baltic Sea (Čepulytė 1971).

Čepulytė was the first to appreciate that 'the role of the Last Glacial was not decisive for the formation of uplifts' (Čepulytė 1958, p. 42). The youngest glacial strata were not the thickest ones. She was also the first to show that the pre-Quaternary surface in the Žemaičiai Upland was more complex than had previously been supposed. In the highest part of the recent upland, the pre-Quaternary surface varied from 10 to 45 m a.s.l. (above sea level). To the NE, the height of the surface reached 80 m a.s.l. near the Latvian border (today it is occupied by a plain). On several occasions Čepulytė confirmed her previous assumption that the Žemaičiai Upland was unre-lated to the pre-Quaternary surface. She deter-mined that 'the basement of Žemaičiai Upland was represented by a morainic upland of Middle Pleistocene age' (Čepulytė 1959, p. 86). In the area between Pavandenė, Varniai and Kaltinėnai the basement rose to 160 m a.s.l and in places to approximately 180 m a.s.l. The summit of the basement coincided with the highest points of the Žemaičiai Upland (Čepulytė 1959, p. 86). Thus, Čepulytė opposed the common opinion of Lithua-nian geologists concerning the influence of neotec-tonics on the formation of the Žemaičiai Upland. She suggested 'that there were no noticeable neo-tectonic movements in the zone of the Žemaičiai Upland during Middle Pleistocene' (Čepulytė 1959, p. 87). Meanwhile, other researchers related the formation of the upland to neotectonics. Čepulytė established that 'the basement of Suvalkai Upland (rising up to 150 m a.s.l) was formed in the Middle Pleistocene' (Čepulytė 1959, p. 86). Other recent hilly terrains were, as a rule, located within zones of Middle Pleistocene uplift (Fig. 4), and only the basement of the Baltic Uplands had already begun to form in the Early Pleistocene. Thus, Čepulytė was the first to establish that the Lithuanian uplands had a binary structure 'the basement of the uplifts [was] . . . composed of the Middle Pleistocene uplift on whose surface different morphogenetic types of

recent relief formed during the last Ice Age' (Čepulytė 1958, p. 32).

Čepulytė was not so successful in solving problems related to the formation of recent relief and her works were criticized because she differed from most other authors in her attitude to degla-ciation. She was an exponent of zonal (areal) deglaciation. Overall, Čepulytė was very consistent in her scientific positions and did not pay any attention to criticism. Even in her last article, 'On the genesis of hilly marginal relief', published after her death in 1988, she wrote that 'the glacier declined areally and a wide glacier periphery called the "marginal zone" melted simultaneously under the impact of ablation' (Čepulytė 1988, p. 43). Čepulytė associated the formation of the relief forms with a passive (negative) glacier, although she did not discount the possibility of the role of active glaciers in some areas. In general, she interpreted geomorphology as a geol-ogist, emphasizing the structure of relief elements to their form. In her geomorphological map she classified the relief of Lithuania into morphogen-etic types and gave them their lithological charac-teristics, distinguishing as many as 12 types of relief. The 'classification was based on the struc-ture of relief elements, lithological composition, morphometric peculiarities, and morphometric indices' (Čepulytė 1957, p. 266). In her opinion, a geomorphologist should not be content with knowledge of the surface of geomorphological regions, but should also take into account the struc-ture, origin and formation patterns of the relief forms (Fig. 5). In other words, in order to under-stand the structure of relief it is also necessary to reconstruct the palaeogeographical conditions of Quaternary stages.

We would like to emphasize the enormous amount of work carried out by Čepulytė in making the geomorphological map of the Lithuania. All the territory of Lithuania and large areas of the sur-rounding countries were investigated by her alone. About 80 000 km were covered during the field-work in Lithuania and about 2000 in neighbouring countries (Čepulytė 1957). Unfortunately, the geo-morphological map was never published and the manuscript has been stored in the archives of the Lithuanian Academy of Sciences, Department of Manuscripts (F 290–310) until now.

Adhering to the principle of areal deglaciation, Čepulytė interpreted the formation of the main Lithuanian uplands differently from other authors, placing emphasis on the processes of passive glacier melting. She considered the Baltic Uplands to be marginal formations of the Pomeranian Stage, which formed in a wide melting zone of a passive glacier. This zone is now regarded as being subdivided into three marginal belts: an

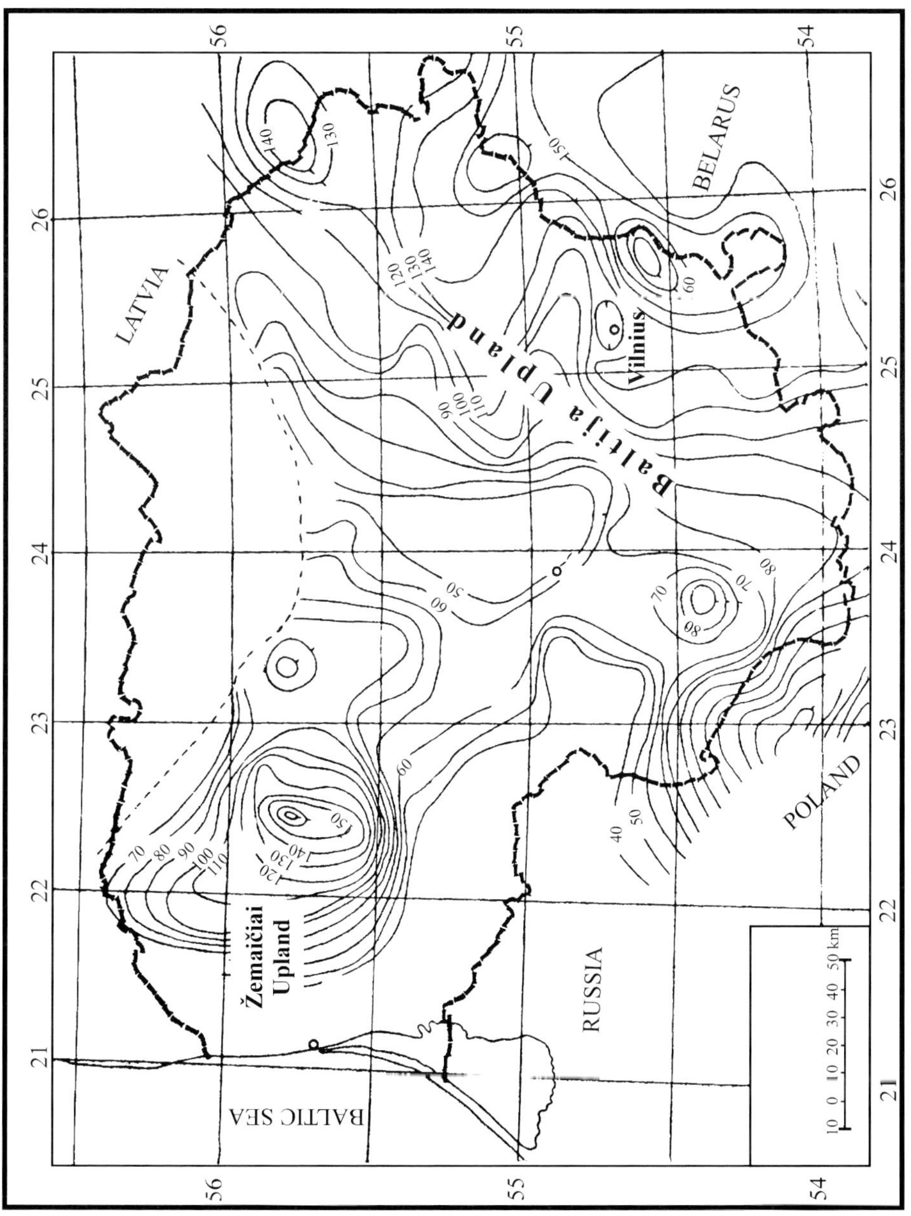

Fig. 4. Contours (m) of Middle Pleistocene till surface relief according to Čepulytė (1959).

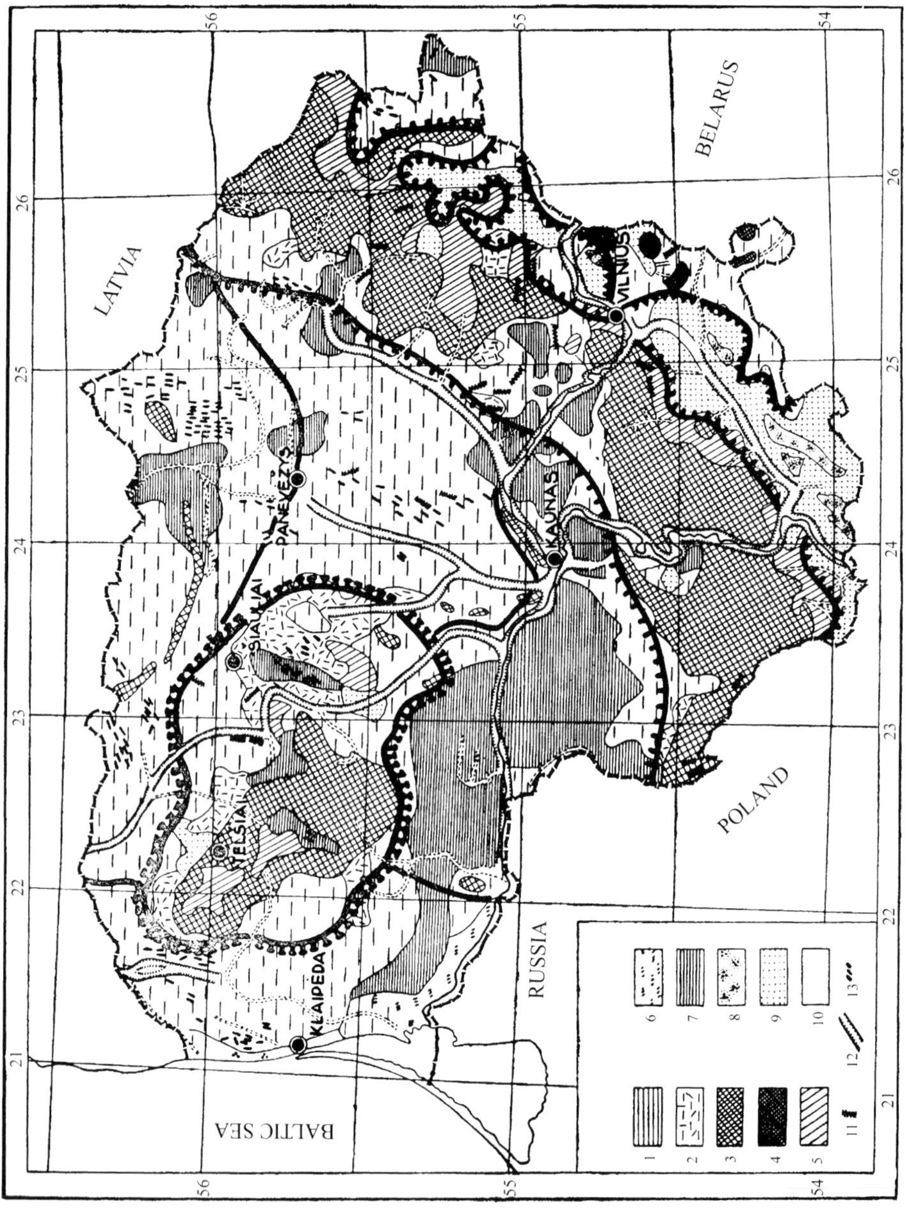

Fig. 5. Morphogenetic scheme of Lithuania according to Čepulytė (1958). 1. Plain of the basic till; 2. slightly undulating relief of the basic till; 3. strongly and moderately undulating relief of the marginal zone; 4. strongly undulating relief, with kames predominating; 5. undulating glacio-fluvial relief; 6. alluvial–deltic plain; 7. glacio-limnic plain; 8. hilly relief of aeolian origin (dunes); 9. glacio-fluvial plain; 10. marine terraces; 11. esker; 12. river valleys; 13. drumlins.

external one (eastern) composed of sandy kames and till cover; a *middle* belt composed of large- and medium-sized kames covered by a 0.5–1.5 m-thick pinkish ablation till; and an *internal or central* belt composed of pinkish till-hills. This zone includes glacier channels and channel lakes arranged almost perpendicularly to the glacier margin. Čepulytė did not recognize the South Lithuanian glacier marginal zone, but considered it to be a local element of the marginal belt. She explained the formation of the Baltic Uplands approximately as follows: 'the advancing Late Pleistocene glacier was active, and, travelling over the mezzo-Pleistocene formations, formed particular elements of the relief. Yet, the entire marginal landscape was formed in the phase of Pomerania–Baltija glacier decay when the ice cover lost its mobility and converted into a 20–40 km width (on average) marginal belt of passive glacier' (Čepulytė 1962a, p. 340).

Čepulytė had a similar view on the formation of the Žemaičiai Upland. She wrote that the hilly morainic and kame relief, characteristic of the whole Žemaičiai Upland, proved the 'territorial torpidity' of a passive glacier. In the territory of the Middle Pleistocene uplifts, the glacier lost its mobility earlier than in the depressions. It melted in the area of the Žemaičiai Upland as a mass of dead ice. The only difference was that the morphogenetic types of the Žemaičiai Upland were arranged concentrically around closed depressions, whereas the hills of the Baltic Uplands were aligned along the elevations. In Čepulytė's opinion, the Vidzemė Upland in Latvia and the Otepė Upland in Estonia formed in a manner or pattern similar to that of the Žemaičiai Upland (Čepulytė 1962b). The Ašmena Uplands formed in a different way. She never associated the Ašmena Upland with the Baltija or Švenčionys Uplands, but maintained that 'the recent relief of the Švenčionys Upland was identical to that of the Baltija Upland and absolutely different from the relief of Ašmena Upland' (Čepulytė 1958, p. 27). According to Čepulytė, the hills of Ašmena Upland formed at the margin of an active glacier.

Čepulytė was the first to point out the incorrect use of the term 'end moraines' (Čepulytė 1961). Thanks to her, the term resumed its original meaning, according to T. C. Chamberlin (1894). This issue claimed much time and effort. She criticized those who applied the term 'end moraines' to the entire marginal zone, and it took time to clarify the fact that the marginal zones included a variety of accumulative elements: terminal moraines, 'thrusted' hills, kames, glaciofluvial and glaciolimnic kames, eskers, and other forms. The term 'end moraines' was designed to name only *some* elements of relief and should be used appropriately. Čepulytė pointed out that imprecise application of the term was confusing because homogeneous or analogous forms could be named differently, and heterogeneous forms could be named similarly. Thanks to her efforts, the term 'end moraines' is now used more carefully and is often replaced by the terms 'edge moraines' or 'marginal zone', as appropriate.

The data obtained by researchers in the last decade have shown that Čepulytė was largely correct in her assumptions. Cosmogenic dating (^{10}Be) of a large number of boulders on the eastern coast of the Baltic Sea (between north Poland and south Finland) has shown that the greater part of the East Baltic region (from the marginal ridge of the so-called Middle Lithuanian Phase to the Salpauselkä marginal moraines in Finland) disengaged from the ice cover at almost the same time, at approximately 13–12.5 ka (Rinterknecht *et al.* 2006). This means that zonal (areal) melting of ice cover took place in the greater part of the Baltic region. Thus, Čepulytė was right, often repeating that 'the glaciers did not race around. [They] ... advanced and melted following a change of climate; but the process of melting was complicated' (Čepulytė pers. commun).

We have mentioned only a few of Čepulytė's assumptions about the Quaternary structure and palaeosurfaces, and their stages in Lithuania. There were other very interesting propositions in Čepulytė's work. But, unfortunately, not all of them can be objectively evaluated, confirmed or rejected. Subjectivity in Quaternary investigations will not be overcome until we have reliable criteria for the stratigraphic analysis of Pleistocene sections. Some subjectivity is, of course, also found in Čepulytė's work. Nevertheless, she made a major contribution to Lithuanian Quaternary studies through her strongly empirical approach, and shed considerable light on a field of study where previously much had been confused and obscure.

There are many more different aspects of Čepulytė's scientific work and civic life that could be discussed. It is owing to her complicated and long life's work that she left us, her colleagues and students, with such a rich heritage, to which we are greatly indebted.

We are grateful to Professors D. R. Oldroyd, R. H. Grapes and A. Grigelis for their comments and suggestions that clarified and improved the article. This study was carried out within the framework of the project 'Indications of climate change in the sediment records of last glacial-interglacial cycle (PALEOKLIMATAS)' and financed by the Lithuanian State Science and Studies Foundation (C-07008).

References

CHAMBERLIN, T. C. 1894. Proposed genetic classification of Pleistocene glacial formations. *Journal of Geology*, **2**, 517–538.

ČEPULYTĖ, V. 1946. *Lietuvos TSR moreninių priemolių stratigrafiniai tipai (Stratigraphic types of the morainic tills of the Lithuanian SSR)*. PhD thesis, Vilnius, Geologijos ir Geografijos Institutas.

ČEPULYTĖ, V. 1955. Pagrindiniai Lietuvos TSR moreninių priemolių horizontai (Basic horizons of the morainic tills in Lithuania SSR). *Lietuvos TSR Mokslų Akademijos Geologijos ir geografijos institutas, Moksliniai pranešimai, Geologija, geografija, Vilnius*, **1**, 139–156.

ČEPULYTĖ, V. 1956. Nekotoryie dannyie o dochetvertichnom reliefe Litovskoi SSR (Some data about the pre-Quaternary relief in Lithuania). *Trudy Akademii Nauk Litovskoi SSR, Serija B*, **2**, 61–72 (in Russian).

ČEPULYTĖ, V. 1957. Geomorfologicheskaya karta Litovskoi SSR (Geomorphological map of the Lithuanian SSR). *Lietuvos TSR Mokslų Akademijos Geologijos ir geografijos institutas, Moksliniai pranešimai, Kvartero geologija ir geomorfologija*, **4**, 265–278 (in Russian).

ČEPULYTĖ, V. 1958. Lietuvos geomorfologiniai rajonai ir jų geologinė raida (The geomorphological districts of the Lithuanian SSR and their geological development). *Lietuvos TSR Mokslų Akademijos Geologijos ir geografijos institutas, Moksliniai pranešimai, Geologija, geografija*, **6**, 23–53.

ČEPULYTĖ, V. 1959. Lietuvos apatinio ir vidurinio pleistoceno bei subkvarterinio reljefo bendrieji bruožai (General features of the Lower and Middle Pleistocene and sub-Quaternary relief in Lithuania). *Lietuvos TSR Mokslų Akademijos Geologijos ir geografijos institutas, Moksliniai pranešimai, Geologija, geografija*, **10**, 81–92.

ČEPULYTĖ, V. 1961. O razlichnom ponimanii termina 'konechnaya morena' (About the different understanding of the term 'marginal tills'). *Trudy Akademii Nauk Litovskoi SSR, Serija B*, **1**, 195–206 (in Russian).

ČEPULYTĖ, V. 1962a. Ledyninės akumuliacijos formų pasiskirstymas Pomeranijos–Baltijos stadijos ruože. *Geografinis metraštis, Vilnius*, **5**, 339–353 (in Lithuanian).

ČEPULYTĖ, V. 1962b. Osnovnyje polosy marginalnych obrazovanij na territorii Litvy i prilegaiuschih rajonov (The main ridges of the marginal formations in Lithuania and surrounding areas). *Trudy Akademii Nauk Litovskoi SSR, Serija B*, **1**, 189–197 (in Russian).

ČEPULYTĖ, V. 1965. Dochetvertichnaja poverchnost' i zakonomernosti raspredeleniya stratigraficheskih gorizontov morennych suglinkov jugo-vostochnoy Litvy (Sub-Quaternary surface and peculiarities of distribution of horizons of moraine tills in southeastern Lithuania). *Stratigrafiya chetvertichnyh otlozeniy i paleogeografiya antropogena jugo-vostochnoy Litvy, Trudi instituta Geologi, vipusk*, **2**, 7–38 (in Russian).

ČEPULYTĖ, V. 1967. K voprosu stratifikacii morennyh suglinkov (About the stratification of the morainic till). *Voprosy geologii i paleogeografii chetvertichnogo perioda Litvy, Trudi instituta Geologi, vipusk*, **5**, 15–39 (in Russian).

ČEPULYTĖ, V. 1968. *Stroeniya, stratigrafiya i paleogeomorfologiya poverhnosti pleistocena Litvy (Composition, stratigraphy and Pleistocene paleogeomorphological surfaces in Lithuania)*. Manuscript of Doctor Habilitation Degree, Vilnius, Geologijos institutas (in Russian).

ČEPULYTĖ, V. 1971. Dochetvertichnaya poverchnost' i raspredeleniya moren niznepleistocenovogo (mindel'skogo) oledeneniya (Pre-Quaternary surface and distribution of the till formations of the Lower Pleistocene (Mindel) Glaciation). *Stroieniya, litologiya i stratigrafiya otlozeniy niznego pleistocena Litvy, Trudi instituta Geologi, vipusk*, **14**, 11–34 (in Russian).

ČEPULYTĖ, V. 1988. Kalvoto marginalinio reljefo genezės klausimu (On the genesis of hilly marginal topography). *Geografinis metraštis*, **24**, 43–48.

DALINKEVIČIUS, J. 1930. Lietuvos ir jos pakraščių pagrindinis (podiliuvinis) reljefas (su žemėlapiu). Lietuvos geologijos keletas morfologijos bruožų (Basic (pre-Quaternary) relief (with the map) of the Lithuania and neighbouring areas. Some morphological aspects of the Lithuania geology). *Kosmos*, **10**, 145–154 (in Lithuanian).

LEWIŃSKI, J. & SAMSONOWICZ, J. 1918. Uksztaltowanie powierzchni, sklad I struktura podloźa dyluwium wschodniej części Niź Polnocno-Europejskiego. *Prace Towarzystwa Naukowego Warszawskiego*, **31**, 172 (in Polish).

RINTERKNECHT, V. R., CLARK, P. U. *ET AL.* 2006. The last deglaciation of the southeastern sector of the Scandinavian ice sheet. *Science*, **311**, 1449–1452.

Early ideas about erratic boulders and glacial phenomena in The Netherlands

FREDERIK R. VAN VEEN

Department of Technical Earth Sciences, Technical University of Delft, The Netherlands
(e-mail: f.r.vanveen@gmail.com)

Abstract: The development of ideas about the origin of erratic boulders in the northern Netherlands is reviewed for the period from 1770 to 1907. A Scandinavian origin of these rocks was recognized at an early stage, but the transport mechanism was not understood. Initially, the Biblical Flood was proposed as a geological agent by Horace de Saussure (1740–1799) in 1780. Charles Lyell (1797–1875) developed a theory of climate change and a 'glacial drift theory' to account for the movement of large boulders in the Alps, and he introduced the term 'drift' in 1840. Several prize contests of the two Dutch Scientific Societies, the Hollandsche Maatschappij der Wetenschappen and the Teyler Genootschap, both at Haarlem, concerned erratics. The competitions of 1827 and 1828 were won by Johann Hausmann (1782–1849) from Göttingen University and Reinhard Bernhardi (1797–1849) from the Forstakademie Hitzacker, respectively. Hausmann assumed that a great freshwater flood, caused by the breakthrough of natural dams in the Scandinavian mountains, swept boulders to the plains of the northern Netherlands. Bernhardi vaguely suggested the possibility of transport by glaciers. The prize for the third contest (1861) was awarded in 1868 to the Swedish geologist Otto Torell (1828–1900). He invoked the land-ice theory, which, as regards The Netherlands, proposed that the boulders had been transported by glaciers descending from the Bothnian Gulf and extending into the northern Netherlands, amongst other areas. However, for reasons unknown, Torell's manuscript was never printed, and he never collected his gold medal and the prize money. At a historic meeting of the Deutsche Geologische Gesellschaft at Berlin in 1875, 7 years after winning the Haarlem contest, Torell managed to convince his audience of the land-ice theory after showing striated rock surfaces at a well-known outcrop at Rüdersdorf near Berlin. Thus, it took about a century from the first speculations in the late eighteenth century about the origin and transport of erratic rocks to about 1880 before the land-ice theory became generally accepted in continental NW Europe.

Early observations

The first academic discussion in the Netherlands on erratic rocks dates from 1771 when the town physicist and rector of Groningen University, Wouter van Doeveren (1730–1783), mentioned his collection of *Hondsrug* (*Hondsrug* = 'hound's ridge' or 'dog's back') rocks in a speech on the occasion of handing over his rectorship to Anthonius Brugmans (van Doeveren 1771). He also announced his intention to write a book on the subject. The Hondsrug is a 20 m-high and 70 km-long sandy ridge, which stretches SSE of Groningen. It is now interpreted as an 'ice-pushed ridge' (*stuwwal*) at the edge of a Saalian till-plane, bordered by an ice-marginal valley. It is the site of many *hunebeds* (megalithic graves) and boulders that were popularly believed to have grown locally in the earth or had fallen from heaven as *Cerauniae* (thunderstones). The ridge (see Fig. 1) remained at the centre of the discussions on erratic rocks in Holland. The book promised by van Doeveren cannot be traced (assuming it was ever published), but the son of Rector Brugmans, Sebaldo Justini, wrote a doctoral thesis on van Doeveren's collection (Brugmans 1781). He was aware of the Scandinavian origin of the Hondsrug erratics and mentioned in passing that no precious stones were found. The means of transport across the Baltic seas was, however, not addressed. Brugmans described the stones according to the classification of the Swedish naturalist Johann Gottschalk Wallerius (1709–1785), who had stated: '[i]f we rightly understand the words of the sagacious Man of God Moses, the greatest *physicus* of all times, then we realise all that geologists want to know about the origin of the world and specially of our planet' (Wallerius 1776; quoted by Bemhardi 1832*a*).

From Genesis to drift

In 1778 the Swiss geologist Jean De Luc (1727–1817) visited Groningen (De Luc 1778/1779). He was impressed by the many large boulders in the fields and believed that their 'mother rock' had been destroyed by a great flood, which had

From: GRAPES, R. H., OLDROYD, D. & GRIGELIS, A. (eds) *History of Geomorphology and Quaternary Geology.*
Geological Society, London, Special Publications, **301**, 159–169.
DOI: 10.1144/SP301.11   0305-8719/08/$15.00 © The Geological Society of London 2008.

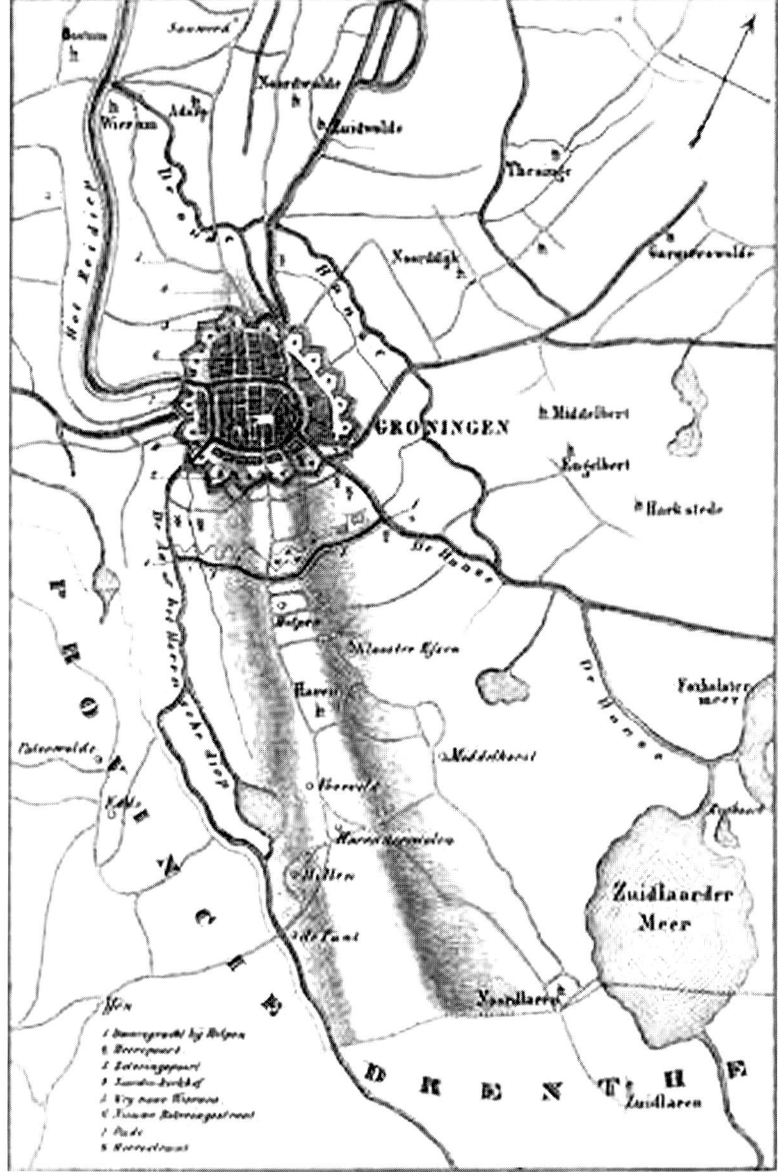

Fig. 1. Map of Hondsrug, from Cohen (1842).

transported the stones and covered them by sand. He 'fruitlessly climbed the Martini tower to see whether he could find in the surroundings any trace of the mother [source] mountains' (Acker Stratingh & Westerhoff 1839). Later, he proposed that violent explosions had hurled the rocks through the air and that the blocks had come sliding across frozen ground. In 1791 De Luc became a corresponding foreign member of the Hollandsche Maatschappij der Wetenschappen. In the eighteenth century, leading persons in civil society in The Netherlands became aware of the importance of the natural sciences. To encourage studies in these fields, private individuals – some of them wealthy – established scientific societies such as the Hollandsche Maatschappij der Wetenschappen (Holland Society of Sciences), founded in Haarlem in 1752. Another one was

'Teyler's Second Society', also founded in Haarlem (1779) 'to promote all the natural sciences'. The publication of scientific treatises and contests with cash prizes or gold medals were considered as means of achieving this goal. The Teyler Society still regularly holds prize contests. An 'offspring' of the Teyler Foundation is the still prominent Teyler's Museum in Haarlem. (Pieter Teyler van der Hulst (1702–1778) was a wealthy cloth merchant and Amsterdam banker of Scottish descent, who bequeathed his fortune for the advancement of art and science.)

During the period 1780–1795 a total of 11 scientific papers were published on the Hondsrug rocks. But no papers are known from 1795 to 1813, the period of the French-inspired Batavian Republic, when Stadtholder William V and several 'Orange' professors went into exile. One of them, the 'universalist' and poet Willem Bilderdijk (1756–1831), moved to Brunswick, where he shared a house with De Luc. On his return, Bilderdijk wrote the first book on geology in the Dutch language (Bilderdijk 1813). He was obviously influenced by De Luc, who now held the Mosaic Flood, which supposedly swept over the Earth, responsible for the erratic boulders.

The catastrophic influence of the Flood had already preoccupied Johannes Silberschlag (1716–1791), also a member of the Holland Society. Silberschlag wrote a three-volume treatise on *Geognosy and the Mosaic Flood* (see Fig. 2). In his description of the Earth, we find the huge subterranean caverns invoked by the ancient cosmogonists and sanctioned by later writers such as Kircher and Leibniz. The ocean could supposedly have drained into these caverns and then been restored when elastic vapours pressed the waters out again. Silberschlag constructed an ingenious machine, which by applying various pressure changes could supposedly simulate the Deluge (see Fig. 3). He was a mining engineer and calculated the pressures required to break up rocks and 'open the earth' to allow the underground waters to flood the Earth. He calculated the volume of subterranean water as 27 409 185 cubic miles and concluded that this volume would be sufficient to flood the entire globe to a level exceeding the height of Mount Ararat. An interesting 'Location Map of Mount Ararat and Paradise according to the Most Recent Accounts of Journeys' indicated the route of Noah's children to Paradise! (Silberschlag 1780) (see Fig. 4).

The English geologist and theologian William Buckland (1784–1856) was the first professor of geology (strictly a 'reader') at Oxford. In his famous book, *Reliquiae diluvianae*, he also preached the idea of the Great Flood as a geological agent to explain the presence of erratic blocks in

Fig. 2. Title page of J. E. Silberschlag's *Geogenie oder Erklärung der Mosaischen Erderschaffung* (1780).

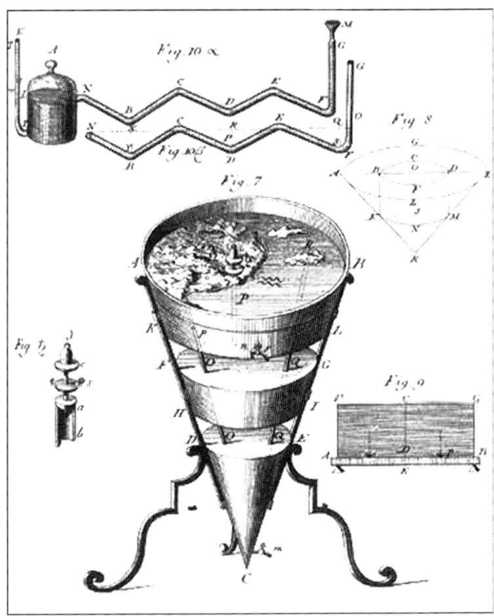

Fig. 3. Figures 7, 8, 9 and 10 from Silberschlag (1780). Apparatus to experiment with subterranean waters.

Fig. 4. The route to Paradise, according to Silberschlag (1780).

Britain, but without the help of experiments (Buckland 1823). Such ideas persisted well into the nineteenth century and his book was translated into German as late as 1877. However, Buckland's opponents joked: 'All was darkness once about the Flood, till Buckland rose and made it clear like mud' (Faber 1949). As used by Buckland (1819), the term *Diluvium*, Latin for flood, became widely accepted, and was distinguished from 'alluvium' – the kind of material deposited by river floods today. In The Netherlands, Buckland's term was first used by a schoolteacher Remmert Venhuis in a Dutch textbook on natural history in 1827 (Venhuis 1829). Meanwhile, mineralogical studies of the Hondsrug rocks, including speculations about their transport from a Scandinavian source, continued at the University of Leiden. Hendrik Karel van der Boon Mesch (1795–1831) proposed in his dissertation that icebergs could have contributed to the transport of Scandinavian rocks, based on observations of the sailor Lucas Fox in 1631 in Hudson Bay of icebergs carrying enormous boulders, estimated to weigh '120 *centenaars*' (van der Boon Mesch 1820). The idea was not wholly original as the German geologist Erhard Wrede (1766–1826) was the first to argue that the transport of large rock masses, such as are scattered on the north German plains, was possible if large icebergs and a sufficiently high sea level were assumed (Wrede 1794).

In the third part of his *Principles of Geology*, Charles Lyell (1797–1875), a pupil of Buckland, presented his glacial drift theory (Lyell 1833) and later formally introduced the term 'drift' as a replacement for Buckland's 'diluvium' (Lyell 1840). Winand Staring (1808–1877) adopted the drift theory in his standard work on the geology of The Netherlands (Staring 1856). He also issued the first geological map of The Netherlands, on which he distinguished two different 'Diluvial' formations, one of Scandinavian origin, the other with a southern source, supposedly transported by the rivers Rhine and Meuse. His map was honoured at the 1862 World Exhibition in London, with a gold medal for the first detailed subdivision of the Diluvium (but see Klemun 2008 in this volume).

Land ice

In 1828, Teyler's Second Society announced an essay competition on geology, proposed by its secretary, the medical doctor and botanist Marinus van Marum (1750–1837). (The polymathic van Marum carried out researches on chemistry, mineralogy and electricity, and also lectured on geology.) The lengthy instructions for the competition started: 'Write a treatise in which is to be presented fully, clearly, and briefly what is now known about geology ... with a detailed description in eighteen points of the subjects to be dealt with' (Forbes 1971). One of the questions asked about the transport of erratic blocks to the northern Netherlands.

Reinhard Bernhardi (1797–1849), originally a theologian and later professor of natural history at the forestry academy of Dreiszigacker near Meiningen, submitted a manuscript in German for the prize, consisting of 353 pages. As a motto for the essay he quoted the Swedish chemist Berzelius (1779–1848): 'Es ist aber einmal unser Loos Allemal auf Unbegreifliches zu stossen, sobald wir uns bemühen alles verstehen zu wollen' ('It is alas our lot that as soon as we try to understand everything, we will be faced with the incomprehensible'). Under Point 17, he proposed that the erratics in northern Holland, Germany and Poland could have been transported by glaciers from Scandinavia. This casual remark apparently went unnoticed, as no reference to it in the Dutch contemporary literature has been found. The time was seemingly not yet ripe for such views. Nevertheless, the essay was awarded the Gold Prize and was published in the Society's *Transactions* (Bernhardi 1832*a*, *b*).

An early suggestion about climate change in the Alps was made by the Swiss engineer Ignaz Venetz (1788–1859) (1822) and about land-ice transport in the Alpine region by 'ungeheure Ausdehnungen der Gletscher in früher Zeit' ('tremendous expansion of the glaciers in former times') (Venetz 1833). In this context, it is interesting that, on 8 January 1820, Goethe (1749–1832) sent a letter to his Grand Duke Carl August von Sachsen-Weimar, informing him of a 'Provisional Communication about the expansion of glaciers in the Swiss Cantons'. He referred to a prize contest issued in 1817 with the question: '[o]b sich früher größere Gletscher durch vorgeschobene Felstrümmer verraten würden' ('whether in former times larger glaciers could be revealed by pushed-forward heaps of stones'). The prize was won by Venetz in 1821 (Seibold & Seibold 2003; see also von Engelhardt 1989). Arthur Wichmann (1851–1927), professor of geology at Utrecht (Wichmann 1914), and Wolf von Engelhardt (1999) have drawn attention to another early mention by Goethe of the new land-ice theory in his novel of 1829, *Wilhelm Meisters Wanderjahre*, where he expressed the idea of rock masses being transported by glaciers into northern Germany. However, after visiting the 10.5 m-high erratic *Grosze Markgrafenstein*, Goethe is said to have remarked: 'Mir macht man aber nicht weis, dasz der Markgrafenstein bei Fürstenwalde weit hergekommen sei; an Ort und Stelle ist er liegen geblieben, als Reste grosser in sich selbst zerfallender Felsmassen' ('I cannot believe that the *Markgrafenstein* could have come from far away. It has remained on the spot and is a remnant of a fallen rock mass') (quoted by Wahnschaffe 1910 and Faber 1949). Venetz's early prize essay has not been located, but it would appear that he had the idea of a former extension of the Alpine glaciers at least as early as 1821.

As is well known, ideas of transport by glaciers during a colder climate were later worked out by Louis Agassiz (1807–1873) for the Alps (Agassiz 1840) and by Otto Torell for Scandinavia and northern Europe (Frängsmyr 1985). From the 1830s onwards, discussions on the transport of erratics increased, but initially the land-ice theory was hardly taken seriously by the geologists of The Netherlands and north Germany. An account of ice coverage of The Netherlands was only first published in an article by Hermanus Hartogh Heys van Zouteveen (1841–1891) as late as 1881 (Hartogh Heys van Zouteveen 1881).

However, in the period 1813–1885, 30 papers were published on Netherlands Diluvium. An interesting paper by Ali Cohen (1817–1889) on sedimentary structures of the Hondsrug presented the first detailed description of the supposed Diluvial sedimentary sequences, accompanied by accurate drawings of geological profiles taken in a canal excavation (Cohen 1842) (see Fig. 5). Cohen observed polished and striated rock surfaces,

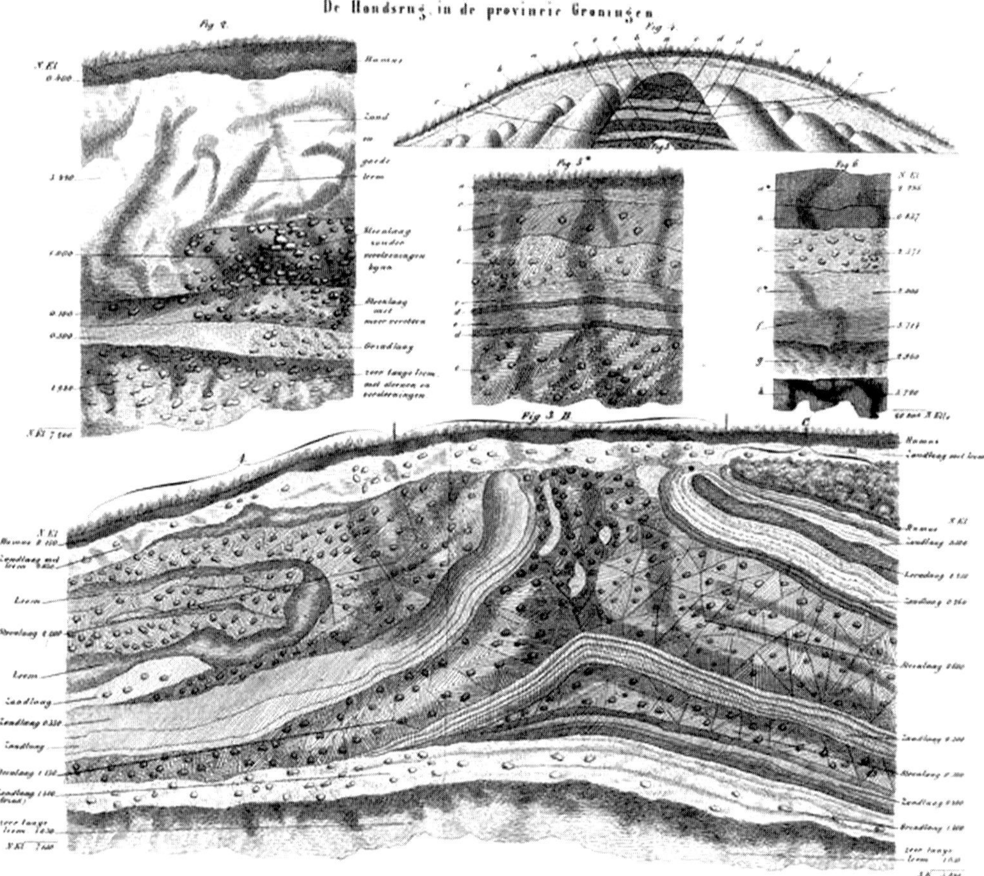

Fig. 5. Sections of the Hondsrug by Cohen (1842).

and also contorted structures that are now interpreted as a result of glacial thrusting. Probably a lack of familiarity with glacial features was the reason why he did not link his observations with a glacial covering.

After 1885, the main interest shifted to the study of the fossiliferous calcareous rocks, the Hondsrug being well known for its abundance of fossil corals, and their source and the presumed direction of transporting ice flows were investigated. Hagen Jonker (1875–1917), palaeontologist at Groningen and Delft universities, became the authority on the determination of these fossils and confirmed Römer's conclusion (Römer 1857, 1858) (see below) that the majority of them derived from the Silurian of Estonia, Gotland and Oesel (now Saaremaa) (Jonker 1905a). In 1904 a heated debate on the direction of ice flows flared up between Jonker and Eugène Dubois (1858–1940), the discoverer of the *Pithecanthropus erectus* (or *Homo erectus*) on Java, who had become

professor of geology and palaeontology at Amsterdam: did they flow parallel to, or at right-angles to, the Hondsrug? Dubois believed in the former existence of a Scottish ice flow, which diverted the Baltic flow along the Hondsrug ridge (Dubois 1905). Jonker (1905b) judged such a deviation to be impossible. These controversies were unfortunately accompanied by a slander campaign, even with accusations of insanity! (Correspondence Dubois–Jonker, 1904: Dubois Archive, Nationaal Natuurhistorisch Museum, Leiden.) It may be remarked that controversy on this matter still continues today.

More prize contests: Hausmann and Torell

Because the geological prize contests organized by the Hollandsche Maatschappij der Wetenschappen played an active role in the erratic boulder debate and influenced the gradual change from Lyell's drift theory to the glacial land-ice theory in The Netherlands, two important contests of the total 79 geological questions in the period 1828–1864 will be discussed in some detail. Sixty-nine of the questions were posed by Jacob van Breda (1788–1867), professor of natural philosophy successively at Franeker, Gand and Leiden universities. He was a teacher of Winand Carel Hugo Staring (1808–1877), and later became secretary of the Haarlem Society and director of Teyler's Museum (De Bruijn 1977; Breure & De Bruijn 1979), and came to be regarded as the founding father of Dutch geological research and mapping.

The 1827–1828 contest asked for suggestions about the origin of the granite and other 'primitive' boulders that are scattered over The Netherlands and north German plains. It was won by Johann Hausmann (1782–1859), professor of mineralogy at the University of Göttingen (Hausmann 1831). Hausmann discussed six different *Gefühle* ('feelings' or 'sentiments') regarding the origin and transportation of the erratic rocks, rejecting (and refuting) the old ideas of local origin, extraterrestrial origin, extrusion by volcanic eruption or the suggestion that all erratics south of the Baltic Sea derived from the Sudeten and Carpathian mountains, as envisaged by Meierotto (1790). Hausmann suggested that the erratics were transported by a huge freshwater flood, which had broken through natural obstructions ('walls') in the Scandinavian mountains. Van Breda translated and commented extensively upon the 100-page octavo contribution, which was written in German. There was no mention of ice ages and glaciers. In his comments, Van Breda rejected the ideas of Cuvier and De Luc that the city of Groningen was built on Tertiary limestone. Rather, he thought it was built on diluvial loam with limestone erratics, and he pointed out that the Dutch Diluvium had two sources: northern Scandinavian and southern Meuse deposits (Breure & De Bruijn 1979). These observations were later worked out by his pupil Staring. Thereafter, serious research on erratics in Holland got underway and large collections of rocks were initiated. Between 1852 and 1864 Staring collected 14 948 geological objects, which were displayed to the public at Haarlem.

The prize competition of 1841 asked: 'What is to be thought of the discovery made by Agassiz, and confirmed by Lyell and Buckland, that glacial moraines occur in northern Europe far from present glaciers?' The contest of 1843 wanted to know: 'What traces of glaciers are present in the Netherlands Diluvium?' But both questions remained unanswered (De Bruijn 1977).

Another important geological contest followed in 1861, asking for a comparison of Silurian boulders from the Groningen Hondsrug with Silurian formations in SW Norway. Staring had meanwhile entrusted his extensive collection of fossiliferous carbonate erratics for identification by Carl Ferdinand Römer (1818–1891), the famous palaeontologist from the University of Breslau. Römer had concluded that they originated from Gotland, Oesel and Estonia (Römer 1857, 1858), which was contrary to a Norwegian source favoured by Dutch scholars.

The Swedish professor of geology and zoology at Lund, Otto Torell (1828–1900) (Fig. 6), was the only one who entered the contest in 1866, and even then only after the submission date had twice been extended. In 1850 Torell had applied Agassiz's glacial theory to explain his own observation of arctic molluscs on the west coast of Sweden, 'but as he saw all the serious objections to this theory by the leading geologists, he resolved to spend some years pursuing his own research to make sure he was correct' (Frängsmyr 1985). From 1856 to 1859 Torell visited Switzerland, Iceland, Spitzbergen and Greenland to study glaciers. In 1861 he headed the first Swedish expedition to the Polar Sea and in 1865 he visited Holland to investigate the erratic rock collections

Fig. 6. Photograph of Otto Torell, from Holmström (1901).

of Hondsrug fossils, and accepted Römer's conclusions about a source for them in the eastern Baltic. In 1867 Torell submitted a separate second part of his entry, in which he made a strong plea or argument for transport by glaciers by land ice. The gold medal, worth 400 Dutch guilders, and prize money of 150 guilders for the contest were awarded to Torell in 1868.

At a meeting of the Royal Netherlands Academy of Sciences at Amsterdam on 28 September 1867, Staring, who was a member of the Holland Society's competition jury, had announced the 'forthcoming publication of an important study by a Swedish scholar in which a totally new light was [to be] shone on the origin of the Scandinavian Diluvium'. Remarkably, however, Torell's manuscript, of more than 300 pages (written in French), was never printed. Because the correspondence between the Haarlem society and Torell is incomplete, we can only speculate about Torell's reasons for not responding to repeated requests by the society's secretary to edit and return his manuscript for publication, and to finalize the stipulated study of Staring's and Cohen's entire rock collections. In the autumn of 1868 the secretary wrote to Torell that he had waited '*tout l'été*' for his arrival and again courteously invited him to come to Haarlem to collect his prize; but again there was no reaction. (Archives Hollandsche Maatschappij van Wetenschappen. HMW inv. 163, 461. Rijksarchief in Noord-Holland, Haarlem.) The manuscript finally arrived only after Torell's death in 1900, by way of the Swedish Geological Survey of which Torell had become Director in 1870. It contained Torell's corrections and also marginal comments in French and Dutch by the jury members Staring, Joseph Bosquet and Van Breda (Oele 2001). The prize was then posthumously presented to Torell's widow.

Could Torell's silence be explained by the request of the jury to complete the study of the large collections of more than 600 erratic specimen at the State Museum of Natural History at Leiden – to be included in his final manuscript? He may have regarded this extra study as irrelevant to his land-ice proposal. A comment in the *Inventory of Prize Essays of the Hollandsche Maatschappij* may hint at this explanation: 'Two advisors regretted that the author did not make use of the rich collection of the Netherlands Diluvium, which is kept at the Leiden museum and which could have thrown much light on the Groningen Diluvium'. In 1868 the essay was evaluated again, this time by a jury now without Van Breda, who had died in the interim. A commentator has considered it curious that the prize was awarded with conditions attached and concluded: '[c]aused by this rare example of censoriousness the Society missed the chance of priority of an epoch-making publication'

(De Bruijn 1977). Frängsmyr (1985) gives another explanation: 'Torell won the prize but he asked to have his manuscript back to complete some details before publication. Unfortunately, due to his undisciplined character, this manuscript was never published'.

In August 1869 Torell was invited by the British Association for the Advancement of Science to form a committee together with the British professors Andrew Ramsey and Hilary Bauerman, to report on 'Ice as an Agent of Geological Change' (Holmström 1901). The committee proposed an 'Investigation on the amount of waste suffered due to glacier action' and recommended national scientific co-operation (Report of the 29th Meeting of the British Association for the Advancement of Science held at Exeter, August 1869. *Geological Magazine*, **7**, 175–178, 1870). This study may also have deflected Torell from completing his study of the Haarlem rock collection, although as stated in the Introduction to the seventh volume of the *Geological Magazine*, the 1870 war between Germany and France was thought to be likely to interfere with the fieldwork required for this major project. Torell eventually published his land-ice theory in the *Proceedings of the Swedish Academy* at Stockholm (Torell 1872, 1873, 1887), written in Swedish. But by then it apparently passed unnoticed by most international scholars. This may have been owing to a 'language problem' but by the 1870s the land-ice theory, promulgated by Ramsay and others in Britain in the 1860s, was no longer a novel topic.

Thus, James Geikie (1839–1915) of the Scottish Geological Survey had become convinced of an Ice Age and his book, *The Great Ice Age* (Geikie 1874), was well received in Britain and internationally. Geikie still used the term 'drift deposits' for the Scottish tills (the Scottish term 'till' is still used today), but in fact he envisaged an ice cover for Britain and Ireland like that suggested by Agassiz for the Alps. His ideas involved global climate changes, possibly caused by slow changes of the shape of the Earth's orbit – an idea based on the work of James Croll (1821–1890), Geikie's colleague at the Scottish Survey. Geikie also recognized a cyclical sequence of the 'Diluvium' of cold glacial and warm interglacial periods, and by 1893 he was suggesting no less than five distinct glacial epochs (Geikie 1893, pp. 319–321), and also mentioned Torell's paper of 1875, referred to below (Geikie 1893, p. 223). Also, in the following year the third, 'largely rewritten', edition of *The Great Ice Age* (Geikie 1894) again mentioned Torell. In this edition, Geikie published maps of the ice cover and flow directions of glaciers in Britain, and the extent of his three different ice ages over the whole of northern Europe.

Fig. 7. Glacial striae at Rüdersdorf, from Holmström (1901).

In the event, Torell delivered his famous lecture on the land-ice theory at a meeting of the Deutsche Geologische Gesellschaft in Berlin on 3 November 1875 (Protokoll der November Sitzung 1875. *Zeitschrift der deutschen Geologischen Gesellschaft*, **27**, 961–962), 7 years after he had submitted his manuscript to the Haarlem Society. The audience was apparently not very receptive initially until Torell referred to the beautifully exposed parallel glacial striae in a well-known Muschelkalk outcrop at Rüdersdorf near Berlin (see Fig. 7). These phenomena had previously been observed as far back as 1836, but their true significance was now pointed out by Torell (Holmström 1901), and he is honoured accordingly (see Fig. 8). Wichmann who attended the Berlin meeting, remembered that 'not everyone was convinced, but at least the ice was broken' (Wichmann 1914). This was another indication that the geologists of northwestern Continental Europe were lagging behind their colleagues elsewhere. The Dutch geology student Jan Lorié (1852–1924), who at the time was studying in Germany where he became a close friend of the glacial geologist Albrecht Penck (1858–1945), considered that 'with the publication of Torell's award-winning manuscript the land-ice victory would have begun at Haarlem' (Lorié 1907). Lorié, later a lecturer at Utrecht, became an authority on the stratigraphy of the Dutch Pleistocene, with many publications to his name. In fact, Geikie (1893, p. 223) stated that a 'great

Fig. 8. Torell Monument at Rüdersdorf, from Holmström (1901).

literature' on the land-ice theory sprang up after Torell's Berlin address.

In 1904 Hagen Jonker studied Torell's rediscovered manuscript (Jonker 1905*a*). He quoted Torell's important passage 'would it be absurd to suppose that a glacial cover similar to the Greenland icecap has covered northern Europe during a period when the marine fauna of Spitzberg lived between 50° and 60° latitude, when *Betula nana* [the cold climate dwarf birch] occurred in Devonshire and reindeer lived in southern France?'. Jonker returned the manuscript to the Holland Society at Haarlem, where it afterwards mysteriously disappeared again (Oele 2001).

The Norwegian geologist Amund Helland (1846–1918) visited Groningen in 1878, exactly 100 years after De Luc's visit, and found glacially striated rocks on the Hondsrug, thus finally settling the Dutch debate on the origin and transport of the boulders in favour of the land-ice theory (Helland 1879). The theory was officially introduced by Friedrich Van Calker (1841–1913) on 17 November 1881 at the Groningen Physical Society during a lecture on the Hondsrug (Van Calker 1884), again exactly 100 years after Brugman's first publication on the enigmatic Hondsrug rocks.

In the period 1885–1906 the number of publications in the Netherlands on the glacial geology of the Diluvium, now called Pleistocene, increased to 37, mainly by Jonker and Lorié (Jonker 1907). Their studies concentrated on the identification of fossils in the Scandinavian erratics and on the subdivision of the Pleistocene.

I am most grateful to Dr S. K. Donovan (Nationaal Natuurhistorisch Museum, Leiden) for a detailed review on an earlier draft of this paper. Ms W. Cawthorne (Geological Society of London), M. van Hoorn (Teylers Museum, Haarlem) and Frank Rietz (Otto von Guericke Universität, Magdeburg) have kindly supplied bibliographical assistance. D. R. Oldroyd has also assisted with editing the paper and made a number of useful suggestions.

References

ACKER STRATINGH, G. & WESTERHOFF, R. 1839. *Natuurlijke historie der Provincie Groningen.* Oomkens, Groningen.

AGASSIZ, L. 1840. *Études sur les glaciers.* Soleure, Neuchâtel.

BERNHARDI, R. 1832*a*. Ueber das, was man von der Geologie jetzt (1830) weiss. *Verhandelingen Teylers Tweede genootschap,* **21**.

BERNHARDI, R. 1832*b*. Wie kamen die aus dem Norden stammenden Felsbruchstücke und Geschiebe, welche man in Norddeutschland und den benachbarten Ländern findet, an ihre gegenwärtigen Fundorte? *Jahrbuch für Mineralogie, Geogenie, Geologie und Petrefactenkunde,* **3**, 257–267.

BILDERDIJK, W. 1813. *Geologie of Verhandeling over de vorming en vervorming der Aarde* [*Geology or Discussion Concerning the Formation and Defirmation of the Earth*]. Wybe Wouters, Groningen.

BOON MESCH, H. C. & VAN DER., 1820. *Disputatio geologica inauguralis de granite.* Hazenberg Jr, Leiden.

BREURE, A. S. H. & DE BRUIJN, J. G. 1979. *Leven en Werken J. G. S. van Breda (1788–1867).* Tjeenk Willink, Groningen.

BRUGMANS, S. J. 1781. *Dissertatio physico mineralogica inauguralis de lapidibus et saxis agri Groningani (juxta ordinem Wallerii digesta, cum synonimis aliorum, inprimis Linnaei et Cronstedii).* Doekema & Muller, Groningen.

BUCKLAND, W. 1819. Description of the Quartz Rock of the Lickey Hill in Worcestershire, with considerations on the evidence of a recent Deluge. *Transactions of the Geological Society, London,* **5**, 516–544.

BUCKLAND, W. 1823. *Reliquiæ diluvianæ: or, Observations on the Organic Remains Contained in Caves, Fissures, and Diluvial Gravel and on other Geological Phenomena Attesting the Action of a Universal Deluge.* John Murray, London.

COHEN, L. A. 1842. Geognostische beschrijving van den *Hondsrug* (Geognostic description of the *Hondsrug*). *Tijdschrift voor Natuurlijke Geschiedenis en Physiologie,* **9**, 17–67, 268–295.

DE BRUIJN, J. G. 1977. *Inventaris van de prijsvragen uitgeschreven door de Hollandsche Maatschappij der Wetenschappen 1753–1917 (Inventory of the Prize Competitions set by the Holland Society of Sciences).* Tjeenk Willink, Groningen.

DE LUC, J. A. 1778/1879. *Lettres physiques et morales sur l'histoire de la terre et de l'homme,* **5**, 247–266, Lettre 128. De Thune, La Haye & Duchesne, Paris.

DOEVEREN, W. VAN 1771. *Academische Rede uit het Latijn vertaald door Matthias van Geuns (Farewell Lecture Translated from the Latin by Mattias van Geuns).* Barlinkhof Bolt, Groningen.

DUBOIS, E. 1905. De geographische en geologische betekenis van de *Hondsrug* en het onderzoek der zwerfsteenen in ons noordsch diluvium. Verslag van de gewone vergadering van 30 September 1905 (The geographical and geological significance of the *Hondsrug* for the investigation of the erratic rocks of our Nordic diluvium). *Verslagen Koninklijke Akademie van Wetenschappen, Wis-en Natuurkundige Afdeling,* **14**, 360–368.

ENGELHARDT, W. VON 1989. *Johann Wolfgang Goethe. Schriften zur Allgemeinen Naturlehre, Geologie und Mineralogie.* Deutscher Klassikerverlag, Frankurt.

ENGELHARDT, W. VON 1999. Did Goethe discover the Ice Age? *Eclogae geologicae Helveticae,* **92**, 123–128.

FABER, F. J. 1949. *Van Zondvloed tot Landijs (From Flood to Land Ice).* Thieme, Zutphen.

FORBES, R. J. (ed.). 1971. *Martinus van Marum. Life and Work,* Volume 3. Tjeenk Willink & Zoon, Haarlem.

FRÄNGSMYR, T. 1985. Otto Torell. Gestalter i Svensk Lärdomhistorie 2 (Swedes in the history of science and learning 2). *Lychnos,* 73–88.

GEIKIE, J. 1874. *The Great Ice Age and its Relation to the Antiquity of Man.* Edward Stanford, London.

GEIKIE, J. 1893. *Fragments of Earth Lore: Sketches & Addresses Geological and Geographical.* John Bartholomew & Co., Edinburgh; Simpkin, Marshall, Hamilton, Kent & Co., London.

GEIKIE, J. 1894. *The Great Ice Age and its Relation to the Antiquity of Man*, 3rd edn. Edward Stanford, London.

HARTOGH HEYS VAN ZOUTEVEEN, H. 1881. Het Diluvium van de nederlandsch–noordduitsche vlakte (The Diluvium of the Netherland–North German Plain). *Weekblad voor Natuurwetenschap Isis*, **10**, 97–112.

HAUSMANN, J. F. L. 1831. Verhandeling over de oorsprong der graniet en andere primitieve rotsblokken die over de vlakten der Nederlanden en van het noordelijk Duitschland verspreid liggen (Vertaald uit het Hoogduitsch door J. G. S. van Breda) (Essay on the origin of granite and other Primitive boulders that are distributed on the plains of The Netherlands and North Germany (translated from German by J. G. S. Van Breda)). *Natuurkundige Verhandelingen Hollandsche Maatschappij der Wetenschappen te Haarlem*, **19**, 271–400.

HELLAND, A. 1879. Über die glacialen Bildungen der nord-europäischen Ebenen. *Zeitschrift der deutschen Geologischen Gesellschaft*, **31**, 63–106.

HOLMSTRÖM, L. 1901. Otto Torell, Minnesteckning (Otto Torell, obituary). *Geologiska Föreningens i Stockholm Förhandlingar*, **23**, 391–455.

JONKER, H. G. 1905*a*. Zwerfsteenen van den ouderdom der oostbaltische zone (The erratics of the eastern Baltic region). *Verslagen der Koninklijke Akademie van Wetenschappen te Amsterdam, Afdeling Wis- en Natuurkunde*, **13**, 550–552.

JONKER, H. G. 1905*b*. De stroomrichting van het diluviale landijs over Nederland (The direction of flow of the diluvial land ice of The Netherlands). *In: Handelingen van het Xde Natuur- en Geneeskundig Congres, gehouden te Arnhem op 27, 28 en 29 April 1905.* Bohn, Haarlem, 459–465.

JONKER, H. G. 1907. Lijst van geschriften welke handelen over de geologie van Nederland (Catalogue of the publications on Netherlands geology). *Verhandelingen Koninklijke Akademie van Wetenschappen*, **13**, 101–134.

KLEMUN, M. 2008. Questions of periodization and Adolphe von Morlot's contribution to the term and the concept 'Quaternär' (1854). *In:* GRAPES, R. H., OLDROYD, D. R. & GRIGELIS, A. (eds) *History of Geomorphology and Quaternary Geology.* Geological Society, London, Special Publications, **301**, 000–000.

LORIÉ, J. 1907. Het interglacialisme in Nederland (Interglacials in The Netherlands). *Tijdschrift Nederlandsch Aardrijkskundig Genootschap*, 2nd Ser., **24**, 406–448.

LYELL, C. 1833. *Principles of Geology, Being an Attempt to Explain the Former Changes of the Earth's Surface, by Reference to Causes now in Operation*, Volume 3. John Murray, London.

LYELL, C. 1840. On the Boulder Formation or Drift and associated freshwater deposits composing the mud cliffs of Eastern Norfolk. *Proceedings of the Geological Society, London*, **3**, 171–179.

MEIEROTTO, J. H. L. 1790. *Gedanken über die Entstehung der Baltischen Länder.* August Milius, Berlin.

OELE, E. 2001. Ontwikkelingen in de Kwartairgeologische opvattingen over ons land na Staring's proefschrift van 1833 [Developments in ideas concerning the Quaternary geology of our country [The Netherlands] post Staring's dissertation of 1833]. *Grondboor en Hamer*, **55**, 16–21.

RÖMER, F. 1857. Über Holländische Diluvialgeschiebe. *Neues Jahrbuch für Mineralogie*, 385–392.

RÖMER, F. 1858. Versteinerungen des Silurisches Diluvialgeschiebes von Groningen in Holland. *Neues Jahrbuch für Mineralogie*, 257–272.

SEIBOLD, E. & SEIBOLD, I. 2003. Erratische Blöcke – erratische Folgerungen: ein unbekannter Brief von Leopold von Buch von 1818. *International Journal of Earth Sciences (Geologische Rundschau)*, **92**, 426–439.

SILBERSCHLAG, J. E. 1780. *Geogenie oder Erklärung der Mosaischen Erderschaffung nach physikalischen und mathematischen Grundsätzen.* Verlag der Buchhandlung der Realschule, Berlin.

STARING, W. C. H. 1856. *De Bodem van Nederland (The Soil of the Netherlands)*, Kruseman, Haarlem.

TORELL, O. 1872. No. 10. Undersökningar öfver istiden (Researches on the Ice Ages). *Öfversigt af Kungliga Vetenskapsakademiens förhandlingar*, 25–66 (Prize essay, Part 1).

TORELL, O. 1873 (No. 1) Undersökningar öfver istiden: Skandinaviska inlandsisens utsträckning under isperioden. *Öfversigt af Kungliga Vetenskapsakademiens förhandlingar*, 47–64 (Prize essay, Part 2).

TORELL, O. 1887. Undersökningar öfver istiden: Om glacialbildningarnns lagerföljd och temperaturen under isidens olika skeden. *Öfversigt af Kungliga Vetenskapsakademiens förhandlingar*, 429–438 (Prize essay, Part 3). (The pagination is continuous in this volume, and there were no numbered parts for 1887.)

VAN CALKER, F. J. P. 1884. Beitrage zur Kenntnis des Groninger Diluviums. *Zeitschrift der deutschen Geologischen Gesellschaft*, **36**, 726–727.

VENETZ, I. 1822. Sur les variations du climat dans les Alpes. *Bibliothèque universelles des sciences, belles-lettres, et arts, faisant suite à la Bibliothèque Britannique. Redigée à Genève par les auteur de ce dernier recueil*, **21**, 77–78. [This was actually a *report* of a paper written by Venetz, which he presented to the *Société Helvétique des Sciences Naturelles* in 1822 on a question posed by the Society; 'Sur les variations du climat dans les Alpes'. It was subsequently published in full in 1833.]

VENETZ, I. 1833. Mémoire sur les variations de la température dans les Alpes Suisses. *Denkschrift Schweizerischer Gesellschaft der Naturwissenschaften*, **1**, 1–38.

VENHUIS, R. A. 1829. *Eene Natuurlijke Historie der Provincie Groningen (A Natural History of the Province of Groningen).* Oomkens, Groningen.

WAHNSCHAFFE, F. 1910. Grosze erratische Blöcke im norddeutschen Flachlande. *Geologische Charakterbilder.* Heft, **2**.

WALLERIUS, J. G. 1776. *Tankar om verldenes, i synnerhet jordenes, danande och ändring; författade af Johan Gotsch. Wallerius.* Henr. Fougt, Stockholm.

WICHMANN, A. 1914. Aus den Kindheitstagen der Glazialtheorie. *Der Geologe*, **12**, 223–229.

WREDE, E. G. F. 1794. *Geologische Resultate aus Beobachtungen über einen Theil der Subbaltischen Staaten.* Rengersche Buchhandlung, Halle.

Planation surfaces in China: one hundred years of investigation

KE ZHANG

Department of Geosciences, Sun Yat-Sen University, Guangzhou, 510275, People's Republic of China (e-mail: eeszke@mail.sysu.edu.cn)

Abstract: By introduction of the erosional cycle theory of W. M. Davis, Bailey Willis, an American geologist, began pioneering work on planation surfaces in northern China between 1903 and 1904. He identified two surfaces. Subsequently, the study of planation surfaces in northern China was continued by both Western and Chinese scientists, and this stimulated similar research in other parts of China. However, in the 1930s and 1940s the study of planation surfaces all but stopped because of the Japanese invasion of China, the civil war that followed, and again during the period of the 'Cultural Revolution' between 1966 and 1976. In the early 1960s the cycle theory proposed by Davis was much criticized, and with it the concept of the planation surface. However, from the 1980s, after careful re-examination of former criticisms, geomorphologists realized that the early concepts did have value and began to restudy planation surfaces over the whole of China. The most interesting research has been on planation surfaces preserved in the area of the Tibetan Plateau, which relates to uplift of the plateau and its impact on the environment.

It is generally accepted that the end result of a long period of erosion under relatively stable tectonic conditions results in a surface of low relief, called a peneplain or pediplain, although it is understood that the end product of long-continued denudation in different climate regimes can result in different landscapes. Recognition of areas (relics) of low relief on hill and mountain tops as low gradient or summit surfaces has been fundamental in the development of the idea of a planation surface, i.e. uplifted and dissected peneplains. In China these planation surfaces are typically an ancient (Late Mesozoic–Tertiary) feature with the potential to provide a 'time base' from which late Tertiary, or Quaternary, development of the present landscape by denudation in relation to climate change and tectonic activity can be understood. In this paper, over 100 years of investigation on such planation surfaces in China is documented that not only reflects developments in scientific thinking but also changes in society.

The early years: Bailey Willis

After the introduction of the theory of the erosion cycle by W. M. Davis, an American geologist Bailey Willis (Fig. 1) began pioneering work on planation surfaces in northern China in 1903–1904 (Willis *et al.* 1907). He was sent by the US Geological Survey to China with specific instructions to examine Cambrian rocks and search for the oldest trilobites, the 'Adam trilobite' he called it, and to solve the mystery of the sudden appearance of trilobites in the fossil record, which was troubling to the evolutionary theory. However, he also undertook fundamental work that significantly promoted and influenced geological studies in China, especially in the fields of stratigraphy, structural and tectonic geology (Oldroyd & Yang 2003). Willis was particularly impressed by the ideas of Davis on the efficacy of fluvial erosion. When he passed through Hebei and Shanxi provinces in northern China, besides investigating the stratigraphy and rock types, he also noted tectonic uplift and recognized that the summit heights of the Mt Wutai (the 'Five Platform Mountain') in Shanxi Province (see Fig. 2(1) for location) were plateau-like and evidently represented an ancient erosion surface. He identified two surfaces that were termed the Beitai and Tangxian surfaces. The Beitai Surface, at 2500–3000 m above sea level, is covered by a reddish sandy clay regolith, and occurs as flat summits on the main mountain ranges in northern China. It is typically represented by the 'Middle Platform' of Mt Wutai. The Tangxian Surface, at 200–1400 m above sea level, is covered by Late Tertiary red clay and often shows broad valley features. The two surfaces are common in northern China, e.g. Mt Xuehuashan (Fig. 3). Willis considered that the Beitai Surface was probably developed in the Late Mesozoic–Early Tertiary (corresponding to the Beitai Stage), while the Tangxian Surface formed in the Early–Late Tertiary (corresponding to the Tangxian Stage).

Period between 1910 and 1945

Between 1910 and the 1940s, especially after the foundation of the Geological Survey of China in

From: GRAPES, R. H., OLDROYD, D. & GRIGELIS, A. (eds) *History of Geomorphology and Quaternary Geology.* Geological Society, London, Special Publications, **301**, 171–178. DOI: 10.1144/SP301.12   0305-8719/08/$15.00

Fig. 1. Photographs of: Bailey Willis (left) (1857–1949), an American geologist; Z. X. Zeng (or C. S. Tseng) (centre) (1921–2007), a famous geomorphologist in China; and Z. J. Cui (right) (1933–), also a famous geomorphologist in China, who has carried out extensive research both on the planation surfaces of the Tibetan Plateau and the adjacent area, and on glacial landforms.

1913 (Li 2003), the northern China planation surfaces were studied by both Western (Andersson 1923; Barbour 1927, 1928) and Chinese (Ye 1920; Wang 1925; Zhao 1931; Sun 1935) scientists, who arrived at both similar and different opinions as to the ages of the surfaces. For example, Barbour considered that the Beitai Surface was uplifted in Late Oligocene–Early Miocene while the Tangxian

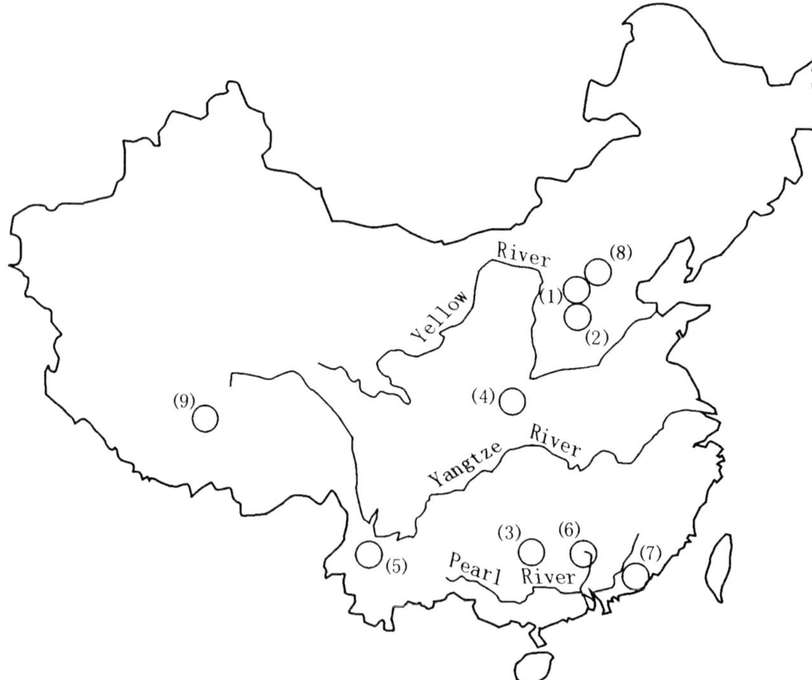

Fig. 2. Locations of planation surfaces in China mainland, mentioned in this paper. (1) Mt Wutai area. (2) Mt Xuehuashan area. (3) Kweilin (Guilin) area, Guangxi Province. (4) The Qinling Mountain Range. (5) The Yunnan Surface in Yunnan Province. (6) Northern Guangdong Province. (7) Eastern Guangdong Province. (8) The typical Dianziliang Surface, Hebei Province. (9) Tibetan Plateau.

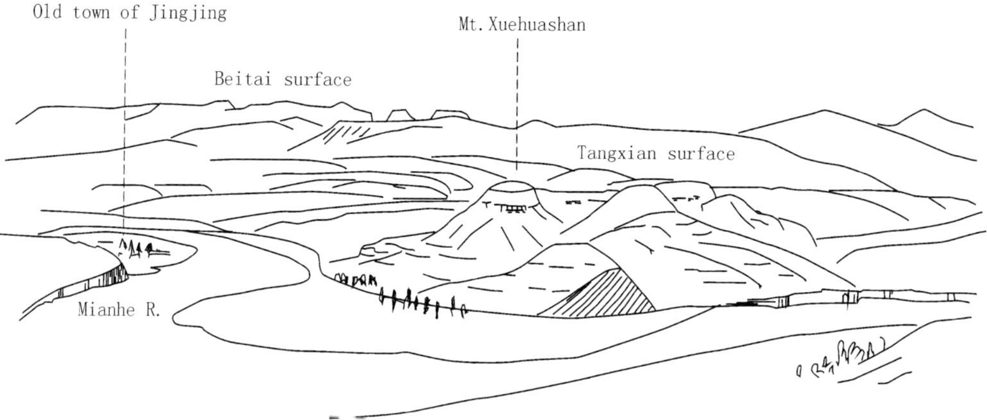

Fig. 3. Two planation surfaces, the Beitai and the Tangxian surfaces, in north China. Note: the top of Mt Xuehuashan is covered by basalt dated at 6.31 Ma, see Figure 2(2) for the location. (Modified after Li 1999.)

Surface was uplifted in the Late Miocene–Pliocene. Zhao argued that the Beitai Stage, referred to as the 'Grand Plain Stage', lasted from the end of the Mesozoic through most of the Tertiary, and the surface was subsequently covered by basalt, which he named as the 'Nanling Lava Stage'. He also considered that as the Tangxian Surface was covered by fossil Hipparion-bearing red clay, it should have been formed at the end of the Tertiary. Following the work of several Chinese scholars, Barbour (1936) revised his former opinion and considered that the Beitai Stage lasted from Late Cretaceous to Eocene, the Nanling Stage between Eocene and Oligocene, and the Tangxian Stage formed from Oligocene to Miocene. The main divergence of opinions related to the age of the Beitai Surface, the corresponding sediments of the Beitai Stage, and the existence of the Nanling Stage proposed by Zhao and Barbour.

The research on planation surfaces in northern China stimulated similar research in other parts of the country, such as NE China (Tan & Wang 1929), central China (Chao 1929; Chao & Huang 1931; Hsieh 1931; Lee 1933; Barbour 1935, 1936), eastern China (Gao 1935), southern China (Teilhard de Chardin *et al.* 1935; Wu & Tseng 1948) and NW China (Young 1931; Young & Bien 1936–1937; Chen 1947). Two or three planation surfaces were identified in the different areas and most could be approximately age-correlated. For example, Teilhard de Chardin *et al.* (1935) identified three planation surfaces, S, S_1 and S_2, in Guangxi Province, south China (see Fig. 2(3) for location). They considered that the oldest surface, or the upper limit of the karst pillar surface S, could be correlated with the Beitai Surface of North China. The lower surface, or the lower marginal or external karst surface S_1, surrounding the high karst masses, and in some places covered by Early Cenozoic red beds, was obviously formed before deposition of the red beds. The youngest surface, or lower internal karst surface S_2, separating the oldest and lower surfaces was correlated with the Tangxian Surface of North China (Fig. 4).

During the invasion and occupation of the eastern part of China by the Japanese in 1937–1945,

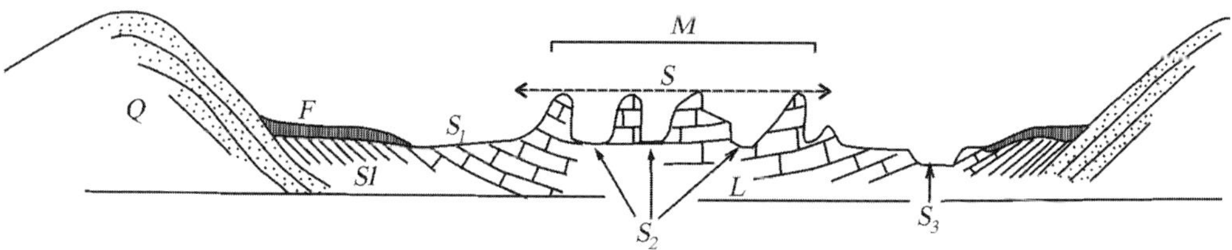

Fig. 4. Diagram illustrating the structure and physiographical elements of basins in the Kweilin (Guilin) area, Guangxi Province, south China. Q, Paleozoic quartzite. Sl, Palaeozoic slates. L, Palaeozoic limestone. F, Late Pliocene erosional fans (laterites). S, upper plane of karst. S_1, lower, external karst surface (covered locally by Lower Cenozoic red beds in the south). S_2, lower, internal karst surface. S_3, lower surface of canyons. M, mass of limestone isolated by S_1. (Redrawn from Teilhard de Chardin *et al.* 1935.)

many institutes were either closed or moved to SW China. Limited investigations on the planation surfaces continued in SW China, e.g.: Hou & Yang (1939), Guo (1946, 1948), and Li *et al.* (1946) in Sichuan Province; Yang (1944) and Si (1945) in Guizhou Province; Bien (1940), Chen (1948) and Fong (1948) in Yunnan Province. For example, Bien (1940), in a paper on the Cenozoic geology of Yunnan, SW China, considered that the affinity of Early Tertiary fauna of Mongolian, northern, central and SW China, indicated the extensive distribution of the different elements of a single faunal unit. This affinity demanded an explanation because there was no effective faunal barrier, i.e. the Qinling Mountain Range (see Fig. 2(4) for location) with an approximately east–west trend, between North and South China during the Early Tertiary. Bien cited the work of Teilhard de Chardin *et al.* (1935), who suggested that the Qinling Mountain Range was largely reduced to a low level at that time and that a single, probably swampy, peneplain environment extended from the Western Gobi south to the present Yangtze Basin. Bien even suggested that this peneplain extended as far south as Burma. The uplifted peneplain in Yunnan was called the Yunnan Surface (see Fig. 2(5) for location). Uplift to form the Qinling Mountain Range, the natural boundary between North and South China, and dissection of the proposed peneplain were the result of the mid-Tertiary Hengyangian tectonic movement, implying that the Yunnan Surface correlates with the Beitai Surface.

Guangdong Province in southern China was one of the first places where planation surfaces were investigated in detail. As early as 1930, Fong (1930) noted that some gorges in Guangdong and Guangxi provinces, such as Feelay Hsia (Feilai Xia; *Xia* = gorge) of Peikang (Beijiang, *bei* = north, *jiang* = river), Lin Yang Hsia (Linyang Xia) of Si Kiang (Xijiang, *xi* = west, *jiang* = river), Nu Tan Hsia (Nutan Xia) of the Liu Kiang (Liujiang) are obviously not simply consequent features because they cut across the structural trends of mountain ranges, i.e. transverse gorges. He speculated that some of the rivers may have be superimposed on an older peneplain topography that naturally truncated all former structures, and which was deeply incised, resulting in formation of the gorges as part of a new cycle of erosion by uplift and rejuvenation. During the Japanese occupation in Guangzhou (Canton, capital city of Guangdong Province) in the early 1940s, Sun Yat-sen University moved to northern Guangdong, where Wu and Tseng (their paper was published several years later in 1948) (Fig. 1) undertook a detailed investigation of the landscape in northern Guangdong (see Fig. 2(6) for location). They recorded that during the Japanese advance into southern China, refugees discovered to their surprise unexpectedly smooth topography on mountain summits, which provided easier ways for them to escape. They also reported that there were more people living on the tops of mountains in that area than at their bases.

Post-World War II

After the end of the war in 1945 and from the end of the civil war in 1949, China entered into a period of relatively stable economic and academic development. Investigation of planation surfaces continued in northern (Zhang 1950, 1959), central (Shen 1965), eastern (Yan 1958) and southern China (SITTBIGG 1962) (Fig. 5a–c). For example, Zhang (1950) noted that many rivers, such as the Yongdinghe, Qingshuihe, Yanghe, Baihe, Chaohe, Luanhe, Sangganhe and Guishuihe rivers, in northern China transect the northeastern mountain ranges at high angles and considered therefore that they must be older than the mountain ranges. He agreed with Barbour's opinion that they should have formed earlier than the Tangxian Surface.

Since the 1960s, the erosion cycle proposed by Davis has been much challenged and criticized. Many, for example Hack (1960), have argued that real landforms as open systems would not evolve as a closed system as envisaged in the erosion cycle theory, and that in an open system similar landforms could have different origins. For example, accordant summit heights may have originated in areas of similar erosion-resistant rocks and drainage density rather than represent remnants of a dissected peneplain. If there are different erosion-resistant bedrocks, there are possibly different levels of accordant summit heights, which are not necessarily explained by multiple erosion cycles. Effected by this idea, some geomorphologists in China began to criticize the concepts of planation surfaces, the physiographical stages and tectonic cyclic uplifts, and to question the correlation of planation surface ages and physiographical stages over the large areas that had been proposed. Others diverted their interest into geo-engineering rather than pure science, especially in the 1950s when the country was beginning its recovery from a long period of turmoil. During the 'Cultural Revolution', between 1966 and 1976, nearly all scientific research in China, including geomorphology, ceased.

Contemporary ideas

Following the end of the 'Cultural Revolution', a period known as the 'second spring of science in China' began. Scientists began to carefully

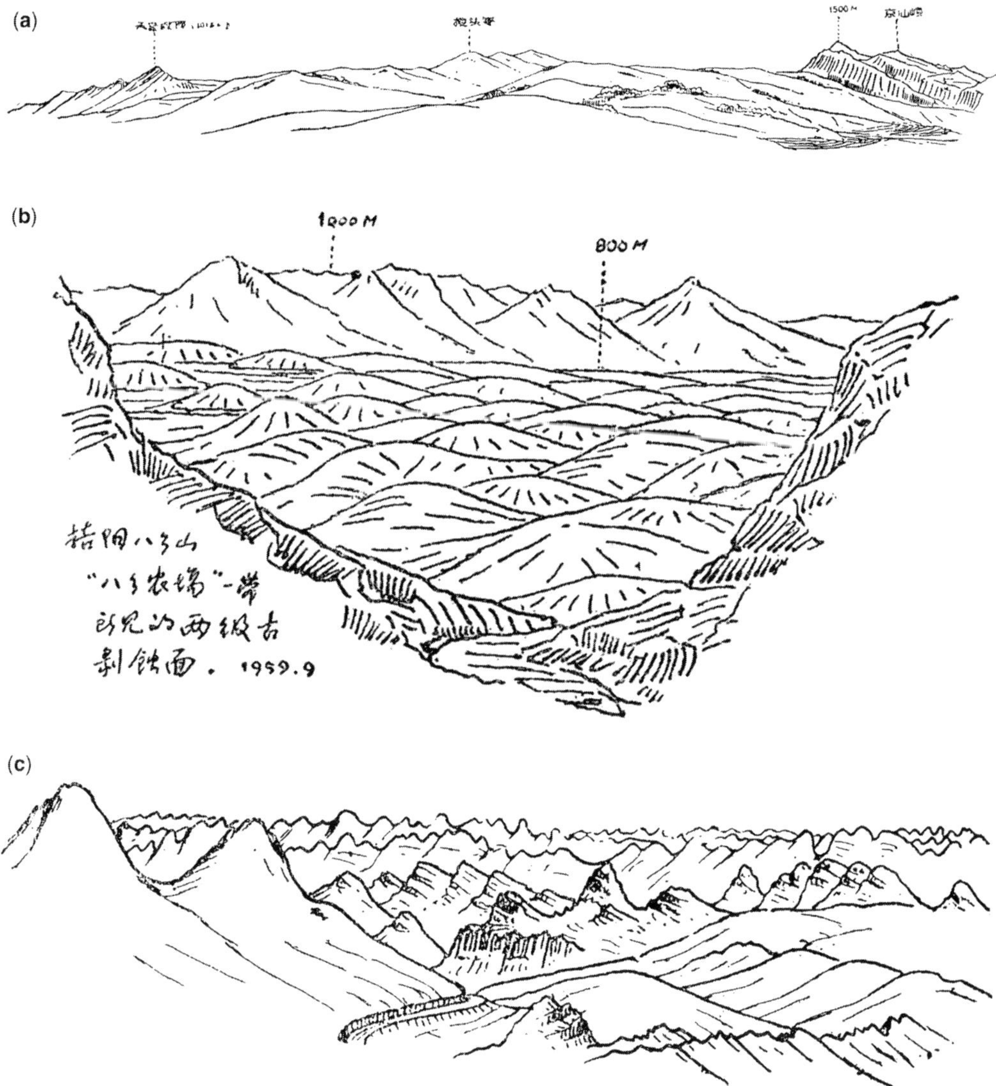

Fig. 5. Sketches of planation surfaces in Guangdong Province, south China. (**a**) Planation surface at 1013–1487 m above sea level, cutting across an anticline structure in the Yaoshan Mountains, North Guangdong (Zeng 1959). (**b**) Two planation surfaces at 1000 and 800 m above sea level in Baxiang, East Guangdong (Huang 1959). (**c**) Karstic planation surface about 600 m above sea level, in Yangshan, North Guangdong (Li 1959). See (a) and (c) in Fig. 2(6), and (b) in Fig. 2(7) for location. (Sketches of Zheng 1959, Huang 1959 and Li 1959 are all from SITTBIGG 1962.)

re-examine their former research and knowledge, and they realized that well-preserved peneplains or planation surfaces both in China and worldwide reflected the existence of periods of tectonic and climatic stability so that they could be considered as approximately closed systems. The erosion cycle theory of planar surface formation thus came to be viewed as a kind of scientific abstraction from the real world (Zhang 1999), and careful identification of the dissected peneplains in China with this understanding would be required in order to elucidate the complex interplay between tectonics, lithology and climate, and their effects on landscape evolution. Careful reinvestigation of planation surfaces over the whole country has only recently started, e.g.: J. J. Li *et al.* (2001) in central China; Huang *et al.* (1999) in eastern China; Zhang & Huang (1995) in SE China; Wang (1998) in NW China;

Chen & Zhao (1988) in SW China; and Cui *et al.* (1996) in the Tibetan Plateau (Fig. 1). Over the last decade, publications on planation surfaces have increased markedly (Qiu 1999). Although research in northern China continued after the 1980s, the most detailed work was carried out in the 1990s by Wu *et al.* (1999), supported by the Natural Science Foundation of China. They identified an additional planation surface, termed the Dianziliang Surface, at 1100–2200 m above sea level (see Fig. 2(8) for location), intermediate between the Beitai and Tangxian surfaces, and considered that the three planation surfaces were formed during the Late Cretaceous–Early Eocene, Oligocene and Late Miocene–Pliocene, respectively. Recent studies indicate that the Tangxian Surface is covered by red clay deposits dated at 8.35 Ma by magnetostratigraphy (Qiang *et al.* 2001) and by basalt dated to 6.31 Ma (Li 1999) (Fig. 2), which supports the conclusions of Wu *et al.* (1999). In SE China Zhang & Huang (1995) undertook further studies on the planation surfaces in northern Guangdong (see Fig. 2(6) for location), and identified four levels. Recently, Zhang (1998) and Zhang & Grapes (2006) have drawn attention to the importance of the distribution of Mesozoic granite, which forms topographic highs that are erosion remnants of former extensive planation surfaces.

The most interesting research on planation surfaces is probably related to the Tibetan Plateau (see Fig. 2(9) for location) and adjacent areas, which closely relates to the question of Cenozoic uplift of the plateau. The Tibetan Plateau has become a focus for geosciences because it is perhaps the best example of ongoing continental collision. The area is not only ideal for testing and developing plate tectonic and new geodynamic theories, but is also one of the agents that has changed both local- and global-scale environments during the Cenozoic (Li & Fang 1999). In the 1970s the Chinese Government sent a scientific investigation team to the Tibetan Plateau (Yang *et al.* 1983). Their work indicated that there was evidence of at least two planation surfaces: the Shanyuan Surface (Summit Surface), at 5000–6000 m above sea level, on mountain summits in general; and the Basin Surface (Main Surface), at 4000–4500 m above sea level, which is widely distributed, well preserved and corresponds to the Tangxian Surface of northern China. Cui *et al.* (1996, 1997) showed that the Shanyuan Surface formed during the Early Miocene, while on the basis of fission-track dating of speleotherm calcite, the Basin Surface was uplifted in the Pliocene. Cui *et al.* (2001, 2002) and D. W. Li *et al.* (2001) considered that the Basin Surface was a type of palaeo-karst (termed a karst planation surface), of subsurface origin and explained its development using the "double planation model" of Budel (1982). By chemical and mineralogical correlation of karst weathering products in the Tibetan Plateau and its eastern area over a distance of 1000 km and a height difference of 5000 m, Cui *et al.* (2001, 2002) and D. W. Li *et al.* (2001) have attempted to demonstrate the existence of an extensive planation surface. The planation surfaces, as 'reference planes', suggest an increasing rate of uplift of the Tibetan Plateau during the Cenozoic. Recent research (Wu *et al.* 2006) indicates that there were large lakes, indicated by lacustrine deposits of the Lower Miocene Wudaoliang Group (23.5–16.0 Ma; 20–5 Ma, Yi *et al.* 2000), the correlative sediments of planation, in the interior of the Tibetan Plateau, suggesting a planation age for the Basin Surface. However, Yi *et al.* (2000) have argued that there are no obvious boundaries between the Shanyuan and the Basin surfaces, between which there is a progressive change in topography. They also argued that there was only one palaeo-planation, the Shanyuan Surface, and that there was no planation surface corresponding to the surface on which the Wudaoliang Group sediments were deposited. Because of minor deformation of the Wudaoliang Group, they considered that if crustal thickening and shortening were the result, the extensive deformation process should have ended prior to 20 Ma. This age appears to be a key time-marker horizon in deciphering the evolution of the Tibetan Plateau, and supports arguments against the idea of rapid uplift of the whole Tibetan Plateau commencing at 3.6 Ma proposed by Cui *et al.* (1997, 2002) and Li & Fang (1999). Thus, current research on planation surfaces in the Tibetan Plateau highlights unresolved problems concerning the interpretation of planation surfaces in the China in general.

I wish to thank both R. H. Grapes and R. F. Diffendal, Jr for their very kind help with the preparation and reviews of this paper, and Q. Z. Gao for providing his photograph of Z. J. Cui. The Natural Science Foundation of China (NSFC) is gratefully acknowledged for financial support (40672124).

References

ANDERSSON, J. G. 1923. Essays on the Cenozoic of North China. *Memoirs of the Geological Survey of China (Series A)*, **3**, 144–155.

BARBOUR, G. B. 1927. Note on correlation of physiographic stages. *Bulletin of the Geological Society of China*, **5**, 279–280.

BARBOUR, G. B. 1928. Post-Mesozoic history of North China. *Bulletin of the Geological Society of America*, **39**, 209–210.

BARBOUR, G. B. 1935. Physiographic history of the Yangtze. *Memoirs of the Geological Survey of China (Series A)*, **14**, 112.

BARBOUR, G. B. 1936. The Paote and Chingshui Physiographic Stages and their significance. *Bulletin of the Geological Society of China*, **9**, 45–48.

BIEN, M. N. 1940. Preliminary observations on the Cenozoic geology of Yunnan. *Bulletin of the Geological Society of China*, **20**, 179–204.

BUDEL, J. 1982. *Climatic Geomorphology*. Princeton University Press, Princeton, NJ, 253–443.

CHAO, Y. T. 1929. Geological notes in Szechuan. *Bulletin of the Geological Society of China*, **8**, 137–150.

CHAO, Y. T. & HUNAG, T. K. 1931. The geology of the Tsinlingshan and Szechuan. *Memoirs of the Geological Survey of China (Series A)*, **9**, 206–215.

CHEN, F. B. & ZHAO, Y. T. 1998. *Neotectonics in Panxi Area*. Chengdu, Science and Technology Press of Sichuan, 1–128 (in Chinese).

CHEN, M. X. 1947. Physiography in central Gansu. *Geological Review*, **12**, 545–556 (in Chinese).

CHEN, S. P. 1948. Physiography of the Tanglongchuan Drainage Basin, Yunnan Province. *Journal of Geographic Science*, **15**, 1–22 (in Chinese).

CUI, Z. J., GAO, Q. Z., LIU, G. N., PAN, B. T. & CHEN, H. L. 1996. Planation surfaces, palaeokarst uplift of Xizang (Tibet) Plateau. *Science in China (Series D)*, **39**, 391–396.

CUI, Z. J., GAO, Q. Z., LIU, G. N., PAN, B. T. & CHEN, H. L. 1997. The initial elevation of palaeokarst and plantation surfaces on Tibet Plateau. *Chinese Science Bulletin*, **42**, 934–939.

CUI, Z. J., LI, D. W., LIU, G. N., FENG, J. L. & ZHANG, W. 2001. Characteristics and planation surface formation environment of the red weathering crust in Hunan, Guangxi, Yunnan, Guizhou and Tibet. *Science in China (Series D)*, **44**, (Supplement), 162–175.

CUI, Z. J., LI, D. W., FENG, J. L., LIU, G. N. & LI, H. J. 2002. The covered karst, weathering crust and karst (double-level) planation surface. *Science in China (Series D)*, **45**, 366–380.

FONG, K. L. 1930. Some problems in geology of Kwangtung and Kwangsi. *Bulletin of the Geological Society of China*, **10**, 127–133.

FONG, K. L. 1948. Geology and mineral resources of Yuhih District, Yunnan. *The Science Reports of National Tsing Hua University (Series C)*, **1**, 263–284.

GAO, P. 1935. Geology in Eastern Zhejiang Province. *Geological Report*, **25**, 49–76 (in Chinese).

GUO, L. Z. 1946. Report on geographic investigation in the Dabashan Mountain (Vol. 1, Landforms). *Geography Special Journal*, **4**, 1–42 (in Chinese).

GUO, L. Z. 1948. Landform in Beipei area. *Geography*, **5**, 1–15 (in Chinese).

HACK, J. T. 1960. Interpretation of erosional topography in humid temperate regions. *American Journal of Science*, **258**, 80–97.

HOU, D. F. & YANG, J. Z. 1939. Several landforms and their history in the Sichuan Basin. *Geological Review*, **4**, 315–321 (in Chinese).

HSIEH, C. Y. 1931. The Yühuatai gravel and its physiographic significance. *Bulletin of the Geological Society of China*, **11**, 61–64.

HUANG, P. H., DIFFENDAL, R. F., YANG, M. Q. & HELLAND, P. E. 1999. Mountain evolution and environmental changes of Huangshan (Yellow Mountain), China. *The Journal of Chinese Geography*, **9**, 25–34.

LEE, C. Y. 1933. The development of the Upper Yangtze Valley. *Bulletin of the Geological Society of China*, **13**, 107–117.

LI, C. S., ZHOU, T. R., GUO, L. Z. & GAO, Y. Y. 1946. Report on geography of the Jialingjiang Drainage Basin (Vol. 1, Landforms). *Geography Special Journal*, **1**, 68–72 (in Chinese).

LI, D. W., CUI, Z. J., ZHANG, Q. & MA, B. Q. 2001. Development features and original altitude of a karst planation surface, Preliminary review. *Chinese Science Bulletin*, **46**, (Supplement), 33–38.

LI, J. J. 1999. In memory of Davisian theory of erosion cycle and peneplain. A Centurial study in China. *Journal of Lanzhou University (Natural Sciences)*, **35**, 157–163 (in Chinese with English abstract).

LI, J. J. & FANG, X. M. 1999. Uplift of the Tibetan Plateau and environmental changes. *Chinese Science Bulletin*, **44**, 2117–2124.

LI, J. J., XIE, S. Y. & KUANG, M. S. 2001. Geomorphic evolution of the Yangtze Gorges and the time of their formation. *Geomorphology*, **41**, 125–135.

LI, X.T. 2003. Research on the history of the bureau of national geological survey of China. *China Historical Materials of Science and Technology*, **24**, 351–358 (in Chinese with English abstract).

OLDROYD, D. & YANG, J. Y. 2003. Bailey Willis (1857–1949), Geological theorizing and Chinese geology. *Annals of Science*, **60**, 1–37.

QIANG, X. K., LI, Z. X., POWELL, C. MCA. & ZHENG, H. B. 2001. Magnetostratigraphic record of the Late Miocene onset of the East Asian monsoon and Pliocene uplift of northern Tibet. *Earth and Planetary Science Letters*, **187**, 83–93.

QIU, W. L. 1999. The history of the research on physiographic states in China. *China Historical Materials of Science and Technology*, **20**, 95–106 (in Chinese with English abstract).

SHEN, Y. C. 1965. *Valley Landforms of the Upper Yangtze River*. Science Press, Beijing, 1–162 (in Chinese).

SI, Y. F. 1945. Landform adjacent to Zunyi. *Geological Review*, **10**, 113–121 (in Chinese).

SUN, J. C. 1935. Geology in Suiyuan and southwest Chahaer. *Geological Special Report (Series A)*, **12**, 6–8 (in Chinese).

SITTBIGG 1962. *Landform Division of Guangdong Province* (restricted publication), Synthetic Investigation Team of Tropic Bio-resources and Institute of Geography of Guangzhou, Chinese Academy of Science, 1–430.

TAN, X. C. & WANG, H. S. 1929. Geology along the Nengjiang River, Heilongjiang Province. *Geological Report*, **13**, 63–81 (in Chinese).

TEILHARD DE CHARDIN, P., YOUNG, C. C., PEI, W. C. & CHANG, H. C. 1935. On the Cenozoic formations of Kwangsi and Kwangtung. *Bulletin of the Geological Society of China*, **14**, 179–210.

WANG, S. J. 1998. *On the Planation Surfaces in Central Asia, Exemplified by Study of the Tianshan Mountain Range*. Science Press, Beijing, 1–128 (in Chinese).

WANG, Z. Q. 1925. Brief history of physiographic development for Shanxi and Shan'xi Provinces. *Science of China*, **10**, 929–938 (in Chinese).

WILLIS, B., BLACKWELDER, E. & SARGENT, R. H. 1907. *Research in China*, Volume 1, part 1. Carnegie Institution of Washington, Washington, DC, 75–83, 203–261, 319–339.

WU, C., MA, Y. *ET AL.* 1999. *Topographic Surface, Physiographic Period and Geomorphic Evolution of the Mountain Area in North China.* Hebei Science and Technology Publishing House, Shi Jiazhuang, 1–285 (in Chinese with English abstract).

WU, S. S. & TSENG, C. S. 1948. On the principle erosion surfaces and transversal gorges of north Kwangtung. *Lingnan Journal*, **9**, 254–260.

WU, Z. H., HU, D. G., ZHAO, X. & YE, P. S. 2006. Features of Early Miocene large paleolakes in the interior of the Qinghai–Tibet Plateau and their tectonic significance. *Geological Bulletin of China*, **25**, 782–791 (in Chinese with English Abstract).

YAN, Q. S. 1958. On geomorphological evolution processes in the areas of Qiantangjiang River, Zhejiang Province and the Taihu Lake. *Geography Symposium of East China Normal University*, Issue 1 (in Chinese).

YANG, H. R. 1944. Landform development in central Guizhou Province. *Journal of Geographic Science*, **11**, 1–14 (in Chinese).

YANG, Y. C., LI, B. Y., YIN, Z. S. & WANG, F. B. 1983. *Tibetan Landform.* Science Press, Beijing, 200–229 (in Chinese).

YE, L. F. 1920. Geological Memoirs of the Peking West Mountain. *Geological Special Report (Series A)*, **11**, 51–63 (in Chinese).

YI, H. S., WANG, C. S., LIU, S., LIU, Z. F. & WANG, S. F. 2000. Sedimentary record of the planation surface in the Hoh Xil region of the northern Tibet Plateau. *Acta Geologica Sinica*, **74**, 827–835.

YOUNG, C. C. 1931. On the Gobi Plane of deflation – the Gobi Erosion Plane. *Bulletin of the Geological Society of China*, **11**, 161–169.

YOUNG, C. C. & BIEN, M. N. 1936–1937. Cenozoic geology of the Kaolan–Yungteng Area of central Kansu. *Bulletin of the Geological Society of China*, **16**, 221–260.

ZHANG, B. S. 1950. Preliminary research on landforms in the Chanan Basin. *Science of China*, **32**, 335–339 (in Chinese).

ZHANG, B. S. 1959. Mountain landform of the Wutaishan Mountain. *Quaternary Sciences of China*, **2**, 39–46 (in Chinese).

ZHANG, K. 1998. The characteristics and origins of the landforms along the coastal area of Southeast China. *Tropical Geomorphology*, **19**, 22–30 (in Chinese with English abstract).

ZHANG, K. 1999. On the equilibrium and evolution of landforms. *Tropical Geography*, **19**, 97–106 (in Chinese with English abstract).

ZHANG, K. & GRAPES, R. 2006. Relationship between large rivers and granite-cored anticlines in the Lower Pearl River System, Southeast China, an example of a long-lived drainage pattern. *Catena*, **66**, 190–197.

ZHANG, K. & HUANG, Y. K. 1995. Research on the planation surfaces in North Guangdong. *Tropical Geography*, **15**, 295–305 (in Chinese with English abstract).

ZHAO, J. K. 1931. Physiography and loess in China. *Bulletin of the Geological Society of Peking University*, **5**, 213–230 (in Chinese).

The Palaeo-Tokyo Bay concept

MICHIKO YAJIMA

*College of Liberal Arts and Sciences, Tokyo Medical and Dental University,
2-8-30 Kohnodai, Ichikawa-shi, Chiba 272-0827, Japan
(e-mail: yajima.michiko@gupi.jp or pxi02070@nifty.com)*

Abstract: Tokyo is situated in the southern part of the Kanto Plain, the largest plain in Japan. It is filled with thick marine Quaternary sediments deposited in the Palaeo-Tokyo Bay. The concept of a Palaeo-Tokyo Bay was proposed by Hisakatsu Yabe in 1913 and 1914, based on molluscan fossils, geography and tectonics. Palaeo-Tokyo Bay used to open to the east and, at the time of the high sea-level phase, perhaps also to the south, whereas the modern Tokyo Bay opens to the south. The Paleo-Tokyo Bay concept is supported by recent sedimentological evidence, sequence stratigraphy and tectonic studies, and is considered to have formed in a forearc basin near the triple junction of plates (the Pacific, Asian and Philippine plates) and trenches (Japan Trench, Sagami Trough and Izu-Ogasawara Trench) in the Quaternary. The bay is unique and shows evidence of repeated differential vertical movements during its formation.

Metropolitan Tokyo is situated on the southern part of the Kanto Plain, the largest plain in Japan. The Kanto Plain and the northern part of the Boso Peninsula (Figs 1–3) are underlain by more than 3000 m of marine Neogene sediments that contain natural gas. Thirty per cent of the Japanese population live on the Kanto Plain and a large earthquake related to subduction of the Pacific Plate is anticipated. The first issue of the *Journal of the Geological Society of Japan* in 2006 was concerned with '[the] geotectonic boundary between Northeast and Southwest Japan concealed beneath the Kanto Plain – [an] integrated approach of surface and subsurface geology', and it provided data concerning the basement rocks of the thick Neogene sediments.

The uppermost Quaternary sediments of the Kanto Plain are considered to have been deposited in a Palaeo-Tokyo Bay, a concept first proposed by Hisakatsu Yabe (1878–1969) in 1913–1914, based on evidence of fossil molluscs, geomorphology and interpretation of the tectonics of the Japanese Archipelago (Yabe 1913, 1914). This concept is now generally accepted, as it has been tested against more recent data from the sedimentology, palaeontology and neotectonics of the area. Of course, Yabe never thought in terms of the Kanto Plain being a forearc basin developed near a triple plate junction during the Neogene, and that it represented the uppermost part of this basin during the Quaternary, so it is interesting to examine how he formulated his concept.

The study of Quaternary geology in Japan in the late nineteenth century

Quaternary deposits in NW Europe and North America are intimately linked with the extension and fluctuation of the northern polar ice sheets, but no similar glaciations affected Japan in the Quaternary, except for small accumulations of ice and the development of cirques in some of the high mountain areas of Hokkaido and central Honshu.

The study of geology in Japan started with the establishment of the Geological Institute and the Geological Survey during the Meiji Era in the late nineteenth century. The study of Quaternary geology in Japan was difficult as Japanese geologists learned their science from European geology, which offers a very different geology to that of Japan. The Japanese geologists had to work out their Quaternary geology with no glaciation 'symptoms', and, needless to say, radiometric and isotopic methods of age determinations were unavailable. New ways of approaching Quaternary geology were needed. The many problems confronting the study of Quaternary geology included such questions as: What is the appropriate boundary between the Tertiary and the Quaternary? Did the distinction between diluvium and alluvium hold in Japan? Which ice age from the Günz, Mindel, Riss and Würm ice ages was represented in Japan? The Palaeo-Tokyo Bay concept was developed in the context of the answers to such questions.

For the early research, fossil molluscs were studied in the Tokyo area, where at the end of the nineteenth century there were many outcrops with fossils, even in Tokyo itself (Fig. 4). The strata were horizontal and soft, and the absence of Palaeozoic and Mesozoic fossils indicated that they were Cenozoic. The scientific study of Japanese fossils began in the Meiji Era, but there were so few books on palaeontology that it was difficult to determine generic and specific names. The molluscan fossils from the Tokyo area were, however,

From: GRAPES, R. H., OLDROYD, D. & GRIGELIS, A. (eds) *History of Geomorphology and Quaternary Geology.*
Geological Society, London, Special Publications, **301**, 179–187.
DOI: 10.1144/SP301.13   0305-8719/08/$15.00 © The Geological Society of London 2008.

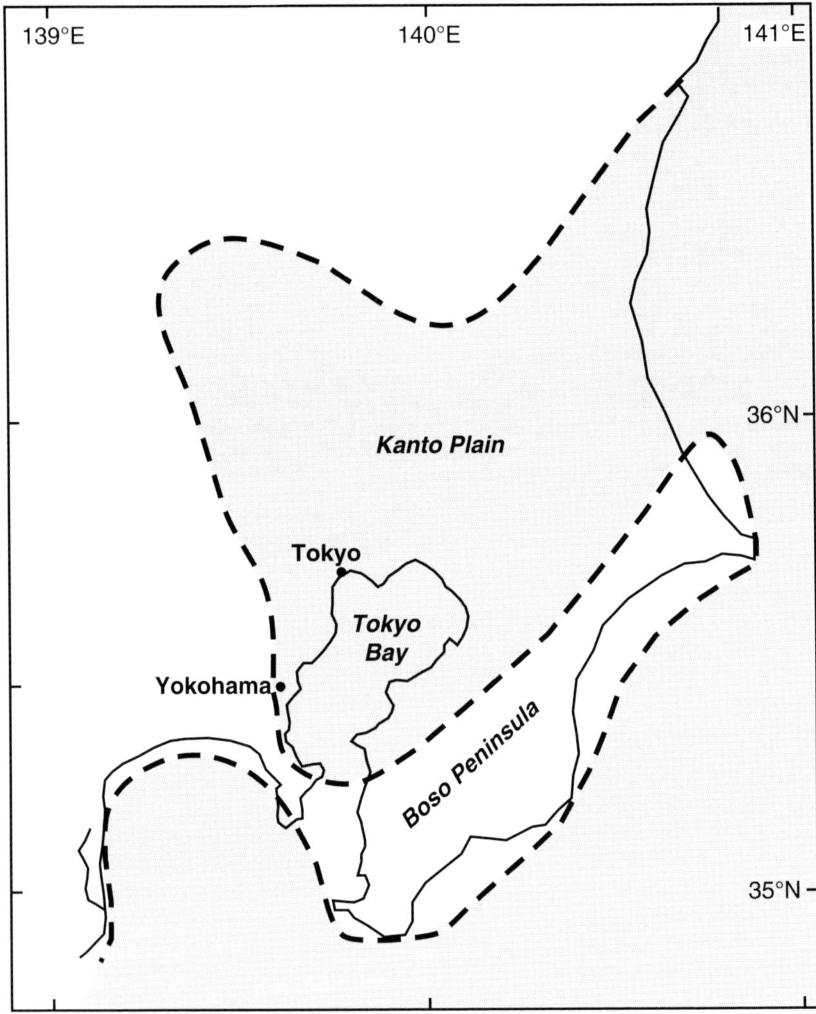

Fig. 1. Palaeo-Tokyo Bay redrawn from Koike (1952). The Recent shoreline is shown by the thin continuous line.

mostly identified as Recent. 'Ecological' information about Recent molluscs was plentiful, as they formed a major part of the Japanese diet. Today, along the Pacific coast of Japan, warm and cold sea currents converge north of Tokyo Bay (Fig. 5). The warm current is called the Japan Current or the 'Black Stream' ('*Kuroshio*'); the cold current is called the Kurile Current or the 'Parent Stream' ('*Oyashio*'); and the cold- and warm-water molluscs are quite distinct. Thus, fossil molluscs in the Cenozoic rocks of the Tokyo area were discussed in terms of being either warm- or cold-water species. Examples are shown in Figure 6.

Fossil shells near Tokyo were first reported by Franz Hilgendorf in 1875 and published in 1876.

This German scientist had been appointed by the Japanese government as the first lecturer of Natural Hstory in the Medical Department of the University of Tokyo. He described fossilized cold-water molluscs such as *Glycymeris yessoensis* (Fig. 6) distributed in northern Japan and interpreted the strata in which they occurred as 'diluvial'. In 1879 Edmund Naumann, the first professor of the Geological Institute of the University of Tokyo, discovered fossilized warm-water shells near Tokyo and concluded that warm water flowed from south to east across the present area of Tokyo. Fossil coral reefs near Tateyama in the southernmost part of the Boso Peninsula were known as early as 1879, but their age could not be determined. Naumann also located the jaw of a

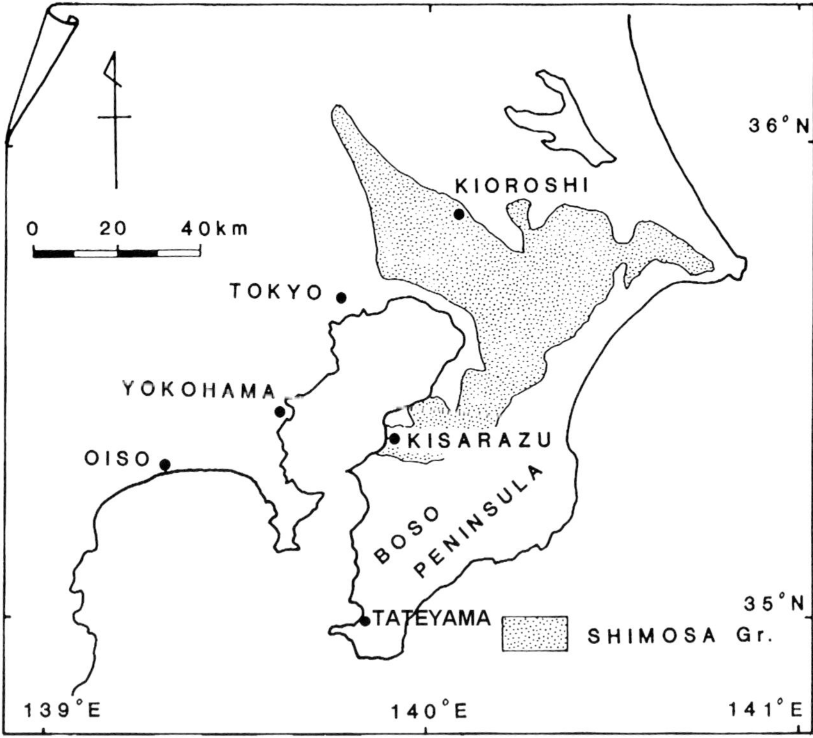

Fig. 2. Distribution of the Shimosa Group (slightly revised from Yajima & Lord 1990).

supposed fossil mammoth near Tokyo in 1881, and suggested that during what we now call the Neogene period the Tokyo area was warm, whereas during 'Diluvium' (or 'Diluvial') times it was cold (Naumann 1881). In 1881 David Brauns, the second professor of the Geological Institute at the University of Tokyo, published descriptions of many molluscan fossils from the environs of

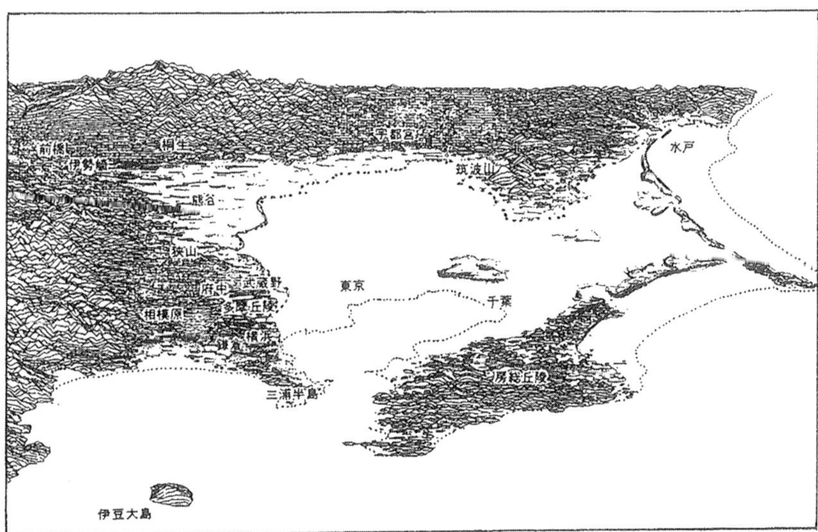

Fig. 3. Reconstruction of the barrier reef in Palaeo-Tokyo Bay (Masuda 1992).

Fig. 4. Sketch of an outcrop of Cenozoic rocks near Tokyo (after Brauns 1881).

Tokyo and maintained that they were all warm-water species. Later, Naumann came to think that his fossil jaw was not that of a mammoth but, rather, that of an Asiatic elephant. So in 1885 he concluded that even during the 'Diluvium' the climate of the Tokyo area was warm (Naumann 1885). In 1888 a 1:200 000 scale geological map and explanation of the Tokyo area was published by the Geological Survey of Japan (Suzuki 1888). Satoshi Suzuki, the third director of the Survey, considered that the fossil shells from the environs of Tokyo were all of Tertiary age because they occurred below alluvial sediments in the valley of Tokyo town and the 'diluvium' loam sediments of higher ground. The question of palaeo-climate was not discussed.

In 1898 Hisakatsu Yabe published a list of fossils from the Tokyo area. Although he was only an undergraduate at the time, he had already started research in palaeontology and geology. He studied molluscan fossils and foraminifera, not only from Tokyo but also from the Boso Peninsula, and proposed that the fossils were Tertiary. Thus, reports of fossil localities increased but only in an unco-ordinated manner.

In 1907 Shigemoto Tokunaga (1874–1940), who was a graduate of the Zoological Institute of the University of Tokyo, published an account (in English) of fossil molluscs from the environs of Tokyo and maintained that they were cold-water species (Tokunaga 1907). However, 4 years later in 1911, Matajiro Yokoyama (1860–1942), the second Japanese professor at the Geological Institute of the University of Tokyo, concluded that the same fossils were all warm-water species. He determined that they were Pliocene, and considered that the fossil coral reefs from Tateyama were of 'Diluvium age'. He thought the North Pole was then located near Europe and that the Japanese Archipelago was near the Equator. He wrote (Yokoyama 1911, p. 3):

So when Japan was opened to international traffic, and geologists, both foreign and native, began to scour the country, they naturally looked for evidence of glaciers. But strange to say, they were nowhere to be found. They were not found in Honshu, nor in Hokkaido, nor even in the cold island of Sakhalin where even in the southernmost part the mean January temperature falls far below freezing point, to -18 degrees C, a temperature that we find in Labrador and Southern Greenland. From this negative evidence they were obliged to conclude that glaciers had never existed in Japan, probably because the climate had never been cold enough to produce them. But why had it not been cold? There was no one who could answer this question.

The Palaeo-Tokyo Bay concept

The concept of a Palaeo-Tokyo Bay was proposed by Hisakatsu Yabe in 1913 and 1914 (see Fig. 1). This idea originated from several facts: modern Tokyo Bay is surrounded by the Boso and Miura peninsulas and opens to the *south* via a deep submarine canyon. Fossil shells from Quaternary sediments of the two peninsulas are mostly cold-water species, and warm and cold sea currents join to the north of Tokyo Bay (Fig. 5). Yabe considered that the ocean-current system along the Pacific

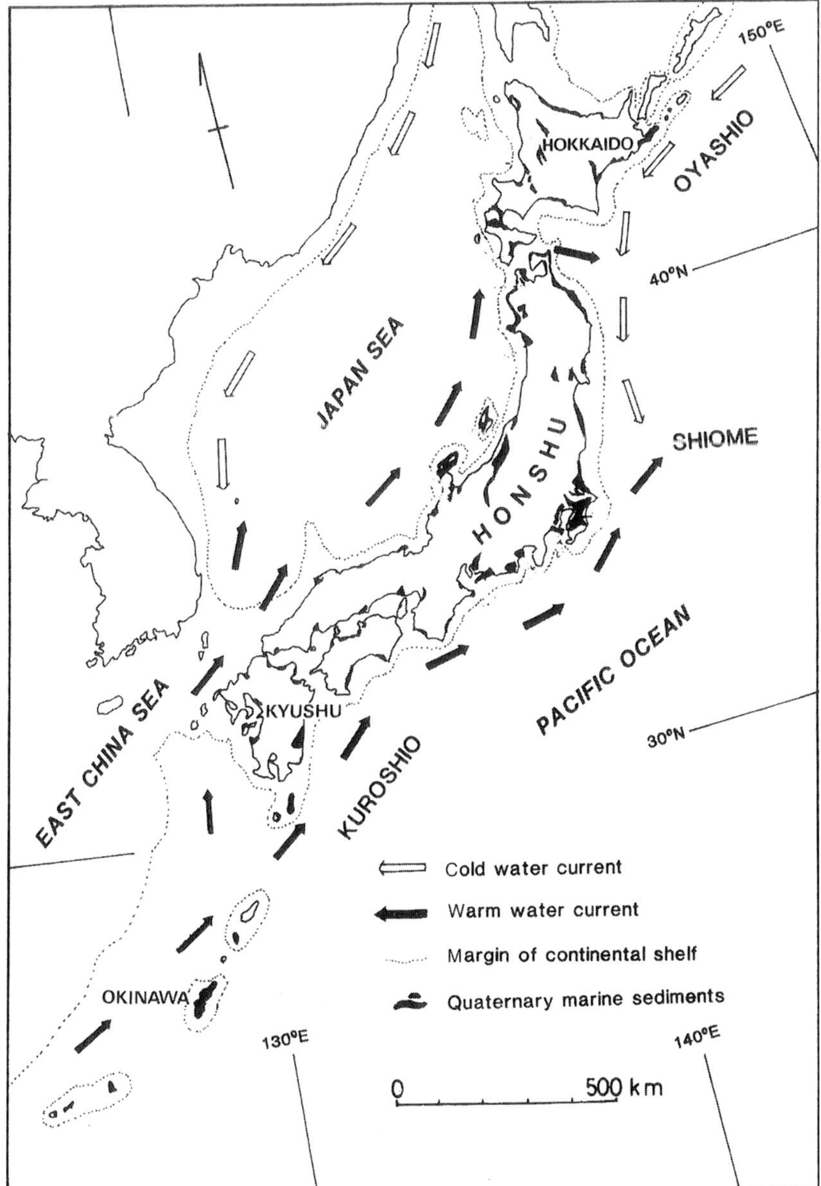

Fig. 5. The Japanese Islands, showing the distribution of marine currents, the continental shelf margin and the distribution of Quaternary marine sediments.

coast of the Japanese Islands was not substantially different from that which existed during the Quaternary, but he also considered that the entrance to Tokyo Bay might have subsequently changed from east to south. He concluded that if Tokyo Bay had formerly opened directly to the east then the cold current would have flowed into the bay, which would account for the presence of the cold-water fossils (Fig. 1).

However, Yabe thought the fossil evidence was insufficient to establish a boundary between the Tertiary and Quaternary deposits because of insufficient data. The extinct/extant ratio was therefore uncertain. Stratigraphic evidence was also poor

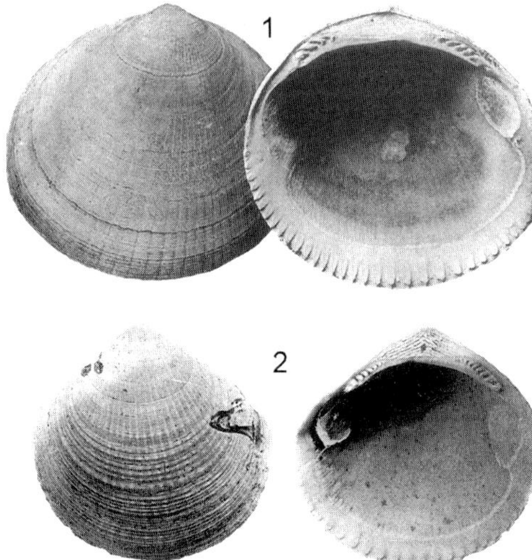

Fig. 6. Examples of warm- and cold-water species of molluscs: 1. Warm-water species *Glycymeris vestita* (Duker); 2. cold-water species *Glycymeris yessoensis* (Sowerby) (after Yokoyama 1922).

because of the presence of many unconformities. However, he surmised that the geographical evidence and the geological sequence, taken together, made a determination possible. At first he separated the fossil coral reefs at Tateyama from the fossil shells in the Tokyo area. The coral reefs supposedly belonged to the 'Alluvium' and indicated a different environment from that of the Palaeo-Tokyo Bay. However, the molluscs belonged to the 'Diluvium' and lived within the Palaeo-Tokyo Bay. Thus, Yabe rejected Yokoyama's (1911) idea of a warm climate in the Pliocene and 'Diluvium', even though Yokoyama was one of Yabe's lectures.

The central parts of the Boso and Miura peninsulas have altitudes of 200–400 m a.s.l. (metres above sea level), and are mainly composed of volcanic rocks that trend east–west and form parallel grabens. This block-faulted structure was recognized by Yabe, who thought that the rocks formed the geological 'backbone' of the Japanese Archipelago, which he discussed after 1920. Yabe needed a large sedimentary basin for the deposition of the Quaternary sediment of the Tokyo area and the northern part of the Boso Peninsula that was derived from this uplifted area of volcanic rocks. A Palaeo-Tokyo Bay opening to the east fulfilled this requirement and explained the presence of cold-water fossil molluscs. The Palaeo-Tokyo Bay concept was thus proposed rather 'instinctively', but it was the first palaeogeographical reconstruction based on neotectonics to be made in Japan.

Hisakatsu Yabe

Hisakatsu Yabe (Fig. 7) was interested in molluscan fossils even in his earliest days and wrote his first paper in 1898. He graduated from the Geological Institute of Tokyo Unversity in 1901, and from 1908 to 1912 he studied in Europe, mostly in Germany and Austria. In 1912 he returned to Japan and established a new geological institute at Tohoku University in Sendai. The first geological institute had previously been founded at Tokyo University in 1877 and the one at Tohoku University was the second.

Yabe carried out research on the geology, geography, stratigraphy and palaeontology of various areas and strata of various ages. His graduate thesis (1901) was concerned with coalfields in Hokkaido under the guidance of Professor Bunjiro Koto (1856–1935), the first Japanese professor of geology at Tokyo University. Yabe identified Cretaceous ammonites, orbitorinas and trigonias; but he was also skilled in both vertebrate and invertebrate palaeontology. In 1904 he proposed a new genus, *Nipponites* for an abnormally-coiled ammonite (Yabe 1904). He conducted research throughout the Japanese Archipelago, in Formosa, Korea and NE China, with the Boso Peninsula being one of his most intensive areas of study. Yabe received the 'Cultural Medal' in 1953 – the highest medal for civilians in Japan, and he is the first and only geologist to have received this award.

Fig. 7. Hisakatsu Yabe (1878–1969) (photograph from the records of attendance at the International Geological Congress in Belgium in 1922).

Yabe belonged to the second generation of Japanese geologists. The first generation, e.g. Koto and Yokoyama, learned geology from foreign teachers in Japan and then went to Europe to further their studies in geology. The second generation learned geology from Japanese teachers but also studied in European countries.

The Palaeo-Tokyo Bay concept in relation to controversies about glaciation in Japan

We now know that the Quaternary in Japan had small accumulations of ice, and cirques were developed in some of the mountains of Hokkaido and Central Honshu as the 'footprint' of glaciation. But the study of the high-mountain geology of Japan started late. In the early days of geological studies in Japan the idea of glaciation was proposed only cautiously, because most of the known Quaternary fossils were warm-water species. Early on the earthquake investigator John Milne (1850–1913) reported, in 1881, that cirques were developed in some of the high mountains in Japan (Milne 1881), but many geologists were not interested and neglected to follow up his observations. In 1902 Naomasa Yamazaki (1870–1929) gave a talk on glaciation in Japan at a meeting of the Tokyo Geological Society (Yamazaki 1902) after having studied glacial geology for 3 years in Germany under Albrecht Penck. Unfortunately, Yamazaki's ideas were rejected by Yokoyama (1911) on the basis that all known Quaternary fossils in Japan were warm-water species. However, Yabe's (1913, 1914) proposal that a Palaeo-Tokyo Bay had been occupied by cold-water organisms initiated a second glaciation controversy, although this quickly ended because the presence of a Palaeo-Tokyo Bay did not require glaciation. It was only in 1931, following the publication of Takuji Ogawa's paper on glaciation, that all Japanese geologists agreed that glaciation had occurred in Japan, with many geologists becoming involved with research in the mountainous areas of Hokkaido and central Honshu.

Successive ideas about a Palaeo-Tokyo Bay

Following the Palaeo-Tokyo Bay proposal, Yabe was involved in research on the geological structure of the Kanto area (1920), from which he proposed a generalized stratigraphy of Japanese Cenozoic strata, suggesting a subdivision into four series: Akitsu, Takachiho, Mizuho and Shikisima (Yabe & Aoki 1926). Yabe aimed at a complete revision of the geological structure of Japan that had been proposed by the first generation of Japanese

geologists. He also advanced the understanding of the tectonics of the Boso Peninsula (Yabe & Tayama 1932). In 1951 Yabe published his reminiscences and was gratified that his Palaeo-Toyo Bay concept had not been rejected, although most of his proposals regarding the geological structure of Japanese Archipelago were forgotten.

Kiyoshi Koike (1926–1957) started an investigation of the geology of the Boso Peninsula for his graduate thesis at the University of Tokyo, and also studied the history of Palaeo-Tokyo Bay; the first map showing the possible extent of the bay (Fig. 1) was produced by him in 1952. Koike was interested in the Cenozoic history of southern Kanto Province and his study concentrated on the basement of Palaeo-Tokyo Bay. He thought that there were three phases in its formation: an anticlinal elevation; a rapid submergence on its southern side; and a gentle, basin-forming, down-warping on its northern side in each phase. He thought that the Palaeo-Tokyo Bay was located in a more southerly area in its early phase and subsequently moved northwards. However, his work was not completed because of his premature death.

Contemporary interpretations of the Palaeo-Tokyo Bay

Yabe's Palaeo-Tokyo Bay concept was based on geographical, geological and palaeontological evidence. In his day there were, as mentioned, many outcrops, even in Tokyo; but in the twentieth century most of these outcrops have disappeared. The centre of the study of Palaeo-Tokyo Bay has therefore shifted to the Boso Peninsula (Fig. 7). A striking characteristic of the Quaternary sediments (Shimosa Group) on the peninsula is that the marine strata of this period can usually be observed on land without any substantial breaks, and the sequence includes many pyroclastic layers that provide datable index horizons.

In the 1970s the Palaeo-Tokyo Bay concept was further supported by stratigraphic work based on pyroclastic marker horizons (e.g. Machida *et al.* 1974; Sugihara *et al.* 1978). Machida *et al.* proposed a thick sequence of at least eight cyclothems during the last 500 000 years, suggesting glacio-eustatic cycles with interglacial sediments marking the base of each cyclothem. The Palaeo-Tokyo Bay was interpreted as being open to the east, but at the highest sea-level phase of the interglacial age it would have been open to the south.

In the 1980s and 1990s the Palaeo-Tokyo Bay concept was further supported by the results of sequence stratigraphy (Ito & Masuda 1989) and studies of fossil ostracods (e.g. Yajima 1982; Yajima & Lord 1990). The Shimosa Group consists of four formations, each showing a tendency towards cyclic sedimentation and reflecting a cycle of glacio-eustatic sea-level change. Marine components dominate in the lower part of each cycle. Ostracod distribution patterns reflect water depth, changes in which were induced by eustatic and isostatic/tectonic movements, and agree well with the major sea-level changes suggested by Machida *et al.* (1974).

Most recently, Palaeo-Tokyo Bay is presently interpreted as being the uppermost part of a forearc basin, near a triple junction of oceanic trenches (the Japan Trench, the Sagami Trough and the Izu-Ogasawara Trench) (Kaizuka *et al.* 2000). At the triple junction, three plates – the Pacific, Asian and Philippine plates – have moved in a complex manner. In the southern part of the Palaeo-Tokyo Bay area, the Mineoka–Hayama Uplift Zone has been active since the late Miocene as the outer ridge of the Japanese Archipelago, followed by uplift of the Miura Peninsula and the southern part of the Boso Peninsula. These uplifts indicate that Palaeo-Tokyo Bay had to open to the east. However, at the same time the Kashima–Boso Uplift Zone, parallel to the Japan Trench, was also active on the east side of Palaeo-Tokyo Bay (Nakasato & Sato 2001), so reconstruction of the bay before 120 000 years BP indicates that, although it opened to the east, there were barrier islands (reefs) in the eastern part of the bay (Fig. 3) (Masuda 1992; Nishikawa & Ito 2000) based on sequence stratigraphy. During the highest interglacial sea-level phase of 120 000 years BP, Paleo-Tokyo Bay opened to the south (Okazaki 2001). The alternate eastern and southern openings of the bay are therefore in accordance with the suggestions first proposed by Hisakatsu Yabe.

I am grateful to T. Yamada (Chiba High School), M. Ito (Chiba University) and H. Ando (Ibaraki University) for their helpful comments and information in relation to tectonics. I am also indebted to D.R. Oldroyd (University of New South Wales, Australia) and R.H. Grapes (Sun Yat-sen University, China) for their language assistance and review that helped improve and clarify some of the ideas reported in the paper.

References

BRAUNS, D. 1881. Geology of the environs of Tokio. *Memoirs of the Science Department, Tokyo Daigaku*, **4**, 1–82.

HILGENDORF, F. M. 1876. Sitzung in Yokohama am 16th October 1875. *Mitteilungen der deutschen Geselschaft fuer Natur und Voelkerkunde Ostasiens*, **1**, 5.

ITO, M. & MASUDA, F. 1989. Petrofacies of Paleo-Tokyo Bay sands, the Upper Pleistocene of Central Honshu, Japan. *In*: TAIRA, A. & MASUDA, F. (eds) *Sedimentary Facies in the Active Plate Margin.* Terra Scientific, Tokyo, 179–196.

KAIZUKA, S., KOIKE, K. & SUZUKI, T. 2000. History of development of geomorphology in the Kanto District. *In*: KAIZUKA, S., ENDO, K, SUZUKI, T., KOIKE, K. & YAMAZAKI, H. (eds) *Regional Geomorphology of the Japanese Islands – Volume 4. Geomorphology of Kanto and Izu-Ogasawara*, University of Tokyo Press, Tokyo, 303–325 (in Japanese).

KOIKE, K. 1952. Structural evolution of Southern Kanto. *Ritti sizen kagaku kenkyujo Hokoku (Reports of the Institute of Natural Sciences and Earth Sciences)*, **10**, 5–11 (in Japanese with English abstract).

MACHIDA, H., ARAI, F., MURATA, A. & HAKAMATA, K. 1974. Correlation and chronology of the Middle Pleistocne tephra layers in South Kato. *Journal of Geography*, **83**, 302–338.

MASUDA, F. 1992, Barrier islands in the Late Pleistocene Palaeo-Tokyo Bay. *Chishitsu News*, **458**, 16–27 (in Japanese).

MILNE, J. 1881. Evidences of the Glacial Period in Japan. *Transactions of the Asiatic Society of Japan*, **9**, 53–86.

NAKASATO, H. & SATO, H. 2001. Chronology of the Shimosa Group and movement of the 'Kashima' Uplift Zone, Central Japan. *Quaternary Research*, **40**, 251–257.

NAUMANN, E. 1879. Ueber die Ebene von Yedo, Eine geographisch–geologisch Studie. *Petermann's Geographische Mitteilungen*, **4**, 121–136.

NAUMANN, E. 1881. Ueber Japanische Elephanten der Vorzeit. *Palaeontographica*, **28**, 1–39.

NAUMANN, E. 1885. *Ueber den Bau und die Entstehung der japanischen Inseln.* R. Friedlaender und Sohn, Berlin.

NISHIKAWA, T. & ITO, M. 2000. Late Pleistocene barrier-island development reconstructed from genetic classification and timing of erosional surfaces, Paleo-Tokyo Bay, Japan. *Sedimentary Geology*, **137**, 25–42.

OKAZAKI, H. 2001. Significance of tectonic movement and eustatic sea-level changes to depositional systems of Palaeo-Tokyo Bay, analyzed from the Pleistocene Shimosa Group, Japan. *In: Geological History of the Boso Peninsula – Studies of Sedimentology and Paleontology. Journal of the Natural History Museum and Institute, Chiba*, **4** (Special Issue), 5–12.

OGAWA, T. 1931. Glaciation in the Diluvium in Central Japan. *Chikyu (The Earth)*, **16**, 321–332, 401–408; **17**, 11–8, 159–170 (in Japanese).

SUGIHARA, S., ARAI, F. & MACHIDA, H. 1978. Tephrochronology of the Middle and Late Pleistocene sediments in the northern part of the Boso Peninsula, Central Japan. *Journal of Geological Society of Japan*, **84**, 583–600.

SUZUKI, S. 1888. *Explanation of the Geological Map of Tokyo, 1/200 000 Scale.* Geological Survey, Tokyo (in Japanese).

TOKUNAGA, S. 1907. Fossils from the environs of Tokyo. *Journal of the College of Science, Imperial University, Tokyo, Japan*, **21**, 1–96.

YABE, H. 1898. The list of Tertiary fossils from environs of Tokyo. *Journal of the Geological Society of Japan*, **5**, 387–395 (in Japanese).

YABE, H. 1901. *General geology of the Middle and the Southern parts of the Ishikari coal-field, Hokkaido.* Graduate thesis, University of Tokyo (in Japanese).

YABE, H. 1904. Cretaceous cephalopoda from Hokkaido. *Journal of the Colleges of Science, Tokyo Imperial University,* **20**, 1–45.

YABE, H. 1913. On Japanese climate in the Pleistocene. *Gendai no Kagaku (Modern Science),* **1**, 552–557 (in Japanese).

YABE, H. 1914. Boso Peninsula. *Gendai no Kagaku (Modern Science),* **2**, 227–234 (in Japanese).

YABE, H. 1920. The geological structure of the northeastern part of the Kanto mountains. *Journal of the Geological Society of Japan,* **27**, 129–149, 187–198, 243–251 (in Japanese).

YABE, H. 1951. On the Palaeo-Tokyo Bay. *Shizenkagaku to Hakubutsukan (Natural Sciences and Museum),* **18**, 42–146 (in Japanese).

YABE, H. & AOKI, R. 1926. The Great Kwanto Earthquake of September 1, 1923, geologically considered.

Annual Report of Saito Gratitude Foundation, **1**, 70–83 (in Japanese).

YABE, H. & TAYAMA, R. 1932. Hojo Trough. *Bulletin of the Earthquake Research Institute,* **9**, 333–339 (in Japanese).

YAJIMA, M. 1982. Late Pleistocene ostracoda from the Boso Peninsula, Central Japan. *In:* HANAI, T. (ed.) *Studies on Japanese Ostracoda, University Museum, University of Tokyo Bulletin,* **20**, 141–228.

YAJIMA, M. & LORD, A. 1990. The interpretation of Quaternary environments using ostracoda: an example from Japan. *Proceedings of the Geologists' Association,* **101**, 153–161.

YAMAZAKI, N. 1902. Glaciation has not appeared in Japan? *Journal of Geological Society of Japan,* **9**, 361–369, 390–398 (in Japanese).

YOKOYAMA, M. 1911. Climate changes in Japan since the Pliocene Epoch. *Journal of the Colleges of Science, Tokyo Imperial University,* **32**, 1–16.

YOKOYAMA, M. 1922. Fossils from the Upper Musashino of Kazusa and Shimosa. *Journal of the Colleges of Science, Tokyo Imperial University,* **44**, 1–200.

Australia – a Cenozoic history

DAVID BRANAGAN

School of Geosciences, University of Sydney, Sydney 2006, Australia
(e-mail: dbranaga@usyd.edu.au)

Abstract: The study of the Quaternary events that shaped the surface of Australia cannot be separated from the previous 60 Ma of the Tertiary. This history has only begun to be clearly understood since about 1950. Explorers from the late seventeenth century to the mid-nineteenth century laid the groundwork for the later understanding by their observations, particularly under difficult conditions, and by their attempts to interpret what they recorded. The western and southwestern coasts provided evidence of relative uplift of land, while the northeastern coast (the Great Barrier Reef) indicated evidence of the opposite. Explorers of various nationalities provided the evidence. Knowledge of the dry inland began to emerge from the 1830s, with puzzlement about 'hard crusts'. Evidence of limited Pleistocene glaciation and relatively young, but quite extensive, volcanic activity took somewhat longer to evaluate. A major interest for European savants was the discovery, in the late 1820s, of cave deposits of extinct vertebrates.

The vast canvas of the island continent, Australia, has seen the widest range of climate and physical conditions during the Quaternary, and indeed throughout the Cenozoic. Popular writing and even more serious works (Feeken *et al.* 1970) often refer to Australia as 'the oldest continent'. This is, of course, completely wrong, but it is almost certainly the result of the appearance of much of the inland, a 'low, flat, arid and old Landscape' as Twidale & Campbell (2005) call it. The old age of the surface, over much of the continent, is being increasingly recognized (Gale 2006). The Quaternary history of the Australian landscape cannot be easily separated from the Tertiary story over much of the continent, although, thanks to the availability of accurate, very detailed maps and modern techniques of age dating and geochemistry, much of the story of the past 20 Ma or so is now being teased out, and presenting some surprises. But these modern results, obtained in the last half century, belong essentially in a different paper.

Not surprisingly, many aspects of the Cenozoic story attracted the attention of the earliest European visitors. Opinions about the landscape, aridity and climate of the coastal regions were recorded as early as the seventeenth century by Dutch and English travellers (e.g Tasman, Dampier, Vlamingh), with increasing attention to scientific description and analysis (von Buch 1814; Fitton 1826*a*, 1826*b*; Darwin 1845; de Strzelecki 1845; Leichhardt 1847, 1855; Dana 1849; Jukes 1847, 1850; *see* also Vallance 1975; Branagan 1984, 1994, 2005*a*). Aspects of Australia's Cenozoic history continued to be studied with diminishing interest by scientists until the 1850s, with a revival of intellectual concern from the 1950s. There were, of course, always some interested individuals undertaking significant research on specific topics in the intervening period. Where Australian geology and geomorphology stood in 1950 can be read in David (1950), an expansion (largely by W. R. Browne) of what David (see below) had written in 1932.

For the purpose of discussion, this vast story can be broken into a number of topics: (1) the coastline including the reef systems; (2) the deserts and internal drainage; (3) climate change, particularly glaciation; (4) tectonism/stability, notably volcanism and earthquakes; and (5) human occupation, particularly Aboriginal, with the coincidental faunal and floral extinctions. This is a huge canvas, which, while I presented it in summarized form, with considerable visual aids, at the Vilnius INHIGEO meeting, cannot be covered adequately in a relatively short paper. Aspects of the history of some specific topics have been covered in earlier studies. See, for instance, concerning: (a) glaciation – Banks *et al.* (1987) and Branagan (1999); (b) volcanism – Branagan (1998); (c) earthquakes – Burke-Gaffney (1952), Day (1966) and Hunter (1991); (d) Aborigines and extinctions – Mulvaney (1994). For a broad cover of the changes that have led to the present Australian continent during the Cenozoic the series of volumes, albeit in popular form, but based on excellent sources, by White (1994, 1997, 2000) is invaluable. A more compressed statement (BMR Palaeogeographic Group 1990), made largely through a series of maps, is almost essential viewing. However, neither of these last named works gives virtually any hint of the workers, particularly the beginners, who began to unravel a very complex story. Some of these workers, their interwoven scientific lives and their contributions have been

From: GRAPES, R. H., OLDROYD, D. & GRIGELIS, A. (eds) *History of Geomorphology and Quaternary Geology.*
Geological Society, London, Special Publications, **301**, 189–213.
DOI: 10.1144/SP301.14   0305-8719/08/$15.00 © The Geological Society of London 2008.

studied, to some extent, by Tate (1894), Dunn (1910), Andrews (1942), Brown (1946), De Jersey (1968), Vallance (1975, which contains a very extensive bibliography), Branagan (1972*a*, *b*, 1983, 1990), Branagan & Townley (1976), Johns (1976), Brock & Twidale (1984), Darragh (1987), Twidale *et al.* (1990), Banks (1994), Organ (1997), Moyal (2003), to mention just a few.

No extensive history of the development of geomorphology in Australia has been written. There are brief, but useful reviews by Young & Twidale (1993), Twidale & Campbell (2005) and, more particularly, by Scott (1977), although the last-named was misled by some inaccurate material that he relied on. However, all three reviews begin essentially at the end of the nineteenth century, paying tribute to T. W. Edgeworth David (1858–1934), who encouraged so many students, of several generations, at the University of Sydney to pursue studies in what he called 'physiography'. They included E. C. Andrews (1870–1948), T. Griffith Taylor (1880–1993), W. R. Browne (1884–1975), W. G. Woolnough (1876–1958) and H. I. Jensen (1879–1966), who all made important contributions to Australian geomorphology (see Branagan 2005*a*; Oldroyd 2008 in this volume). J. Jutson (1874–1959), mainly in Western Australia, carried out superb work under difficult conditions (Jutson 1914, 1934; Brock & Twidale 1984). Walter Howchin (1845–1937), active in South Australia from 1881 until his death, was enthusiastic about landscape studies (Howchin 1913) and clashed with Taylor on a number of matters. J. W. Gregory's brief stay in Australia (1900–1904) produced a school textbook that was influential for several decades, to be followed by the even more influential books of E. S. Hills (1906–1986) (1940, 1975). This brings us to the 1950s and the acceleration of research, which was to reveal so much of the story of Australia's last 60 Ma or so. One other late nineteenth-century pioneer should be mentioned. This is Baron von Mueller (1825–1896). Himself a noted explorer of Australia, particularly taking part in A. C. Gregory's North West Australia expedition of 1855–1856, and earlier in the Australian alpine region, von Mueller encouraged other explorers, such as William Gosse (1842–1881) and Ernest Giles (1835–1897), to venture into the difficult inland and to make their discoveries known. He enthused businessmen and politicians to support 'geography', founding the Victorian branch of the then influential Geographical Society of Australasia in 1883, and ensuring that 'geography' had its place in the proceedings of the newly formed Australasian Association for the Advancement of Science from its inception in 1888. Furthermore, his studies of fossil plants were to prove important in the subdivisions of the Cenozoic.

The Australian landscape

Twidale & Campbell (2005) in an excellent volume *Australian Landforms*, including a long section on the Quaternary, subtitle their book: *Understanding a Low, Flat, Arid and Old Landscape*. Australia, like South Africa, did not 'enjoy' a recent Ice Age. There was, to be sure, glaciation in the eastern highlands, albeit very limited according to some authorities (e.g. Galloway 1963) and Tasmania (Lewis 1944; Paterson 1966; Branagan 1999). Some regions had frozen soils (permafrost) for limited periods, but for Australia the Ice Age (comprising several separate phases) produced two 'negatives'. The first was lowered sea level, variably so, but at its greatest extent perhaps 150 m lowering, which accelerated erosion of the coastal valleys and their surrounds, and then, as the sea finally rose to its present level, choked the eroded valleys with sediment, which over the past century and a half has caused many headaches for engineers trying to bridge or dredge the coastal waterways (Branagan 1995, 1996*a*). There were probably also repercussions for the inland waterways as well.

The more important 'negative' of the almost 'non' Ice Age was the lack of deposition of glacial debris across the land. Unlike the vast western plains of Canada and the USA, Russia and much of northern Europe, the older rocks (and even the younger ones) of Australia were not covered by nutrient-rich material during the Pleistocene. Charles Darwin (1836) told his sister Susan: 'the country is so dry and the soil [so] light that the aspect even of the better parts is very miserable'. It turned out that much of the Australian inland consisted of rocks capped only by thin residual soils, often mono-mineralic 'sheets' of calcium-, silica- or iron-rich material (which became known collectively as 'duricrust' (Woolnough 1918, 1927, 1930).

The lowered sea levels also, in many regions, left exposed for long periods unconsolidated sands that were largely the source of various dune systems which were, in time, found to cover much of the inland.

The main period of concern of this paper deals with the time from earliest European contacts to the gold rushes of the 1850s. This first extensive metal mining period took what attention there had been away from the young 'skin' of Australia to the older rocks containing the eagerly sought 'mother-lodes', once the limited finds of gold in the shallow alluvial deposits had been largely exhausted.

Even in Europe, pre-1850, the idea of an Ice Age was only just emerging. Nevertheless, it is interesting to find a hint of such in the observation of the mariner, Thomas Raine (1793–1860), following his visit to Macquarie Island in the Southern Ocean in 1822, writing to the merchant Edward Wollstonecraft (1783–1832): 'on top of the island are many lakes ... probably glacial ... evidence everywhere that the island has been covered by ice in the past' (Goddard 1940). However, Banks *et al.* (1987) suggest that the observations were most likely to have come from another on board, Dr David Ramsay, a Scot, who was a medical graduate from Edinburgh University, and who was possibly influenced by James Hutton and John Playfair.

This paper, then, reviews the history of exploration and understanding of the Cenozoic history of Australia from the earliest European records, reaching briefly to the late nineteenth century. It considers the work of the pioneers to about 1860, but mostly earlier, with just a few comments on the more recent work that has built on that original platform. In a sense these early observers, those concerned about 'geology', while aware of present natural conditions that might be changing the environment were more interested in what they regarded as past events. Thus, we read constant references to 'Tertiary' and 'Diluvium' (Vallance 1975, p. 29), which suggests that geologists continued to use the term 'diluvium' as essentially synonymous with what we call the 'Pleistocene' long after the term had become dissociated with the Flood concept. We see also both 'catastrophist' and 'uniformitarian' ideas. In the botanist Baron Karl von Hügel's record of his 10-month visit to Australia in 1833–1834, in conjunction with his perceptive observations on rock types, their relation to the overlying soils and the plant variations met with, we find him commenting on 'the small number of *revolutions* in New Holland, which explains why no rivers of any magnitude are to be found' (Clark 1994, p. 248). von Hügel attributed the formation of deep beds of rivers, and even of the smallest brooks to the astonishing 'floods ... associated with the almost tropical rains [which] are the most terrible phenomena imaginable' (Clark 1994, p. 248).

Geology or geognosy?

In a perceptive paper, Martin Rudwick (1996) has pointed out two reasons that Karl von Zittel had for naming the early nineteenth century the 'heroic period of geology'. They were 'the feats of some naturalists in exploring dangerous and remote places' and the rejection of large-scale theorizing in favour of 'the collecting of facts'. Nevertheless, as Rudwick points out, 'the search for so-called 'facts' was still guided by theoretical goals, albeit more limited ones'.

Rudwick goes on to suggest that many of the workers of this period were 'geognosts' rather than geologists, essentially interested in the (three-dimensional) structural features of particular regions, rather than contributing to any rather speculative 'theory of the earth'. He argues further that true geohistory only began when concepts developed in the *human* sciences [his emphasis], and 'specifically from human historiography' were adopted by geologists. He suggests the significance of fossils in the work of William Smith has been overplayed, in that his mapping still 'fell clearly into the tradition of geognostic description ... [in the case of Smith finding] by careful fieldwork the structural order of the formations in a specific region'. While Rudwick's review deals with the first 20 years of the nineteenth century we can, perhaps, see something of the 'geognostic' concept lingering longer in the work of the early Australian researchers.

First European observations

It should not be forgotten that the Aboriginal precursors of the European 'discovery' and later occupation of Australia had (and indeed still have) a close affinity with the land; see, for instance, Dawson (2007). While this has been recognized by many anthropologists (McBryde 1963; McCarthy 1964; Mulvaney 1969; Flood 1984, 2006; Rosenfeld 1994), the degree of understanding of geology by Aboriginal people is still in the process of investigation (Branagan 1977), and will not be developed in this paper, which begins with the first European visitors.

The observations of Willem Janzoon (1606), Dirk Hartog (1616), Frederik Houtman (1618), Jan Carstensz and Willem van Colster (1623), and Pieter Nuyts (1627) provided an outline of the western and southwestern coastline of what was later to be named Australia, but not much geological or geomorphological information can be derived from the records of these voyages. Neither did the now well-known tragic expedition of the *Batavia*, led by François Pelsaert (1629), yield a great deal for the Earth sciences. However, it is appropriate to begin the consideration of this small part of the history of Australian geomorphology on the western part of the continent.

William Dampier, an English privateer, visited the area twice, in 1688 and 1699, and was not impressed, glimpsing a 'long series of reefs and shoals behind which lay sandhills and barren country' (recorded in a series of simple topographical sketches), 'without water and inhabited by the miserablest People in the World' (Dampier 1703 published in 1939).

Nevertheless, Dampier recorded wind patterns along the western coastline, the beginning of a long history of the study of wind in Australia, a natural phenomenon that played a major part in forming the landscape of much of inland Australia. It should not be forgotten, either, that Dampier identified, named and partly mapped Shark Bay, including noting a 'shoal of coral rocks', being thus probably the first European to see, but not understand, the extraordinary Quaternary stromatolite 'colonies' that puzzled a great many fine naturalists in the coming centuries and which, in time, proved a Recent analogue of the beautifully preserved Archaean stromatolite occurrences in Western Australia (McNamara 1988).

In 1696–1697 just before Dampier's second visit, the Dutchman, Willem de Vlamingh, also untrained in science, found the western coast a 'barren, bare desolate region', and, naturally enough, he did not apply his understanding beyond the coastal region. Vlamingh's work, now available in English (Schilder 1985) is supplemented by a series of fine watercolours by Victor Victorszoon, which show many of the coastal features observed, and discussed in some detail, 120 years later, by P. P. King (King 1826; see also Branagan & Moore 2008). Sadly, the various members of this Dutch expedition seem to have had little or no interest in (or knowledge of) geological matters, except to remark on various 'duny' land, particularly south of Jurien Bay where 'the coast here resembles the high dunes of the Vlie' (Fig. 1), and to comment that on Rottnest Island the soil 'is white sand and rocky ... [and] unsuitable for cultivation'. Although rocky shores were frequently mentioned, there was no attempt to identify the rocks. However, Nicolaes Witsen (1641–1717), one of the main promoters of Vlamingh's voyage, in his 'summary' of the expedition made an interesting interpretation of some of the data, perhaps by analogy with his experience of Holland: 'in many places this Hollandia Nova seems to be drowned land. The soil is mostly salty, so that sweet drinking water is difficult to get in the holes which the natives had dug'.

Although Vlamingh's voyage and its observations have been largely ignored by scholars, this idea of at least a partly drowned coastline was the opposite of what would be proposed in later, better known, voyages. The expedition took back to Europe a small collection of shells (Schilder 1985, appendix 14), probably only living forms, several of which drew the attention of Witsen, and which he sent to the Englishman Martin Lister (1638?–1712), writing:

[T]he master hath brought me two shells from that country, hitherto not sufficiently known, which he found on the sea side, and I make bold to send you the draught of them, the shells themselves being twice as long as broad as the draught, and indeed I could not bestow them better, than to one who hath the best knowledge of these, and all other sea products.

(Shilder 1985, appendix 16, p. 221)

Witsen further wrote in a footnote:

the two shells herin [*sic*] mentioned, are printed in my Synopsis Conchyliorum, the one is the 1st nautilus, the other the concha persica clavicula radiata.

(Schilder 1985, appendix 16, p. 222)

In view of the variable time spent, and limited locations visited, such as safe harbours, it is not surprising that these early visitors often made both observations and interpretations hastily and that they were, in some cases, quickly refuted by later visitors. However, some ideas hung on for relatively long periods.

Sea-level changes

As we have seen above, Witsen was already thinking about sea-level change for Australia before 1700. Vallance (1975, 1983) has discussed aspects of relative sea-level change, beginning with the visit of George Vancouver (1757–1798) to King George's Sound in 1791. Vancouver relied on the presence of what he believed was coralline limestone on hills well above present sea level, and suggested that the elevation had occurred quite recently. This story, relating to the 'coastal limestone', is dealt with, in part, in this volume by Mayer (Mayer 2008), but perhaps some comments may be made here.

Baron von Hügel (1795–1870) visiting the Swan River settlement in 1833, in the area walked over by Vlamingh and his companions, 150 years earlier, commented on the fact that the:

[local] limestone takes a most extraordinary form. On the rises the limestone assumes the shapes of the living trees so distinctly up to a height of 8 to 10 feet that I conceived the theory that it might actually have been formed by the vegetation. Why should plants, in certain circumstances, not have the same capacity as the lower forms of animal life to transform water, particularly seawater, into a substance similar to limestone? [Banksia?] and Eucalyptus tree trunks are so clearly recognizable in these shapes that there can scarcely be any doubt that this is how they originated, particularly [as] in places where the sea has eaten into the land, and at many spots on the banks of the Swan River, these tree trunks may be distinctly seen rising from the similarly petrified roots.

(Clark 1994, p. 29)

Of his visit to Vancouver's site at King George's Sound in March 1836, Darwin wrote in his Diary:

One day I accompanied Captain Fitz Roy to Bald Head, ... where some imagined that they saw corals, and others that they saw petrified trees, standing in the position in which they had grown. According to our view, the beds have been formed by the wind having heaped up fine sand, composed of minute rounded particles of shells and corals, during which process branches and roots of trees, together with many land-shells, became enclosed. The whole then became consolidated by the percolation of calcareous matter; and the cylindrical cavities left by the decaying wood were thus filled up with a hard pseudo-stalactitical stone. The weather is

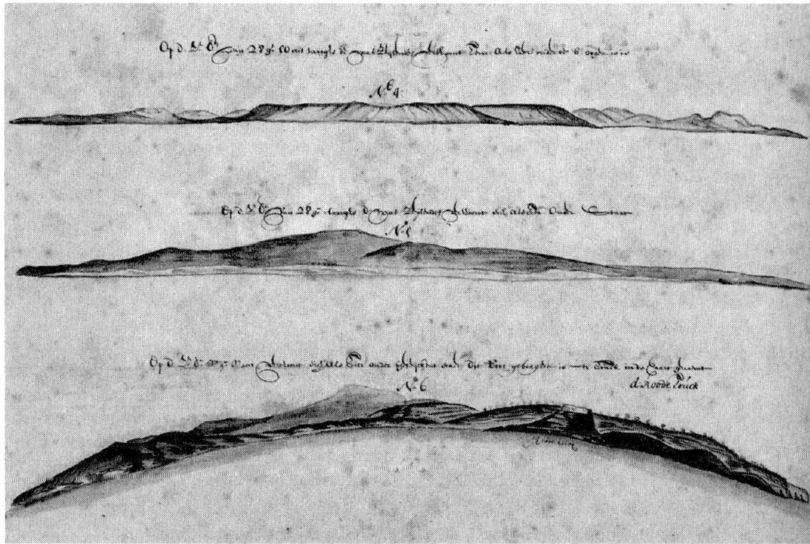

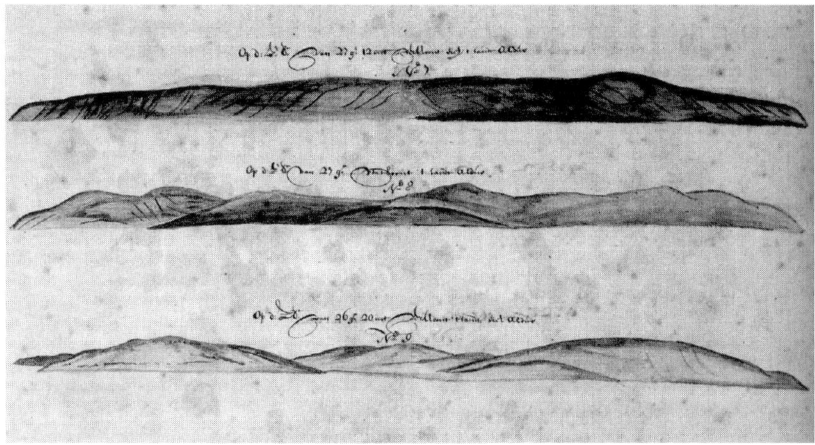

Fig. 1. Sketches by V. Victorzoon of the west coast of Australia, 1696–1697 (Schilder 1985). (Originals in colour: Prins Hendrik Maritime Museum, Rotterdam.)

now again wearing away the softer parts, and in consequence the hard casts of the roots and branches of the trees project above the surface, and, in a singularly deceptive manner, resemble the stumps of a dead thicket.

(Cited in Nicholas & Nicholas 1989, p. 166)

But Joseph Beete Jukes (1811–1869), visiting Swan River a little later, would have none of it. In the same area that von Hügel had examined, Jukes saw things differently:

In a little cliff near Fremantle, however, near the entrance of the Swan, I saw some of these dendritic masses fully exposed, and from their peculiar structure and conformation I believed them to be nothing more than *stalactites formed in the sand* by the percolation of rain water dissolving and taking up the carbonate of lime found in the sand, and re-depositing it in fantastic forms wherever a predisposing cause happened to determine it. I believe the limestone in these sands likewise to be formed in the same way, as the bedding had frequently a rather highly inclined or contorted dip, evidently not due to movements of elevation, but the result of their original formation.

(Jukes 1850. p. 61)

McNamara (in Laurent & Campbell 1987) comes down in support of Darwin's idea that the redeposition of calcium carbonate essentially forms 'fossilized roots'. However, Armstrong (1985), quoted by Nicholas & Nicholas (1989), and Semeniuk & Meagher (1981), propose that an upward capillary activity is involved in the process.

Jukes went on to describe a rather argillaceous and marly red sandstone (containing some hard grey 'gritstone' and ferruginous concretions) lying conformably below the sands and limestone near

Fremantle. Although it resembled the 'new red sandstone' of England, he found no fossils in it and its close relation with the overlying units which were 'certainly tertiary, as they in some places contained shells of the genera *arca* and *venus* apparently of species now existing on the coast, and certainly of very recent and unaltered aspect', made him 'inclined to believe ... [they] may in reality be a lower part of the same formation' (Jukes 1850, p. 62).

The southern cliffs

The cliffs of the Great Australian Bight were first recorded as early as 1623 when the Dutch ship *Gulden Zeepard* travelled almost to its eastern end (St Francis Island in Nuyts Archipelago, SSW of Ceduna) and named the region Pieter Nuyt's land after, a Dutch East Indies official on board. (Most maps have the date of discovery, 16 January 1627, marked on them.) Emanuel Bowen's 'complete Map of the Southern Continent' (1744) (reproduced in Clancy & Richardson 1988) has appended, in addition: 'This is the country seated according to Coll: Purry in the best climate in the world'. This was a reference to the work of Jean Pierre Purry, born in Neuchâtel in 1667, who is said to have visited South Australia about 1717 (!) and who published a pamphlet in 1718 extolling the virtues of living within the latitudes of 30° to 33°, in either the north or south hemisphere (Rowland *et al.* 1996). This is perhaps the reason why the region of the Bight was the place chosen by Dean Jonathan Swift to have Gulliver shipwrecked, and tied down by human creatures only a few centimetres high. Dutton (1967) comments on the irony that, even today, humans are dwarfed by the scale of the landscape. But Purrry's recommendation or prophesy was not fulfilled: the famous Nullarbor Plain is, except for a few days a year, a treeless plateau almost entirely bereft of standing surface water.

Others after Nuyt were to marvel at the cliff line's impressive nature, but it was not until Matthew Flinders was charged with surveying the coastline of Australia that details began to emerge. He wrote:

[T]he height of this extraordinary bank is nearly the same throughout, being nowhere less by estimation than 400 feet, not anywhere more than 600 . . .. This equality of elevation for so great an extent, and the evidently calcareous nature of the bank, at least in the upper 200 feet, would bespeak it to have been the exterior line of some vast coral reef, which is always more elevated than the interior parts, and commonly level with high water mark. From the gradual subsiding of the sea, or perhaps from some convulsion of nature, this bank may have attained its present height above the surface, and however extraordinary such a change may appear, yet when it is recollected that branches of coral still exist, upon Bald Head [King George Harbour], at the elevation of 400 feet or more, this supposition assumes a degree of probability, and it would farther seem that the subsiding of the waters has not been at a period very remote, since these frail branches have yet neither been all beaten down nor mouldered away by the wind and weather . . .. The bank may even be a narrow barrier between an interior and the exterior sea, and much do I regret the not having formed an idea of this probability at the time, for notwithstanding the great difficulty and risk, I should certainly have attempted a landing upon some part of the coast, to ascertain a fact of so much importance.

(Flinders 1814, Vol. 1, p. 97)

Flinders made sketches of the cliff section, showing the general twofold character of the rocks that comprised it (see Fig. 2).

More details did not emerge until the extraordinary journey in 1841 led by Edward John Eyre (1815–1901) from Fowler's Bay in South Australia

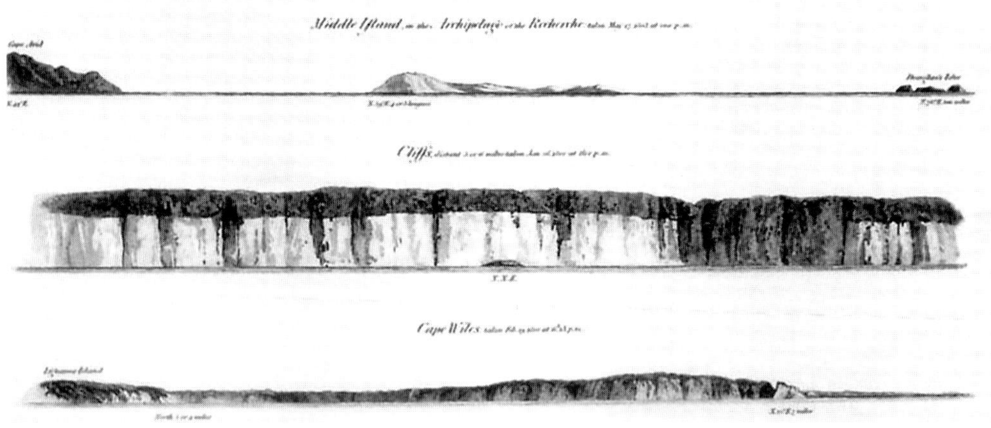

Fig. 2. Cliffs of the Great Australian Bight (William Westall 1802, in Flinders 1814).

to King George's Sound (Western Australia), a distance of about 1400 km. Despite severe privations, particularly the sparse availability of water, Eyre was still able to make useful observations of the geology (Eyre 1845):

Being now at a part of the cliffs where they receded from the sea, and where they had at last become accessible, I devoted some time to an examination of their geological character. The part that I selected was high, steep and bluff towards the sea, which washed its base, presenting the appearance described by Captain Flinders ... by crawling and scrambling among the crags I managed, at some risk, to get at these singular cliffs. The brown or upper portion consisted of an exceeding hard, coarse, grey limestone, among which some few shells were embedded, but which from the hard nature of the rock, I could not break out; the lower or white part consisted of a gritty chalk, full of broken shells and marine productions, and having a somewhat saline taste; parts of it exactly resembled the form that I had found up to the north, among the fragments of table land The chalk was soft and friable at the surface, and easily cut with a tomahawk; it was traversed horizontally by strata of flint, ranging in depth from six to eighteen inches, and having varying thicknesses of chalk between the several strata. The chalk had worn away from beneath the hard rock above, leaving the latter most frightfully overhanging, and threatening instant annihilation to the intruder. Huge misshapen masses were lying with their rugged pinnacles above the water in every direction at the foot of the cliffs, plainly indicating the frequency of a falling crag; and I felt quite a relief when my examination was completed, and I got away from so dangerous a post.

(Eyre 1845, pp. 338–339)

A less dramatic description by Eyre added just a little further information, in descending order:

At the head of the Bight: 1. Oolitic limestone in a crust probably of no great thickness, 2. Hard concrete sand with pebbles and marine shells, 3. Hard coarse grey limestone, 4. White substance like chalk. West of longitude 126° the section was a little different 1. An upper crust of oolitic limestone with shells, 2. Coarse grey hard limestone, 3. Alternate strata of white and yellow limestone in horizontal layers.

(Jukes 1850, p. 57; derived from Eyre 1845, pp. 284–285, 302–303)

The whole succession was occasionally observed by Eyre to be underlain by 'granite'. All the observers from the sea noted the succession as thinning out westerly at the Salt Lakes, east of Cape Le Grand, and from there to the west Eyre described sandstone and ironstone resting on granite. Jukes (1850) was 'strongly disposed to look on this sandstone and ironstone as belonging to the same [T]ertiary formation as the fossiliferous limestone [described by Sturt on the Murray; see below], that the two kinds of rock are either different parts of the same formation, or that they replace each other in the same geological horizon' (Jukes 1850, p. 60). This seems to have been an early suggestion of facies change, a topic suggested about the same time by Ludwig Leichhardt for the somewhat older rocks he examined on the coastline between Sydney and the Hunter River (Leichhardt 1855; Branagan 1972a, 1994).

Woods (1862), commenting on Eyre's work during his 'terrible and disastrous journey', agreed with Jukes on correlation of the cliffs of the Bight with those on the Murray, and included also the limestone which occurred around Mount Gambier. He thought 'the upper and lower deposits ... strongly resemble the mode in which the Pleiocene [*sic*] Crag occurs at home', adding 'thus a geological period which has left but slender records in Europe, is largely represented in Australia, and forms a very large portion of its continent' (Woods 1862, p. 387).

Commenting on Flinders' 'probable origin of the great sea wall, which appeared to him to be of calcareous formation', Sturt wrote:

Flinders ... concluded that it had been a coral reef raised by some convulsion of nature. Had Capt. Flinders been able to examine the rock formation of the Great Australian Bight, he would have found that it was for the most part an oolitic limestone, ['ooolitic' here is probably meant in purely compositional terms, rather than as an age (Jurassic) interpretation] with many shells imbedded in it, similar in substance and in formation to the fossil bed of the Murray, but differing from it in colour.

(Sturt 1849, vol. 2, pp. 131–132)

The main point was that Eyre's observations finally laid to rest the 40-year old idea of Flinders that the cliff was a vast uplifted coral reef. Of course, it had been uplifted, but it had not been a fringing reef, like the still-existing Great Barrier Reef. Rather, it was originally an extensive broad shallow sea. Jukes had no hesitation in assigning the age of the rocks of the Bight to the tertiary [*sic*]. He was inclined to believe that Australian geology was marked by a long break in deposition between the palaeozoic [*sic*] and the tertiary, so that there seemed no evidence of Secondary deposits.

Jukes, quoting Darwin at Hobart, Tasmania, mentions two small patches of Tertiary travertinous limestone, one tilted by an intruding 'mass of trap', the other nearly horizontal and 'but little elevated above the sea'. He went on to say 'there are very thick masses of gravel, consisting of pebbles as large as the fist, accumulated on the sides of the Derwent River at some places, and Count Strzelecki mentions great accumulations of loose sand, from beneath which he procured a large Cypraea, this was at Newton, a short distance from Hobarton [Hobart Town]' (Jukes 1850, p. 15).

Jukes on the Great Barrier Reef

Jukes (1847) wrote: 'on January 7 1843 I landed for the first time on a coral island. It was First Bunker's Island of the Capricorn Group on the north-east coast of Australia at 152° longitude on the Tropic of Capricorn'. He went on to give quite detailed descriptions of various parts of what became known as the 'Great Barrier Reef' and the adjacent coast. Even when under pressure Jukes proved himself a true geologist, noting when being attacked by Aborigines, that they were 'pelting us with large blocks of rough basalt'!

After describing the coastal journey of HMS *Fly*, in a masterful chapter of 37 pages Jukes presented a summary of his observations and ideas about the Reef and its formation (see Fig. 3). He confessed to having a problem with terminology, suggesting that the term 'reefery' might be useful to comprise a group of individual reefs.

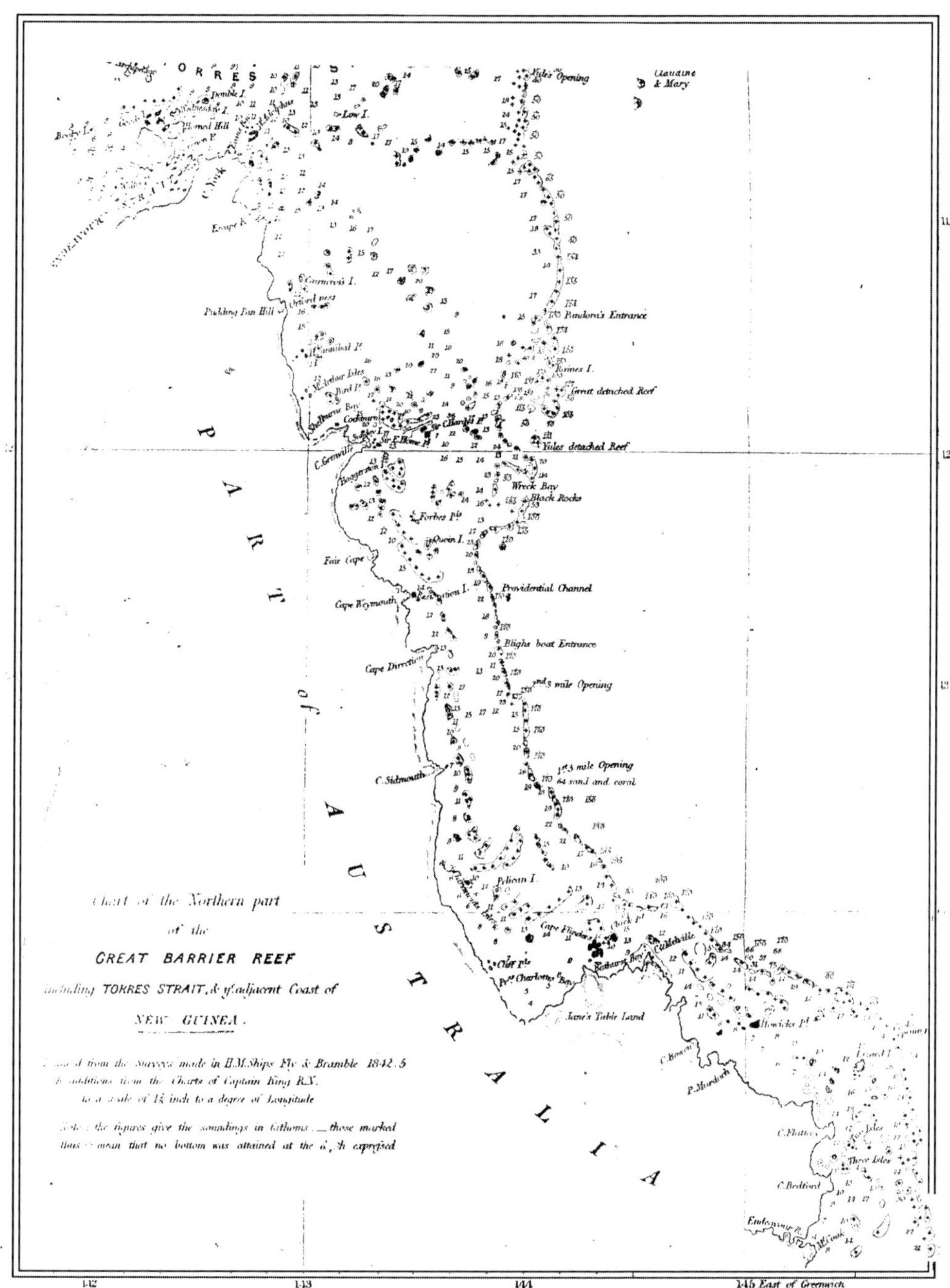

Fig. 3. Great Barrier Reef (Jukes 1850).

The outer edge of the 'Barrier' consisted essentially of linear reefs, with a few detached reefs outside. There was a variety of inner reefs. He felt the north-east coast and its barrier reefs had been 'stationary for a long period of time (say two or three thousand years) or have even been elevated a little in particular locations. The present is, so to speak, the *coral reef age of the globe*'. Jukes seemed to think that reef-forming corals were absent in earlier geological periods.

'Speaking generally', Jukes continued:

[T]he outline of the Great Barrier reef [*sic*] is parallel to the outline of the north-east coast ... [C]urves and flexures show the two are connected ... [C]ircumstances that modified the coast modified the general outline of the reefs, that is, the outline of the reefs depends on the depths of the water ... imagine the coast extending [formerly] to the present barrier position. The reef begins in shallow water, then there is slow and gradual depression.

(Jukes 1847, p. 314)

Jukes said he had no trouble accepting Darwin's theory, extending it from atolls (as first postulated) to fringing reefs. He felt the 'theory rises beyond a mere hypothesis into the true theory of coral reefs'. Having expended so many words on the geology of the Reef his summary of the geology of Australia, published just 3 years later (Jukes 1850) devoted little space to the Reef proper, but expanded somewhat on the northern extension and the changes that occur towards Torres Strait and New Guinea.

Although visits by E. C. Andrews (1870–1948) and C. Hedley (1862–1926) some years earlier added useful information about the Great Barrier Reef, it was not until 1923 that attention turned to making a long-term serious study of it by the formation of the Great Barrier Reef Research Committee, strongly supported by Edgeworth David and under the Chairmanship of H. C. Richards (1884–1947). Even as late as 1938, Richards (1938) was to write that Jukes' 1847 writing was still the best description of the complex.

J. D. Dana, who had visited the east coast in 1839–1840, stressed the variable changes (both rises and lowering) in sea level which he felt had occurred around the Australian coasts:

[Since the Tertiary] subsequent changes have taken place, as is apparent in the coral reefs of the shores north of the parallel of twenty-eight degrees, in the tertiary of the southern and western shores, and certain terraces and shell deposits along the coast.

The coral reefs indicate an extensive subsidence along the east and northeast coasts of New Holland, the amount of which we have had no means of definitely ascertaining. That it must have been great is evident from the wide channel and deep waters between the outer reef or barrier and the shore, the distance, as we have stated, being, in some parts, fifty or sixty miles, and the depth sixty to eighty fathoms. We cannot believe the subsidence to have been less than the depth of the inner channel, or five

hundred feet. The forms and extent of Ports Jackson, Broken Bay, Macquarie and Stephens have been alluded to as proofs of subsidence, probably the same that is indicated by the coral reefs.

(Dana 1849, pp. 533–534)

Dana continued: 'calcareous deposits on the southern and western shores, first noticed by Flinders, appear to indicate, in some parts, a considerable rise of the land. For facts of reference to this coast, we must refer to other authors, and especially Fitton's appendix to Captain King's Voyage [see Branagan & Moore 2008], Mr. Darwin's Volcanic Islands, and Strzelecki's New South Wales [Strzelecki 1845]. The formation appears to be of recent origin' (Dana, 1849, pp. 533–534).

Dana then cautioned about accepting evidence of uplift from shell deposits as they are 'often heaped up in great quantities by the natives of the country, who subsist generally to a great extent on the species of the coast'. However, he described features on the Illawarra coast, south from Sydney, writing 'the form of these shores leaves little occasion for doubting that the upper ridge is actually the summit of an ancient beach ... a large portion of Illawarra has been but lately reclaimed from the sea'. Although he was less certain, Dana also suggested that some shell beds along the Hunter River appeared to support the idea of a 'rise of land', a matter confirmed much later by David (1907).

The coastal dunes of Botany Bay, near Sydney, in the 1840s had reminded Ludwig Leichhardt of the Brandenburg region of northern Germany, a product of the Great Ice Age. Along the coast near Newcastle (north of Sydney) he saw something of the coastal lakes formed by growing sand spits, and of the effects of drowning by rising sea-level, but this phenomenon was not studied in Australia in any detail until the late 1800s.

Strzelecki (1845. p. 142) also had a few words to say about the recent geology – the events of his Fourth Epoch – suggesting there were three specific types of material deposited during this period: *'loose gravel or sand, ... elevated beaches*, [and] the *osseous breccia* at Wellington [Caves, inland from Sydney]'. However, he felt 'the greater part lie in confused masses, and in a state of partial decomposition, either filling the bottoms, or lodged against the sides of the valleys'. Strzelecki felt there was evidence that the 'loose ... substances' gave evidence that 'the surface of the colonies has been gradually rising, and had been for some time exposed to the attritive action of shallow water, before it arrived at its present height above the sea' (pp. 142–143). He continued:

Elevated beaches are disposed, at wide intervals, along the present coast of the two colonies [New South Wales and the then Van Diemen's Land (Tasmania)]: they present commonly horizontal beds, and occur at various heights above the existing sea; some showing marks of greater antiquity than others. Thus the elevated beaches at Lake King (Gipps Land) are seventy feet above the sea: they are composed of an indurated reddish clay and calcareous paste, containing *ostrea* and *anomia*, which are different from the existing specie ... agglutinated by a gritty paste. The elevated beach which forms Green Island, in Bass's Straits, is again but a comminuted mass of shells, and rises to the height of 100 feet: that of the south-west point of Flinders Island exhibits the same character. The two last beaches are abutted against granite, sienite and greenstone.

(Strzelecki 1845, p. 143)

Strzelecki also described beaches in Tasmania near Cape Grimm and at Table Cape approaching 'in structure to a coarse and porous sandstone', and listed the fossils found at the Table Cape beach. He concluded:

The character of these elevated beaches, and their occurrence in localities widely separated, furnish important additions to the evidence collected in other parts of the world, not only respecting the agencies which still operate in uplifting the earth's surface, but to the local and confined manifestations of such upheavings.

(Strzelecki 1845, p. 144)

A shoreline of sorts?

The relatively well-defined scarp marking the eastern edge of the Blue Mountains at Lapstone, some 50 km west of Sydney, called up numerous ideas in those minded to wonder about its presence. Catastrophic concepts had more than their day in the new colony of New South Wales. Both H. J. Antill (1779–1852) and the visiting Russians, scientist F. I. Stein (d. 1845) and artist E. Karneev, accompanied by Allan Cunningham in the Blue Mountains, were influenced by the wild scenery (Vallance & Branagan 1976; Branagan 2005*b*). The age of the rocks of this plateau began to be argued about when some of the earlier visitors claimed there was a close relation between the fossils found in the rocks and the present eucalypt forests. This alleged similarity took some years to dispel.

Baron von Hügel saw that beyond the scarp 'the most fertile black soil is to be found at the very highest point and on the slopes ... here the only theory that can be entertained is that the soil was formed before the geological revolution which elevated these hills' (Clark 1994, p. 248).

Charles Darwin, possibly reflecting his experience of seeing an emerged coast in southern South America, seems to have thought the Blue Mountains edge, west of Sydney, might have been a relatively young former coastline, and that the rocks (now regarded as Triassic) were also quite young. In his *Journal of Researches* (1845 edition) he drew an analogy between what had been observed in the West Indies and the Blue Mountains:

[In the] West Indies sediment is not in a uniform heap, but heaped around submarine rocks and islands ... waves have the power to form high and precipitous cliffs, even in land-locked harbours, I have noticed in South America. To apply these ideas to the sandstone platforms of New South Wales, I imagine that the strata were heaped by the action of strong currents, and of the undulations of an open sea, or an irregular bottom; and that the valley-like spaces thus left unfilled had their steeply sloping flanks worn into cliffs, during a slow elevation of the land; the worn-down sandstone being removed, either at the time the narrow gorges were cut by the retreating sea, or subsequently by alluvial action.

(Darwin 1845, p. 327)

The Lapstone landscape is not a shoreline, but is a tectonic feature (Branagan & Pedram 1990), as recognized by Richard Taylor (1838 p. 282), Thomas Mitchell and others (Vallance & Branagan 1976). It is a minor active scarp (see Oldroyd 2008).

The inland

Perhaps the first 'serious' geological observations of the younger rocks of the inland were made by Charles Sturt (1795–1869). The important earlier explorers, George Evans (1780–1852), John Oxley (*c*. 1785–1828) and Allan Cunningham (1791–1839), did not travel through localities where such rocks occurred to any great extent (that is, apart from volcanic rocks, for which they had, at the time, few means of dating).

Although untrained in geology, Sturt's work as a military officer had given him skills in observation and recording (particularly sketching), which he put to good use in his two widely separated periods of exploration. There were two expeditions between 10 November 1828 and 25 May 1830, by which he explored the Murray–Darling river system, including its important tributaries, the Murrumbidgee and Lachlan rivers. The second period extended from 10 August 1844 to 19 January 1846, into the dry interior north of the present city of Broken Hill in search of an inland sea that was thought by many, including Sturt, to exist (Gibbney 1967). Sturt based his idea of a possible sea on the results of Oxley's earlier excursions ending up, on both the Lachlan and Macquarie rivers, in extensive reedy swamps. He wrote 'it became, therefore, a current opinion, that the western interior of New Holland comprehended an extensive basin, of which the ocean of reeds which had proved so formidable to Mr. Oxley, formed most probably the outskirts (Sturt 1833, vol. 1, p. lxxx).

Sturt also wrote:

my impression, when traveling the country to the west and N.W of the marshes of the Macquarie, was, that I was traversing a country of comparatively recent formation. The sandy nature of its soil, the great want of vegetable decay, the salsolaceous character of its plants, the appearance of its isolated hills and flooded tracts, and its trifling elevation above the sea, severally contributed to strengthen these impressions on my mind.

(Sturt 1833, vol. 1, p. 160)

Sturt's instructions for his first expedition from Governor Ralph Darling (18 November 1828) told him to 'be particular in describing the general face of the country ... the rivers ... velocity, breadth, and depth, are carefully to be noted ... you will, as far as may be in your power, attend to the animal, vegetable, and mineral productions of the country, ... preserving specimens as far as your means will admit' (Sturt 1833, vol. 1, Appendix, p. 187).

It is interesting, in view of the above instructions, that only geological specimens were afforded special attention by Sturt (1833) in his Appendix 4 of his first volume. The older rocks (which included what he called 'Old Red Sandstone') will not be considered, but the possible younger rocks should be noted. They were:

Breccia – pale ochre colour, silicious cement, extremely hard. Cellular, and sharp edges to the fractured pebbles. Has apparently undergone fusion. Occurs in the bed of the Darling in one place only .... Chrystallized Sulphate of Lime – found imbedded in the alluvial soil forming the banks of the Darling river. Occurring in a regular vein, soft, yielding to the nail; not acted upon by acids. – See Plate [a selenite] .... Sandstone varieties – Colour dull red and muddy white; appears like burnt bricks; light, easily frangible; adheres to the tongue; occurs in large masses in the bed of the Darling; probably in connection with the rock-salt of the neighbourhood, which, from the number of brine springs discovered feeding the river, must necessarily exist. Jasper and Quartz – showing itself above the surface of a plain.

(Sturt 1833, vol. 1, pp. 199–200)

This last was a foretaste of important observations some 15 years later.

Sturt went on to write:

It is a remarkable fact, that not a pebble or stone was picked up during the progress of the expedition, on any one of the plains; and that after it left Mount Harris [north of the present township of Warren] for the Castlereagh, the only rock-formation discovered was a small Freestone tract near the Darling river. There was not a pebble of any kind either in the bed of the Castlereagh, or in the creeks falling into it.

(Sturt 1833, vol. 1, p. 200)

Sturt's second expedition took him SW from Sydney. His appendix to the volume recording his travels referred to the geological specimens he found (pp. 249–256, with three plates). In this case, only two sites are relevant: (1) a selenite occurrence, 'found imbedded in the deep alluvial soil in the banks of the Morumbidgee [sic] River', similar to that found on the Darling; and, from a geological history point of view; and (2) a more important and, at the time, relatively surprising outcrop on the bank of the Murray (see below).

Concerning the Murrumbidgee, Sturt (1833, vol. 2, p. 252) wrote that the river 'may be said to have entered the almost dead level of the interior ... a coarse grit occasionally traversed the beds of the rivers, and their lofty banks of clay or marl appear

to be based on sandstone and granitic sand. The latter occurs in slabs of four inches in thickness, divided by a line of saffron-coloured sand, and seems to have been subjected to fusion, as if the particles or grains had been cemented together by fusion'.

Sturt described the second site in some detail: 'on the right bank of the Murray, a little below the junction of the Rufus with it'. However Sturt's map, with the label 'Remarkable Cliffs', suggests that the site was *upstream* of the junction with the Rufus (see Fig. 4) (as Meinicke, see later, also interpreted).

Sturt continued:

a cliff of from 120 to 130 feet in perpendicular elevation here flanks the river for about 200 yards, when it recedes from it, and forms a spacious amphitheatre that is occupied by semicircular hillocks, that partake of the same character as the cliff itself; the face of which shewed the various substances of which it was composed in horizontal lines, that if prolonged would cut the same substances in the hillocks. Based upon a soft white sandstone, a bed of clay formed the lowest part of the cliff; upon this bed of clay, a bed of chalk reposed; this chalk was superseded by a thick bed of saponaceous earth, whilst the summit of the cliff was composed of a bright red sand. Semi-opal and hydrate of silex [flint or quartz] were found in the chalk, and some beautiful specimens of brown melenite [?melanite (garnet)] were collected from the upper stratum of the cliff.

A little below this singular place, the country again declines, when a tertiary fossil formation shews itself, which, rising gradually as an inclined plane, ultimately attains an elevation of 300 feet. This formation continues to the very coast, since large masses of the rock were observed in the channel of communication between the lake and the ocean; and the hills to the left of the channel were based upon it. This great bank cannot, therefore, average less than seventy to ninety miles in width. At its commencement, it strikingly resembled skulls piled one on the other, as well in colour as appearance. This effect had been produced by the constant rippling of water against the rock. The softer parts had been washed away, and the shells (a bed of turritella) alone remained.

(Sturt 1833, vol. 2, pp. 252–253)

Sturt included identification of a number of fossils, which he assigned to the Tertiary (presumably in terms of Lyell's classification). There is no indication of any other author but Sturt, but it seems unlikely that he had the expertise to make the identifications himself. The plate (Fig. 5) is marked by the initials J. D. C. S. This was almost certainly James de Carle Sowerby (1787–1871), the naturalist and artist who 'devoted himself to describing and illustrating specimens for others' (Cleevely 1983). Sturt (1833, vol. 2, p. 255), noted that the various genera 'are scarcely ever, and some of them not at all, found in any but tertiary formations'. He added: 'all appear to belong to the newer tertiary formations'. He also noted a 'Limestone Flustra, and their corallines, probably tertiary, from the mouth of the Sturt, on the coast

Fig. 4. Cliffs on the Murray River ('Sunset on the Murray': Sturt 1849, vol. 1, opposite p. 64).

line, nearly abreast of Mount Lofty'. Interestingly, 30 years later the different London publisher of the work of another pioneer geologist, the Reverend J. E. Tenison Woods (1862), writing on essentially the same region seems to have cut up Sturt's original plate to illustrate the fossils described by Woods!

An 1837 review

In a fascinating review of Australian geology, C. E. Meinicke (1837) began with a strong condemnation of some ideas which had been proposed up to that time, writing:

Various hypotheses concerning the geological age of Australia have been advanced but we shall put them aside as they lack profound scientific basis. The view, based on Vancouver's so-called corals of Bald Head and Flinders' survey of the Great Australian Bight, that the land only recently emerged from the ocean, and in essence, is no more than a vast coral reef, has collapsed following a closed examination of the facts. [Some had interpreted this uplift as a violent one] Recent emphatic claims of a greater antiquity, on the basis of bones found in limestone caves in the Wellington Valley, require a more detailed study of those organic remains before they can be assessed. We confine ourselves to a review of scattered geological facts.

(unpublished translation by J. Slanska & T. G. Vallance, 1970s, which also documented Meinicke's footnotes, which mainly give his sources)

Meinicke went on to comment:

we know so little about the structure of the country can be explained by the slight attention colonists pay to this aspect of Australian nature. And yet the little we do know is sufficient for us to recognise striking similarities in the geological formation of various parts of the country despite many differences in

detail' .... 'As a general rule it can be shown that there is a direct contact of basement with Tertiary rocks here; intermediate members seem to be either suppressed or altogether lacking.

(Meinicke 1837, translation by Slansky & Vallance 1970s)

This last point, mentioned earlier, was not to be abandoned until the 1860s (Clarke 1867), although the study of Strzelecki's fossils (1845) by Lonsdale and Morris already placed some doubt on the idea (Branagan 1986).

Although he did not dwell on the matter, Meinicke related the presence of intrusive porphyry (post coal measures, according to Buckland 1836, vol. 2, fig. 1) to 'the concept of uplift of the Australian highlands', a problem that still exercises the minds of researchers. In Meinicke's time, Élie de Beaumont was suggesting 12 periods of (?universal) uplift.

Meinicke then turned his attention to the inland:

Because of the predominance of flat country in Australia it is quite natural that the greater part is covered by drift. For instance, the south-eastern plains thus far explored consist of a very parched, loamy and clayey red-coloured sand, in many places containing mica. After rain it soaks up moisture rapidly and soon becomes a bog. That is the awful thing about this stretch of country; it is either an impenetrable bog or a waterless desert.

In this diluvium are found Tertiary rocks like those that were its source. The young sandstone along the Darling and the Murray belongs here as do the gypsum formation and, especially, those rocks of the first cliffs on the lower Murray above its junction with the Rufus River that have so much in common with similar formations observed in England and France. The alluvial formation of calcareous sandstone [in Australia] usually called limestone, along the coasts, corresponding to these youngest formations. It has always been of the greatest interest to travellers.

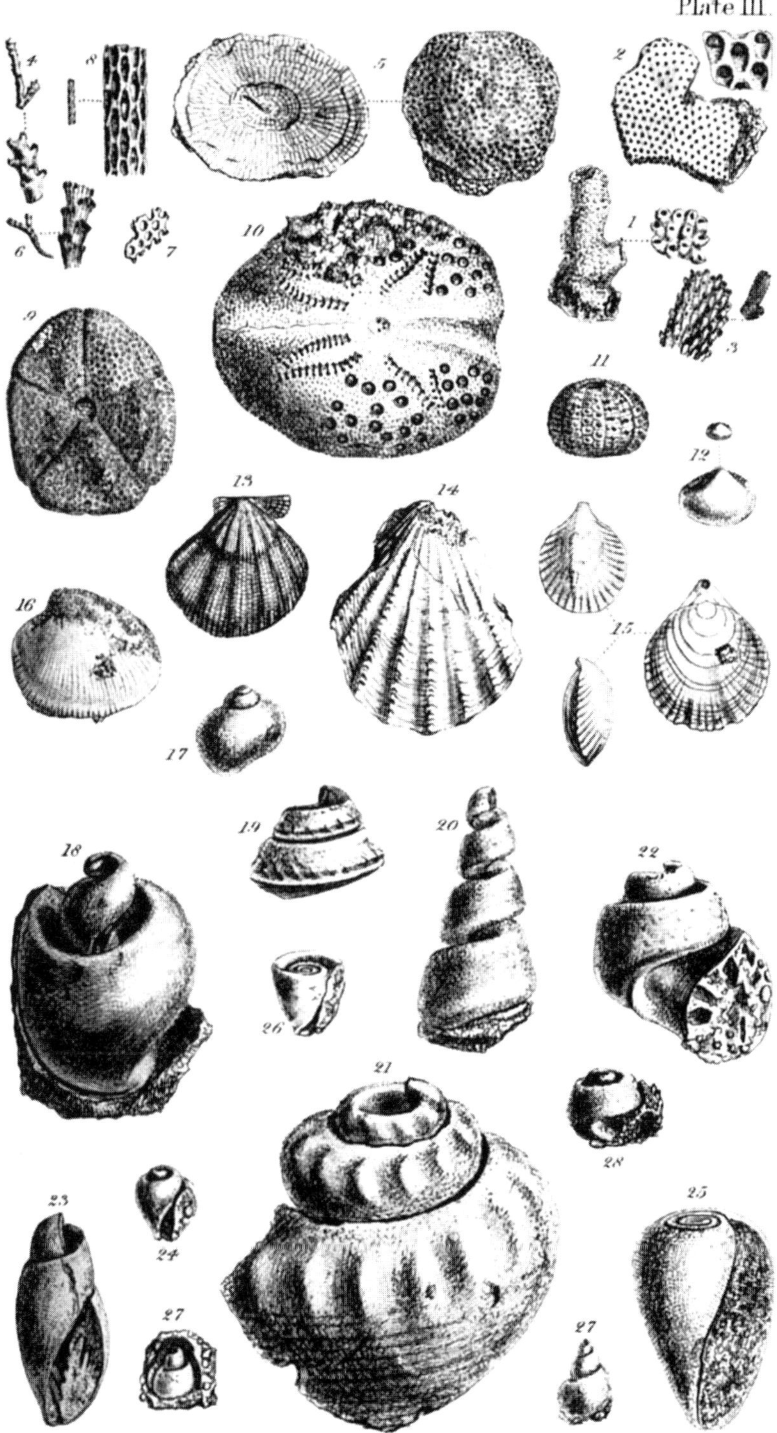

Fig. 5. Fossils collected on the Murray River by Charles Sturt (Sturt 1833, vol. 1, plate III).

It occurs along the entire southern coast [west] from Preservation Island as far as North West Cape on the west coast and forms a base for all the coastal dunes. King found it also at Cygnet Sound; Flinders had it in abundance in the Gulf of Carpentaria. On the north-east coast its place is taken by the madrepore limestone of the coral reefs; it seems to be missing from the east coast and from Van Diemen's land. Although found in many places about the Mediterranean Sea, in the West Indies and in South Africa, the formation is nowhere more extensive than in Australia. It consists of isolated fragments of older rocks and of quite various organic masses bound by a calcareous cement. Hence the rock is sometimes called limestone, sometimes sandstone. Under the influence of sea-water this rock is still steadily being formed along the coast. What aroused such great interest in the rock were the peculiar forms, similar to organic structures, found abundantly in it. These were held, wrongly, to be either corals or fossil plants although it does enclose undoubted organic masses (trees) in no small number, as at Recherche Archipelago, on the Swan River and at Sharks Bay.

So far this formation has been observed on the sea coast only but during his remarkable journey on the Murray Sturt found that it reaches more than 100 miles inland near Encounter Bay. The river has cut this extraordinary deposit, particularly marked here by the abundance of shells. Its exposed banks present an unusual aspect that eventually will be significant in the study of these rock formations.

(Meinicke 1837, translation by Vallance & Slansky 1970s)

Seeing the desert

Although the idea of an inland sea was becoming less believable, Sturt finally had the opportunity to venture much further into the interior of the continent, 14 years after his examination of the Murray–Darling river system.

This epic interior journey in 1844–1846 introduced the 'stony deserts' and expanded on the knowledge of the salt lakes that dot the continent's interior, first noted during an earlier expedition by Eyre. There was a link also back to observations that Sturt had made on his earlier travels – the presence of silicification and ferruginization in otherwise soft and 'young' sediments (see for instance Sturt's description of breccia 'Siliceous cement of the most *westerly* of the hills between the Lachlan and Macquarie Rivers'). Sturt illustrated the curious hard flat-topped caps that dotted the interior, and which were to take more than 100 years to explain as a process of 'duricrusting' mainly by the upwards migration of silica (or in other cases calcium or iron, depending mainly on climatic conditions). The stony deserts resulted from the decay of these hard surfaces and the spread of the broken fragments over lower plains. All these features were related to the problem of the so-called 'Desert Sandstone', first raised by Richard Daintree (1868) and R. L. Jack (1879), which seemed to be a major stratigraphic unit of uncertain age, but is now mainly considered Tertiary (Branagan 2004).

The only fossils collected on this expedition of Sturt were again Tertiary shells similar to those found in the Murray River cliffs 15 years earlier. There were also numerous samples of quartz and silicified materials related to the idea of the Desert Sandstone problem mentioned above. Sturt (1849, vol. 2, p. 62) wrote: 'on the Stony Desert the fragments of rock, with which it was covered, were composed of indurated quartz, rounded by attrition, and coated with oxide of iron'. On p. 60 he mentioned only one apparently recent deposit, 'Soapstone', found at Carnapaga on the first creek to the NW of the Darling River:

The valley of Cooper's Creek was, however, bounded by low quartzose hills, covered with sand. The general level of the interior was otherwise ferruginous clay, on which the long sandy doones [*sic*] or ridges rested, excepting where their regularity was broken by flooded plains. The clay rested on sandstone, which, with a few exceptions, where fossil tertiary limestone occurred, similar to that of the Murray cliffs, was ferruginous sandstone, at the depth of two feet and a half or three feet.

(Sturt 1849, vol. 2, appendix, p. 62)

We now jump 15 years for a final comment on the desert-like inland. It is by John McDouall Stuart (1815–1866), who had been one of Sturt's companions on the difficult 1845–1846 expedition, and who had already examined much of the region around Lake Eyre in 1859 and ventured onto the lake 'but could only get about two miles; it became so soft that I was sinking to the ankles, and the clay was so very tenacious that it completely tired me before I got back to the horses' (Mudie 1968, p. 87).

Stuart recognized, and named, some of the springs of the region that were later to be identified as outlets of the huge 'Great Artesian Basin', which later provided water for a very thirsty land. In 1860 he successfully reached the geographical centre of the continent (as it was then thought), naming the closest hill after Sturt, but it was later renamed after Stuart. On this expedition he named the Finke River, recognizing, even in its mainly stagnant waterholes, its importance as a drainage feature. On 6 April 1860 he noted a 'remarkable hill, which at this distance has the appearance of a locomotive with its funnel'. He found that it was 'a pillar of sandstone, standing on a hill upwards of 100 feet on height. From the base to the top it is 105 feet, twenty feet wide by ten feet deep, and quite perpendicular, with two peaks at the top'. He named it Chambers Pillar after one of the Adelaide backers of his expeditions (see Fig. 6). Stuart continued: 'to the N. and N. E. of it are a number of remarkable hills, which have the appearance of old castles in ruins, and are standing in the midst of sandhills' (Mudie 1968, p. 100). These relate closely to what Sturt had illustrated earlier and relate directly to the longstanding enigma of the Desert Sandstone.

Fig. 6. Chambers' Pillar, Central Australia, named by J. McDouall Stuart (McDouall Stuart 1866).

Sturt also reminded his readers:

[T]he desert of Australia is not more extensive than the deserts in other parts of the world. Its character constitutes its peculiarity, and that may lead to some satisfactory conclusion as to how it formed, and by what agent the sandy ridges which traverse it were thrown up. I repeat that I am diffident of my own judgment.

(Sturt 1849, vol. 2, p. 134)

Ice Age evidence

The Reverend W. B. Clarke (1798–1878), a great pioneer of Australian geology, in 1852 thought there had been glaciation in the highest Australian range surrounding Mount Kosciuszko (named by Strzelecki), but he produced only a little evidence (Clarke 1853). While the idea of Raine (or Ramsay) about glaciation a considerable distance away from the mainland has been previously mentioned, the first suggestion of glaciation on the Australian continent seems to have been made by Murray (1843) (also by quoted Vallance 1975; see also Banks *et al.* 1987) in the western highlands of Victoria, WNW from Melbourne. Although dismissed by Vallance and others, it is possible that Murray was seeing glacial evidence, not of Pleistocene age, but of late Palaeozoic age (Branagan 1999). Alfred Selwyn recognized definite evidence of glaciation near sea level in South Australia in 1859, and believed it was also Pleistocene. It was later recognized as late Palaeozoic. Clarke in 1861 wrote more fully about possible glaciation in the Australian alpine region in a letter to Charles Darwin, who incorporated some of this information in the third edition of his *Origin of Species* (see Moyal 2003, vol. 1, pp. 560–562).

Definite recognition of Pleistocene glaciation only came in the 1880s. The Austrian zoologist, Robert von Lendenfeld, presented good evidence, which was later supplemented by more detailed work by Edgeworth David, Richard Helms and others. W. R. Browne took up the work during the middle 1900s (Branagan 1999). The extent of glaciation in the Kosciuszko region began to be challenged in the 1950s, with geomorphologists such as R. W. Galloway and A. Costin arguing that true glaciation was relatively limited in extent, but that there had been widespread periglaciation in the eastern highlands (Galloway 1963).

Branagan and several others have suggested that the region was also affected by the Late Palaeozoic glaciation, and that similarities in glacial effects can cause confusion. The superposition of the two separate glacial epochs can be clearly recognized in Tasmania, where there is also evidence of multiple Pleistocene glaciation in studies first made by Paterson (1966).

Volcanism

Volcanic activity was observed by William Dampier in the New Guinea region in 1700 (Dampier 1703, reprinted 1939; Branagan 1998). This could have suggested to many that one might expect volcanoes on the continent. James Cook, Flinders, George Bass, John Oxley and others were on the lookout (Branagan 1998; Branagan & Moore 2008). An important observation was the recording from the *Lady Grant*, captained by James Kent in 1803 of Mounts Schank and Gambier in SE South Australia. It was not until 1836 that this region was visited by a scientific observer, when the surveyor and explorer Thomas

Fig. 7. Important workers: (**a**) Matthew Flinders; (**b**) Charles Sturt; (**c**) Edward J. Eyre; (**d**) Thomas Mitchell; (**e**) J. Beete Jukes; and (**f**) J. Mcdouall Stuart.

Mitchell was captivated by the perfect shape of Mount Napier, in Victoria, and found on climbing to its top a circular vent containing cellular rocks – lava and scoria: 'The igneous character of these was so obvious that one of the men thrust his hand into a chasm to ascertain whether it was warm'. Mitchell felt that 'some might conclude that the volcano had been in activity at no very remote period' (Mitchell 1838). Work by Burr (1846) in South Australia, Westgarth (1846, 1853) in Victoria, followed by Selwyn & Ulrich (1866) brought out details of the volcanic history of what became known as the Western Victoria Volcanic Province, with three ages of volcanism extending from pre-Miocene to Present (possibly as recent as 4000 years).

Aware of some of this work Dana (1849) designated the western Victoria region 'the single volcanic region on the Australian Continent', but neither he nor Jukes (1850) seems to have known of the observations in (present) Queensland by Leichhardt (1847) of the Peak Range resembling 'very much the chain of extinct volcanoes of Auvergne', and along a tributary of the Burdekin River, 'where the whole appearance ... showed that the stream of lava was of much more recent date than the rock of the tableland' (Leichhardt 1847, pp. 126 and 244). By 1862 the western Victoria volcanic region was sufficiently well known to be listed by Scrope in his 1862 inventory of world volcanoes (see Branagan 1998).

According to Jukes (1850, p. 31) concerning the three islands Murray, Darnel and Bramble Key (or Cay), off the farthest north-eastern Australian coast: 'the rocks in these three localities are volcanic, consisting partly of sandstone and conglomerate made of pebbles of lava and coral limestone, with some beds of finer tuff, and partly of large masses of dark heavy hornblendic lava. The eruption of these volcanic rocks though probably of comparatively modern origin, geologically speaking'. Jukes added:

before quitting the eastern coast of Australia I must mention one singular circumstance, namely the occurrence throughout its whole extent from Bass's Straits to Torres Straits of pebbles of pumice, strewed over the flats just behind the beaches to the height of a few feet above high water mark. These pebbles were always well rounded. Varying in size from that of nuts to pieces as big as the fist, and are frequently found in considerable abundance. They have never been seen floating nor left on the actual beach by the wash of the sea, except in a few instances where they may have been washed down from above by the rains. It is difficult to conjecture their origin, but they may have proceeded from some old very violent eruption of the volcanoes of New Zealand, or of one or more of the foci of the volcanic band between that country and New Guinea. Their occurrence above high water mark does not necessarily involve the supposition of the elevation of the coast; as a wave of sufficient magnitude to float them into their present position may have been the result of one or more earthquakes which probably accompanied the exhibition of the volcanic violence by which they were originally produced.

(Jukes 1850, pp. 34–35)

However, the Reverend C. P. N. Wilton (1832) had already suggested that they originated from eruptions of White Island, off the North Island of New Zealand.

In terms of age of eruption, Jukes (1850, p. 37) noted: 'Seen about Melbourne was a heavy hornblendic trap or lava, sometimes very heavy and compact, but often cellular and even scoriaceous, with all the appearance of recent subaérial lava ... in some instances this lava seems to have flowed down the existing valleys of the country'.

Vallance (1975, pp. 23–24) has discussed the interest of the French scientists in the Australian volcanic story and the acceptance by non-visiting Germans, such as Zimmerman (1810), of possible volcanism on the continent, based possibly on Peter Pallas's scheme, which demanded volcanic activity in the South Seas.

Vertebrate palaeontology and the 'Diluvium'

In the perceptive preliminary chapter in his first volume, Sturt (1833) gave an interesting account of his observations on the close relations between the geology and vegetation of much of New South Wales. His brief statement of the geology extending from Sydney to Wellington, some 260 km NW, could still be largely accepted today. In his mention of the limestone caves of 'Moulong' (the present Molong) and Wellington, Sturt gave an indication that he had observed caves in Europe in earlier years. He commented: 'A close examination of these caves has led to the discovery of some organic remains, bones of various animals imbedded in a light red soil; but I am not aware that the remains of any extinct species have been found ... There can be little doubt but that the same causes operated in depositing these mouldering remains in the caves of Kirkdale and those of Wellington Valley' (Sturt 1833, p. xxxvi). The Kirkdale Caves in Yorkshire had become famous after their discovery in 1821 by William Buckland (1784–1856), of Oxford, who showed that many of the fragmentary bones discovered there were of extinct animals (Buckland 1823; Rudwick 1985).

In 1826 Fitton had regretted there was little or no information about the Australian diluvial deposits. But he did not have too long to wait. In fact, Charles Frazer had already found red earth cave deposits at Limekilns, north of Bathurst. But it was the finds of fossil bones in similar deposits, just 3 years later at Wellington Caves, New South Wales, which

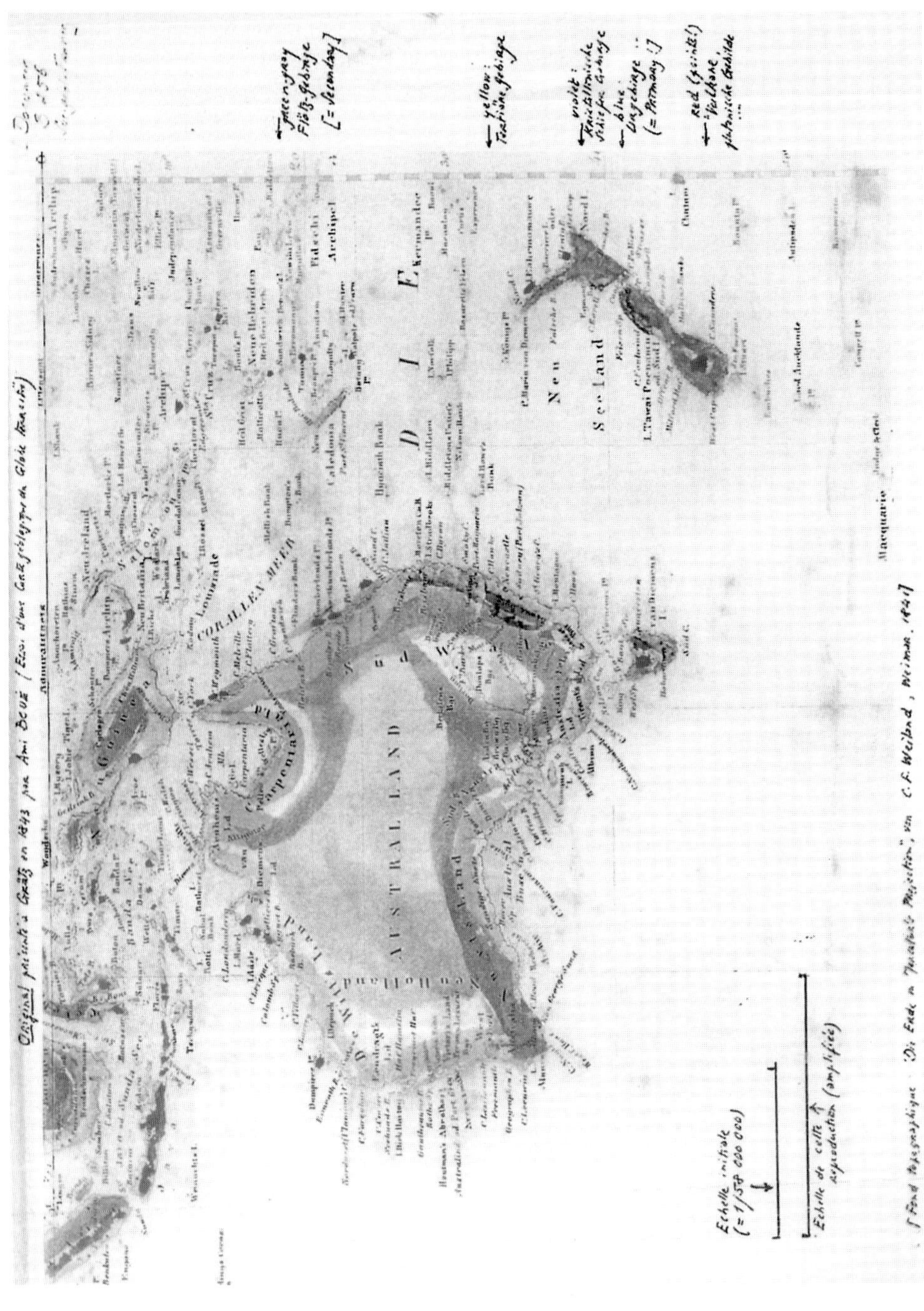

Fig. 8. Geological map of Australia by Ami Boué (Société Géologique de France, Paris 1843).

Fig. 9. Geological map of Australia by Jules Grange (Grange 1850).

stirred European interest (Holland 1992; Oldroyd 2007). J. D. Lang took specimens to Robert Jameson in Edinburgh in 1830, which were sent on to Cuvier, but arrived after his death and were studied by Pentland (1830, 1832, 1833). Thomas Mitchell arranged for material to go to William Buckland (1830), and this material was also sent on to France.

The relations between the fossil vertebrate faunas and the diluvium concept attracted the attention of many continental scientists. Publications by Melleville (1842), d'Archiac (1847, 1865) and Gervais (1854, 1869, 1870) on aspects of the Australian vertebrate fossil fauna testify to continuing interest in this topic in France. It would have been further stimulated by displays of relevant fossil material at the 1867 Paris Exposition (Krefft 1867).

The involvement of French savants in the study of early man was also considerable. The Aboriginal people of Australia were, at that time, regarded as the most primitive race known. Consequently there was continuing interest in the possibility of finding evidence of fossil Man, and likely associated extinct vertebrates, in Australia.

A magnificent collection, with more than 1000 'partly determined specimens', of the remains in the Wellington Caves was made in 1870 by Thomson & Krefft, and sent to Professor Owen, who used it to advance the knowledge of Australian vertebrate palaeontology (Owen 1877, p. 182). Cuvier, of course, had been one of the major influences on Owen, in his earliest years as an anatomist (Branagan 1992). However, it was the Australian megafauna that directly attracted Owen into palaeontology and which was the basis of his career as creator of the Museum of Natural History (Rupke 1992).

Summary/conclusions

As indicated in the previous pages, by 1850 a considerable amount of factual information had accumulated about the landscape of Australia through the work of researchers such as those shown in Fig. 7. But despite attempts to summarize the results on geological maps such as those by the Frenchman Ami Boué (1843) and Jules Grange (1850) (see Figs 8 and 9), who relied on the information from others, and the Pole Strzelecki (1845) and the Englishman Jukes (1850), who saw something of the continent for themselves, there was still a long way to go to explain the geology and the landscape. Both Boué and Grange indicated extensive Tertiary rocks in most of the regions discussed in this paper. Strzelecki's map, which is more restricted in extent, showed 'Fourth Epoch' rocks in the Gippsland region of eastern Victoria,

and smaller areas in both Tasmania and in the Hunter River region, north of Sydney. Jukes (1850) attempted greater sophistication, showing five relevant units: Alluvial; Coral Reefs; Tertiary Rocks; Ages unknown but supposed to be Tertiary; Basalt, lava and Tertiary or modern igneous rocks.

All these ideas formed a fascinating platform on which to build, but the need to feed and clothe the growing population, and produce coal and metals for export and profit, meant that geological attention did not turn quickly to problems that many would have thought, at the time, were purely of academic interest. We know now that the nation might have avoided many large-scale problems had attention been directed much earlier to the continent's fragile surface, its river systems and climatic changes. A small band of dedicated workers has done an extraordinary job on such matters in the past 50 years (see for instance Williams *et al.* 1991), but that is another story.

In some ways, the unravelling of the youngest events in Australian geology has probably proved a much slower process than the study of more ancient events.

My thanks to A. Grigelis for encouragement with my paper presented at the INHIGEO meeting in Vilnius, August 2006. I owe a considerable debt to my late colleague, T. Vallance, who gave me much unpublished research material on early Australian geology and which deserves to see the light of day. Numerous researchers, referred to in the paper, have provided information on which the paper is based. A. Mason first made me aware of the geological map by Ami Boué. I am grateful also to R. H. Grapes, who accommodated my plea for the late submission of this paper. My thanks go also to the helpful reviews of the first version of this paper by R. Twidale and B. Joyce, both of whom have made major contributions to the study of Australian geography in the past 50 years. I express my admiration and thanks to the pioneers; they noted the facts; they pondered the reasons.

References

ANDREWS, E. C. 1942. The heroic period of geological work in Australia. *Journal and Proceedings of the Royal Society of New South Wales*, **76**, 96–128.

ARCHIAC, A. D.' 1847. *Histoire des progrès de la géologie de 1834–1845.* Société Géologique de France, Paris.

ARCHIAC, A. D'. 1865. *Leçons sur la faune quaternaire (paléontologie stratigraphique).* Société Géologique de France, Paris, 263–274.

ARMSTRONG, P. 1985. *Charles Darwin in Western Australia: A Young Scientist's Perception of an Environment.* University of Western Australia Press, Nedlands.

BANKS, M. R. 1994. Geological torchbearers in Tasmania, 1840 to 1890—two amateurs and a professional. *In*: BRANAGAN, D. F. & MCNALLY, G. H. (eds)

Useful and Curious Geological Enquiries beyond the World. The 19th International INHIGEO Symposium, Sydney. Conference Publications, Springwood, New South Wales, 214–230.

BANKS, M. R., COLHOUN, E. A. & HANNAN, D. 1987. Early Discoveries XXXIV: Early discoveries of the effects of ice action in Australia. *Journal of Glaciology*, **33**, 231–236.

BMR PALAEOGRAPHIC GROUP (co-ordinated by H. I. M. Struckmayer & J. M. Totterdell). 1990. *Australia, Evolution of a Continent*. Australian Government Publishing Service, Canberra.

BOUÉ, A. 1843. *Essai d'une carte géologique du globe terrestre (presenté à Grätz en 1843)*. Société Géologique de France, Paris, Document, **B 256 bis**.

BRANAGAN, D. F. 1972a. *Geology and Coal Mining in the Hunter Valley 1791–1861*. Public Library, Newcastle, Newcastle Historical Monograph, **6**.

BRANAGAN, D. F. 1972b. Words, actions, people: 150 years of the scientific societies in Australia. *Journal and Proceedings of the Royal Society of New South Wales*, **104**, 123–141.

BRANAGAN, D. F. 1977. Aborigines in the landscape. *In*: STANBURY, P. J. (ed.) *The Moving Frontier: Aspects of Aboriginal–European Interaction in Australia*. Reed, Terrey Hills, NSW, 38–48.

BRANAGAN, D. F. 1984. The history of geological mapping in Australia. *In*: BORCHARDT, D. H. (ed.) *Some Sources for the History of Australian Science*. University of New South Wales, Historical Bibliography Monograph, **12**, 33–46.

BRANAGAN, D. F. 1986. Strzelecki's geological map of southeastern Australia: an eclectic synthesis. *Historical Records of Australian Science*, **6**, 44–58.

BRANAGAN, D. F. 1990. Alfred Selwyn – 19th century trans-Atlantic connections via Australia. *Earth Sciences History*, **9**, 143–157.

BRANAGAN, D. F. 1992. Richard Owen in the antipodean context [a review]. *Journal and Proceedings of the Royal Society of New South Wales*, **125**, 95–102.

BRANAGAN, D. F. 1994. Ludwig Leichhardt: geologist in Australia. *In*: LAMPING, H. & LINKE, M. (eds) *Australia: Studies on the History of Discovery and Exploration. Frankfurter Wirstchafts-und Sozialgeographische Schriften*, **65**, 105–122.

BRANAGAN, D. F. 1995. The Sydney–Newcastle railway: 19th century engineering geology? *In*: BOYD, R. F & MCKENZIE, G. A. (eds) *29th Sydney Basin Symposium*. Department of Geology. University of Newcastle, 125–132.

BRANAGAN, D. F. 1996a. Bricks, brawn and brains – two hundred years of geology and engineering in the Sydney region (presidential address). *Journal and Proceedings of the Royal Society of New South Wales*, **129**, 1–32.

BRANAGAN, D. F. 1996b. John Alexander Watt, geologist on the Horn Expedition. *In*: MORTON, S. & MULVANEY, D. J. (eds) *Exploring Central Australia, The Environment and the Horn Expedition*. Surrey, Beatty and Sons, Chipping Norton, 42–58.

BRANAGAN, D. F. 1998. Streams of fire, burning mountains, pumice and volcanic activity in no very remote period: volcanism in the Australasian region: the search for understanding: to 1860. *In*: MORELLO, N.

(ed.) *Volcanoes and History. Proceedings of the 20th INHIGEO Symposium (Italy)*, Brigati, Genoa, 1995, 23–45.

BRANAGAN, D. F. 1999. Antipodean ice ages. *Eclogae geologicae Helvetiae*, **92**, 327–338.

BRANAGAN, D. F. 2002. Australian stratigraphy and palaeontology: the nineteenth century French contribution. *Comptes Rendurs Palevol*, **1**, 657–662.

BRANAGAN, D. F. 2004. The Desert Sandstone of Australia, a late nineteenth century enigma of deposition, fossils, and weathering. *Earth Sciences History*, **23**, 208–256.

BRANAGAN, D. F. 2005a. *T. W. Edgeworth David: A Life*. National Library of Australia, Canberra.

BRANAGAN, D. F. 2005b. Two Russian mineralogists 'lost' in the Australian bush. *Journal of the Russian Mineralogical Society*, **N1, 2005**, 45–52.

BRANAGAN, D. F. & LIM, E. 1985. J. W. Gregory – traveller in the Dead Heart. *Historical Records of Australian Science*, **6**, 71–84.

BRANAGAN, D. F. & MEGAW, J. V .S. 1969. The lithology of a coastal aboriginal settlement at Curracurrang, NSW. *Archaeology and Physical Anthropology in Oceania*, **14**, 1–17.

BRANAGAN, D. F. & MOORE, D. T. 2008. W. H. Fitton's Geology of Australia's Coasts, 1826. *Historical Records of Australian Science*, **19**, 1–51.

BRANAGAN, D. F. & PEDRAM, H. 1990. The Lapstone Structural Complexi—New South Wales. *Australian Journal of Earth Sciences*, **37**, 23–36.

BRANAGAN, D. F. & TOWNLEY, K. A. 1976. The geological sciences in Australia – a brief historical review. *Earth Sciences Reviews*, **12**, 323–346.

BROCK, E. J. & TWIDALE, C. R. 1984. J. T. Jutson's contribution to geomorphological thought. *Australian Journal of Earth Sciences*, **31**, 107–121.

BROWN, I. A. 1946. An outline of the history of palaeontology in Australia. Proceedings of the Linnean Society of New South Wales, **71**, v–xviii.

BUCH, L. VON. 1814. Einige Bemerkungen über die geognostiche Constitution von Van Diemens Land. *Magazin Gesellschaft naturfunde Freunde zu Berlin*, **6**, 234–240.

BUCKLAND, W. 1823. *Reliquiae diluvianae; or Observations on the Organic Remains Contained in Caves, Fissures, and Diluvial Gravel, and on Other Geological Phenomena, Attesting the Action of a Universal Deluge*. John Murray, London.

BUCKLAND, W. 1830. Sur les ossemens découvertes à la Nouvelle Hollande. *Bulletin de la Société Géologique de France*, **1**, 227.

BUCKLAND, W. 1836. *Geology and Mineralogy Considered in Relation to Natural Theology*. 2 Vols. William Pickering, London.

BURKE-GAFFNEY, T. N. 1952. The seismicity of Australia. *Journal and Proceedings of the Royal Society of New South Wales for 1951*, **85**, 47–52.

BURR, T. 1846. *Remarks on the Geology of South Australia*. Murray, Adelaide.

CLANCY, R & RICHARDSON, A. 1988. *So Came They South*. Shakespeare Head Press, Silverwater, Sydney.

CLARK, D. (translator & editor). 1994. *Baron Charles von Hügel New Holland Journal November 1833–October*

1834. Melbourne University Press, Miegunyah Press, Melbourne.

CLARKE, W. B. 1853. Southern goldfields report from Eden, Twofold Bay, 6 March 1852. On the geology of the south-east parts of the County of Wellesley, with remarks on Maneroo generally [Report 12]. *Papers Relative to Geological Surveys, New South Wales: New South Wales Parliamentary Blue Book*, **16**, 66–71.

CLARKE, W. B. 1867. Remarks on the sedimentary formations of New South Wales, Illustrated with references to other provinces of Australia. *In: Catalogue of the Natural and Industrial Products of New South Wales, Forwarded to the Paris Universal Exposition*. Government Printer, Sydney, 65–80.

CLEEVELY, R. J. 1983. *World Palaeontological Collections*. British Museum (Natural History) & Mansell, London.

DAINTREE, R. 1868. *Report of the Cape River Diggings and the Latest Mineral Discoveries in Northern Queensland*. Government Printer, Brisbane.

DAMPIER, W. 1939 [1703]. WILLIAMSON, J. A. (ed.) *A Voyage to New Holland . . .*. Argonaut Press, London.

DANA, J. D. 1849. *United States Exploring Expedition during the years 1838, 1839, 1840, 1841, 1842 under the Command of Charles Wilkes, U.S.N. Vol. X. Geology*. C. Sherman, Philadelphia, PA.

DARRAGH, T. A. 1987. The Geological Survey of Victoria under Alfred Selwyn, 1852–1868. *Historical Records of Australian Science*, **7**, 1–25.

DARWIN, C. 1836. Letter to Susan Darwin (from Sydney) 28 January, 1836. *In:* BARLOW, N. (ed.) *Charles Darwin and the Voyage of the Beagle*. Pilot Press, London (1945), 132.

DARWIN, C. 1845. *Journal of Researches into the Natural History and Geology of the Countries visited during the Voyage of H.M.S. Beagle round the World, under the Command of Capt. Fitz Roy, R. N.* 2nd edition. John Murray, London, 327.

DAVID, T. W. E. 1907. The geology of the Hunter River coal measures . . . . *Memoir of the Geological Survey of New South Wales*, **4**. Government Printer, Sydney.

DAVID, T. W. E. 1932. *Explanatory Notes to Accompany a New Geological Map of the Commonwealth of Australia*. Commonwealth Council for Scientific and Industrial Research, Melbourne.

DAVID, T. W. E. 1950. *The Geology of the Commonwealth of Australia (Edited and Much Supplemented by W. R. Browne)* (3 vols). Edward Arnold, London.

DAWSON, K. 2007. From the red earth to Paris. Obituary of Dawson (*c.* 1935–2007). *Sydney Morning Herald*, 9 February, 16.

DAY, A. A. 1966. The development of geophysics in Australia. *Journal and Proceedings of the Royal Society of New South Wales*, **100**, 33–60.

DE JERSEY, N. J. (ed.) 1968. Palaeobotany and palynology in Australia: a historical review. *Review of Palaeobotany and Palynology*, **6**, 111–136.

DUNN, E. J. 1910. Biographical sketch of the founders of the Geological Survey of Victoria. *Bulletin of the Geological Survey of Victoria*, **23**.

DUTTON, G. 1967. *The Hero as Murderer: The Life of Edward John Eyre Australian Explorer and Governor of Jamaica 1815–1901*. Collins Cheshire, Sydney.

EYRE, E. J. 1845. *Journals of Expeditions of Discovery into Central Australia and Overland from Adelaide to King George's Sound in the Years 1840–1*, Volumes 1 and 2. T. W. Boone, London.

FEEKEN, E. H. J., FEEKEN, G. E. E. & SPATE, O. H. K. 1970. *The Discovery and Exploration of Australia*. Nelson, Melbourne.

FITTON, W. H. 1826*a*. Appendix C. An account of some geological specimens collected by P. P. King, in his survey of the coasts of Australia, and by Robert Brown, Esq., on the shores of Carpentaria, during the voyage of Captain Flinders. *In:* KING, P. P. (ed.) *Narrative of a Survey of the Intertropical and Western Coasts of Australia Performed Between the Years 1818 and 1822*, 566–622.

FITTON, W. H. 1826*b*. An account of some geological specimens, collected by Captain P. P. King, in his survey of the coasts of Australia, and by Robert Brown, Esq., on the shores of the Gulf of Carpentaria, during the voyage of Captain Flinders. *Philosophical Magazine and Journal*, **68**, 14–34, 132–147.

FLINDERS, M. 1814. *A Voyage to Terra Australis: Undertaken for the Purpose of Completing the Discovery of that Vast Country and Prosecuted in the Years 1801, 1802, and 1803 . . .* (2 vols). G. W. Nicol, London.

FLOOD, J. 1984. *Archaeology of the Dreamtime*. Collins, London.

FLOOD, J. 2006. *The Original Australians: Story of the Aboriginal People*. Allen & Unwin, Sydney.

GALE, S. J. 2006. Long-term landscape evolution in Australia. *Earth Surface Processes and Landforms*, **17**, 323–343.

GALLOWAY, R. W. 1963. Glaciation in the Snowy Mountains: a reappraisal. *Proceedings of the Linnean Society of New South Wales*, **88**, 180–198.

GERVAIS, F. L. P. 1854. *Histoire naturelle des mammifères avec l'indication de leurs moeurs, et de leurs rapports avec les arts, le commerce, et l'agriculture* (2 vols). L. Curmer, Paris.

GERVAIS, F. L. P. 1869. Mémoire sur les formes cérébrales propres aux marsupiaux. *Nouvelles archives du Muséum d'Histoire Naturelle*, **5**, 229–251, plates 13 and 14.

GERVAIS, F. L. P. 1870. Sur les formes cérébrales des mammifères marsupiaux, édentés et carnivores. *Bulletin de la Société Géologique de France*, **28**, 130–131.

GIBBNEY, H. J. 1967. Sturt, Charles. *In:* PIKE, D. (ed.) *Australian Dictionary of Biography*, **2**, 495–498.

GODDARD, R. H. 1940. Captain Thomas Raine of the *Surry*, 1795–1860. *Journal and Proceedings of the Royal Australian Historical Society*, **26**, 277–317.

GRANGE, J. 1850. *Géologie, mineralogie et géographic physique du voyage au Pole Sud et dans l'Océanie, sous le commandement de Dumont d'Urville* (2 vols with atlas). Folios 1848–1854. Gide et Cie, Paris.

HILLS, E. S. 1940. *The Physiography of Victoria*. Whitcombe & Tombs, Melbourne.

HILLS, E. S. 1975. *The Physiography of Victoria*, 5th edn. Whitcombe & Tombs, Melbourne.

HOLLAND, J. 1992. Thomas Mitchell and the origins of Australian vertebrate palaeontology. *Journal and Proceedings of the Royal Society of New South Wales*, **125**, 103–110.

HOWCHIN, W. 1913. The evolution of the physiographic features of South Australia. *Australasian Association for the Advancement of Science*, **14**, 148–177.

HUNTER, C. 1991. *The Earth was Raised up in Waves like the Sea: Earthquake Tremors Felt in the Hunter Valley since White Settlement*. Hunter House, Newcastle, Australia.

JACK, R. L. 1879. *Report on the Bowen River Coal Field*. Government Printer, Brisbane.

JOHNS, K. (ed.) 1976. *History and Role of Government Geological Surveys in Australia*. Government Printer, Adelaide.

JUKES, J. B. 1843. Art[icle] 1, A few remarks on the nomenclature and classification of rock formations in new countries. *The Tasmanian Journal of Natural Science*, **2**, 1–12.

JUKES, J. B. 1847. *Narrative of the Surveying Voyage of H.M.S. Fly in Torres Straits, New Guinea, and other Islands of the Eastern Archipelago, during the years 1842–1846*. (2 vols). T. & W. Boone, London.

JUKES, J. B. 1850. *A Sketch of the Physical Structure of Australia, so far as it is at Present Known*. T. & W. Boone, London.

JUTSON, J. T. 1914. Outline of the physiographical geology, physiography of Western Australia [with bibliography]. *Western Australian Geological Survey Bulletin*, **61**, 229.

JUTSON, J. T. 1934. The physiography (geomorphology) of Western Australia. *Geological Survey of Western Australia*, Bulletin, **95**. Government Printer, Perth.

KING, P. P. 1826. *Narrative of a Survey of the Intertropical and Western Coasts of Australia. Performed Between the Years 1818 and 1811* (2 vols). John Murray, London.

KREFFT, G. 1867. Australian vertebrates, fossil and recent. *In: Catalogue of the Natural and Industrial Products of New South Wales, Paris Universelle Exposition*. Government Printer, Sydney, 90–110 (also issued as a separate pamphlet, Sydney, 1871).

LANGFORD-SMITH, T. (ed.) 1978. *Silcrete in Australia*. Department of Geography, University of New England, Armidale, New South Wales.

LAURENT, J. & CAMPBELL, M. 1987. *The Eye of Reason: Charles Darwin in Australia*. University of Wollongong Press, Wollongong.

LEICHHARDT, F. W. 1847. *Journal of an Overland Expedition in Australia*. Boone, London.

LEICHHARDT, F. W. 1855. *Beitrage zur Geologie von Australien von Ludwig Leichhardt. Herausgaben von Professor H. Gerard*. Halle, H. W. Schmidt.

LEWIS, A. N. 1944. Pleistocene glaciation in Tasmania. *Papers of the Royal Society of Tasmania*, **41**, 41–56.

MAYER, W. 2005. Deux géologues français en Nouvelle-Hollande (Australie): Louis Depuch et Charles Bailly, membres de l'expedition Baudin (1801–1803). *Travaux du Comité Francais d'Histoire de la Géologie (COFRHIGEO), Series 3*, **19**(695), 95–112.

MAYER, W. 2008. Early geological investigations of the Pleistocene Tamala Limestone, Western Australia. *In*: GRAPES, R. H., OLDROYD, D. R. & GRIGELIS, A. (eds) *History of Geomorphology and Quaternary Geology*. Geological Society, London, Special Publications, **301**, 275–289.

MCBRYDE, I. 1963. An unusual series of stone arrangements. *Oceania*, **34**, 137–146.

MCCARTHY, F. 1964. The Archaeology of the Capertee Valley, N.S.W. *Records of the Australian Museum*, **26**, 197–246.

MCNAMARA, K. M. 1988. Stromatolites, the ultimate living fossils. *Australian Natural History*, **22**, 476–480.

MEINICKE, C. E. 1837. *Das Festland Australien, eine geographische Monographie* (2 vols). F. W. Kalkerberg, Prussia (translation of Meincke (vol. 1, chapter 4, section) titled *Geology of Australia* by J. Slanska & T. G. Vallance, unpublished).

MELLEVILLE, M. 1842. *Du Diluvium: recherches sur les dépôts auxquels on doit donner ce nom et sur las cause qui les a produits*. Lonlois et Leclercq, Paris.

MITCHELL, T. L. 1838. *Three Expeditions into the Interior of Eastern Australia* (2 vols). Longman, London.

MOYAL, A. 2003. *The Web of Science: The Scientific Correspondence of the Rev. W. B. Clarke, Australia's Pioneer Geologist* (2 vols). Australian Scholarly Press, Melbourne.

MUDIE, I. 1968. *The Heroic Journey of John McDouall Stuart*. Angus & Roberston, Sydney.

MULVANEY, D. J. 1969 (1975). *The Prehistory of Australia*. Thames & Hudson, London.

MULVANEY, D. J. 1994. Australian anthropology since Darwin: models, foundations and funding. Dominions apart: reflections on the culture of science and technology in Canada and Australia 1850–1945. *Scientia Canadensis*, **17**, 155–184.

MURRAY, R. D. 1843. *A Summer at Port Phillip*. William Tait, Edinburgh.

NICHOLAS, F. W. & NICHOLAS, J. M. 1989. *Charles Darwin in Australia*. Cambridge University Press, Cambridge.

OLDROYD, D. R. 2007. In the footsteps of Thomas Livingstone Mitchell (1792–1855): soldier, surveyor, explorer, geologist, and probably the first person to compile geological maps in Australia. *In*: WYSE JACKSON, P. (ed.) *Four Centuries of Geological Travel: The Search for Knowledge on Foot, Bicycle, Sledge and Camel*. Geological Society, London, Special Publications, **287**, 343–373.

OLDROYD, D. R. 2008. Griffith Taylor, Ernest Andrews *et al.*: early ideas on the development of the river systems of the Sydney region, eastern Australia and subsequent ideas on the associated geomorphological problems. *In*: GRAPES, R. H., OLDROYD, D. R. & GRIGELIS, A. (eds) *History of Geomorphology and Quaternary Geology*. Geological Society, London, Special Publications, **301**, 239–274.

ORGAN, M. 1997. *Reverend W. B. Clarke 1798–1878: Chronology and Calendar of Events*. M. Organ, Wollongong.

OWEN, R. 1877. *Researches on the Fossil Remains of the Extinct Mammals of Australia with a notice of the Extinct Marsupials of England* (2 vols). J. Erxleben, London.

PATERSON, S. J. 1966. Glacial landforms in Tasmania. *The Australian Geographer*, **11**, 374–389.

PENTLAND, W. 1830. Communication verbal sur les ossemens trouvés dans une brèche calcaire sur la Rivière de Hunter (N. S. Wales). *Bulletin de la Société Géologique de France*, **1**, 144.

PENTLAND, W. 1832. On the fossil bones of Wellington Valley, New Holland. *Edinburgh New Philosophical Journal*, **12**, 301–308.

PENTLAND, W. 1833. Observations on a collection of fossil bones, sent to Baron Cuvier from New Holland. *Edinburgh New Philosophical Journal*, **28**, 120, plate 5.

RICHARDS, H. C. 1938. Some problems of the Great Barrier Reef. *Journal and Proceedings of the Royal Society of New South Wales*, **71**, 68–85.

ROSENFELD, A. 1994. Prehistory to 1788. *In*: BAMBRICK, S. (ed.) *The Cambridge Encyclopedia of Australia*. Cambridge University Press, Cambridge, 60–64.

ROWLAND, L. S., MOORE, A. & ROGERS, G. C. 1996. *The History of Beaufort County, South Carolina*. University of South Carolina Press, Columbia, SC.

RUDWICK, M. J. S. 1985. *The Meaning of Fossils*, 2nd edn. University of Chicago Press, Chicago, IL.

RUDWICK, M. J. S. 1996. Cuvier and Brongniart, William Smith, and the reconstruction of geohistory. *Earth Sciences History*, **15**, 25–36.

RUPKE, N. 1992. Richard Owen and the Victorian museum movement. *Journal and Proceedings of the Royal Society of New South Wales*, **125**, 133–136.

SCHILDER, G. (ed.) 1985. *Voyage to the Great South Land: Willem de Vlamingh 1696–1697*. Royal Australian Historical Society, Sydney.

SCOTT, H. I. 1977. *The Development of Landform Studies in Australia*. Bellbird Press, Sydney.

SCROPE, G. P. 1862. *Volcanos: their share in the structure and composition of the surface of the globe and their relation to its internal forces, with a descriptive catalogue of all known volcanoes and volcanic formations*. Longmans, Green, Longman & Roberts, London.

SELWYN, A. R. & ULRICH, G. H. 1866. Notes on the physical geography, geology and mineralogy of Victoria. *In*: *Intercolonial Exhibition Essays*. Government Printer, Melbourne.

SEMENIUK, V. & MEAGHER, T. D. 1981. Calcrete in Quaternary coastal dunes in southwestern Australia: a capillary rise phenomenon associated with plants. *Journal of Sedimentary Petrology*, **51**, 47–68.

STUART, J. M. 1864. *Explorations in Australia: the journals of Stuart, John McDouall during the years 1858, 1859, 1860, 1861, and 1862, when he fixed the centre of the continent and successfully crossed it from sea to sea, edited from Mr. Stuart's manuscript by W. Hordman with maps, a photographic portrait of Mr. Stuart, and engravings drawn on wood by G. F. Angas, from sketches taken during the different expeditions*. Saunders, Otley, London.

STURT, C. 1833. *Two Expeditions into the Interior of South Australia during the Years 1828, 1829, 1830, and 1831 with observations on the soil, climate, and general Resources of the Colony of New South Wales* (2 vols). Smith, Elder & Co., London.

STURT, C. 1849. *Narrative of an Expedition into Central Australia, performed under the Authority of Her Majesty's Government, during the years 1845, 5, and 6. Together with a notice of the Province of South Australia in 1847*. T. & W. Boone, London.

STRZELECKI, P. E. DE. 1845. *Physical Description of New South Wales and Van Diemen's Land. Accompanied by a Geological Map, Sections, and Diagrams, and Figures of the Organic Remains*. Longman, Brown, Green & Longmans, London.

TATE, R. 1894. Century of scientific progress. *Report of the Meeting of the Australasian Association for the Advancement of Science*, **5**, 1–69.

TAYLOR, R. 1838. Visit to Harpur's Hill. *Journal of Richard Taylor*. Unpublished manuscript. The Alexander Turnbull Library, Wellington, 282.

THOMSON, A. & KREFFT, G. 1870. The geology of the Wellington Caves. *In*: *New South Wales Legislative Assembly, Votes and Proceedings, Session 1870–71*. Government Printer, Sydney, 1182–1184.

TWIDALE, C. R. 1988. Geomorphology. *Australian Geographic Studies*, **26**, 5–20.

TWIDALE, C. R., PARKIN, L. W. & RUDD, E. A. 1990. C. T. Madigan's contribution to geology in South and Central Australia. *Transactions of the Royal Society of South Australia*, **114**, 157–167.

TWIDALE, C. R. & CAMPBELL, E. M. 2005. *Australian Landforms: Understanding a Low, Flat, Arid and Old Landscape*. Rosenberg, Dural, New South Wales.

VALLANCE, T. G. 1975. Presidential address: origins of Australian geology. Proceedings of the Linnean Society of New South Wales, **100**, 13–43.

VALLANCE, T. G. 1978. Pioneers and leaders – a record of Australian palaeontology in the nineteenth century. *Alcheringa*, **2**, 243–250.

VALLANCE, T. G. 1981–1982 (1983). Lamarck, Cuvier and Australian geology. *Histoire et nature*, **19–20**, 133–136.

VALLANCE, T. G. & BRANAGAN, D. F. 1968. New South Wales geology: its origins and growth. *In*: *A Century of Scientific Progress*. Royal Society of New South Wales, Sydney, 265–279.

VALLANCE, T. G. & BRANAGAN, D. F. 1976. *Beginnings of Geological Knowledge in New South Wales*. 25th International Geological Congress Excursion, Guidebook, **4B**.

WESTGARTH, W. 1846. Observations on the geology and physical aspect of Port Phillip. *Tasmanian Journal of Natural Science*, **2**, 402–409.

WESTGARTH, W. 1853. *Victoria, Late Australia Felix, or Port Phillip District of New South Wales*. Oliver & Boyd, Edinburgh.

WHITE, M. E. 1994. *The Greening of Gondwana*, 2nd edn. Reed, Chatswood, New South Wales.

WHITE, M. E. 1997. *Listen … Our Land is Crying*. Kangaroo Press, Sydney.

WHITE, M. E. 2000. *Running Down – Water in a Changing Land*. Kangaroo Press, Sydney.

WILLIAMS, M. A., DECKER, P. J. DE & KERSHAW, A. P. (eds) 1991. *The Cainozoic in Australia: A Re-appraisal of the Evidence*. Geological Society of Australia, Special Publications, **18**.

WILTON, C. P. N. 1832. Australian geology: a sketch of the geology of country about the River Hunter. *In*: *Australian Almanack for 1832*. Ralph Mansfield, Sydney, 179–188.

WOODS, J. E. T. 1862. *Geological Observations in South Australia*. Longman, Roberts & Green, London.

WOOLNOUGH, W. G. 1918. The physiographic significance of laterite in Western Australia. *Geological
Magazine*, Decade 6, **5**, 385–393.

WOOLNOUGH, W. G. 1927. The duricrust of Australia.
*Journal and Proceedings of the Royal Society of
New South Wales*, **58**, 24–53.

WOOLNOUGH, W. G. 1930. The influence of climate and
topography in the formation and distribution of
products of weathering. *Geological Magazine*, **67**,
123–132.

YOUNG, R. W. & TWIDALE, C. R. 1993. History of
geomorphology in Australia. *In*: WALKER, H. J. &
GRABAU, W. E. (eds) *The Evolution of Geomorphology: A Nation by Nation Summary of Development*.
Wiley, New York, 29–43.

ZIMMERMAN, E. A. W. 1810. *Australien in Hinsicht:
der Erd-, Menschen- und Produktenkunde nebst einer
allgemeinen Darstellung des grossen Oceans auf den
Handel und die Politik* (2 vols). Friedrich Perthes,
Hamburg.

The study of desert dunes in Australia

C. ROWLAND TWIDALE

School of Earth and Environmental Sciences, Geology and Geophysics, University of Adelaide, Adelaide 5005, Australia (e-mail: rowl.twidale@adelaide.edu.au)

Abstract: Most of the early explorers were bewildered by the features they encountered in the Australian deserts, but Sturt's observations led him to speculate on the origin of the sand ridges that were so much a part of his desert experience. Scientific investigations of the dunes, however, awaited the twentieth century. In the 1930s Madigan made signal contributions to the understanding of the features, but he also raised as many problems as he resolved. Post-war investigations by King and those due to Wopfner, initially related but incidental to the search of oil and gas, have done much to clarify the dynamics of dune development. More recently, luminescence dating has allowed the sand ridges, as well as periods of lake fill and alluviation, to be dated with some confidence. Chronological research has been extended to include the major palaeodunefields of southern Australia.

We must examine old ideas, old theories, although they belong to the past, for this is the only way to understand the importance of the new ones and the extent of their validity.

(Einstein & Infeld 1938)

Australia is the driest inhabited continent. Climatic statistics show that approximately one third of the continent is arid, and one third semi-arid. The character of the arid plains varies, with many salinas, alluvial tracts and gibber plains – the 'stony desert' of Sturt (1849) and Carnegie's (1898) 'desert of gravel'. But fields of linear, longitudinal or seif dunes – the *reg* of the North African deserts – dominate the desert landscapes. Older inherited dunefields are prominent in the southern, presently semi-arid, areas, although there parabolic forms also are well represented. Yet, not even in the hyperarid desert around Lake Eyre, with an annual average rainfall of 125 mm or less, are there extensive fields of dunes devoid of vegetation. Monstrous piles of bare moving sand, such as occur in the Arabian Sand Sea, in Namibia, and in the Atacama and Sahara deserts, are few and ephemeral in central Australia (Wopfner & Twidale 1988) (Fig. 1). Moreover, although repeat photography and eyewitness accounts show that some of the dunes of the Simpson Desert (*sensu lato*) advance downwind from time to time, their movement is spasmodic.[1] Most frequently only the crests of the

ridges are mobile. Such limited movement is favoured by extreme conditions of drought or by vegetation depletion or other causations, as for example the rabbit plagues of the 1920s and 1930s (e.g. Ratcliffe 1936, 1937). Paradoxically the vegetated dune slopes are modified during and following rains and floods, when seepage and standing water cause sapping, undermining and collapse, wash and gullying of the dune flanks (e.g. Twidale 2002).

The reason the dunefields of central Australia have developed in partial denial of the regional climate, in some degree results from monsoonal incursions and westerly depressions that bring occasional rains to all the inland deserts. Also, vegetation established in former humid periods persists. But geological conditions are the major factors, for Lake Eyre is located in a structural basin that has been subsiding for some tens of millions of years and is still sinking. The bed of the salina stands 15–16 m below sea level. As was demonstrated by Dulhunty (1987), the precise depth and the exact location of the lowest point vary in time. Nevertheless, the lake bed, which is the base level for rivers draining some $1.2-1.3 \text{ km}^2$ of central and northeastern Australia, stands below sea level. Not only do subsurface waters migrate toward this structural low located in the southwestern quarter of the basin, but many large rivers, including some that rise in the monsoonal north, run and flow towards the lowest and also, as it happens, the driest part of the continent. If there are heavy

---

[1]The term 'Simpson Desert' was coined by Madigan in 1928 (see Madigan 1930) to denote the dunefields of the Lake Eyre topographic basin, in central Australia. It was named in recognition of the support accorded his explorations in central Australia, and particularly the 1939 expedition, by A. A. Simpson, a prominent Adelaide businessman and then President of the local geographical society. Subsequently three subregions

have been recognized within the Simpson – the Tirari dunefield east of Lake Eyre; Sturts Stony Desert; and the Strzelecki dunefield of the southern areas – but their boundaries are gradational and informal.

From: GRAPES, R. H., OLDROYD, D. & GRIGELIS, A. (eds) *History of Geomorphology and Quaternary Geology*. Geological Society, London, Special Publications, **301**, 215–239.
DOI: 10.1144/SP301.15   0305-8719/08/$15.00 © The Geological Society of London 2008.

Fig. 1. Active sand dune with east-facing avalanche slopes, Sturts Stony Desert, some 75 km west of Cordillo Downs H. S., South Australia (photograph by H. Wopfner).

rains in consecutive years soils reach field capacity and channels are cleared of debris in the first wet year. In the second and subsequent rainfall events there is runoff and the rivers flow to or towards Lake Eyre. Some precipitation and runoff are dissipated in the soils and sediments of the arid centre or are lost to evapotranspiration, but from time to time the rivers reach the lower part of the catchment and the corridors between sand ridges are inundated (Fig. 2). Sediment is deposited in playas and channels, and in the interdune corridors. Water fills lagoons and lakes, which occasionally overflow, but which on drying leave behind a complex of channels and algal-coated flats (Fig. 3). Lake Eyre may live up to its name.

Lake Eyre fills three or four times a century, the most recent occasion being in 1974 (Kotwicki 1986; see also Bonython & Mason 1953; Peake-Jones 1955). In the brief humid phases, animals awake from aestivation and breed. The desert quickly blooms as plants that survive by tapping shallow subterranean waters flower and seed after rains. All but the most active dunes carry some vegetation and most of the desert is held under

Fig. 2. Part of the Simpson Desert in flood, 1998 (photograph by P. Canty).

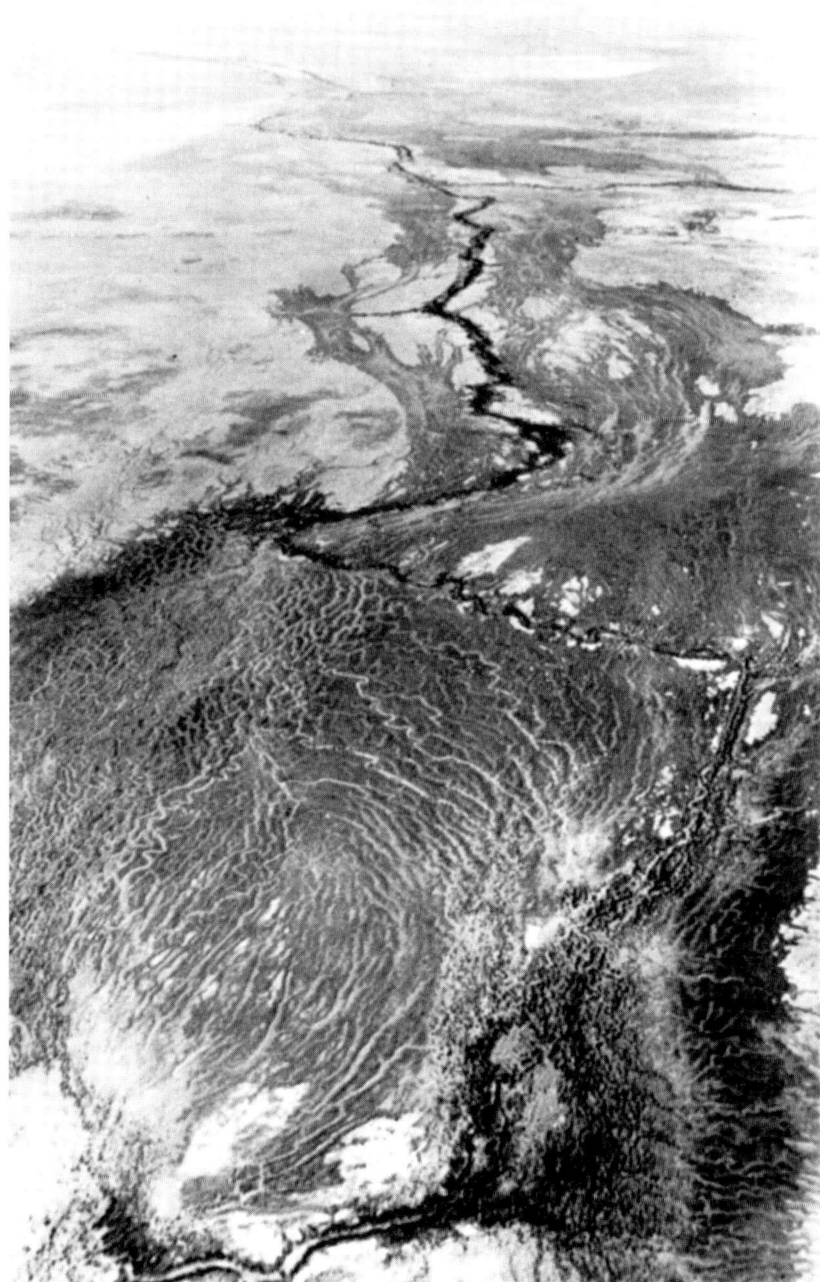

Fig. 3. Flood-out or overflow depression of the Diamantina River near Birdsville, SW Queensland.

pastoral lease. During the 1974 lacustrine interlude the lakes came alive as fish and other water-loving creatures magically reappeared. A regatta was held on Lake Eyre. The coastline underwent remarkable and rapid morphological changes, with depositional features such as beaches, spits and bars forming in a matter of weeks. In the light of these developments questions arise as to whether old high-level beach ridges (Dulhunty 1975) are related to secular climatic changes or reflect short-term 'catastrophic' or 'storm' events like that of 1974. Such modern lake-full periods are brief, although long-term lacustrine phases have been demonstrated for the later Quaternary (Magee *et al.* 2004).

Early explorers and the shattering of a dream

What did the early explorers make of this strange desert? The early European explorers of Australia had no perception of aridity and what it implied for the landscape. In September 1840, for example, Edward John Eyre encountered scrubby sandy ridges and a streamless terrain in what he called 'a perfect desert' on northwestern Eyre Peninsula (Eyre 1845, p. 199). The region he described is semi-arid and is today, although marginally, part of the Australian wheat–sheep belt. Indeed, so unreliable is its precipitation that some argue that the area ought to have been left in its virgin state rather than cleared for agricultural and pastoral use. During the Late Pleistocene, however, it was a genuine desert (see later).

Several explorers of the inland, including Eyre, glimpsed from afar the white salt crusts of salinas and thought they were lakes. Some explorers and surveyors were additionally deceived by the ephemeral pools and running rivers resulting from short-lived local rains. This led to a belief that the centre of Australia was well watered, with lakes and lush pastures. Henry Freeling was sent out to explore this 'promised land' with a portable boat, the easier to cross the water obstacles that he was expected to encounter. But all were disappointed to find only a harsh desert (see Mincham 1964).

Later travellers in the centre and west of the continent commented on the sandplains and dunes, and cursed the scarcity of water and the hard-going in the scrubby sand country (e.g. Warburton 1875; Giles 1889). Ernest Giles lamented his choice of horses rather than camels during his first or westerly journey from the centre to the west coast: a mistake he rectified for the return trip. At no time did he make the error perpetrated by John Charles Darke, who during his 1844 expedition into the Gawler Ranges relied solely on oxen as draft animals with neither horses nor camels to provide easier travel and more importantly the capacity to reconnoitre (see Twidale 1974). But whether on foot or with camels, all the explorers appreciated their good fortune when their overall plan allowed them to travel along the interdune corridors parallel with the dune trend, as did Giles when he crossed from west to east through what he called the Great Victoria Desert (Giles 1889, vol. 2, p. 202; he also named the Gibson Desert after a lost companion). Carnegie's plans, on the other hand, called for him to trek northward in the

same dunefield, taking him and his party over ridge after countless sand ridge (Carnegie 1898, p. 249 et seq.).

All the travellers of the inland noted dune trend and height but, with rare exceptions, they were concerned mainly with survival. With the comment that they are 'blown, I suppose' Carnegie ventured the opinion that the dunes were of aeolian origin (Carnegie 1898, p. 178), but 50 years earlier, and despite the constant discomfort and danger of his situation, Charles Sturt had given serious thought to the origins of the landscapes over which he had travelled in central Australia. He was undoubtedly the greatest of the early explorers who ventured into the Australian inland. He was the first to penetrate into the arid centre and bequeathed to posterity unforgettable accounts of the searing heat and bleak landscapes that he and his companions experienced (Sturt 1849). Who, on reading Sturt's account of the inescapable heat he and his companions endured, cannot still be moved and horrified by: 'the mean of the thermometer for the months of December, January and February, had been 101°, 104° and 101° [degrees Fahrenheit] respectively, in the shade. Under its effects every screw in our boxes had been drawn, and the horn handles of our instruments, as well as our combs, ... were split into fine laminae. The lead dropped out of our pencils .... Our hair ... ceased to grow' (Sturt 1849, vol. 1, pp. 305–306).

Sturt encountered dunes of red sand and the 'adamantine' and 'gloomy stone clad plains' of the almost featureless stony desert (Fig. 4). He noted 'A number of sandy ridges ... abutted upon, and terminated in, this plain like so many head lands projecting into the sea' (Sturt 1849, vol. 1, p. 372). He thought they had been promontories or islands in a sea in which a current flowed transverse to their trend. As for the stony desert, either it had been swept clean of sand or the force of the current had prevented deposition (Sturt 1849, vol. 1, p. 381). He frankly admitted (Sturt 1849, vol. 1, p. 373) that he was 'ignorant of the existence of a similar geographical feature in any other part of the world' and was 'at a loss to divine its nature'.

As for the sand ridges, Sturt expressed his amazement that they extend unbroken and stand in parallel and in the same orientation over vast distances, and he entertained various ideas concerning their origin. He vehemently denied the possibility that the ridges were of aeolian origin. 'No! Winds may indeed have assisted in shaping their outlines. But I cannot think, that these constituted the originating cause of their formation. They exhibit a regularity that water alone could have given' (Sturt 1849, vol. 1, pp. 380–381). Thus, he misread the landscapes he so arduously traversed,

Fig. 4. Sturts Stony Desert west of Bedourie, SW Queensland.

but the fact that he found the time and detachment to consider such esoteric questions speaks volumes for his mental toughness and sense of curiosity.

Some of the early explorers reported and inadvertently drew attention to an aspect of dune dynamics that seemed puzzling at the time. Sturt, travelling with John Browne between the Stony Desert and Eyre's Creek in September 1846, recorded that: 'about 9 o'clock in the morning, we distinctly heard a report as of a great gun discharged, to the westward, at distance of half a mile' (Sturt 1849, vol. 2, p. 24). During the Burke and Wills expedition, William John Wills was with John King near Coopers Creek and at 'about 11 A.M. both of us heard distinctly the noise of an explosion, as if of a gun, at some considerable distance' (Wills 1863, p. 228, W. J. Wills' diary entry for 24 May 1861; see also Moorehead 1963, p. 128). But the sounds that Sturt and Wills reported could not have been caused by the discharge of guns, for no Europeans were in the vicinity at the times in question.

Sandy beaches and dunes that generated noises have been reported for many centuries from SW Asia, Europe and North America (Bagnold 1941, pp. 250–256; Twidale 1968, for references; see also Gibson 1946, p. 41; see also Tey 1953). The sounds have been described variously as like that produced by a whistling kettle just before the water is thoroughly boiled; something like the rustling of rain on an iron roof; or the sound produced by sand falling. Under certain conditions saltating grains of quartz sand can produce noises on

beaches and on dunes, but louder noises involve the collapse of a sandmass. Travelling in the Arabian Sand Sea, Wilfred Thesiger (1964, p. 166) reported hearing a low vibrant hum that 'sounded as though an aeroplane was flying low over our heads'. By deliberately disturbing the crest of a dune and setting sand in motion down the flank, he demonstrated that the noise was due to slippage of sand on slip faces of dunes. The inclination of dune slope is a function of the angle of rest of the sand, which depends on grain size, shape and sorting, as well as the presence of fines (clay, silt) and moisture. It is a delicate balance that is readily disturbed by heating and desiccation, as well as additional loading; as, for instance, by an animal hopping on to the slope near the dune crest (or Thesiger's foot and weight being imposed).

Madigan and a scientific approach

In 1936 a pastoralist, E. A. Colson, crossed the Simpson Desert in the latitude of Birdsville on camel, accompanied by a young Indigenous man; but the scientific exploration of the dunes awaited Cecil Thomas Madigan.

Madigan (1889–1947) (Fig. 5) was an archetypal officer and gentleman of the old school. Educated in the University of Adelaide, he went to Oxford as a Rhodes Scholar, participated in Mawson's 1911 Antarctic expedition and served with distinction in France in World War I (Twidale *et al.* 1990). He worked as a geologist

Fig. 5. C. T. Madigan aboard a 'ship of the desert', Simpson Desert, 1939 (photograph by E. A. Rudd).

for 2 years in Sudan before taking up an appointment in his *alma mater*, where he remained for the rest of his life, except for a spell in the Army as an instructor during World War II.

War service in France alerted Madigan to the use of aircraft for reconnaissance purposes, and both before the war and in the 1920s photographers like Herbert Wilkins, and archaeologists like Sir Henry Welcome and O. G. S. Crawford, had used aerial photography in their work. In 1929, with the co-operation of the military, Madigan made nine flights over the Macdonnell Ranges and the dunefields of what he then named the Simpson Desert. He used the photographs he took as base maps for plotting topographical and geological data (Madigan 1930, 1931, 1932*a*, *b*, 1938),[2] but became obsessed with the Simpson dunefield in its various aspects. After completing several field seasons in central Australia he wrote a well-known paper on the sand-ridge deserts (Madigan 1936) and

set about organizing a multidisciplinary expedition that would cross the Simpson from west to east with camels and in radio contact with the outside world. The expedition left Adelaide by train on 25 May 1939, disembarked at Charlotte Waters, travelled to Andado Station (i.e. Homestead) by truck and then set out on camels across the desert. They were fortunate in travelling east as the SSE–NNW-trending sand ridges of the dune field are asymmetrical in cross-section, with the western slope most often the gentler. The party left Andado on 28 May and arrived in Birdsville on 6 July (Fig. 6). After a pause the party turned south and then SW, travelling parallel with the Warburton River. On 8 August 1939 they arrived at Marree, on the rail line, whence they returned to Adelaide on The Ghan (the Alice Springs–Adelaide train).

Madigan realized that the lower Lake Eyre basin is a depocentre into which many large rivers have in the past debouched and delivered sediment. Occasionally, they still do, overflowing their banks, filling lagoons and flooding the interdune corridors (Fig. 2). As stated earlier, rivers occasionally run and Lake Eyre carries water. The importance both in the past and at present of the paradoxical relationship between floods and dunes cannot be overstated, for river channels and playas or local depocentres are the major source of quartz sand from which the sand ridges or longitudinal dunes of the Simpson Desert have been shaped by the wind (Figs 7–9). It is also the source of fines – clay and silt – which are deposited with the sand grains, but are later washed deeper into the dune ridges to form cores that harden on drying. Most of the calcareous matter found in some dunes and commonly in interdune corridors is also introduced on the wind.

Madigan (1946*a*, p. 62) concluded that the sand of which the dunes are constructed is 'almost entirely alluvial'. He referred to dunes built up by 'sand driven upwards and northwards rising to summits' (Madigan 1946*a*, p. 61). Mineral analyses of the dune sands revealed that heavy minerals derived from the Precambrian and Palaeozoic rocks exposed in the Macdonnell and adjacent ranges are more prominent in the north than in the south (Carroll 1944), but all are alluvial in origin in an immediate sense. Thus, there is a paradoxical but vital link between rivers and floods and the sediment they transport, and the construction of desert dunes.

He recorded that 'the great majority of sand-moving winds are southerly' (Madigan 1946*a*, p. 59); that is, the wind regime is unidirectional (see also Aufrère 1928; Madigan 1930) and basically from the SSE, but with occasional cross-winds that account for the asymmetry of the sand ridges in cross-section, and for the characteristic tuning-fork

---

[2]This 1938 paper is the published version of the Clarke Memorial Lecture delivered to the Royal Society of New South Wales on 31 March 1938. It was published in the Society's *Journal and Proceedings* for 1937, which was issued on 17 March 1938. The paper included (pp. 506–513) a detailed history of the exploration of the area in general and the dunefield in particular.

Fig. 6. Members of the 1939 Simpson Desert expedition, at Birdsville after successfully crossing the dunefield. Note Crocker (standing, extreme left) and Madigan (fourth from left) (*The Advertiser*, Adelaide, South Australia).

or Y-junctions. In this Madigan differed from R. A. Bagnold, who attributed sand ridges to the effect of cross-winds on barchans (but see Lancaster 1980; Tsoar 1989), examples of which are rare and ephemeral in the Australian deserts. He noted the relationship between dune height and density (i.e. number of dunes per unit distance normal to dune trend) for some sectors have many closely spaced low dunes, by contrast with others with a few widely spaced but high sand ridges (Wopfner & Twidale 1967; Twidale 1981).

Madigan noted downwind changes in the colour of the dune sand. In the source areas (channels, depressions) the sand is white, but the quartz

Fig. 7. Typical Simpson Desert dune (CSIRO).

Fig. 8. Simpson Desert dunefield with sandy corridors.

grains acquire a red patina of iron oxides (hematite, goethite) derived from the weathering of clay particles trapped amongst the sand grains as they are blown downwind (Wopfner & Twidale 1967; Walker 1979; Nanson *et al.* 1992). The reddening takes place *in situ* and is not inherited from pre-existing lateritic detritus as suggested by Folk (1969, 1976). Madigan noted that the sand is deflected as it passes over the crests of existing dunes (which stand up to 38 m above the adjacent corridors) but he considered that the dunes are relic except near the Diamantina and Cooper where there are ample supplies of fresh sand.

Nearing Birdsville, Madigan's party encountered rain and boggy interdune corridors, driving home the message that no desert is rainless and

Fig. 9. Eastern Simpson Desert dunefield with dunes migrating over a riverine plain.

that even hyperarid regions occasionally experience precipitation, floods and mud.

Post-war dune research and Don King

World War II broke out less than a month after Madigan and his team left the desert and returned to Adelaide on 'The Ghan' (train). Understandably, little or no new work was undertaken while the war lasted (although many Australians compulsorily gained first-hand experience of the North African deserts). But some pre-war research was published while the authors were on active service. Thus, Peel (1941) published an important paper demonstrating the significance of running water in shaping the Libyan desert landscape, an aspect frequently overlooked even more than 50 years later. In Australia, Hills (1940) published a paper on lunettes, which, it transpired, play a significant role in dune generation (see later), but it was not until after World War II that the landform received further attention (Stephens & Crocker 1946). Crocker, who was botanist on Madigan's 1939 Simpson Desert expedition (Fig. 6), published a paper on the old dunes of the southeastern part of South Australia in 1941 but it was not until 1946 that his papers on the palaeodunefield of Eyre Peninsula appeared in print (Crocker 1946*a, b*).

The war also delayed reports on the work of the 1939 Simpson Desert expedition. Madigan was engaged in military duties until 1943, when he returned to his university and could again give his attention to the publication of its scientific work. Eventually nine papers were produced. They included Madigan's (1946*a*) seminal dune paper, which quite fortuitously benefited from the delay in writing and publication because in the meantime Bagnold's magisterial book on sand movement and dunes had appeared (Bagnold 1941). His published papers and book (Madigan 1936, 1946*a, b*) inspired many to pursue some of the dune problems that he had identified.

The demands of war had demonstrated how little was known of the interior and north of Australia, and the late 1940s saw the instigation of systematic geological, topographic and land surveys, as well as the establishment of agricultural research centres. They have benefited many facets of the natural sciences including the study of desert dunes.

One of the many participants in this blossoming of interest in the interior deserts was Don King, then a geologist with the South Australian Geological Survey, who in 1956 and 1960 published papers that provided a different and challenging perspective on the Australian desert dunes and dunefields. King (1926–1989) was born and educated in Adelaide, and made his name as a distinguished exploration geologist (Goscombe 1989, pp. 34–36). He was perhaps best known for the discovery of coal in the Bowen Basin, in east central Queensland, but his two early dune papers proved controversial and stimulating for they truly set the cat amongst the scientific pigeons!

Madigan (1936, 1946*a*) and many others considered the linear dunes of central Australia and other mid-latitude deserts to be of aeolian and depositional origin. King decided to test this explanation by investigating the internal structure of dunes. He augered into two sand ridges adjacent to Lake Eyre North, with what are, even today, surprising results. He found that wind-blown sand formed a veneer, some 3 m thick on crests but less than 1 m in the intervening corridors, overlying an irregular erosional topography cut in shelly gypseous clays that overlie the gypcrete which caps the cliffs bordering the salina on its western side (Figs 10 and 11). King concluded that the dunes are windrift forms caused by the erosion of unconsolidated sediments: they are yardangs (Hedin 1903; Blackwelder 1934; Bobek 1969) with a mantle of wind-blown sand. King's field evidence was irrefutable and stood in contrast with the conclusion of overseas scientists working in the Sahara, for example, and, closer to home, with the conclusions of Madigan and his colleagues. Yet, such was the esteem and respect in which Madigan was held by his students, current and former, that King made the curious claim (King 1956, p. 93) that the results of his investigation 'all ... substantiate earlier work of the late Dr. C. T. Madigan'.

King's interpretation was based on the investigation of only two ridges, but his evidence was convincing and he extrapolated his findings to the entire Australian dunefield (King 1960). He was unfortunate, however, in that the dunes he examined near Lake Eyre are atypical of linear dunes in general and Australian sand ridges in particular. Indeed, the ridges he excavated may well be the only windrift dunes in Australia. Even at this distance of time it is almost impossible to believe that he could have been so unlucky as to select for study the only erosional sand ridges in the entire continent; yet, he did.

There are still those who think of dunes as eroded by the wind. McTainsh & Leys (1993, p. 190), for example, relate 'dunes and present-day eroding winds', but the constructional nature of the longitudinal dunes of the Simpson Desert was demonstrated beyond doubt a few years later when an east–west seismic line was cut across the Simpson Desert during the geophysical exploration of the region during the search for oil and gas (Sprigg 1963; Wopfner & Twidale 1967), and when the internal structure of hundreds of dunes

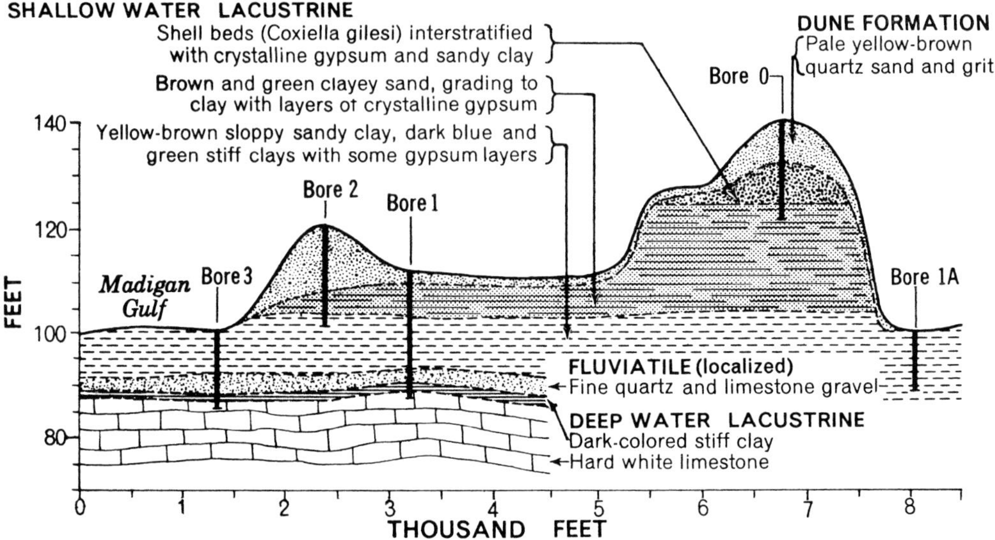

Fig. 10. Section showing windrift dunes adjacent to Lake Eyre (after King 1956; see also Twidale 1972).

was exposed. All were revealed to be entirely built of cross-bedded sands deposited by the wind in a bidirectional regime (Wopfner & Twidale 1967). In confirmation, Mabbutt & Sullivan (1968) augered into dunes in the northwestern part of the Simpson Desert and demonstrated that they are constructed entirely of wind-blown sand.

This general conclusion was in keeping with studies overseas (e.g. McKee & Tibbitts 1964). That the dunes are occasionally mobile is demonstrated by repeat photography, the repeated diversion of station tracks around extending noses of dunes and by the migration of dunes on to the south shores of playas (Fig. 12). The regularity of dune pattern is interrupted near sources of abundant sand, for instance adjacent to sandy channels and flood plains, but overall parallel sand ridges dominate the desert plains.

Folk (1971) and Pell *et al.* (1999, 2000) have argued that, as in many other dunefields, the sand of the Simpson Desert dunes is derived locally. But the suggestion has been refuted by Wopfner

Fig. 11. Gypsite-capped cliff on the western side of Lake Eyre.

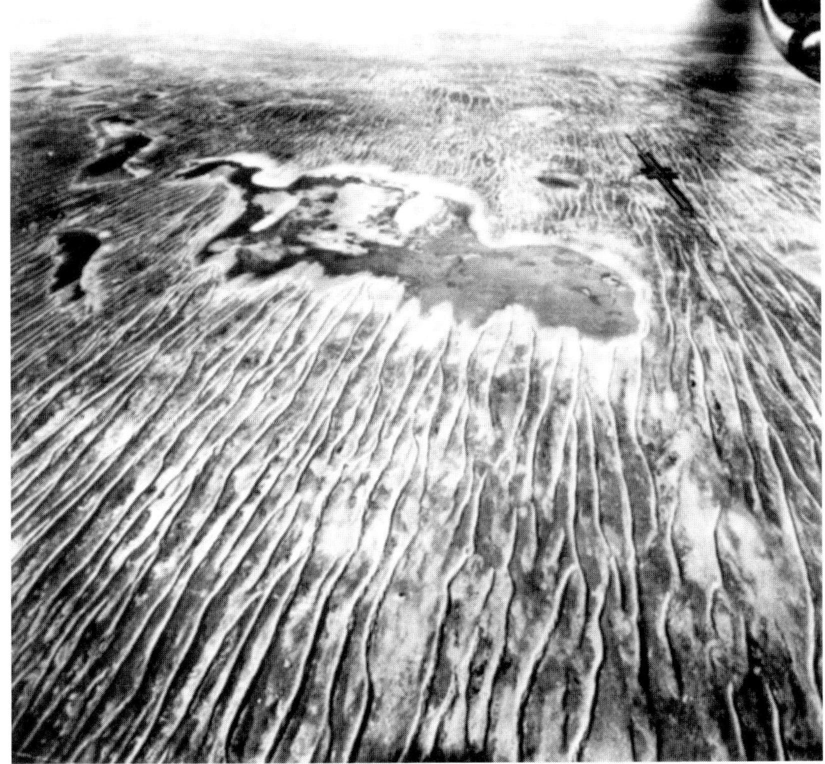

Fig. 12. Wartime low-oblique photograph of the Simpson Desert, SE of Lake Eyre, seen in the left distance. Note the salinas that are preserved in a dismembered old river channel, the dunes advancing on to the southern margin of the salt lake, the lunette of white sand on the lee shore and new dunes, numerous and small, immediately behind the mound but coalescing into fewer but larger ridges beyond (photograph by the RAAF).

Fig. 13. Dunes advancing to the NNW over the Diamantina flood plain north of Birdsville, SW Queensland.

& Twidale (2001), who, although conceding that the alluvia have complex origins, nevertheless present evidence showing that there has been long-distance travel of dune sand in some areas (Fig. 13).

Regional survey and related investigations

Central Australia

The 1950s and 1960s saw increased interest in all the Australian dunefields. Partly this reflected extension of air photography to cover the entire continent. Over the years this was repeated, at various scales, and then in colour for most areas. This in turn demonstrated the extent of the dune-fields, as well as providing the basis for autostereo-scopic mapping, so that many dune areas are now covered by detailed (1:50 000 and/or 1:100 000) topographic maps.

A Land Research and Regional Survey (CSIRO) team investigated the northwestern part of the Simpson as part of the 'Alice Springs' regional survey (e.g. Mabbutt 1962, 1965, 1988). Mabbutt attributed the sandplains (as opposed to dunefields) to the binding effect of clay mixed with the near-surface sand (Mabbutt 1968; see also Jackson 1962). In the Simpson Desert some dunes stand on an alluvial substrate, and at some sites the uncon-formity between dune sand and alluvium stands higher than the alluvia exposed in the intervening dune corridors.

The early reconnaissance surveys that estab-lished the geological structures beneath the dune-field were undertaken in light aircraft by Rudi Brunnschweiler, Reg Sprigg and Heli Wopfner. Wopfner (Fig. 14), a former Luftwaffe pilot and a graduate of the University of Innsbruck, emigrated to join Sprigg's Geosurveys company on the Ero-manga project before joining the South Australian Geological Survey. He was at the Gidgealpa site when the first gas strike was made. He later returned to academic life in Germany, but his frequent visits demonstrate a long-term and active interest in the Australian deserts, happily to the present time and continuing. Although Wopfner is one of the last 'rounded' geologists with wide-ranging interests and expertise, he has, nevertheless, made the study of various aspects of the Australian dunefields his life's work. With Madigan and King he makes up the trio of key contributors to our knowledge of the Australian desert dunes.

The search for oil and gas in the Eromanga Basin, which underlies the Simpson Desert, was successful, as the mining company SANTOS's operation based at Moomba demonstrates (e.g. Wopfner 1990). But it also had significant inciden-tal effects for dune research. Numerous geophysi-cal surveys entailed the bulldozing of long straight tracks through the dunefields, in some instances obliquely to the dune trend and so expos-ing their internal structure. The cross-bedding demonstrated construction by winds blowing from the SE and SW, the SSE–NNW dune being

Fig. 14. Heli Wopfner in the Stony Desert, mounted on a more modern 'ship of the desert', a battered, dusty, but trusty, Land Rover.

Fig. 15. Lunette bordering north shore of Lake Eyre (photograph by J. A. Dulhunty).

a vector or resultant of this regime. The usual asymmetry in cross-section of the sand ridges, with the eastern slope the steeper, indicates a prevalence of sand-moving winds from the SW quarter, although repeat photography shows that the sense of asymmetry changes in time (Wopfner & Twidale 1967, 1988). Field relationships strongly suggest that transverse obstacles such as lunettes (lee-side mounds or source bordering dunes; see, e.g. Dulhunty 1983) (Fig. 15) or river channels are critical to dune development because they are topographic obstacles that induce turbulence and deflection. Sand is lifted from source areas such as playas and river channels. Small dunes form in the lee of obstacles, and then coalesce and grow into fewer but larger ridges downwind (Twidale 1972) (Figs 12 and 16–18). Thus, the volume of sand

Fig. 16. Dunes generated on the bed of Goyders Lagoon (cf. Fig. 3) in the lee of negative obstacles, in these instances depressions that are dry water holes (billabongs).

Fig. 17. Small dunes developing in the lee of a playa depression with lunette, Simpson Desert (photograph by H. Wopfner).

per unit area is constant. Ridges can be traced across transverse obstacles such as river channels, but they are offset (Fig. 19). Reticulate or box-like patterns are found where sand is abundant, as for instance adjacent to major rivers, such as the Finke.

From the beginning of scientific study of desert dunes the relationship between dune form and trend, on the one hand, and wind regime, on the other, has been a matter of controversy and debate. Using the basic topographic and meteorological data amassed during the 1950s and 1960s, Muriel Brookfield (1970) made a detailed study of the relationship in the Simpson Desert. She concluded that summer cyclone winds are responsible for sand movement and shaping the dunes – not winter anticyclones as had been supposed. She found no consistent correlation between mean strong-wind direction and the apparent vector or resultant indicated by dune alignments and morphology. According to Brookfield there have been no latitudinal shifts in the pressure belts responsible for wind patterns. Others, however, recognize a discordance between dune trend and contemporary wind patterns, and conclude that conditions have changed although the source and nature of the

wind data, like the nature of the perceived discordance – whether in terms of a unidirectional or a bidirectional regime – is not stated (McTainsh & Leys 1993, p. 190). Nanson *et al.* (1995) also disagree, claiming discrepancies between past and present sand movement and a northward shift of wind pattern during the last glacial period of 100–150 km or 1.0°–1.5° of latitude.

The nature of the airflow patterns responsible for shaping the dunes has also given rise to discussion. Dune orientation is in places diverted by local topographic effects on airflow, as, for example, in the Rumbalara area (Mabbutt 1968), and climbing dunes were noted in uplands within the dunefields (Madigan, cited in Mabbutt 1965), as, for instance, on the borders of the Ooraminna Ranges.

The major dunefields, however, are developed on plains or flats. By contrast with those workers who attribute asymmetry to a bidirectional wind regime, others suggested that it is caused by roller vortices (Bagnold 1941; Mabbutt *et al.* 1969; Folk 1971; Tseo 1990). Tseo used kites to detect local wind behaviour over existing ridges. And the wind-rift forms demonstrated by King surely argue erosion by roller vortices rather than a bimodal wind regime?

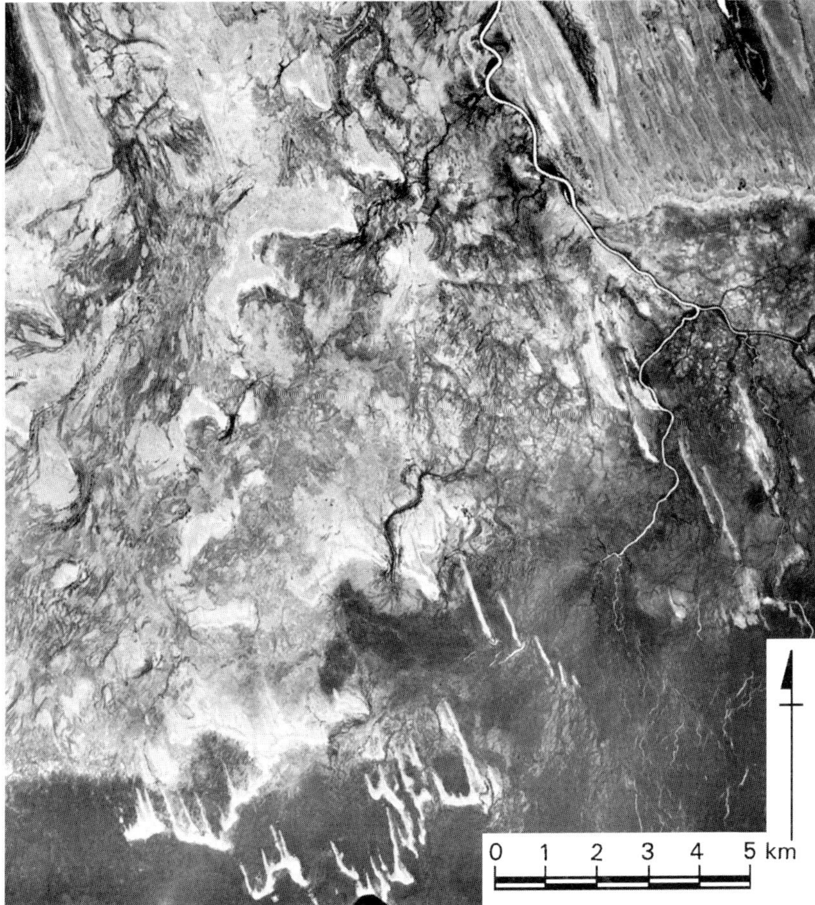

Fig. 18. Extract of a vertical air photograph of part of Goyders Lagoon, showing dunes originating in the lee of dry billabongs, and in the lee of a long mound at the northern margin of the Lagoon (photograph by S. A. Lands Department).

Whether linear dunes migrate laterally as well as downwind is also controversial, with some arguing that it does, indeed, occur (Rubin & Hunter 1987; Rubin 1990; see also Hesp *et al.* 1989; Bristow *et al.* 2000), while others have maintained that it does not, if only because of the vegetation that stabilizes the dune flanks (e.g. Nanson *et al.* 1992). Rubin (1990) has noted anomalous cross-bedding patterns in sand ridges in the Strzelecki field that clearly indicate sideways shifts in position of the dune forms, and in north China (Hesp *et al.* 1989) lateral movement of linear dunes of up to 3 m per annum has been demonstrated by repeat air photography. In the Simpson Desert, however, significant lateral migration of the linear dunes is unlikely and not only for the reason cited by Nanson *et al.* Where the interdune corridors are underlain by sand

(as opposed to alluvium or lacustrine beds) the strata typically consist of horizontally disposed laminae (Wopfner & Twidale 1967). Where, as is commonly the case, the corridors are underlain by alluvium (e.g. Twidale *et al.* 2001) dunes stand on an alluvial plinth the surface of which is 1 m or so higher than that of the adjacent plain: clearly, there is no evidence of dunes other than in the positions they now occupy.

But recent flume experiments, as well as field work in the Sinai and NW China, point to lateral migration (e.g. Rubin & Ikeda 1990; Tsoar *et al.* 2004; Rubin *et al.* 2008). It may be that wind regimes, vegetation cover, and consolidation by calcrete and clay accumulations have rendered stable most Australian sand ridges, but that in other deserts, where different conditions prevail, lateral movement has occurred.

Fig. 19. Deflection of sand ridges by a negative obstacle, namely the channel of the Diamantina River, NE of Birdsville, SW Queensland.

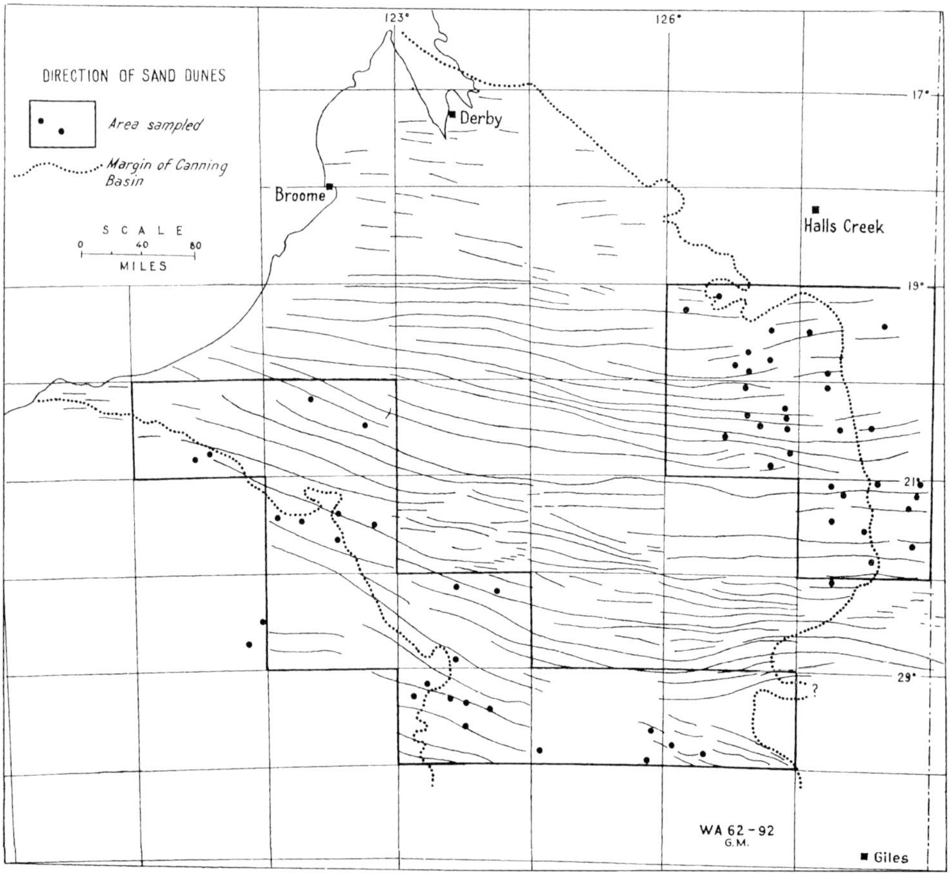

Fig. 20. Dune trends in the Great Sandy Desert, Western Australia (Veevers & Wells 1961, p. 202).

Great Sandy Desert

Many parts of the western deserts of Australia were investigated as part of programmes involving regional geological mapping (e.g. Veevers & Wells 1961; Daniels 1974) and land surveys (e.g. Mabbutt 1963). Veevers & Wells (1961) mapped long linear and chain-like dunes in the course of their geological investigations of the Canning Basin and the Great Sandy Desert that is virtually coincident with it. Fields of seif or linear dunes and sand plains are common. The sand is mainly quartzitic and is coarser than most of the Simpson Desert sand. It is derived mostly from underlying strata, although there was some riverine redistribution prior to aeolian intervention. In the south, alluvium from the dismembered river system now represented by the inverted 'Z' composed of the Percival lakes and Lake Tobin, and lakes Auld, George, Blanche and Dora may also have contributed.

Up to 20 m high, the dunes are asymmetrical in cross-section with the northern face the steeper. Y-junctions are common. The ridges vary in detailed morphology with chain-like forms rather than simple ridges in places, and reticulate forms in areas of abundant sand. In places, sand shadows have formed in the lee of strike ridges aligned obliquely to the dune trend, producing a local box-work pattern. The ridges trend WNW–ESE, tending to west–east in the east (Fig. 20), and on the basis of admittedly sparse wind data were considered to have formed under a bidirectional regime, with winds from the east and SE, in a period of more arid climate prior to the present amelioration (Veevers & Wells 1961, p. 211). As in central Australia, only the crests of dunes are devoid of vegetation, but a few dunes display slip faces. Although the eastern ends of some of the ridges are scoured and there is evidence of deposition at the western ends, the dunes are essentially fixed. In support of this contention, Veevers & Wells (1961, pp. 199–200) pointed out that there are few exposed tree roots, few sand shadows and no ventifacts.

Salinaland

To the south, in the catchment of the Murchison River and at the margin of Jutson's (1914) Salinaland, Mabbutt (1963) described what he termed Wanderrie banks and flats, low (*c*. 1 m) sand rises alternating with alluvial flats. Of various orientations (parallel with, transverse or oblique to local drainage lines), they were attributed to a combination of wind and wash, and relic from an earlier drier period when wind action was more pronounced.

Blackstone Ranges area

Daniels (1974) described the dunefields of the Blackstone region in the interior of Western Australia. The dunes are developed around and amongst uplands. As in the Great Sandy Desert, the dunes are linear (up to 18 m high), asymmetrical in cross-section with the steeper side on the left looking downwind, and with chain-like forms as well as simple ridges. Daniels noted sand wedges between the main ridge and what he terms blebs or 'swellings' in some of the dunes in plan, implying local narrowing of the interdune corridors. Daniels attributed them to unidirectional winds from the NW and NE. He considered that in the south where the field overlaps with the Great Victoria Desert and where the sand ridges run in parallel east–west and some 300 m apart (see Wasson *et al.* 1988), the dunes are shaped by westerly winds. In the north of his study area, however, Daniels attributed the dunes to easterlies. In addition, the complex intermontane patterns were attributed partly to topographic interference and modification of airflow (Fig. 21), although he mentioned the possibility of seasonal changes in wind direction.

A continental pattern of dunes?

Apart from questioning the very nature of desert dunes, King's foray into dune morphology had another significant effect. His map of Australian linear dunes (King 1960; see also Jennings 1968; Wasson *et al.* 1988; Nanson *et al.* 1992) suggested a continental pattern of sand ridges related to an overall pressure and wind system: what became known as the 'anticyclonic swirl' with ridges (and sand-moving winds) trending ESE–NNW in the Simpson Desert, then turning to the SE–NW and then ESE–WNW in the west (Fig. 22). The Canning Basin dunefield is influenced by easterly and southeasterly winds, but to the south, in the vicinity of Minilya where the main highway cuts through dunes in small fields inland from the Exmouth Gulf and the adjacent coast, sand drift varies seasonally. (As with the barchans of the Atacama Desert in Peru (Finkel 1959), the direction of sand and dune movement may change, and even be reversed, with the seasons.) The linear ridges of the Gibson and Great Victoria deserts are shaped by northwesterly and westerly winds, but then fall under the influence of south and southeasterlies so that the ridge system links up with the Simpson fields.

Thus, viewed continentally, an overall regional pattern is apparent, but it implies that the complex of dunefields has been shaped by a unified wind

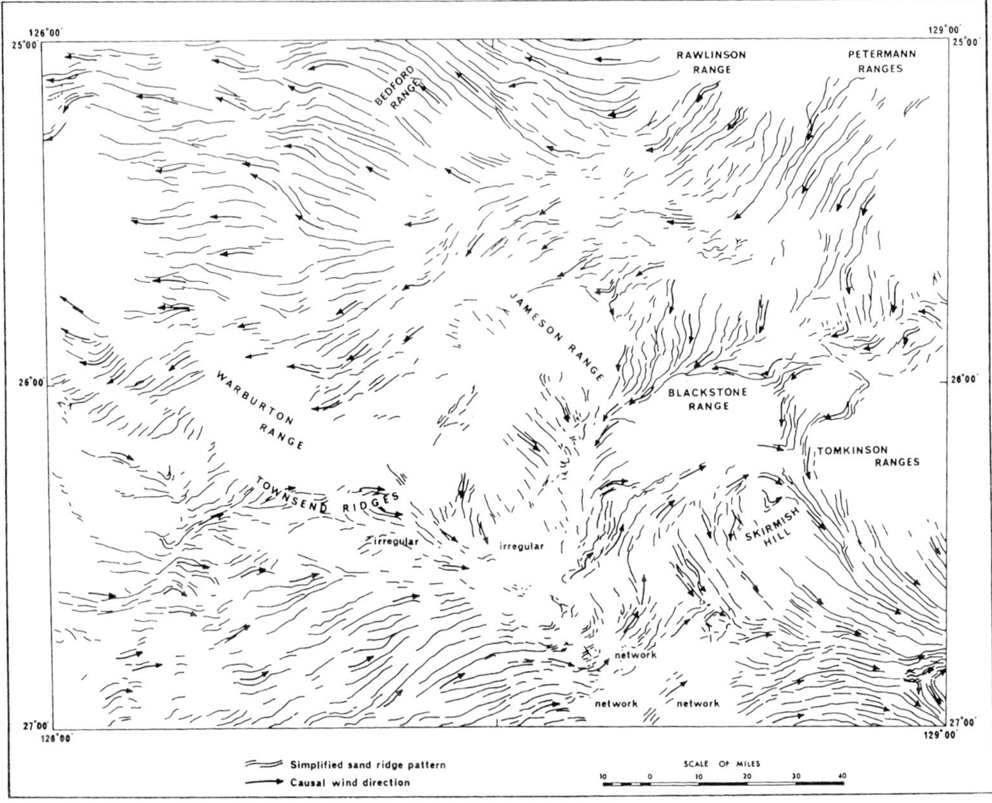

Fig. 21. Interference in dune pattern by topography (uplands) in the Blackstone Range area, Western Australia (Daniels 1974, p. 207).

system and, moreover, that all the dunes are of similar age (see, for example, Wasson *et al.* 1988, p. 96). Neither seems likely.

Palaeodunefields

Most of the Australian dunefields are at most spasmodically and partly active, with only the crestal areas prone to change. In addition to the recognized deserts, fields of such linear dunes, mainly fixed but with crests that are occasionally mobile, occupy most of the Lake Torrens plains, the western Gawler Ranges and sundry relatively small areas. The latter include seifs near Swan Hill, in northern Victoria, of which some are aligned meridionally, others west–east. Phases of sand deposition, plant colonization and weathering, followed by truncation and burial, have been identified (Churchward 1963). The dunes have not been dated, however, and their location suggests that, although the west–east sand ridges are desertic, the others may

be associated with the flow and flooding of the nearby River Murray.

Large areas of South Australia and northwestern Victoria are occupied by palaeodunes: that is sand ridges and mounds that are vegetated throughout and are immobile save where there has been human interference resulting in the depletion or destruction of the protective cover and the exposure of bare sand surfaces, as for instance where roads or tracks have been cut through sand ridges.

At the eastern end of the field, the northern dunes of the Great Victoria Desert turn abruptly NE and then NNW to merge with the sand ridges of the Simpson Desert. In the southern sectors of the Great Victoria Desert, however, the sand ridges continue into what was termed the Nullarbor–Spencer field (Crocker 1946a, b), which extends across northern Eyre Peninsula and includes Eyre's 'perfect desert'. This dunefield appears to be part of a larger but complex ancient dune desert (Fig. 22), that has been termed the

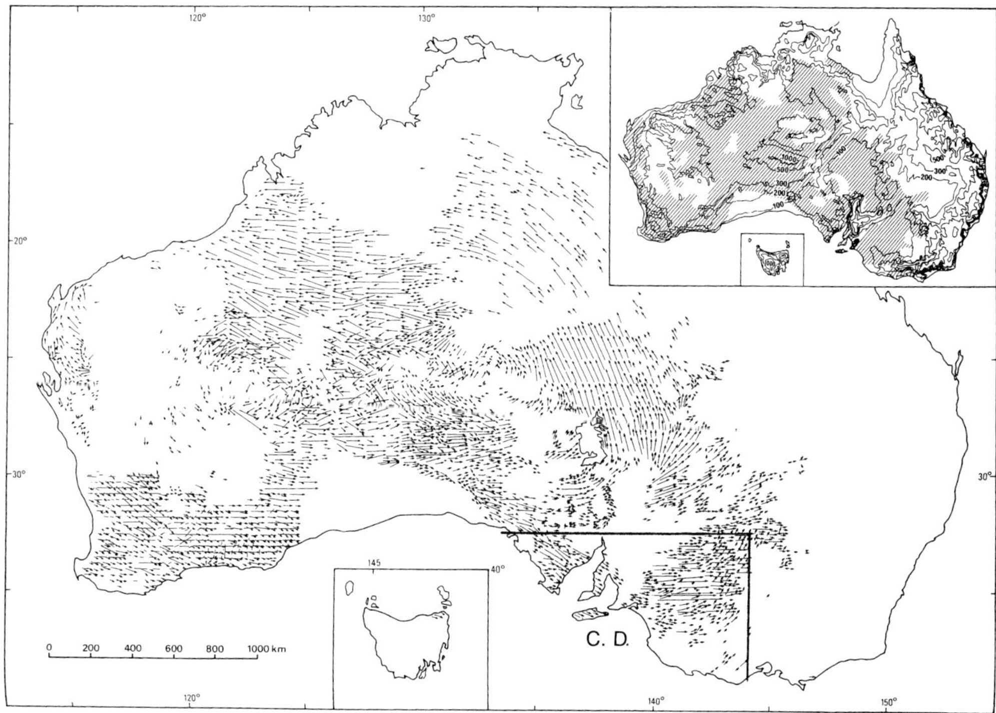

Fig. 22. Map of Australian sand ridges and sand-moving winds. Inset shows topography with areas of dunes shaded (after Wasson *et al.* 1988, p. 96). The area of Crocker Dunefield (C. D.) is indicated.

Crocker Dunefield, which occupies an area of some 200 000 km^2 of southern Australia.[3] The component ridges of the Crocker Dunefield extend across northern Eyre and Yorke peninsulas, the northern Adelaide Plains, and into the Murray Basin in South Australia and northwestern Victoria. In the Murray Basin there is a mix of linear and parabolic forms, with the latter dominant in southern areas but also occurring in the north, amidst the linear forms and in association with outcrops of arenaceous bedrock.

The calcrete cores of linear dunes extend below low-tide level both on the Eyre Peninsula coast and the opposed Yorke Peninsula shore (Jessup 1967; Van Deur 1983). In most instances sand from the NW–SE-trending dunes of northern Eyre Peninsula impinge on the west shores of salinas (e.g. Agars Lake, near Minnipa) indicating an easterly movement of sand. The sandy lunettes deposited on the eastern shore of Lake Gairdner are consistent with this pattern. But a Late Pleistocene lunette located on the southern shore of

Kappakoola Swamp, near Wudinna on northern Eyre Peninsula, must have formed under the influence of a regime dominated by northerly sand-moving winds. Furthermore, an active but immobile dune has accumulated to the south as a result of clearing of the lunette, indicating strong northerly wind activity in recent times (Smith *et al.* 1975). As in central Australia, uplands such as the Darke Range (Fig. 23) interfered with the airflow and disturbed the regularity of the dune pattern (Crocker 1946*b*).

Separate from the Crocker Dunefield both in space and time, a field of white, west–east trending, predominantly linear, climbing and falling dunes (although with some transverse forms) and carrying a scatter of vegetation (Fig. 24) has penetrated the hilly terrain of the western Gawler Ranges (Campbell *et al.* 1996). Sand has been cleared from most hillcrests but the mallee scrub prevents significant dune migration. To the south, the Corrobinnie Depression is a fault-line valley that was exploited by exoreic streams during the Eocene and Pliocene. It is now, and long has been, a depocentre characterized by salinas and an abundant supply of sand from which have developed complex parabolic forms (Bourne *et al.* 1974; Binks & Hooper 1984). It is also the source of

---

[3]The Crocker Dunefield is named after R. L. Crocker who, as has been implied, was a pioneer in the investigation of these ancient deserts (Crocker 1941, 1946*a, b*).

Fig. 23. Deflection of palaeodunes by Darke Range, northern Eyre Peninsula, South Australia.

sand from which are derived the easterly trending dunes of the western Gawler Ranges.

Dune chronology

Dating any landform or event is useful not only because it allows landscape components to be seen in sequence, but also places them in regional or even global context. It also offers the possibility of resolving problems of dune genesis and climatic change. Thus, in the Murray Basin linear dunes dominate northern areas, and parabolic forms the southern. Is this pattern caused by contrasted sand supply, with linear forms developed in areas of relatively sparse sand and parabolics in areas of

Fig. 24. Essentially stationary falling dunes, western Gawler Ranges, South Australia. Note the prominent planation surface, of Early Cretaceous age, preserved on the summits of the bornhardts.

abundant sand, or do the two fields reflect processes generated in different climatic regimes prevalent at different times? Are the southern dunefields of the same age as or older than those of central Australia? Are the various regional and morphological components of the Crocker Dunefield all of the same age range? Did the implied zone of aridity expand and contract or has it migrated latitudinally? Such problems would be clarified and possibly resolved if the ages were known of sufficient dunes to provide a meaningful regional pattern.

Early attempts at dating Australian dunes involved stratigraphy (e.g. Wopfner & Twidale 1967) and ^{14}C estimates based on calcrete and organic material found in the dunes (e.g. Wasson 1983). The dating of calcrete, however, provides only a minimum age for the dunes in which the calcareous horizons occur for the latter accumulated in pre-existing sand accumulations (e.g. Twidale *et al.* 1976; Wasson 1983). Using ^{14}C to date organic remains preserved in sediments, Jennings (1975) dated dunes adjacent to the Fitzroy estuary in the NW of Western Australia as 6000–8000 years old.

Luminescence dating (Aitken 1970, 1985, 1997) revolutionized the investigation of dune chronology, for most dunes are constructed predominantly of quartz sand (although not all quartz sand, for in Tanzania there are barchans of olivine derived from the weathering of basalt, just as some Hawaiian beaches are black and basaltic in origin) and quartz is eminently suitable for the application of luminescence procedures. It has the added advantage of being the most common of all rock-forming minerals. Even dunes constructed predominantly of other minerals, for example calcarenitic forms, are susceptible to dating by this method because, although mainly calcareous, the sand contains some quartz crystals and clasts (cf. Sprigg 1952), and the coastal dunes of western Eyre Peninsula were dated using the uranium series method (Wilson 1991).

Crocker (1946*b*) thought that the dunes of northern Eyre Peninsula were 3000–4000 years old, and were formed in what was then termed the Great Australian Arid Period. Later work based on numerical dating procedures and particularly luminescence dating has destroyed that hypothesis, for major dune developments in Australia can be traced back almost half a million years. Dating has so far been concentrated in the Lake Eyre Basin, in the Tirari and Strzelecki dunefields. But as mentioned earlier, the area is geologically and geomorphologically complex, and the depositional processes active at any given time do not necessarily reflect local or regional climatic conditions and changes. Nevertheless, the chronology of filling and drying of such playas as Lake Eyre not only provides a record of climatically generated events, and particularly of monsoon impacts, but also clearly indicates when sand was readily available for dune construction.

The Simpson Desert and environs has been the subject of several major luminescence studies in the last 30 years. Working in the Birdsville area, using thermoluminescence (TL) and obtaining samples by augering, Nanson *et al.* (1992), obtained dates of up to 78 ka. In the western Simpson Desert, in the vicinity of the Finke River, Nanson *et al.* (1995) obtained TL dates of 30–12, 17–9 and 5–0 ka for the sand ridges.

Around Lake Eyre Magee *et al.* (2004; see also Hesse *et al.* 2004) recognized five periods of lake-full stage, taking the Australian Height Datum as zero, 120 ka (+/−2.9 ka) the lake level stood at approximately +10 m; 85.8 ka (+/−1.5 ka) it was at +3–4 m; 63.5 ka (+/−1.3 ka) at −4 m; approximately 40 ka about 10 m; and 3–10 ka about −9 m. At its maximum Lake Eyre was three times as extensive as it is at present (DeVogel *et al.* 2004). Between the lake-full periods aridity prevailed and dunes formed. Sand ridges have also been initiated during the Holocene (e.g. Twidale *et al.* 2001) and, in addition, the crestal zones of older dunes have continued to be reworked. In the Lake Gregory catchment of northern Western Australia, also in the monsoon zone, Wyrwoll & Valdes (2003) recognized high-water periods at 250–375, 180–225, 50–140 and 0–20 ka.

Sheard *et al.* (2006) obtained dates around 200 ka (215 ka +/−15 ka; 197 +/−14 ka) for dunes in the southeastern Great Victoria Desert. In the western Gawler Ranges the quartz has a very low radioactive content but gave a TL date of some 4000 years (Campbell *et al.* 1996). In the Crocker Dunefield, Gardner *et al.* (1987), working on dunes in the southern Murray Basin in South Australia, obtained TL dates of 8–34.5 ka (19.5–28 ka using ^{14}C). Using optically stimulated luminescence (OSL) and sampling from backhoed trenches Twidale *et al.* (2007) dated linear red dunes near Waikerie with ages of 151–25.3 ka; 157–33.3 ka; and a dune with basal age of 59.6 ka. From the quartz but lime-rich linear dunes of the central Murray Basin Lomax *et al.* (2007) obtained ages as old as 360 and 440 ka. Younger dates were obtained for parabolic dunes further south; but this statement is subject to revision as the dune bases were not exposed and thus not sampled. To the west a red linear dune from Yorke Peninsula gave a basal date of 260 ka with OSL, with an electron spin resonance (ESR) date of 230–380 ka from the same sample. A dune exposed on the coast gave a basal date of about 125 ka (D. Beng pers. comm. 2000). Eyre

Peninsula dunes have provided maximum ages of some 290 Ka (A. Hilgers pers. comm. 2007).

Clearly, desert conditions prevailed when Indigenous peoples entered the Murray Basin 40 000–60 000 years ago (Bowler *et al.* 2003). Sand movement in historical times noted in the Murray Basin may have coincided with droughts or could well be linked to land clearance.

Concluding comment

Much has been achieved in less than a century of scientific investigation of the Australian desert dunes. But they still offer many research opportunities and challenges, particularly in the field of dating. Yet, it has to be borne in mind that despite the aridity of the region the sand ridges are odd in that all but the crests carry a sparse, yet protective, cover of plants that mostly inhibit sand movement. As usual, Australia is unusual.

References

AITKEN, M. J. 1970. Dating by archaeomagnetic and thermoluminescent methods. *Philosophical Transactions of the Royal Society of London*, **A269**, 77–88.

AITKEN, M. J. 1985. *Thermoluminescence Dating*. Academic Press, London.

AITKEN, M. J. 1997. *An Introduction to Optical Dating*. Oxford University Press, Oxford.

AUFRÈRE, L. 1928. L'orientation des dunes et la direction des vents. *Comptes rendus de l'Académie des Sciences à Paris*, **187**, 833–835.

BAGNOLD, R. A. 1941. *The Physics of Blown Sand and Desert Dunes*. Methuen, London.

BINKS, P. J. & HOOPER, G. J. 1984. Uranium in Tertiary palaeochannels, 'West Coast Area', South Australia. *Proceedings of the Australasian Institute of Mining and Metallurgy*, **289**, 271–275.

BLACKWELDER, E. 1934. Yardangs. Bulletin of the Geological Society of America, **24**, 159–166.

BOBEK, H. 1969. Zur Kenntnis der Sudlichen Lut. Ergebnisse einer Luftbildanalyse. *Mitteilungen der Österreichischen Geographischen Gesellschaft*, **111**, 155–192.

BONYTHON, C. W. & MASON, B. 1953. The filling and drying of Lake Eyre. *Geographical Journal*, **119**, 321–330.

BOURNE, J. A., TWIDALE, C. R. & SMITH, D. M. 1974. The Corrobinnie Depression, Eyre Peninsula, South Australia. *Transactions of the Royal Society of South Australia*, **98**, 139–152.

BOWLER, J. M., JOHNSTON, H., OLLEY, J. M., PRESCOTT, J. R., ROBERTS, R. G., SHAWCROSS, W. & SPOONER, N. A. 2003. New ages for human occupation and climatic change at Lake Mungo, Australia. *Nature*, **421**, 837–840.

BRISTOW, C. S., BAILEY, S. D. & LANCASTER, N. 2000. The sedimentary structure of linear sand dunes. *Nature*, **406**, 56–59.

BROOKFIELD, M. 1970. Dune trends and wind regime in Central Australia. *Zeitschrift für Geomorphologie Supplementband*, **10**, 121–153.

CAMPBELL, E. M., TWIDALE, C. R., HUTTON, J. T. & PRESCOTT, J. R. 1996. Preliminary investigations of dunes of the Gawler Ranges Province, South Australia. *Transactions of the Royal Society of South Australia*, **120**, 21–36.

CARNEGIE, D. W. 1898. *Spinifex and Sand*. Pearson, London.

CARROLL, D. 1944. The Simpson Desert Expedition, 1939, Scientific Reports: No. 2 Geology – Desert sands. *Transactions of the Royal Society of South Australia*, **68**, 49–59.

CHURCHWARD, H. M. 1963. Soil studies at Swan Hill, Victoria, Australia. IV. Groundsurface history and its expression in the array of soils. *Australian Journal of Soil Research*, **1**, 242–245.

CROCKER, R. L. 1941. Notes on the geology and physiography of south-east South Australia, with reference to late climatic history. *Transactions of the Royal Society of South Australia*, **65**, 103–107.

CROCKER, R. L. 1946a. An introduction to the soils and vegetation of Eyre Peninsula, South Australia. *Transactions of the Royal Society of South Australia*, **70**, 83–106.

CROCKER, R. L. 1946b. *Post-Miocene Climatic and Geologic History and its Significance in Relation to the Genesis of the Major Soil Types of South Australia*. Commonwealth for Scientific and Industrial Research, Bulletin, **193**.

DANIELS, J. L. 1974. *The Geology of the Blackstone Region, Western Australia*. Geological Survey of Western Australia, Bulletin, **123**.

DEVOGEL, S. B., MAGEE, J. W., MILLER, G. H. & MANLEY, W. F. 2004. A GIS-based reconstruction of late Quaternary palaeohydrography: Lake Eyre, arid central Australia. *Palaeogeography, Palaeoclimatology, Palaeoecology*, **204**, 1–13.

DULHUNTY, J. A. 1975. Shoreline shingle terraces and prehistoric fillings of Lake Eyre. *Transactions of the Royal Society of South Australia*, **99**, 183–185.

DULHUNTY, J. A. 1983. Lunettes of Lake Eyre North, South Australia. *Transactions of the Royal Society of South Australia*, **107**, 219–222.

DULHUNTY, J. A. 1987. Salina bed instability and geodetic studies at Lake Eyre, South Australia. *Transactions of the Royal Society of South Australia*, **111**, 183–188.

EINSTEIN, A. & INFELD, L. 1938. *The Evolution of Physics*. University Press, Cambridge.

EYRE, E. J. 1845. *Journals of Expeditions of Discovery into Central Australia and Overland from Adelaide to King George's Sound, in the Years 1840–41* (2 vols). T. and W. Boone, London.

FINKEL, H. J. 1959. The barchans of southern Peru. *Journal of Geology*, **67**, 614–647.

FOLK, R. L. 1969. Grain shape and diagensis in the Simpson Desert, Northern Territory. *Geological Society of America Abstracts with Programs for 1969*, **7**, 68–69.

FOLK, R. L. 1971. Longitudinal dunes of the northwestern edge of the Simpson Desert, Northern Territory, Australia. 1. Geomorphology and grainsize relationships. *Sedimentology*, **16**, 5–54.

FOLK, R. L. 1976. Reddening of desert sands: Simpson Desert, Northern Territory, Australia. *Journal of Sedimentary Petrology*, **46**, 604–615.

GARDNER, G. J., MORTLOCK, A. J., PRICE, D. M., READHEAD, M. L. & WASSON, R. J. 1987. Thermoluminescence and radiocarbon dating of Australian desert dunes. *Australian Journal of Earth Sciences*, **34**, 343–357.

GIBSON, E. S. H. 1946. Singing sand. *Transactions of the Royal Society of South Australia*, **70**, 35–40.

GILES, E. 1889. *Australia Twice Traversed: The Romance of Exploration* (2 vols). Sampson, Low, Marston, Searle and Rivington, London.

GOSCOMBE, P. 1989. Obituary: Donald King AO, MSc. *Bulletin and Proceedings of the Australasian Institute of Mining and Metallurgy*, **294**, 34–36.

HEDIN, S. 1903. *Central Asia and Tibet* (2 vols). Hurst and Blackett, London.

HESP, P., HYDE, R., HESP, V. & ZHENGYU, Q. 1989. Longitudinal dunes can move sideways. *Earth Surface Processes and Landforms*, **14**, 447–451.

HESSE, P., MAGEE, J. W. & VAN DER KAARS, S. 2004. Late Quaternary climates of the Australian Arid zone: a review. *Quaternary International*, **118–119**, 87–102.

HILLS, E. S. 1940. The lunette, a new landform of aeolian origin. *Australian Geographer*, **3**, 15–21.

JACKSON, E. A. 1962. *Soil Studies in Central Australia: Alice Springs–Hermannsburg–Rodinga Areas.* CSIRO Australia Soil Publication, **1**.

JENNINGS, J. N. 1968. A revised map of the desert dunes of Australia. *Australian Geographer*, **10**, 408–409.

JENNINGS, J. N. 1975. Desert dunes and estuarine fill in the Fitzroy estuary (north-western Australia). *Catena*, **2**, 215–262.

JESSUP, R. W. 1967. The late Quaternary eustatic and geologic history of northern Yorke Peninsula, South Australia. *In*: MORRISON, R. B. & WRIGHT, H. E. (eds) *Means of Correlation of Quaternary Successions. Volume 5, Proceedings of the VII Congress INQUA.* University of Utah Press, Salt Lake City, UT, 395–406.

JUTSON, J. T. 1914. An Outline of the Physiographical Geology (Physiography) of Western Australia. *Geological Survey of Western Australia*, Bulletin, **61**.

KING, D. 1956. The Quaternary stratigraphic record at Lake Eyre North and the evolution of existing topographic forms. *Transactions of the Royal Society of South Australia*, **79**, 93–103.

KING, D. 1960. The sand ridge deserts of South Australia and related aeolian landforms of the Quaternary arid cycles. *Transactions of the Royal Society of South Australia*, **83**, 99–108.

KOTWICKI, V. 1986. *Floods of Lake Eyre*. Engineering and Water Supply Department, Adelaide.

LANCASTER, N. 1980. The formation of seif dunes from barchans – supporting evidence for Bagnold's model from the Namib Desert. *Zeitschrift für Geomorphologie*, **24**, 160–167.

LOMAX, J., HILGERS, A., TWIDALE, C. R., BOURNE, J. A. & RADTKE, U. 2007. Treatment of broad palaeodose distributions in OSL dating of dune sands from the western Murray Basin, South Australia. *Quaternary Geochronology*, **2**, 51–56.

MABBUTT, J. A. 1962. Geomorphology of the Alice Springs area. *In*: PERRY, R. A. (compiler) *General Report on Lands of the Alice Springs Area, Northern Territory, 1956–57.* CSIRO, Melbourne, Land Research Series, **6**, 163–184.

MABBUTT, J. A. 1963. Wanderrie banks: microrelief patterns in semiarid Western Australia. *Bulletin of the Geological Society of America*, **74**, 529–540.

MABBUTT, J. A. 1965. Landscapes of arid and semi-arid Australia. *Australian Natural History*, **15**, 105–110.

MABBUTT, J. A. 1968. Aeolian landforms in central Australia. *Australian Geographical Studies*, **6**, 139–150.

MABBUTT, J. A. 1988. Australian desert landscapes. *GeoJournal*, **16**, 355–369.

MABBUTT, J. A. & SULLIVAN, M. E. 1968. The formation of desert dunes: evidence from the Simpson Desert. *Australian Geographer*, **10**, 483–487.

MABBUTT, J. A., WOODING, R. A. & JENNINGS, J. N. 1969. The asymmetry of Australian desert sand ridges. *Australian Journal of Science*, **32**, 159–160.

MADIGAN, C. T. 1930. An aerial reconnaissance into the south-eastern portion of Central Australia. *Proceedings of the Royal Geographical Society of Australasia (South Australian Branch)*, **30**, 83–108.

MADIGAN, C. T. 1931. The physiography of the western Macdonnell Ranges, central Australia. *Geographical Journal*, **78**, 417–433.

MADIGAN, C. T. 1932a. The geology of the western Macdonnell Ranges, central Australia. *Quarterly Journal of the Geological Society of London*, **86**, 672–711.

MADIGAN, C. T. 1932b. The geology of the eastern Macdonnell Ranges, central Australia. *Transactions of the Royal Society of South Australia*, **56**, 71–117.

MADIGAN, C. T. 1936. The Australian sand-ridge deserts. *Geographical Review*, **26**, 205–227.

MADIGAN, C. T. 1938. The Simpson Desert and its borders. *Journal and Proceedings of the Royal Society of New South Wales*, **71**, 503–535.

MADIGAN, C. T. 1946a. The Simpson Desert Expedition, 1939. Scientific Reports: No. 6. Geology–the sand formations. *Transactions of the Royal Society of South Australia*, **70**, 45–63.

MADIGAN, C. T. 1946b. *Crossing the Dead Heart*. Georgian House, Melbourne.

MAGEE, J. W., MILLER, G. H., SPOONER, N. A. & QUESTIAUX, D. 2004. Continuous 150 ky monsoon record from Lake Eyre, Australia: insolation forcing implications and unexpected Holocene failure. *Geology*, **32**, 885–888.

MCKEE, E. D. & TIBBITTS, G. C. 1964. Primary structures in a seif dune and associated deposits in Libya. *Journal of Sedimentary Petrology*, **34**, 5–17.

MCTAINSH, G. H. & LEYS, J. 1993. Soil erosion by wind. *In*: MCTAINSH, G. H. & BOUGHTON, W. C. (eds), *Land Degradation Processes in Australia.* Longman Cheshire, Melbourne, 188–233.

MINCHAM, H. 1964. *The Story of the Flinders Ranges.* Rigby, Adelaide.

MOORHEAD, A. 1963. *Cooper's Creek.* Hamish Hamilton, London.

NANSON, G. C., CHEN, X. Y. & PRICE, D. M. 1992. Lateral migration, thermoluminescence chronology and colour variation of longitudinal dunes near Birdsville in the Simpson Desert, central Australia. *Earth Surface Processes and Landforms*, **17**, 807–819.

NANSON, G. C., CHEN, X. Y. & PRICE, D. M. 1995. Aeolian and fluvial evidence of changing climate and wind patterns during the past 100 ka in the western Simpson Desert, Australia. *Palaeogeography, Palaeoclimatology, Palaeoecology*, **113**, 87–102.

PEAKE-JONES, K. (ed.). 1955. *Lake Eyre, South Australia. The Great Flooding of 1949–50*. Royal Geographical Society of Australasia (South Australian Branch), Adelaide.

PEEL, R. F. 1941. Denudation landforms of the central Libyan Desert. *Journal of Geomorphology*, **4**, 3–23.

PELL, S. D., CHIVAS, A. R. & WILLIAMS, I. S. 1999. Great Victoria Desert: development and sand provenance. *Australian Journal of Earth Sciences*, **46**, 289–299.

PELL, S. D., CHIVAS, A. R. & WILLIAMS, I. S. 2000. The Simpson, Strzelecki and Tirari deserts: development and sand provenance. *Sedimentary Geology*, **130**, 107–130.

RATCLIFFE, F. N. 1936. *Soil Drift in the Pastoral Areas of South Australia*. CSIR Pamphlet, **64**.

RATCLIFFE, F. N. 1937. *Further Observations on Soil Erosion and Sand Drift, With Special Reference to Southwest Queensland*. CSIR Pamphlet, **70**.

RUBIN, D. M. 1990. Lateral migration of linear dunes in the Strzelecki Desert, Australia. *Earth Surface Processes and Landforms*, **15**, 1–14.

RUBIN, D. M. & HUNTER, R. E. 1987. Bedform alignment in directionally varying flows. *Science*, **237**, 276–278.

RUBIN, D. M. & IKEDA, H. 1990. Flume experiments on the alignment of transverse, oblique, and longitudinal dunes in directionally varying flows. *Sedimentology*, **37**, 673–684.

RUBIN, D. M., TSOAR, H. & BLUMBERG, D. G. 2008. A second look at western Sinai seif dunes and their lateral migration. *Geomorphology*, **93**, 335–342.

SHEARD, M. J., LINTERN, M. J., PRESCOTT, J. R. & HUNTLEY, D. J. 2006. Great Victoria Desert: new dates for South Australia's ?oldest desert dune system. *MESA Journal*, **42**, 15–26.

SMITH, D. M., TWIDALE, C. R. & BOURNE, J. A. 1975. Kappakoola dunes – aeolian landforms induced by man. *Australian Geographer*, **13**, 90–96.

SPRIGG, R. C. 1952. *The Geology of the South-east Province, South Australia, With Special Reference to Quaternary Coastline Migrations and Modern Beach Developments. Geological Survey of South Australia*, Bulletin, **29**.

SPRIGG, R. C. 1963. Geology and petroleum prospects of the Simpson Desert. *Transactions of the Royal Society of South Australia*, **86**, 35–66.

STEPHENS, G. C. & CROCKER, R. L. 1946. Composition and genesis of lunettes. *Transactions of the Royal Society of South Australia*, **70**, 302–312.

STURT, C. 1849. *Narrative of an Expedition into Central Australia Performed under the Authority of Her Majesty's Government during the years 1844, 5, and 6* (2 vols). Boone, London.

TEY, J. 1953. *The Singing Sands*. Macmillan, New York.

THESIGER, W. 1964. *Arabian Sands*. Penguin, Harmondsworth, Middlesex.

TSEO, G. 1990. Reconnaissance of the dynamic characteristics of an active Strzelecki Desert longitudinal dune, southcentral Australia. *Zeitschrift für Geomorphologie*, **34**, 19–35.

TSOAR, H. 1989. Linear dunes – forms and formation. *Progress in Physical Geography*, **13**, 507–528.

TSOAR, H., BLUMBERG, D. G. & STOLER, Y. 2004. Elongation and migration of sand dunes. *Geomorphology*, **57**, 293–302.

TWIDALE, C. R. 1968. Singing sands. *In*: FAIRBRIDGE, R. W. (ed.), *The Encyclopedia of Geomorphology*. Reinhold, New York, 994–996.

TWIDALE, C. R. 1972. Evolution of sand dunes in the Simpson Desert, central Australia. *Transactions of the Institute of British Geographers*, **56**, 77–109.

TWIDALE, C. R. 1974. J. C. Darke's North Western Exploratory Expedition of 1844. *Journal of the Royal Australian Historical Society*, **60**, 28–47.

TWIDALE, C. R. 1981. Age and origin of longitudinal dunes in the Simpson and other sand ridge deserts. *Die Erde*, **112**, 231–247.

TWIDALE, C. R. 2002. A note on the development of wash aprons in dune sand, central Australia. *Zeitschrift fur Geomorphologhie*, **46**, 455–460.

TWIDALE, C. R., BOURNE, J. A. & SMITH, D. M. 1976. Age and origin of palaeosurfaces on Eyre Peninsula and the southern Gawler Ranges, South Australia. *Zeitschrift für Geomorphologie*, **20**, 28–55.

TWIDALE, C. R., BOURNE, J. A., SPOONER, N. A. & RHODES, E. J. 2007. The age of the palaeodunefield of the northern Murray Basin in South Australia–preliminary results. *Quaternary International*, **166**, 42–48.

TWIDALE, C. R., PARKIN, L. W. & RUDD, E. A. 1990. C. T. Madigan's contributions to geology in south and central Australia. *Transactions of the Royal Society of South Australia*, **114**, 157–167.

TWIDALE, C. R., PRESCOTT, J. R., BOURNE, J. A. & WILLIAMS, F. M. 2001. Age of desert dunes near Birdsville, southwest Queensland. *Quaternary Science Reviews*, **20**, 1355–1364.

VAN DEUR, W. J. 1983. *Submerged dunes of northeastern Eyre Peninsula*. Unpublished M.A. thesis, University of Adelaide, Adelaide.

VEEVERS, J. J. & WELLS, A. T. 1961. *The Geology of the Canning Basin, Western Australia. Bureau of Mineral Resources*, Bulletin, **60**.

WALKER, T. R. 1979. Red color in dune sand. *In*: MCKEE, E. D. (ed.) *A Study of Global Sand Seas*. United States Geological Survey, Professional Paper. **1052**, 61–81.

WARBURTON, P. E. 1875. *Journey Across the Western Interior of Australia*. Sampson Low, London.

WASSON, R. J. 1983. The Cainozoic history of the Strzelecki and Simpson dunefields (Australia) and the origin of desert dunes. *Zeitschrift für Geomorphologie Supplementband*, **45**, 85–115.

WASSON, R. J., FITCHET, K., MACKAY, B. & HYDE, R. 1988. Large-scale patterns of dune type, spacing and orientation in the Australian continetal dunefield. *Australian Geographer*, **19**, 89–104.

WILLS, W. (ed.).1863. *A Successful Exploration Through the Interior of Australia, From Melbourne to the Gulf of Carpentaria; From the Journals of William John Wills*. Bentley, London.

WILSON, C. C. 1991. *Geology of the Quaternary Bridgewater Formation of southwest and central South Australia*. PhD thesis, Flinders University of South Australia, Adelaide.

WOPFNER, H. 1990. Subsurface geology and hyrocarbon deposits. *In*: TYLER, M. J., TWIDALE, C. R., DAVIES, M. & WELLS, C. B. (eds) *Natural History of the North East Deserts*. Royal Society of South Australia, Adelaide, 27–44.

WOPFNER, H. & TWIDALE, C. R. 1967. Geomorphological History of the Lake Eyre Basin. *In*: JENNINGS, J. N. & MABBUTT, J. A. (eds) *Landform Studies from Australia and New Guinea*. Australian National University Press, Canberra, 117–143.

WOPFNER, H. & TWIDALE, C. R. 1988. Formation and age of desert dunes in the Lake Eyre depocentres in central Australia. *Geologische Rundschau*, **77**, 815–834.

WOPFNER, H. & TWIDALE, C. R. 2001. Australian desert dunes: wind rift or depositional origin? *Australian Journal of Earth Sciences*, **48**, 239–244.

WYRWOLL, K.-H. & VALDES, P. 2003. Insolation forcing of the Australian monsoon as controls of Pleistocene mega-lake events. *Geophysical Research Letters*, **30** CLM, 6-1-4.

Griffith Taylor, Ernest Andrews *et al.*: early ideas on the development of the river systems of the Sydney region, eastern Australia, and subsequent ideas on the associated geomorphological problems

DAVID R. OLDROYD

School of History and Philosophy, The University of New South Wales, Sydney 2052, Australia (e-mail: doldroyd@optushome.com.au)

Abstract: Thomas Griffith Taylor was one of Australia's leading geoscientists in the early twentieth century. He also developed ideas on race and environment at Sydney University, supposing that brachycephalic people had greater intellectual capacity than dolichocephalics, so that Chinese settlement in Australia was desirable. In consequence, Taylor's promotion at Sydney University was blocked. He therefore moved to a chair at Chicago, and thence to Toronto. He retired to Sydney in 1951 and in his old age published a classic study of the landforms of the region: *Sydney-side Scenery* (1958, published by Angus & Robertson, Sydney). This little book, although scientifically somewhat dated even at the time of its publication, summarized Taylor's earlier geomorphological ideas about Sydney and its hinterland. It applied Davisian ideas to the Sydney region, harking back to the work that Taylor and others had done in the early twentieth century. Taylor's ideas about the Sydney region's river patterns are described and their relationship to supposed Earth movements. In particular, Davisian ideas about river capture and antecedent drainage, and the topography of the Blue Mountains, are discussed in relation to the empirical information and theoretical ideas available in the early twentieth century. Taylor's ideas, as described in 1958, seem(ed) plausible, but they were subsequently thrown in doubt or invalidated by consideration of the form of diatremes in the Blue Mountains, by new geological theory, new data about the ages of rocks and minerals and estimates of the timing of Earth movements, closer mapping of structures, etc. The case exemplifies both the explanatory power, and the weaknesses, of Davisian geomorphology. E.C. Andrews preceded Taylor in introducing American geomorphological ideas into Australia and propounded the idea of Late Tertiary–Pleistocene uplift in eastern Australia, which was held responsible for many of the features that Taylor described. Other important contributors were W. G. Woolnough (1876–1958) and T. W. E. David (1858–1934), who taught Woolnough, Andrews and Taylor. A consensual view of the geomorphological and tectonic history of the Sydney region has still not been achieved.

There is not a single main river in the Sydney region which behaves normally.
(Taylor 1911*a*, p. 23)

Thomas Griffith Taylor (1880–1963)

Thomas Griffith Taylor was born in England but migrated to Australia in 1893, when his father was appointed Government Metallurgist for New South Wales. Taylor studied geology and geomorphology (called physiography for the first-year course) under Professor Edgeworth David (1858–1934) at Sydney University and also some physics and engineering subjects, graduating in 1904, following which he took further studies in the Mining School. After a short stint teaching at a secondary school, Taylor was appointed as a demonstrator in David's department, and gave lectures on 'Commercial Geography' to evening students studying Economics. In 1907 Taylor took up a

Great Exhibition scholarship at Emmanuel College, Cambridge, working on archaeocyathids from South Australia, for which he was awarded a BA (research) degree. In 1910, following his return to Australia, he was appointed to the post of 'physiographer' to the Commonwealth Meteorological Bureau in Melbourne, but was almost immediately seconded to Scott's second Antarctic Expedition, to which he was recommended through his Cambridge connections. Sydney University awarded him a DSc in 1916 for his Antarctic work.

After his return from Antarctica, and during World War I, Taylor served as a senior scientist in the Australian Weather Service, and taught meteorology at the flying school at Laverton, Victoria. In the years that followed, Taylor became increasingly interested in questions of climate and race, and the development of agricultural production and settlement in Australia in relation to soils and climate

From: GRAPES, R. H., OLDROYD, D. & GRIGELIS, A. (eds) *History of Geomorphology and Quaternary Geology.*
Geological Society, London, Special Publications, **301**, 241–277.
DOI: 10.1144/SP301.16   0305-8719/08/$15.00 © The Geological Society of London 2008.

and geography more generally, rather than geology. He also became adept at preparing handdrawn maps and diagrams, which imparted ideas about regions' geology and geography, their suitability for settlement according to temperature, rainfall, soiltype, and so on.

In 1920 Taylor was appointed Associate Professor and head of a new Department of Geography at Sydney University, and began lecturing in 1921. But some of his ideas were 'eccentric' in that, for example, he supposed that brachycephalics had greater intellectual capacity than dolichocephalics, in consequence of which he believed that the migration of 'Asiatics' (especially Chinese peoples) to Australia should be encouraged. This suggestion was put forward at the height of the 'White Australia' policy, and Taylor, who was teaching such ideas at Sydney University (with his students busily engaged in craniometrical studies!), came in for public criticism, although his department thrived under his leadership. Even so, his idiosyncratic anthropological ideas coupled with his writings on the limits to population in Australia (in terms of soils, rainfall, etc.), in opposition to the high hopes of the many 'boosters' in the country in his era, made him unpopular in some quarters in Sydney. So, finding his promotion to a full professorship blocked, Taylor moved to the United States in 1928 to take up a chair at Chicago (at a much better salary than Sydney could offer), from whence he later moved to Toronto where he established the first Canadian geography department.

Taylor retired to Sydney (with its congenial climate) in 1951 and in his old age (see Fig. 1) he prepared a delightful book, *Sydneyside Scenery* (1958), which described the physical structure and history of the city and its environs, and its urban development, all cogently explained in terms of the topography and rocks and soils of the region. Among the discussions, considerable attention was given to the question of the development of the river systems of Sydney and its hinterland and this topic provides the focus of the present paper. But the ideas on drainage, landforms and crustal movements expressed in 1958 were those that Taylor had developed back in the early years of the twentieth century.[1]

Fig. 1. Taylor at work in his study in Sydney, during his retirement (Taylor 1958*b*, facing p. 32).

Geology and geomorphology of the Sydney region and hinterland: historical and topographical background

A general map of the geology of the Sydney region, drawn by Taylor in the back of his notes on 'physiography', taken from David's lectures at Sydney University (1899), is reproduced in Figure 2. The general features of the topography of the Sydney region are represented in Figure 3, and the river systems are delineated in Figure 4.

As suggested by Taylor in the quotation at the start of this paper, the rivers of the Sydney region have unusual features. The city has its famous and quite deep harbour, but the Parramatta (or Paramatta) River that feeds it is small and its headwaters are not far inland, in the large area of flat ground 'behind' Sydney–the Cumberland Plain. The main waterway of the region enters the sea about 30 km north of Sydney Harbour, with a wide estuary (Broken Bay), fed by a large river: the Hawkesbury. But parts of the drainage system of this river rises on the *western* side of the famous Blue Mountains, the eastern edge of which forms a scarp that runs parallel to the coast and about 50 km inland from

---

[1]On Edgeworth David, see Branagan (2005). For Taylor's career, see Taylor (1958*a*), Powell (1979) and Sanderson (1988). For his anthropological hypotheses, see Oldroyd (1994).

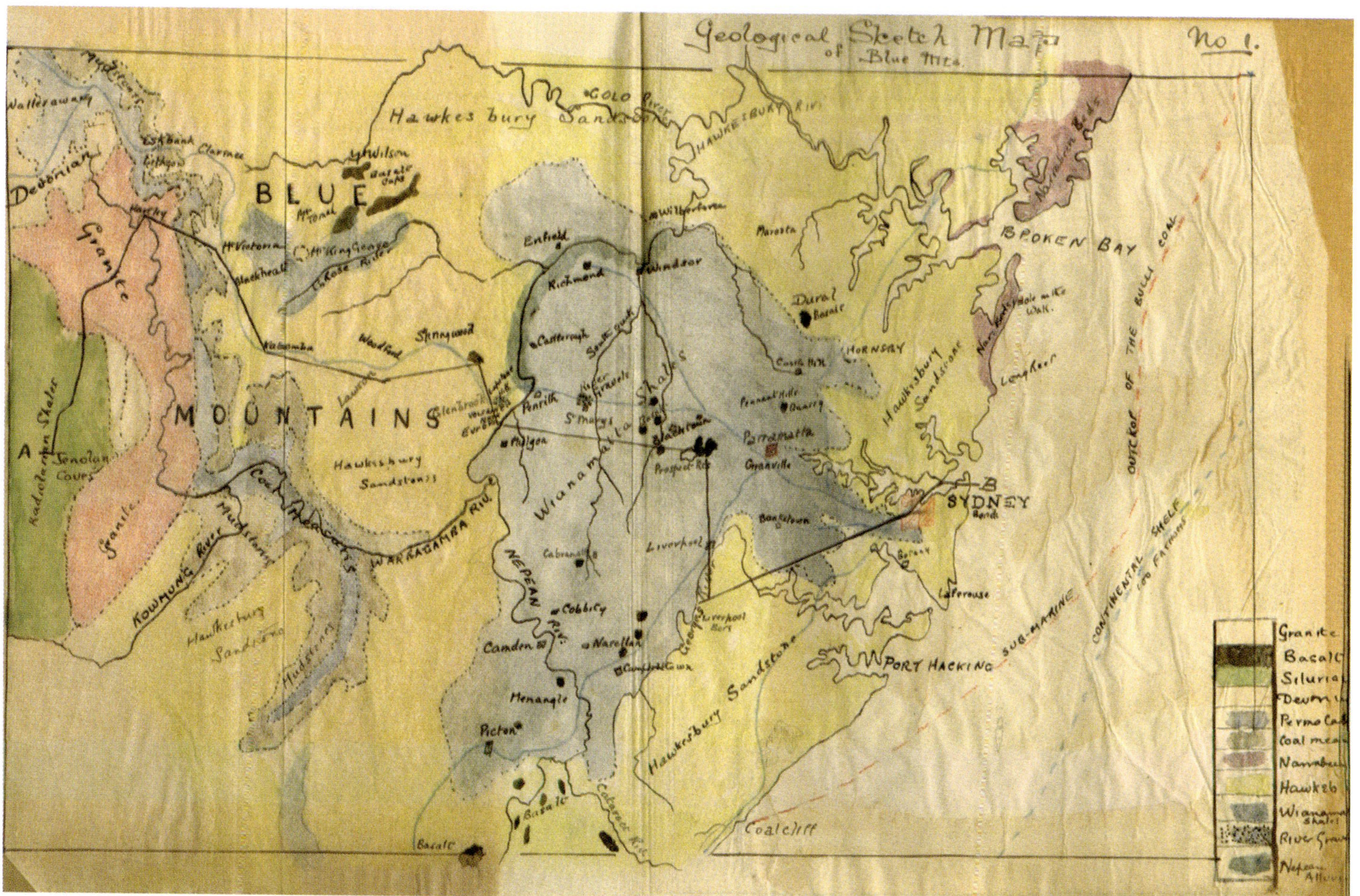

Fig. 2. Geological map of the Sydrey region, according to Taylor/David (1899). From Personal Archives of Thomas Griffith Taylor, Series 1, Lectures and Working Notes, Item 1, p. 196. Reproduced by permission of Norman Taylor and Archives and Records Management Services, The University of Sydney.

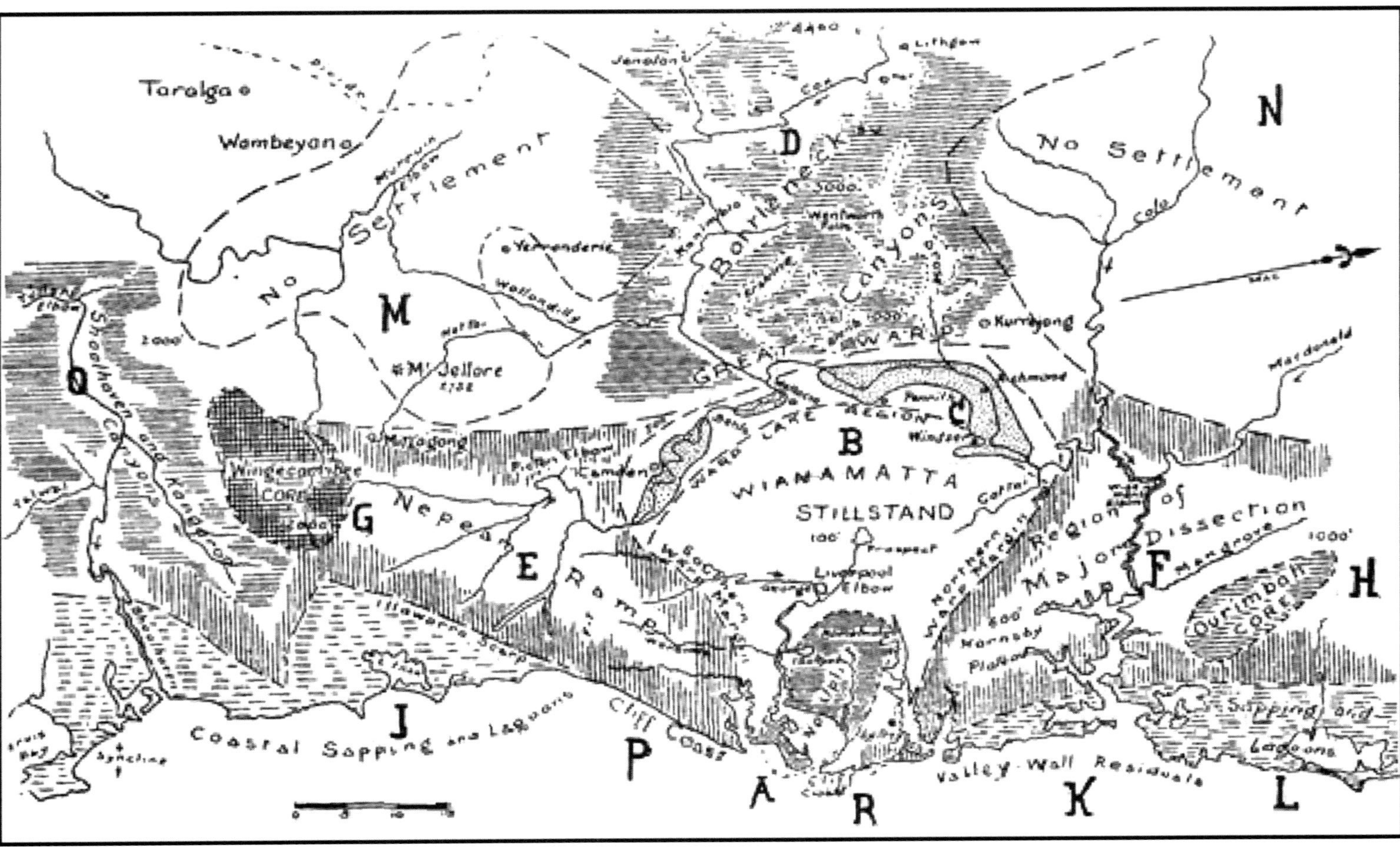

Fig. 3. Geographical regions around Sydney, according to Taylor (1923a, p. 64).

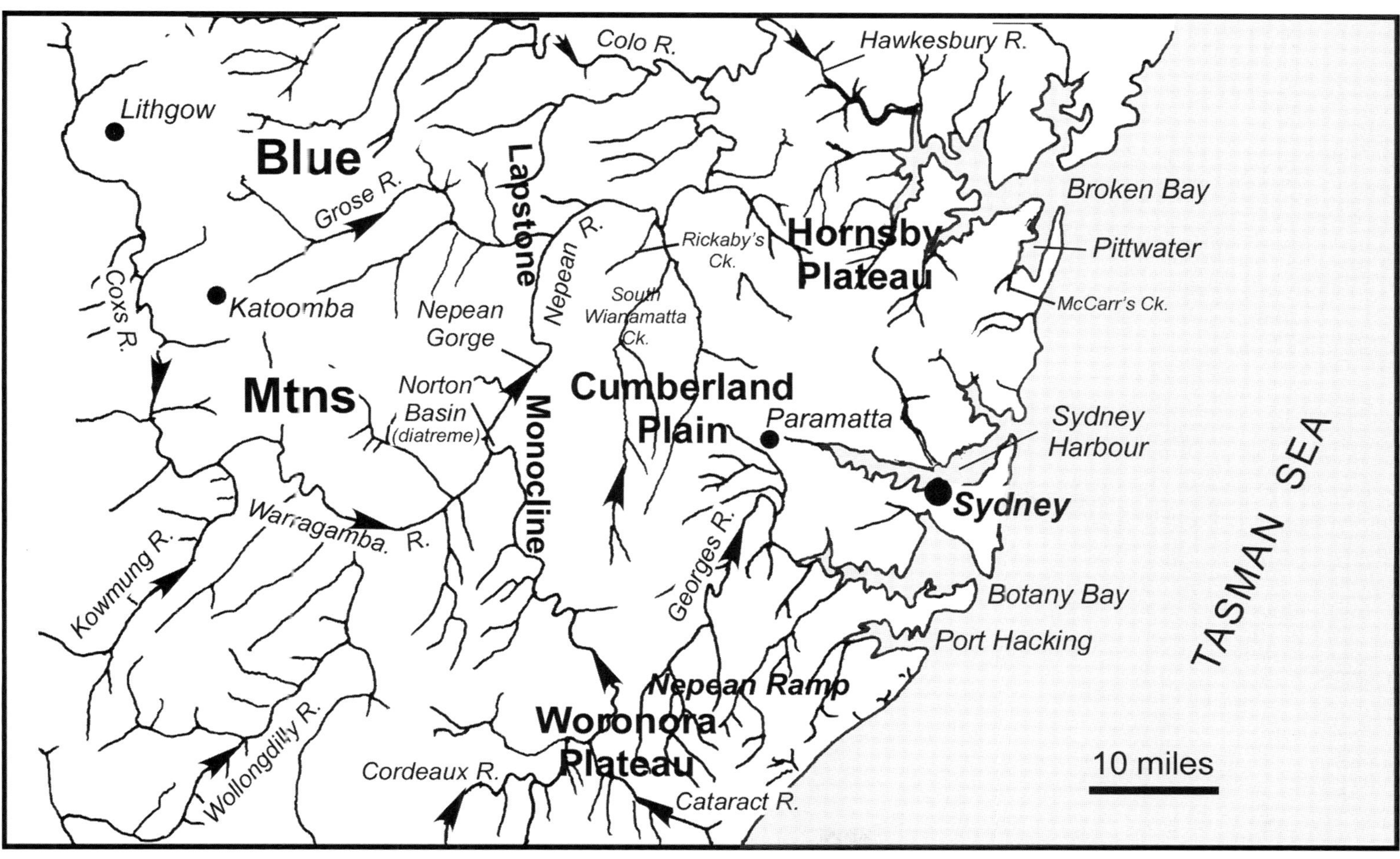

Fig. 4. Sketch of the main rivers of the Sydney region.

Sydney's city centre.[2] The scarp has a monoclinal structure, and is known as the Lapstone Monocline.[3] Upstream, the *same* river as the Hawkesbury has a different name – the Nepean. The river thus named runs northwards, approximately parallel to the foot of the scarp, but sometimes it enters the higher and harder ground of the Lapstone Monocline–Blue Mountains. It takes on its downstream name, the Hawkesbury, near the township of Richmond, at the junction where the Grose River flows out from its beautiful and much admired valley in the Blue Mountains and joins the main waterway.[4] At Richmond, the river turns east and then north, passing *off* the Cumberland Plain and *into* the hilly country north of Sydney, where it is joined by the Colo River from the rugged country that forms the northern part of the Blue Mountains. The enlarged Hawkesbury waterway then follows a *meandering* course, through a valley incised in hard sandstone, before reaching the sea at Broken Bay.

Further upstream, the Nepean is fed by the Cox's and Wollondilly rivers, which unite to form the Warragamba River that follows a straight gorge through the Blue Mountains to reach the Nepean from the *west*.[5] So, for example, water from the quite small Cox's River flows from the comparatively low-altitude country to the west of the Blue Mountains and through the mountains, before turning north to flow mostly near the foot of the Lapstone Monocline. It then enters the area of high ground north of Sydney before turning east to reach the sea. Water also enters the Hawkesbury–Nepean system from the *SE* from the Cataract, Cordeaux and Avon rivers, which have their headwaters on the high ground to the south of Sydney, as does the Nepean itself. To the west of the Blue Mountains, with intervening lower ground, lies the 'Main Divide': a long, mostly quite low, range of 'hills' and modest mountains, chiefly made up of Palaeozoic rocks of complex structure, running more or less south–north and approximately parallel to the coast, from Victoria to North Queensland.

To the south of Sydney's centre, there is the large, approximately circular, Botany Bay, which is fed by the relatively small George's River. The form of the Bay might suggest (as Taylor thought) that it is an area where there was formerly a meandering river, nearly at sea level, which became inundated fairly recently by a sea-level rise – presumably after the last Pleistocene glaciation. Sydney Harbour can be interpreted as a drowned river valley also. Further south, the Shoalhaven River, like the Nepean, also runs northwards and parallel to the coast for some of its course, before turning sharply to the east and flowing to the sea.

From these brief remarks, it would appear that the Sydney region has been affected by substantial Earth movements (explaining what would appear to be antecedent drainage phenomena), and that there have been 'river captures' (in the language of the American geomorphologist William Morris Davis (1850–1934)). Good accounts of the history of geomorphological studies of the Sydney region have been published by Scott (1977), Young (1978) and Bishop (1982).

The nature and form of the strata

As Taylor's maps (Figs 2 and 3) show, Sydney is located near the centre of a basin.[6] There is rough, dissected sandstone country to the north and south (the Hornsby Plateau and the Nepean Ramp–Woronora Plateau, respectively). To the west, there is what Taylor called the 'Great Warp' (or Lapstone Monocline), which marked the frontier of the dissected plateau of the Blue Mountains.[7]

---

[2]Today, urban Sydney covers much of the intervening ground.

[3]More recent publications refer to it as the Lapstone Structural Complex (see, for example, Branagan & Pedram 1990). In the present paper I shall use the older term: Lapstone Monocline. However, the structure has various associated subfolds and faults, and is by no means a simple, smooth and isolated fold.

[4]The different names were coined because, soon after European settlement began in 1788, different parts of the Nepean–Hawkesbury waterway were explored and named at different times and by different people; and the fact that the Nepean and Hawkesbury rivers were all part of one system was not immediately realized.

[5]The gorge follows an anticlinal structure in the hard Triassic sediments that form the sides of the gorge.

[6]As a geological structure, the Sydney Basin is made up of the Permian, Triassic and younger rocks that are discussed in the present paper. But one can also say that the Sydney region and the inland Cumberland Plain near Parramatta have the form of a basin. Thus, the rocks of the Blue Mountains are in the Sydney Basin (geologically speaking), but they are not in 'Sydney's basin'. The 'Sydney Basin' (geologically speaking) is a larger entity than the 'Sydney basin' (topographically speaking).

[7]Taylor always objected to the name of Blue Mountains, as he (rightly) contended that they consisted of a plateau region, dissected by various rivers and do not resemble an ordinary mountain range. For this reason, he advocated the name 'Blue Mountains Plateau', but this suggestion did not catch on. Seen from any lookout on the top of the 'mountains' the area looks like a dissected plateau, but from the bottom of its

He did not regard the low-lying area of the Cumberland Plain as one of subsidence: it was, he thought, *relatively* low because of the *uplift of the adjacent land*. So the supposedly static area he called the 'Wianamatta Stillstand'.[8]

The Triassic rocks of the Sydney Basin, which contain rather few fossils, are subdivided into three main units:

- Wianamatta Group – chiefly *soft* laminated shales, with poor-quality coal in places; dark when freshly exposed but turning brown and friable on weathering. The Group has several constituent members and has yielded some plant remains and amphibia;

overlying

- Hawkesbury Sandstone – massive *hard*, cross-bedded cream-coloured sandstone, with occasional shale or clay lenses. Fish and amphibia have been found in some of the lenses;

overlying

- Narrabeen Group[9] – several units, some of them *soft* clay-stones or shales with ripple-marks and mud-cracks; elsewhere, especially in the Blue Mountains, *hard* massive sandstones and conglomerates, weathering to golden yellow. The group contains red beds and 'chocolate shales' and has yielded amphibia, fish and plants.

These units form much the greatest areas of outcrop in the Sydney region. Below them occur generally *soft* Permian strata, containing good-quality coal. The whole structure is basin-like, so that coals crop out to the north and south of Sydney, where they have given rise to the industrial cities of Newcastle to the north and Wollongong to the south. Coals also crop out at the western margin of the Blue Mountains near Lithgow (see Figs 3 and 4), and they could also be (and at one time were) mined by sinking deep shafts near the centre of Sydney – for the coal deposits run from Newcastle to Wollongong, right under the city. Insofar as there is good agricultural land near Sydney, it occurs chiefly in the area of the Wianamatta Stillstand, and to the south in the Wollongong region, where the rocks of the underlying Shoalhaven Group (Permian) are exposed and weathered volcanics

yield rich soils. Behind Wollongong, runs the 'Illawarra Scarp', formed in hard Triassic sandstones. There are also quite numerous volcanic necks (diatremes[10]) in the Sydney region and there is a small laccolith at Prospect in the centre of the 'Stillstand' (see Fig. 2). Diatremes occur in the Blue Mountains also, and there are outcrops of basalt at Mount Tomah, by the road to the mining town of Lithgow (see Fig. 3), at nearby Mount Wilson and several other localities, which also give good soils; and outcrops of Wianamatta Shales east of Mount Tomah provide soil suited to fruit growing. There are vineyards to the north of Sydney in the good soils of the Hunter Valley, inland from Newcastle.

Early studies of the geomorphology of the Blue Mountains and related issues

In 1896 Edgeworth David read a paper to the Royal Society of New South Wales, in which he described the structure of the Lapstone Monocline (which, as said, forms the eastern scarp of the Blue Mountains, to the west of Sydney), and differentiated between the three Triassic units mentioned above (David 1897). He made special mention of river gravels found at the *top* of the monocline, as well as gravels found on the plain below to the east.[11] Their presence suggested that there had formerly been a river that ran roughly parallel with the present Nepean River; and/or that the same river had changed its course over time. This ancient ('high-level') river had, thought David, cut into the Hawkesbury Sandstone that forms the monocline, but not deeply. It predated the present deep gorges that some rivers (such as the Warragamba or the Grose) have cut into the mountains, forming deep valleys perpendicular to the mountain range and the Lapstone Monocline.

Given the existence of the monocline and the high-level gravels, the question immediately presented itself: had there been uplift to the west or downwarping to the east, where subsidiary folds are displayed on the lower ground, parallel to the

---

valleys, most notably the Grose Valley, there appear to be peaks, like those of an 'orthodox' mountain range.

[8]The (Aboriginal) name Wianamatta was taken from that of a small tributary of the Hawkesbury–Nepean River, which flows northwards across the area of the Stillstand. This tributary is also called South Creek: see Figs 2–4.

[9]Narrabeen is a seaside suburb north of Sydney, where these rocks are well exposed in the sea-cliffs.

[10]Diatreme = διάτρημα or 'through hole': a long breccia-filled volcanic pipe formed by the explosive emission of gas-rich magma. The explosive nature of the eruption may be attributed to the passage of the magma through water bearing sediments.

[11]These gravels had earlier been mentioned (and partly mapped) by the Reverend W. B. Clarke (1878), the so-called 'father of Australian geology'. They were later called the Rickaby's Creek Gravels for the exposures on the plain below the monocline by Valerie Gobert (1976), a geologist with the Geological Survey of NSW.

monocline? David opted for the downwarping alternative, having regard to the evidence of submergence of the coastal ground 'in Tertiary or Post-Tertiary time', such as the 'drowned' Sydney Harbour.

Within the area of the Blue Mountains there are major wide and deep valleys, walled by celebrated orange/yellow cliffs (Narrabeen rocks) (for an account of early ideas on the origin of the valleys of the Blue Mountains, see Young 2007). These valleys have *narrow* 'necks' where their rivers cut through to the coastal plain. David rejected Darwin's hasty suggestion that the valleys were the product of marine erosion[12] and that the monocline marked a recent shoreline. The form of the valleys ('box canyons') was explained by the nature and form of the strata that made up the mountains. The upper layers are hard sandstone, whereas the underlying strata belong to the relatively soft lower Narrabeen rocks and Permian coal measures. But, by virtue of the monocline, the vertical thickness of hard rock to be cut through by the rivers is greatest near the line of the monocline; so the valleys narrow there. Further west, once the streams have cut through the hard sandstone to the softer rocks below, they have been able to widen their valleys by undercutting the sandstones, from which slabs fall off from time to time (chiefly along the lines of joints), leaving the great vertical cliffs of sandstone. Although David did not state the matter exactly like this, he seemingly had the general idea. Without focusing on the geomorphological influence of the monocline on the form of the mountains' rivers and valleys, he emphasized the general eastward dip of the strata, which led to the softer, erosion-prone, rocks being at a higher altitude to the west as compared with the east.

In his 1897 paper, David also suggested that there was a modest *west*-dipping fold directly to the west of the Blue Mountain scarp. But soon thereafter he produced another paper (David 1902) showing that the Lapstone Monocline did not simply flatten out beyond the west of the small west-dipping fold. He now suggested that the eastern margin of the Blue Mountains was a modified long, narrow anticline, with its western limb down-faulted in places. This fault, on the western side of the structure, David named the Kurrajong Fault, Kurrajong Heights being located at the top of the scarp about 15 miles (24 km) north of Glenbrook. David also noted faulting at Glenbrook.

Fig. 5. Photo-portrait of Ernest Andrews. Andrews papers, Box 2, not dated. Reproduced by courtesy of the Basser Library, Australian Academy of Science, Canberra.

Another important early paper was published by an up-and-coming young geologist, Ernest C. Andrews (1870–1948) (1903a), who joined the NSW Geological Survey in 1899 (see Fig. 5).[13] He was initially an autodidact and unqualified schoolteacher, but managed to enrol for teacher training at the Sydney Teachers' College, where he could study science, including geology, and thereby he came in contact with Edgeworth David. After transferring to a degree course, Andrews graduated in 1894 with a 'Second' in Mathematics and was subsequently appointed to a school at Bathurst, a town on the farming land west of the Blue Mountains. From this centre he began to study the geology for many miles around. He also worked with David on his important studies of the coal fields in the Hunter Valley north of Sydney; and David arranged for him to

---

[12]Darwin did not see the narrow necks during his brief traverse of the mountains in 1836; see Nicholas & Nicholas 1989.

[13]On Andrews, see: Anon. (1952) and Scott (1977). There are also two unpublished autobiographical documents in the Andrews papers in Canberra.

travel to Fiji and Tonga on behalf of J. L. R. Agassiz of the Harvard Museum to study coral reefs, following which he took more courses under David at Sydney University. His appointment to the Survey was achieved by success in a public examination, although perhaps assisted by David as his backer.

Andrews' work with the Survey soon took him to the plateau country of New England (northern New South Wales), and he also visited the Great Barrier Reef off the Queensland coast in 1901, in company with the Australian Museum's conchologist Charles Hedley (1862–1926). This northern excursion gave Andrews the opportunity to visit the fertile plateau country of North Queensland, behind Cairns (the Atherton Tablelands).

Working in New England, Andrews (1903*b*) was readily convinced that the gorges of that area, which cut into the region's tableland, were (in the language of W. M. Davis) 'youthful'. Andrews saw the upland country of New England and northern New South Wales as representing a late-Tertiary erosion surface. By contrast, David regarded the New England plateau as an ancient surface left by a receding sea; and he did not accept Andrews' suggestion of relatively recent uplift. Nevertheless, Andrews (1903*a, b*) began to deploy Davis's ideas in Australia, and particularly the idea of geomorphological cycles. Like Davis, Andrews envisaged successive periods of uplift, followed by subaerial erosion, so that one got benches and valleys-within-valleys. Naturally, he also accepted Davis's idea of 'rejuvenation' and the well-known stages of a river's lifehistory or of a landscape: youth, maturity, oldage/senescence. Andrews' (1903*b*) paper on the highlands of New England tentatively proposed that the last uplift for the region occurred in the Pliocene.

Andrews (1903*a*) attempted to account for the geomorphology of the Sydney and its surrounds in Davisian terms, in ways that were to a considerable extent taken over by Taylor. Andrews envisaged three peneplains in the mountains, arising from a succession of uplifts, giving: the Lithgow Plain (3100 ft, most recent); the Blue Mountain Plain (now at 3500 ft); and the Stony Ridges Plain–Jenolan Plain (now at 4100–4300 ft, earliest) (forming part of the Main Divide) (but these were also called 'cycles'!). The lowest Lithgow surface was thought to have been folded thereby creating the Lapstone Monocline. In making these suggestions Andrews was identifying what we now think are, in many cases, depositional surfaces as erosional plains (or planes) albeit sometimes slightly inclined. But despite the objections of his colleague at the Survey, Joseph Edward Carne (1855–1922), Andrews' ideas on a succession of elevations and peneplanations exerted their influence for many years.[14]

Andrews (1903*a*, p. 806) also suggested that 'the old Hawkesbury River bed at Lapstone Hill' belonged to what he called the 'early cañon cycle'. The gravels near Glenbrook at the top and side of the Lapstone Monocline might represent the relics of a 'Miocene (?)' stream that once flowed on the old 'Lithgow Plain', which was formerly not much above sea level.[15] The subsequent elevation had left these gravels 'high and dry' (as I might put it). But the present Nepean River had, nevertheless, managed to cut down in places at about the same rate as that of the elevation, so that it preserved its old meanders 'in to' and 'out of' what is now the scarp of the Lapstone Monocline (see Figs 8b, 15 and 18 later in this paper). Further downstream (but to the NE) the Hawkesbury was an 'obsequent [antidip] stream' (cf. Davis 1895, p. 134),[16] taking an unlikely course to the sea through the hard sandstone country of the Hornsby Plateau.

Like (or maybe from?) David, Andrews understood the significance of the layers of underlying soft and overlying hard strata in the formation of the topography of the Blue Mountain valleys. But his terms 'plateau cycle' – for the formation of the several envisaged plain surfaces ('peneplains', as we would say) – and 'cañon cycle' – for the formation of the gorges and canyons of the Blue Mountains – was not perhaps particularly felicitous, as we would regard the formation of these features as components of a cycle, not a cycle per se. Davis (1899) had written of the evolution of a landscape from youth to old age as a 'cycle'. But for Andrews it seemed as if a part of a landscape cycle were a cycle. It was only in 1905, in a school textbook, that he made it clear that he

---

[14]Carne (1908, p. 12) objected to the idea of the successive formation of peneplains at sea level and successive uplifts; and also to the fact that Andrews kept on changing his thinking about the specifics of his theory in different publications. But Carne's voice, and his suggestion that the Blue Mountains had essentially been created by a single uplift, fell on deaf ears.

[15]Lithgow is an old industrial and mining town situated below the western scarp of the Blue Mountains. In a later publication, Andrews (1905, p. 64) changed the term Lithgow Plain to Blue Mountains Plain, which was an improvement.

[16]Twidale & Campbell (2005, p. 191) have rightly remarked that the old terminology of consequent, subsequent, obsequent, resequent and insequent streams is most unsatisfactory. They recommend the use of the terms dip, strike, anti-dip, fault, anticlinal, synclinal or monoclinal as appropriate. The term 'insequent' was a 'negative ragbag' for streams that did not seem to belong to any other category.

regarded 'any great natural work, whether of elevation, sinking, sedimentation, erosion, volcanicity or earth folding, as a cycle' (Andrews 1905, p. 58). This clarified his usage of the term 'cycle' in his memoir of 1903, but also shows that it was idiosyncratic.

In the years that followed, Andrews began to correspond with the eminent American geologist Grove Karl Gilbert (1843–1918) and was invited by him to visit California in 1908. They also travelled together to Arizona. At the end of the year, Andrews met many of the leading American geologists at the American Association for the Advancement of Science (AAAS) meeting in New York. Following his visit to the New World, he sailed for Britain, where he also met leading figures, including W. M. Davis, who was then in England. Andrews opined that 'English geologists [were] much behind American and Canadian field men and original workers' (Andrews not dated, p. 82)! He returned to Australia after 18 months overseas.

On his return to Australia, Andrews worked up an important paper in which he sought to give a *unified* account of the geology/geomorphology of eastern Australia, based on the supposition that there was a single Miocene peneplanation and that the peneplain surface was then differentially elevated (Andrews 1910). That is, he drew attention to the existence of plateau surfaces in Queensland, various parts of New South Wales and Victoria, and suggested that the whole of the eastern part of the country had been *uplifted* in geologically quite recent times in what he called the *Kosciusko Period*[17] (late Tertiary–Pleistocene), and resultant the land surface had subsequently been subjected to weathering and erosion, giving rise to the analogous youthful topographies encountered in various parts of the three states. The different elevations of the plateaus in different regions might be attributed to faulting. So Andrews attributed the features of the landscape of the Sydney region to more general Earth movements that were linked to the elevation of the chief mountain system of the continent – which he supposed to have taken place remarkably recently. In such a manner, the tectonic ideas that had been expounded by David on a geological basis were developed and modified by Andrews by the study of physiography (or geomorphology as we might say). He acknowledged that what he had earlier taken to be separate

peneplanation surfaces were now to be interpreted (without much independent evidence) as the product of block faulting. The following year, David cited with approval Andrews' work – with the mapping of faults on physiographical evidence – in his synthetic study of the 'chief tectonic lines' of Australia' (David 1912 for 1911).

It should be remarked that by the time of publication of his 1910 paper, Andrews was evidently under the spell of American geologists and geomorphologists, and – rather extraordinarily – imagined that the geology of eastern Australia was analogous to that of western side of the United States. For the Pacific coast and neighbouring ranges, American geologists envisaged peneplanation in the Tertiary, followed by extensive uplift and rejuvenation of topography in the Late Pliocene–Pleistocene, forming the Sierras (Chamberlin & Salisbury, 1906, Vol. 3, pp. 311–318). This uplift, which was accompanied by substantial igneous activity, supposedly led to active erosion, rejuvenation and canyon formation in the Pleistocene. Andrews' model for eastern Australia largely used this scheme as an analogue. But, unfortunately, he was deceived by palaeobotanical evidence found beneath the basalts, which in places had evidently flowed into the supposed newly eroded valleys cut into the uplifted plateaus of eastern Australia. For in Victoria the botanist Ferdinand von Mueller (1825–1896) (1874, First and Second Decades) had identified plant remains found underneath the basalts of the 'Newer Volcanic Period' in Victoria as Upper Pliocene, and Andrews supposed that this held also for the presumed analogous situation in New South Wales. Thus, the Kosciusko Period (or Uplift) and consequent topographic rejuvenation were deemed to be remarkably recent.

As regards the rivers of the Sydney region and surrounds, Andrews supposed the Hawkesbury–Nepean and its headward tributaries had maintained their courses during the period of uplift: he was, he believed, dealing with a manifest instance of antecedent drainage.[18] This meant that the present approximately south–north alignment of the Nepean must have been in existence prior to the uplift. But that arrangement would have been owing to an earlier trend of folding in eastern Australia; otherwise the rivers would be orientated west–east, carrying water directly from the mountains to the sea. The supposed 'youth' of the monoclinal fold of the eastern flank of the Blue Mountains was evidenced by the fact that its

---

[17]Mount Kosciuszko (south of Canberra – formerly spelled Mount Kosciusko) is the highest mountain in Australia (2228 m), and the centre of the only area of the mainland that was glaciated during the Pleistocene. The term Koskiusko Uplift was later used by many writers.

[18]But the upper reaches of the Nepean and its tributaries (the Cataract, Cordeaux and Avon rivers) appear to be dip and joint controlled, as they drain water from what Taylor called the Nepean Ramp.

consequent streams were but little developed. Andrews opined that the Australian (and New Zealand) mountains resulted from vertical uplift, not lateral compression. However, he attributed to Taylor the idea that 'in late Tertiary time ... a force acted from the Antarctic region towards the north-east probably carrying the continent bodily with it from the Antarctic direction' (Andrews 1910, p. 464) – a prescient suggestion considering modern ideas about the separation and northward drift of Australia from Antarctica.

Andrews' Davisian ideas were also deployed by another former David student, Carl A. Süssmilch (1875–1946) (1910 for 1909, p. 348, 1911), lecturer in geology, mineralogy and mining at the Sydney Technical College from 1903 to 1914, who laid special emphasis on the role of block faulting. For Süssmilch (1910 for 1909, p. 353), a 'cycle' involved both a phase of elevation and one of denudation.

Further useful work, relevant to the 'river problem', was published by the Danish-born geologist Harald Ingemann Jensen (1879–1966) (1912 for 1911), who traced out what appeared to be a former course of the Nepean River on the Cumberland Plain, near Penrith – a town by the foot of the Lapstone Monocline – by examining a line of gravels (the Rickaby's Creek Gravels) apparently similar to those found on the top and side of the monocline.[19] It seemed to Jensen that there had been a subsidiary uplift, approximately perpendicular to the axis of the monocline, which might have produced some diversion of the Nepean's course. Be this as it may, there was now firm evidence for the existence of similar gravels at the top, side and foot of the monocline, and northeastward across the Cumberland Plain.[20]

From a remark in Andrews' autobiographical notes (Andrews not dated, pp. 99–100), we are told that his paper of 1910 was 'torn in pieces' by (unnamed) 'seniors', although he recorded that he gained support from biologist friends and some colleagues at the Survey (but probably not Carne). I surmise that Edgeworth David was one of the 'seniors', as he had attributed the form of the Lapstone Monocline and the Cumberland Basin to subsidence, not elevation – although, as said, David referenced Andrews' work in print without negative comment.

Another important contributor to the development of geomorphological ideas for the Sydney area was Walter George Woolnough (1876–1958), who was also one of David's students at Sydney University, and a most distinguished one (see Raggatt 1959). Graduating in 1898, he too did fieldwork with David in the Hunter Valley coalfields, and also worked with him on an attempt to drill through a Pacific atoll (Funafuti) to try to confirm Darwin's theories about coral reef formation (for details see Branagan 2005, chap. 5). Woolnough was appointed a demonstrator at Sydney University, but obtained a lectureship at Adelaide University in 1904, where he worked for 3 years before being lured back to Sydney by David, for whom he deputized while he was away on his Antarctic investigations. In 1913 Woolnough was appointed to the first Chair of Geology at the University of Western Australia (Perth) and in

---

[19]Jensen studied at Sydney University and served there for a time as a demonstrator under Edgeworth David. At the time his observations on the Cumberland Plain were made, he was employed as a soil scientist with the New South Wales Department of Agriculture. Subsequently, he served as Director of Mines and Chief Geologist of the Northern Territory from 1912 to 1916, and was Geologist for the Queensland Government from 1917 to 1922. He was also active in left-wing politics and fell out with the authorities in the Northern Territory.

[20]The outcrops of gravels were clearly shown on a beautiful geological map of the Sydney region, 'prepared under the direction of E. C. Andrews, BA' by the NSW Department of Mines Geologist, Thomas Lindsay Willan (1895–1940) (1925). It seems clear that they are relicts of a river (very likely the Nepean itself) that formerly ran NW from near Penrith to Windsor, approximately along the line of the present

Rickaby's Creek (see Fig. 4), which does not, however, extend today as far as Penrith. Gravels crop out at the top of the Monocline near Glenbrook, and – a little to the north of there – on the side of the monocline in Knapsack Gulley – one of the few consequent streams flowing off the monocline to the east. The gravels there could have been reworked from the top of the monocline, although they also appear again on the side of the monocline further north near Hawkesbury Lookout, but at a lower altitude and in this case parallel with the strike of the fold. In addition, they crop out below the monocline near Mulgoa Creek, on the plateau south of Glenbrook, and further south again near Norton's Basin (see p. 243, 253 and 265), where exposures are indicated by Willan on both the side and the foot of the monocline. The outcrops can reasonably be joined up to mark the course of a former northward-flowing river, with approximately the same course as the Nepean, except for the 'short-cut' taken near the present Rickaby's Creek between Penrith and Windsor (whereas the present Nepean follows an arc parallel with the foot of the higher ground in the NW corner of the Cumberland Plain). But temporal correlation of the different gravel outcrops remains insecure.

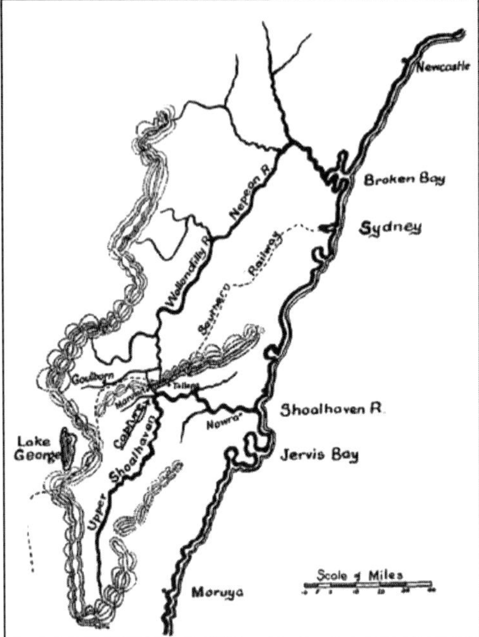

Fig. 6. The relations between the Shoalhaven,
Wollondilly, Nepean and Hawkesbury Rivers:
Woolnough & Taylor (1907–1908 for 1906, p. 547).
(The Hawkesbury runs out to the sea at Broken Bay and
is not labelled on this map.)

1927 he was appointed geological adviser to the
Federal Government.[21]

In 1906 Woolnough (then an assistant lecturer at
Sydney University) and Taylor (then an assistant
demonstrator) collaborated on a paper that discussed the remarkable form of the Nepean, Wollondilly and Shoalhaven rivers (see Figs 6 and 7). The
rivers' forms suggested that there had once been a
continuous south–north river, flowing parallel
with the coast and what is now the upland region
of New South Wales. The Shoalhaven's sharp
swing east, inland from Nowra – the so-called
'Tallong Bend' (see Fig. 7), as it occurs near the
small village of Tallong – was presumed to be
owing to river capture. Woolnough and Taylor
reached this conclusion independently, but then
undertook two joint excursions to check their
ideas and write a paper (Wodnough & Taylor
1907–1908). Their explanatory scenario was that
there had been a peneplain created in the region,
with a substantial northward-oriented river

flowing over it. Then uplift had occurred, encouraging the enlargement of relatively small streams
that flowed directly to the coast. One of these
(today the Lower Shoalhaven) had supposedly cut
back westwards and captured the waters of the old
Upper Wollondilly (now Upper Shoalhaven),
thereby forming the present river pattern, including
most notably the Tallong Bend. A persuasive piece
of evidence cited in the paper was the discovery of
the relicts of an old watercourse through which the
present Upper Shoalhaven's waters might formerly
have flowed to the present Upper Wollondilly.
It was said to be full of abandoned pebbles or
boulders,[22] which, from their composition, had
apparently been transported long distances, there
being, it was asserted, no corresponding rock in
the neighbourhood. In this paper we see Taylor's
early propensity to seek to solve tectonic problems
on the basis of geomorphological evidence, particularly that available from contoured maps and river
patterns.[23]

Thus, Davis's ideas were enthusiastically
applied in New South Wales by Andrews, Woolnough and Taylor. Davis and Taylor were in correspondence not long after the publication of the
Woolnough & Taylor (1907–1908) paper, and the
two met in Europe in 1908, when the American
organized a group to study Alpine landforms.
(Taylor mentions the correspondence in his
autobiography.[24]

---

[21]Woolnough became a formidable linguist, being able to
read French, German, Spanish, Portuguese, Italian,
Russian, Swedish, Danish and Norwegian. He could
also speak the Fijian language.

[22]But, perhaps, these were Permian glacials from the
neighbouring Permian conglomerates.

[23]It should be remarked that the Woolnough & Taylor
paper marked part of the alleged course of the
abandoned river very clearly, but I (in company with
David Branagan) have not been able to find the river
pebbles in places mapped by Woolnough & Taylor.
Also, the present inclination of the ground is contrary
to what might be anticipated if the postulated river
joining the Shoalhaven and the Wollondilly had
flowed as hypothesized. To deal with this problem,
Woolnough & Taylor (1907–1908) suggested the area
might have been warped since the river flowed. But
lacking any independent supporting evidence this
suggestion was a classic ad hoc hypothesis. See
further on this in the next section of this paper. While
admitting that not all authorities agreed with him,
Taylor still advanced the old river-capture hypothesis
in *Sydneyside Scenery* (Taylor 1958*b*, pp. 184–185).

[24]Taylor 1958*a*, p. 61), but failed to say that Davis was in
fact firmly *rejecting* Taylor's idea that the courses of
rivers might be curved or deflected as a result of the
Earth's rotation. Davis visited Australia in 1914 for
the meeting of the British Association for the
Advancement of Science, and his presence there
probably furthered the wide acceptance of his ideas in
Australia.

Fig. 7. View of the Shoalhaven or Tallong Bend, looking east. The Shoalhaven River is flowing northwards at the right of the picture and then makes a bend to the east, running towards the sea at the top of the picture. Photograph by D. R. Oldroyd.

Taylor (1911*a*) considered the geomorphological situation in the Sydney region in a wider discussion of Australian physiography, and adumbrated the idea that there had been a *reversal* of drainage owing to the uplift. In Figure 8a and b we see his sketch map of the river systems and the line of the monoclinal (Lapstone) fold, with the Nepean 'ducking into' the hard rock of the fold at one point, and then out again, But more interestingly, it seemed to Taylor that the general drainage pattern had, at one time, been towards the *west*, for it appeared from its form that, before its capture by the Shoalhaven River, the Kangaroo River (see Figs 8a and 17 later in this paper) might formerly have flowed westward – into what is now the Murray–Darling Basin, perhaps reaching the sea in what is now South Australia! This idea was developed further in later writings and reached its final form in *Sydneyside Scenery*. Taylor (1911*a*) suggested (e.g. on the basis of the valley of the Kangaroo River) that there had formerly been quite a wide strip of land to the east of the present coastline, which had subsided relatively recently. And the occurrence of 'boathook bends' or 'barbed drainage' at several river junctions was taken to indicate the disruption of 'normal' river forms arising from earth movements. (Examples can be seen in Fig. 8a; see also Fig. 17 later.) His idea of a former eastern extension of the Australian coastline meshed with the views of his former teacher Edgeworth David. Taylor also overtly aligned his thinking with that of Davis:

It is to the labours of American geologists and geographers – notably Professor Davis, of Harvard – that we owe the methods whereby *the forms of river valleys* have to a certain extent, taken the place of fossils when we are investigating the movements of the earth's crust in the later periods of geological time.

(Taylor 1911, p. 7; italics in original)[25]

The idea of rivers to the east of the NSW mountains having, in the past, run in a meridional course was hinted at in Figure 8a and b, and in other diagrams

---

[25]Bishop (1982) has contrasted this mode of analysis of physiographical history with that which utilized a 'geological' approach, based on, for example, the determination of stream directions by examining sedimentary structures in gravels preserved under datable lava flows.

D. R. OLDROYD

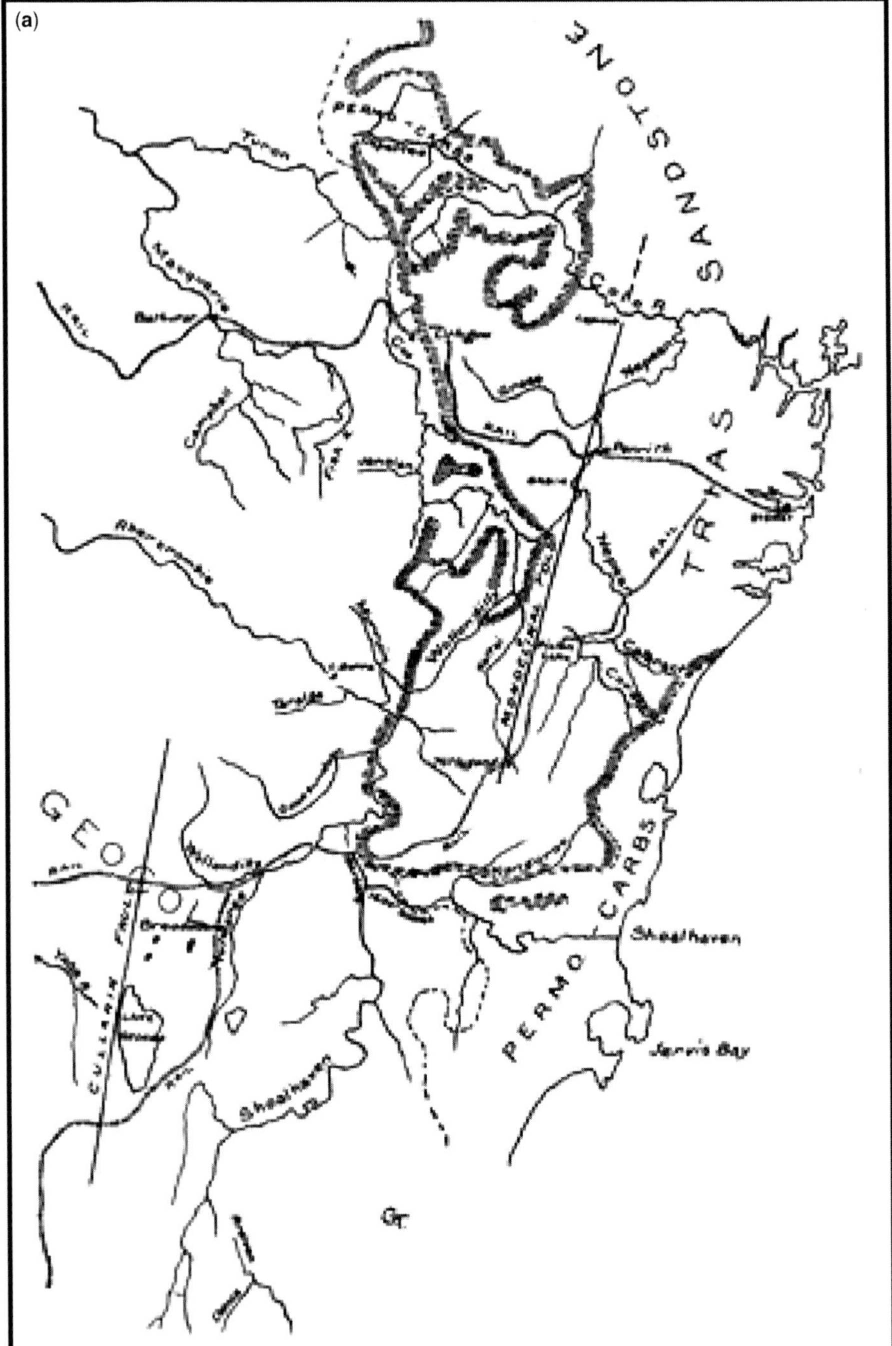

Fig. 8. (**a**) Rivers of the Sydney region, as represented by Taylor (1911*a*, p. 12), with the line of the Lapstone Monocline indicated. The 'dashed' line marks the broad division between the outcrop of the hard and soft rocks, but without showing the soft Wianamatta Shales at the centre of the Sydney Basin. (**b**) Enlargement of a portion of (a); from Jose *et al.* (1912, p. 89).

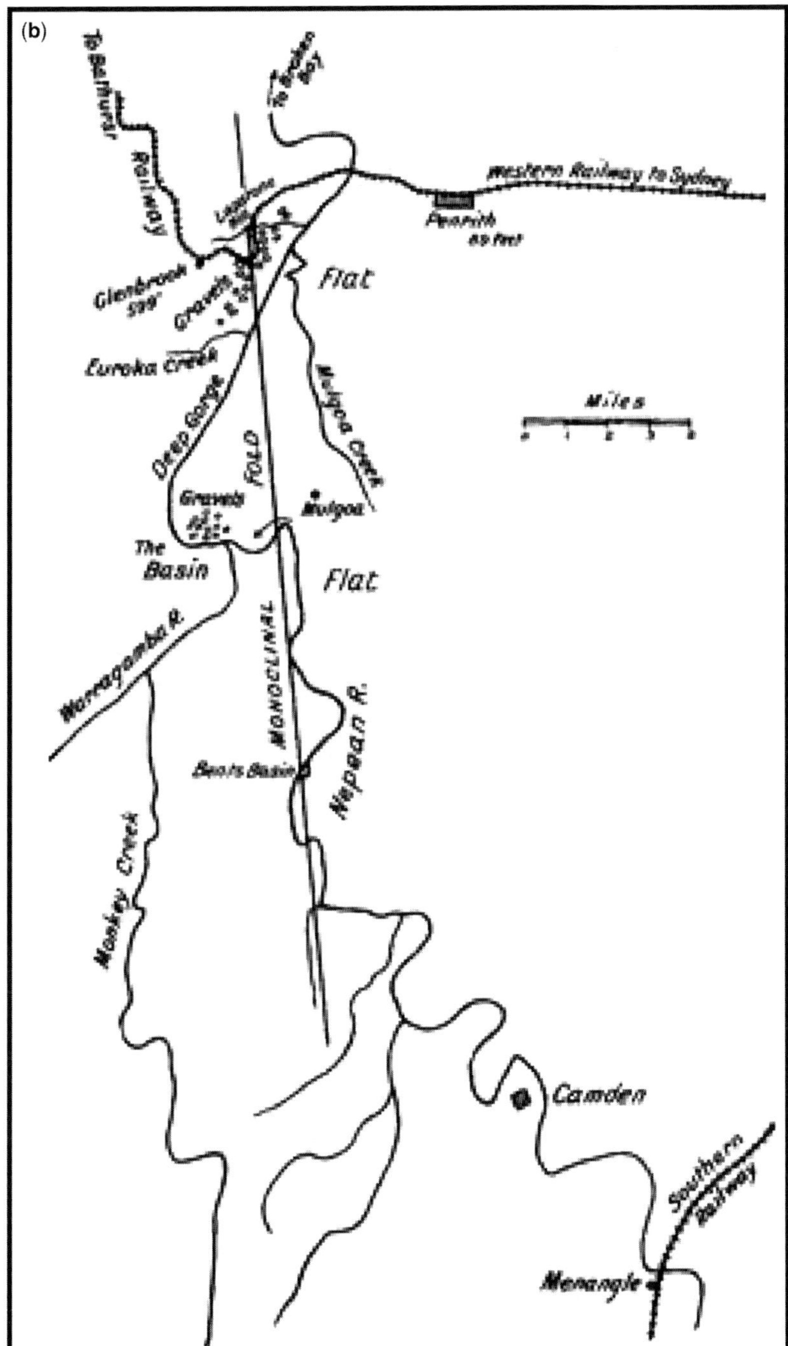

Fig. 8. (*Continued*).

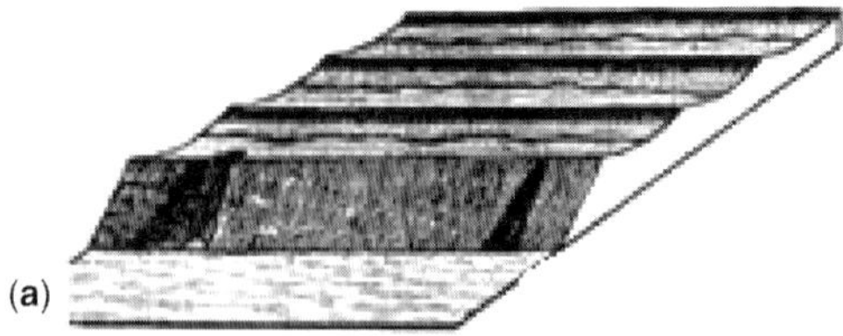

(a)

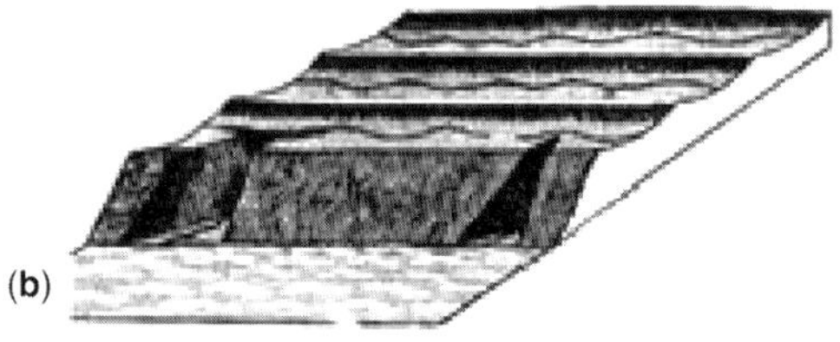

(b)

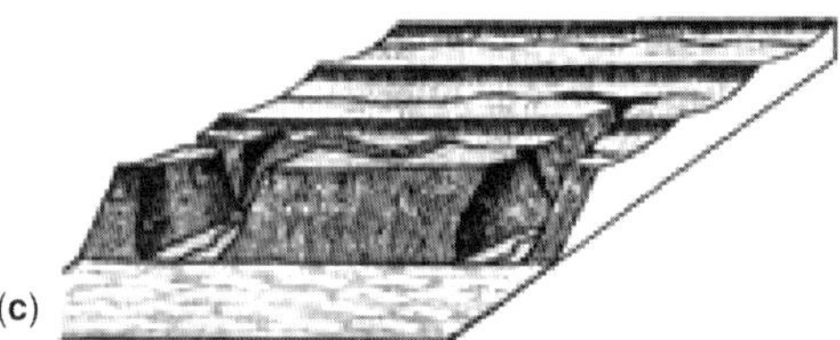

(c)

Fig. 9. Development of rivers systems, where original waterways flow in folds parallel to a coastline; from Jose *et al.* (1912, pp. 80 and 81).

drawn by Woolnough for the Jose *et al.* volume of 1912 (see Fig. 9).

Of course, for drainage to have been from east to west presumably required there to have been high land at some time somewhere out in the region of the present Tasman Sea, which could have provided a source of sediment and of river channels. But Taylor had no *independent* evidence to support the former existence of such high ground. It could have been there, in terms of general 'Lyellian tectonic theory', but *that* did not, in itself, provide direct empirical support for the idea. Nevertheless, in Lyellian style, Taylor (1911*a*, p. 13) hopefully 'invoked' a land area to the east, 'perhaps one or two hundred miles wide, which had subsided lately beneath the waves'.

As is well known, Davis and his disciples frequently deployed anthropomorphic language in their discussion of the histories of landforms. Such diction was enthusiastically employed in Australia, a good example being provided by, for example, Taylor's friend Charles Hedley in his Presidential Address to the Linnean Society of New South Wales in 1911 (Hedley 1912). He had his own preferred model for the development of the

Sydney area's river systems: initially two separate rivers flowing from west to east – the Shoalhaven and the Hawkesbury. Following folding, these joined up to form a single south–north river – the 'Shoalhawke' as he called it – and it was the Hawkesbury part of this that carried the combined waters to the sea. Subsequently, what is now the Shoalhaven part of the system was supposedly 'captured' by a fairly small stream cutting westwards from the coast, so that the major 'right-turn' of the Shoalhaven (Tallong Bend) was created, as previously envisaged by Woolnough & Taylor (1907–1908).

But Hedley's diction is what attracts us more particularly here. Concerning the river capture investigated by Woolnough & Taylor, he wrote:

Perhaps the most *interesting tale* yet told of the physiography of New South Wales is the *vivid story* by Dr. Woolnough and Mr. T. G. Taylor of how the Upper Shoalhaven River formerly flowed into the Wollondilly, thence into the Nepean and so into the Hawkesbury. Thus it reached the sea after *following a course* of about a hundred and sixty miles, roughly parallel to the coast and distant from it about forty miles. A *crisis in its history occurred.* Not only did *a pirate stream*, the Lower Shoalhaven, *behead* the former Wollondilly, but a *further capture* of Wollondilly water *is imminent* in the near future. In the past the Moruya and the Tuross Rivers have each *taken a length from* the old river. No marginal stream could *have the power to excavate and capture* [that is] possessed by a radial [one;] hence the former *must always fall a victim to the latter. These threats and captures* are *attempts and successes to proceed* from marginal to radial drainage, to *progress* from the *normal to the abnormal.*

(Hedley 1912, p. 23, emphases added)

In 1911 Taylor published a text based on his series of lectures given at Sydney University on the Commercial Geography of Australia, in which we can discern his already growing interests in the relationship between economic and social development and geology, topography, soils and climate (Taylor 1911*b*). In the third edition (1916) of this successful textbook he again wrote of a reversed drainage, with the main divide having been shifted to the west in the southern parts of New South Wales, and a former eastern extent of Australia. A figure (see Fig. 10) showing the outcrops of granites and volcanics in the eastern part of the continent was intended to indicate the former western flow of the rivers that presently cut through the Blue Mountain Plateau to reach the coast: 'if the present divide along the New England "massif" be produced to the south, it very probably marks off the region (between Newcastle and Kosciusko) which originally drained west, and indicates what an enormous extent of land has sunk beneath the sea in late geological times' (Taylor 1916, p. 78). (Here he was again thinking about a former eastern extension of Australia.)

Following his return to Australia from the Antarctic and his appointment as Head of the new Department of Geography at Sydney University,

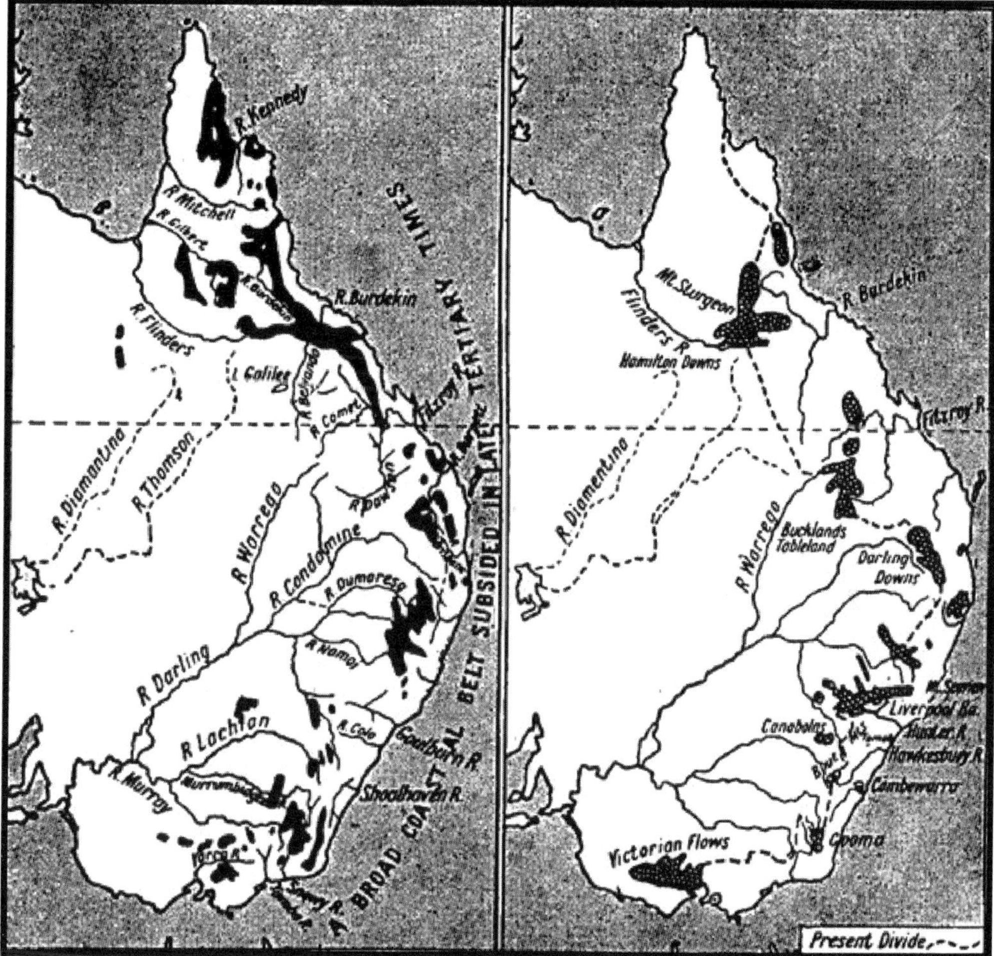

Fig. 10. Outcrops of granitic and volcanic rocks in eastern Australia, according to Taylor (1916, p. 65).

Taylor taught his students how to make large models of various regions of New South Wales, including, of course, the Sydney region and the Blue Mountain Plateau. The making of such models was a significant part of the teaching pro gramme.[26] But, as mentioned, in the next few

years Taylor gave increasing attention to anthropological matters and supposed relationships between climate, race, skull shapes and mental capacities (Oldroyd 1994). He wrote many newspaper articles on such topics, and lectured on them in various places in Sydney, thereby increasing his income; but such issues need not be considered here.

However, in 1923 the Pan-Pacific Science Congress met in Sydney and Taylor prepared a guidebook for an excursion to the Blue Mountains,

---

[26]Models were made of the Kosciusko area, the Jenolan Cave area (west of the Blue Mountains), the Wentworth Falls (over which water flows stepwise from the Blue Mountains Plateau into their major valley), Stanwell Park (a coastal area between Sydney and Wollongong, where the harder Triassic and softer Permian rocks meet), Kurrajong (at the top of the Lapstone Monocline and near the Kurrajong Fault, investigated by David), the Blue Mountain Plateau and gorges, and Port Jackson (i.e. Sydney Harbour). The

last two were geologically coloured. See Taylor Papers, National Library of Australia, MS1003/4/489. Details of the model-making processes were given in Taylor & Taylor (1925), a collaborative teaching booklet prepared by Taylor and his sister Dorothy who worked as a demonstrator in his department.

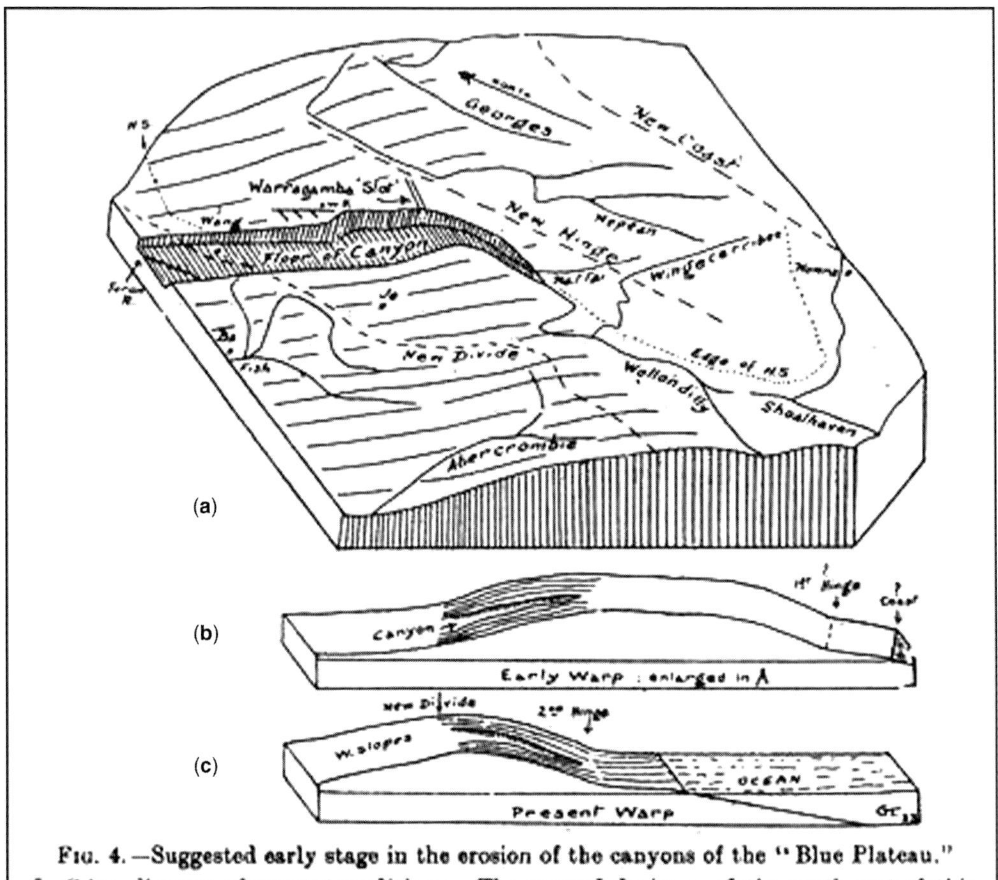

FIG. 4.—Suggested early stage in the erosion of the canyons of the " Blue Plateau." In C is a diagram of present conditions. The general drainage of the south-east of this State is to the north west—and was here also, as the upland streams at Wentworth Falls still indicate. The great uplift of late Tertiary times advanced as an earth-wave (see B), giving rise for a long period to conditions shown in A. The new warp seems to have been oblique and produced a new divide, new hinge, and new coast (all running north-south), as indicated by the broken lines in A.

Fig. 11. The evolution of the plateau, canyons, and river systems of the Blue Mountain region, according to Taylor (1923*b*, p. 29).

Jenolan Caves and Lithgow (Taylor 1923*b*). In that guidebook he first published a diagram (reproduced unchanged in *Sydneyside Scenery*, 1958*b*, p. 119), which sought to explain the supposed *reversal* of drainage by the westward movement of an 'earth-wave', as Late Tertiary uplift occurred (see Fig. 11).

This was not an unreasonable idea, given that the small Cox's River (the head stream of the Warragamba–Nepean–Hawkesbury River) is an insignificant waterway, running in a large and wide valley (the Kanimbla–Megalong–Cox's) on the western side of the Blue Mountains. Seemingly, such a river could not have produced the valley in which it presently runs, according to its present size and the disposition of the mountains. And it could not have excavated its present course right through the Blue Mountains to the eastern plains and thereby join the Hawkesbury–Nepean River.[27]

Taylor outlined his theory as follows:

[W]e know that before the Kosciusko uplift (at the dawn of the Pleistocene[28] the drainage of the whole country was to the north-west or north, *e.g.*, the Upper Nepean, Upper Shoalhaven, South Creek (in the 'Stillstand'), Nattai, 'Wingecarribee, Lachlan, and Macquarie. We … [see] that the relics of the ancient upland

---

[27]However, it must be noted that Cox's River flows over and cuts into different rocks in its valley from those that form the hard obstacle of the Blue Mountains.

[28]In a newspaper article published not long before, Taylor implausibly suggested that uplift occurred only 100 000 or 200 000 years previously (Taylor 1921).

rivers [on the Blue Mountains plateau], as at Wentworth Falls, show the same direction.

… I postulate the coast buckling under stresses acting across the meridional drainage, and advancing gradually in a wave to the west. As Andrews, Jensen, and myself have suggested this occurred in earlier geological time also. The continent was much wider then, and its condition is shown in Fig. 4 [which is Fig. 11 in this paper] at B. Heavy rains fell on the wide coastlands. The main stream of the region developed *along the edge* of the resistant Hawkesbury Sandstone. The ancient Shoalhaven–Wollondilly–Cox's River then cut out a huge valley, which ate back, by headward erosion from the western plains, as is suggested in Fig. 4, A [= Fig. 11, Diagram **A**] … .

The later phase of folding then occurred, whereby the earth-fold advanced perhaps 100 miles to the west. Judging by the change in the slopes, this new axis was somewhat oblique to the old S.W.–N.E. axis, and ran north and south. Much of the old land sank beneath the sea, and all the rivers east of the new divide were reversed. The small river left in the great valley was captured by the headwaters of the Warragamba from the east. This indeed still occupies a narrow slot in the very wide Cox's–Wollondilly valley.[29] In my opinion this narrow slot belongs to the same 'cycle' as the narrow 'slots' which are all that the tributary streams of to-day are cutting in the giant walls of the broad valley which was developed in the 'Proto-uplift period'.

(Taylor 1923*b*, pp. 28–29; italics in original)

In the same year, Taylor published an essay entitled 'The warped littoral around Sydney' (Taylor 1923*a*). It was intended to be the first of a series of papers describing and discussing the topography round Sydney, with a classification of the various features, their geohistorical explanation, and a consideration of the relation of the growth of the economy and population to the geology and geomorphology of the area. But the series got no further than its programmatic outline, although if it had been completed I dare say it might have looked rather like *Sydneyside Scenery*. So we can see that book as the completion of a project that was long in gestation. Presumably, it got set aside in the 1920s when Taylor became so deeply involved in anthropological matters, and polemical debates about the 'carrying capacity' of Australia and its suitability for extended population, Asian immigration, economic, and agricultural and industrial growth. And then he went off to the United States.

The paper (1923*a*) had one particularly interesting concept, however. Taylor wrote, in relation to the area of the Wingecarribee Swamp on the upper part of his so-called 'Nepean Ramp':

This oval area, about 20 miles long … illustrates clearly a stage in the dissection of an uplifted peneplain … It is obvious that

immediately after its uplift a peneplain still exhibits a senile topography, though it is in a 'precarious' position and will soon yield to the headward erosion attacking it on all sides. This stage I have been accustomed [presumably in Taylor's lectures] to name a '*besieged peneplain*' For long ages ago the *central portion* of the besieged peneplain remains in much the same condition, and the Wingecarribee swamps near Robertson illustrate the relics of such a besieged peneplain. For such areas, I suggest the word 'core'.

(Taylor 1923*a*, p. 69; italics in original)

This text nicely illustrates Taylor's understanding of Davisian thinking and his Davisian linguistic style.

But the hypothesis of the Wianamatta Stillstand has an inherent problem. The form of the Cumberland Basin might well be intelligible if there had been a downwarping or sagging of the upper crust in that region. But it is difficult to imagine any system of upward forces that might have generated the Woronora Plateau, the Hornsby Plateau and the Lapstone Monocline at one and the same time. Yet, this seems to have been implicit in Taylor's notion of a 'Stillstand'. The only way round the difficulty is to hypothesize (at least) two episodes of elevation. The Woronora and Hornsby plateaus might have been formed at essentially the same time, and then the Monocline; or vice versa.[30] The former possibility seems the more plausible from a geometrical point of view; but one cannot be certain. The issue seems to have been little discussed; but see Walker (1960), who suggested more than one movement, while accepting the reality of the 'Kosciusko Uplift'.

Another useful idea to be found in the Excursion Guide-Book for the 1923 Pan-Pacific Science Congress was by the NSW Government geologist Thomas Willan (see p. 249). He suggested that the deepest area of the Sydney Basin during Hawkesbury–Narrabeen times was somewhere north of the Hawkesbury River, which was why that waterway had flowed northwards (Willan 1923, pp. 24–25). Later it became an entrenched antecedent river as the Hornsby Plateau was subsequently elevated.

Taylor himself did no further original writing on the development of Sydney's river systems before he left for Chicago, although he did publish some interesting articles on the topic in the *Sydney Morning Herald*. One of these (Taylor 1927*a*) adumbrated a theory that he later developed in *Sydneyside Scenery*. The suggestion was that there had formerly been a south–north running river, which ran approximately along the line of what is the

---

[29]This area is no longer accessible to the public as it lies within the area dammed by the Warragamba Dam: Sydney's main source of water. However, I am informed by David Branagan that pre-dam photographs reveal that the Warragamba outlet was not an excessively narrow gorge.

---

[30]Or the two plateaus might not necessarily have been formed at just the same time, although they are commonly thought of as 'mirror images' or the matching sides of 'Sydney's basin'.

present coastline near Sydney. After the proposed uplift, the sea had broken into the region of the former winding river as sea levels rose after the Ice Age. It 'thus produced the curious series of "residuals"' of headlands and beaches that we find well developed to the north of Sydney. More will be said about these when discussing *Sydneyside Scenery*. It may be remarked here, however, that in the newspaper article of 1927 Taylor had in mind a much earlier date for the Earth movements than that which he had suggested previously: 1 or 2 Ma ago.

Taylor's last published statements on New South Wales topography, before he departed for Chicago, appeared in the *Official Year Book of the Commonwealth of Australia* for 1927 (Taylor 1927*b*). Here he again spoke about uplift, but dated it to 'many thousand years ago'; so he was, it seems, quite uncertain about dates for Earth movements in the Sydney region. But he reiterated his belief in a former south–north drainage system for the Sydney area, and discussed a tectonic lineament that ran right down to Bass Strait (between the Australian mainland and Tasmania). He also suggested, with little direct evidence, that the huge scarp, inland from Wollongong and separating the coastal plain from the plateau country, was the mark of a substantial fault line. This was assuredly mistaken. The line of the scarp is not really straight (but see Fig. 17 later in this paper, where Taylor represented it as having an essentially linear form); it carries no signs of slickensiding; and no relatively displaced strata can be matched up. Such faults as are shown on modern maps are approximately perpendicular to the scarp (i.e. approximately east–west) and do not run far inland.

In addition, the *Year Book* entry (Taylor 1927*b*) made reference to current fieldwork in the area of the Cox's River Basin that supported the idea of a reversal of drainage for the river systems of the Blue Mountain Plateau. Taylor was evidently thinking of the work of his research student at Sydney University, Frank Alfred Craft (1906–1973), who published his results the following year in 1928 stating that the fieldwork was undertaken in 1926 under Taylor's supervision. Later, after Taylor had left Australia, Craft carried out independent work in the Shoalhaven and Monaro areas.

Craft's paper of 1928, which evidently gave expression to Taylor's ideas, contained two figures that represented (hypothetically) the old peneplain surface of the area of the Blue Mountains and the country to the east and west (Fig. 12a), the same area subsequent to uplift and the formation of the supposed Old Blue Mountain Anticline (Fig. 12b), and a simplified diagram showing the existence of a large NW-flowing western river (made up of what are now the Wollondilly, Kowmung and Cox's rivers) prior to the reversal

of drainage (Fig. 12c). This reversal supposedly occurred when the Warragamba River, rejuvenated by further uplift, cut back to the supposed 'western' river, thereby allowing it easy access to the sea through the modern Warragamba Gorge. So river capture and drainage reversal occurred. Taylor reproduced Figure 12c (redrawn) in *Sydney Scenery* (see Fig. 13) (Taylor 1958*b*, p. 121), and referred favourably to Craft's (early) work.[31]

There was one other intended geomorphological publication that might have appeared from Taylor's pen about 1930, but didn't. For many years, Edgeworth David had been collecting materials for a general book on the geology of the Commonwealth of Australia, but he was getting old by the early 1930s and the job was still not completed.[32] David's correspondence, held at Sydney University, reveals that he enlisted Taylor's assistance to write a chapter on 'Physiography' for the book. It appears from a letter from Taylor to David, dated 19 September 1927, that Taylor did write a contribution and sent it to David.[33] In the event, it never got used, but a typed outline of the chapter and its sections is in the letter-box along with the handwritten letter to David. For the section on the region around Sydney, the outline for the proposed chapter had the following headings: Blue Mountain Plateau Region; Pre-Uplift Drainage; Ancestor of Nepean; Kurrajong Fault; Warped Littoral; Sydney Coast Line; Hornsby Warp; Nepean Warp; Southern Blue Plateau; Blue Plateau East; Jenolan Plateau[34]; Western Wollondilly; Evolution of Blue Plateau. Broadly speaking, these headings fit well with most of the chapters eventually published in *Sydneyside Scenery* (Taylor 1958*b*).

Rivers in Taylor's *Sydneyside Scenery*

We now jump to 1958, the year that Taylor's *Sydneyside Scenery* was published. The original proof copy is held in the Taylor papers at the National Library of Australia, but its emendations are only

---

[31]Taylor omitted to mention that Craft's later papers, written in the early 1930s after Taylor had departed for Chicago, began to back away from Taylor's idea of drainage reversal. For discussion, see Bishop (1982) and the next section of this paper.

[32]The book was eventually published posthumously in 1950, as a sterling effort, by his former colleague W. R. Browne; but by then it was somewhat dated.

[33]See David Papers, Box 44; I thank D. Branagan for drawing my attention to this document.

[34]Jenolan Caves, in old limestone rocks to the west of the Blue Mountains, are well known to Sydneysiders as a tourist attraction; they lie outside the Sydney Basin.

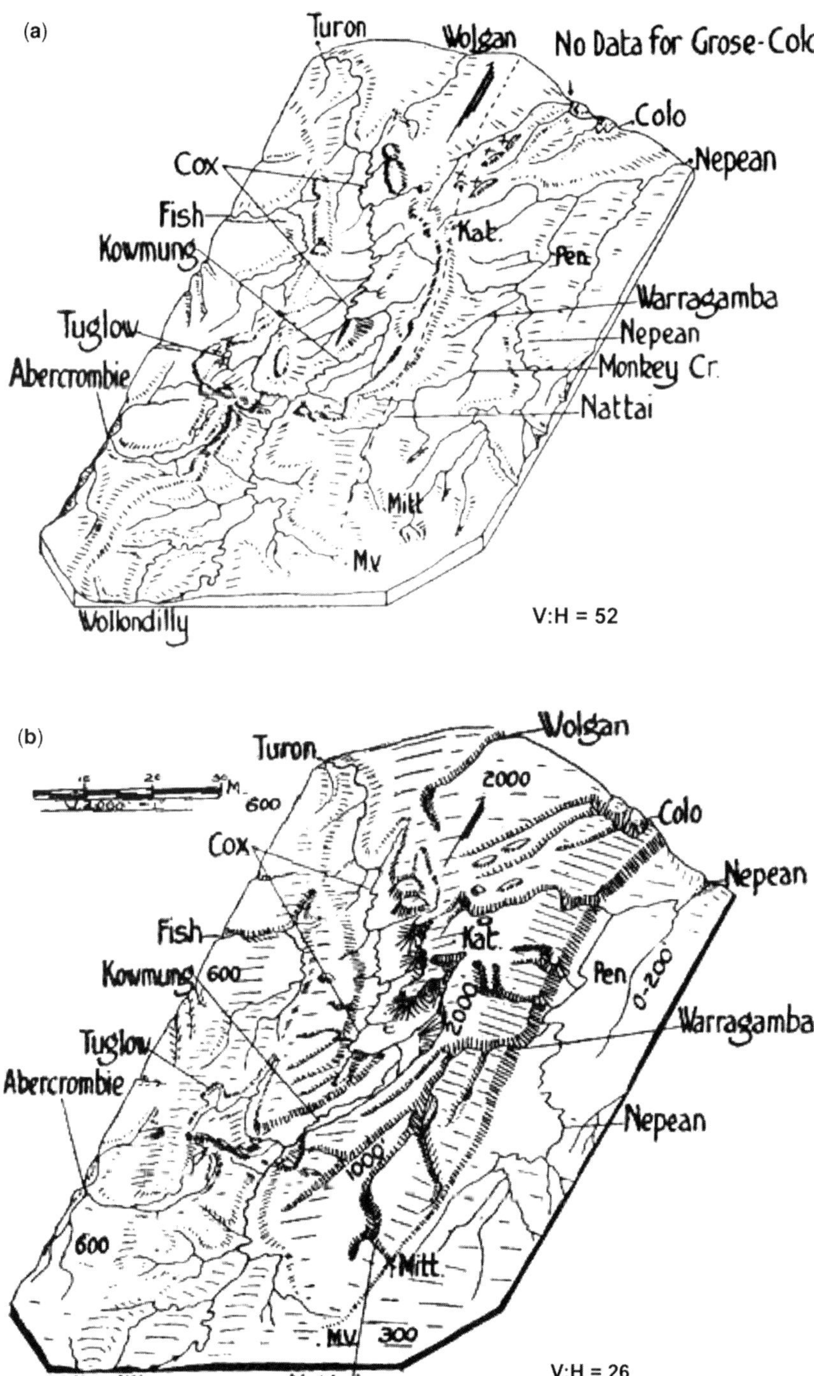

Fig. 12. (**a**) Topography of the area of the modern Blue Mountains and adjacent area and waterways: the former peneplain. Kat., Katoomba; Pen., Penrith; Mitt., Mittagong; MV, Moss Vale; from Craft (1928, p. 241). (**b**) The 'Old Blue Mountain Anticline': abbreviations as for (a); from Craft (1928, p. 243). (**c**) Corresponding fold systems, showing the hypothesized large NW-flowing river; from Craft (1928, p. 244). The dotted line marks what Craft unhelpfully labelled the 'Cox divide' (rather than 'Cox catchment area').

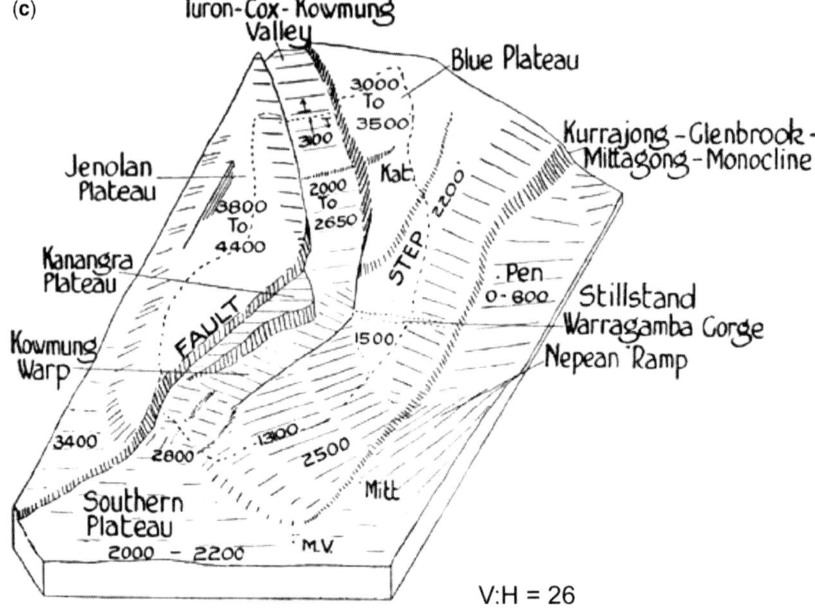

Fig. 12. (*Continued*).

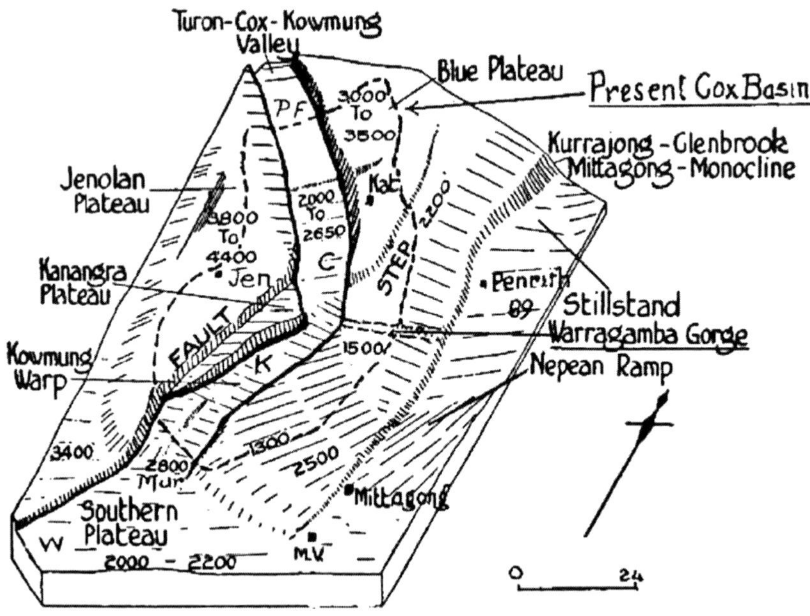

Fig. 13. Taylor's reproduction of Figure 12c; partly redrawn by him. W, Wollondilly Valley; K, Kowmung Valley; C, Cox Valley; PF, Piper's Flat (which lies near the modern mountain divide) (Taylor 1958*b*, p. 121). The area marked by a closed dashed line delineates the catchment area for the present Cox's River. The straight course of the Warragamba Gorge, also shown in Figure 12c, is marked by two parallel dotted lines, and marks the passageway for the Cox's waters to the east arising from the hypothesized concomitant reversal of drainage. (The line of the Warragamba Gorge is today thought to be structurally controlled.)

stylistic and throw no light on the development of Taylor's thoughts. Indeed, by that time his ideas were pretty well fixed and for the most part they can be traced back to the 1920s and beyond. For example, the illustration from his Pan-Pacific Guide (1923*b*) (see Fig. 11) was taken over directly into *Sydneyside Scenery* (p. 119). (The copy of the Guide in the Taylor papers in Canberra has the illustration clipped out and the reason why is obvious.) Taylor failed to mention the fact that his former student, Craft, had begun to move away from his ideas after his teacher departed for North America.[35]

But there was one suggestion about rivers in *Sydneyside Scenery* that was an interesting extension of earlier ideas. The coastal area immediately north of Sydney (called the Northern Beaches) presents some interesting and puzzling features – perhaps the topic attracted Taylor in his old age as his place of retirement was quite close to the Northern Beaches. There is an alternation of rocky headlands (Narrabeen and Hawkesbury Sandstone rocks) and sandy beaches (see Fig. 14a and b). Immediately to the *west* of this area is a large area of salt water, running south–north – called Pittwater – which connects to the estuary of the Hawkesbury River as a lateral branch, but is fed by only a minor freshwater stream (McCarr's Creek, see Fig. 4).

Taylor accounted for this arrangement by proposing an *analogy* between the structures created by the supposed interaction of the Nepean River and the emerging Lapstone Monocline and the structure and origin of the area of the Northern

Beaches. The relationship of the Nepean relative to the monocline had long been explained in terms of the idea of antecedent drainage. Now Taylor supposed that there had formerly been a river (which he called 'River X') that had flowed south–north, approximately along the line of the present coastline. But its valley had been invaded by rising seas, following the end of the Ice Ages, giving rise to the structures as sketched by Taylor, with the eastern side of the river valley having been eroded away by the encroaching sea (see Figs 15 and 16).

Another little diagram shows the supposed evolution of the rivers and landforms, with the folding of land and the westward migration of the warp, as previously discussed in relation to the supposed reversal of drainage over the Blue Mountains. Taylor supposed that the 'primeval' drainage of the region ran south–north, in accordance with the supposed original fold system of the Sydney area, and approximately parallel to the coast of eastern Australia. The three main initial rivers, supposedly flowing northwards in parallel, were: the Nepean; South Creek (= Wianamatta Creek), which still flows over the Wianamatta Stillstand in the course shown in the left-hand diagram of Figure 16 and presently joins with the Nepean near where it turns east (perhaps owing to river capture[36] like the Shoalhaven, studied by the young Taylor in collaboration with Woolnough), with water flowing to the sea via the Hawkesbury River and Broken Bay; and the hypothetical 'River X', which only exists as the relics indicated in Figure 16.

Thus, in his old age, Taylor synthesized his understanding and interpretation of the geomorphology of the Sydney region and its drainage system, including the area of the Blue Mountains. It will be remarked that much of the explanation depended on the idea of the westward-shifting warp, and a geologically quite recent Kosciusko Uplift and formation of the Lapstone Monocline.

Success of the *Sydneyside Scenery* Synthesis? More recent interpretations and plate-tectonic considerations

A major synthesis of New South Wales geology was published by the Geological Society of Australia in 1969 (Packham 1969). The section on geomorphology (pp. 559–580) was written by Edgeworth David's previously mentioned former co-worker and colleague at Sydney University, the elderly

---

[35]Craft (1931) explicitly rejected the idea of Woolnough & Taylor (1907–1908) that the Shoalhaven Bend (Tallong Bend) was the result of river capture, but was, he thought, owing to older structural features. As previously mentioned, the supposed abandoned river bed sloped the wrong way for water to have formerly flowed from the Shoalhaven into the Wollondilly catchment – although this had been explained away ad hoc by Woolnough & Taylor's suggestion of a later warping of the surface. More importantly, Tertiary basalts were found to occupy an old valley running *eastwards*, with a 'persistent eastern drainage over a long period' (p. 127), discounting Taylor's idea that the Kangaroo River formerly flowed westwards for an extended distance and time prior to uplift. Thus, Craft was resisting his former supervisor's idea of a reversal of drainage. Craft (1933) wholly rejected Taylor's idea of high ground moving westward so as to cause a reversal of the drainage system ('the Main Divide . . . [is] a rather accidental feature of essentially non-tectonic origin' (p. 439); but he retained the Davisian notion of peneplanation.

---

[36]There may, however, be an underlying structural control, accounting for the Hawkesbury's eastward turn.

Fig. 14. (**a**) 'Islands' (headlands and bays) forming the coastline (Northern Beaches), north of Sydney: looking south; viewed from the hill indicated above 'Avalon' in Taylor's lower drawing (of Fig. 15). Photograph by D. R. Oldroyd. (**b**) The same peninsula viewed from the Pittwater side, looking SW; viewed from the headland above the 'w' of the name 'Pittwater' in Taylor's lower drawing (of Fig. 15). Photograph by D. R. Oldroyd.

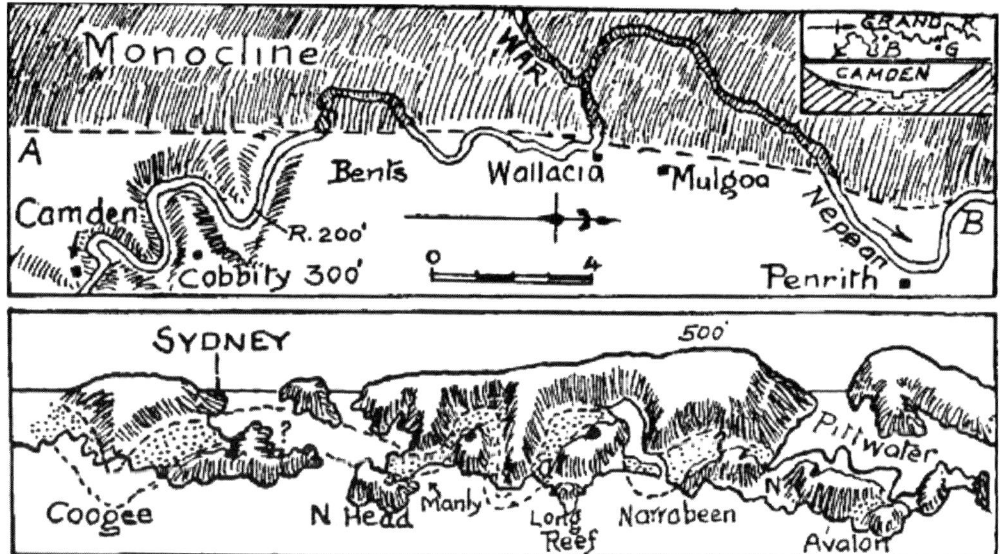

Fig. 15. Sketches from Taylor's *Sydneyside Scenery*, p. 86. The upper sketch shows the meanders of part of the Nepean River in relation to the Lapstone Monocline. The lower sketch suggests that the small 'rock islands' on the Narrabeen (Northern Beaches) coastline are similar bisected meanders. WAR, Warragamba River/Gorge. Taylor supposed that there might have been either down-faulting to the east or folding, or both, to form the coastline.

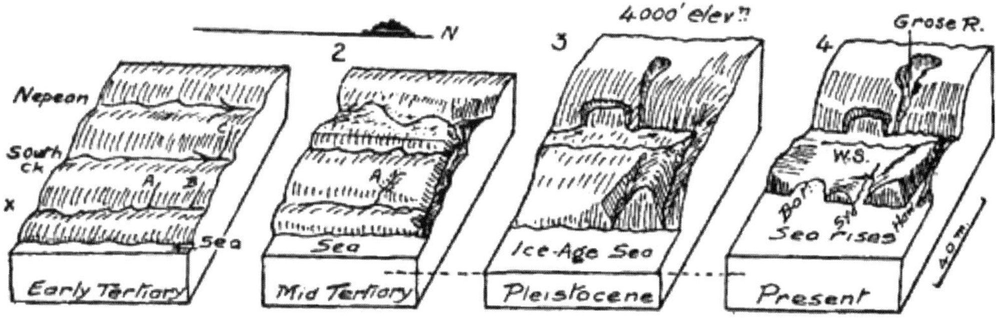

Fig. 16. Diagrams suggesting the evolution of the rivers and the topography of the Sydney region, according to Taylor (1958*b*, p. 22). Bot., Botany Bay; Syd., Sydney Harbour; Haw., Hawkesbury River; WS, Wianamatta Stillstand. A tributary (A) of 'River X' could have been the ancestor of the present Sydney Harbour.

William R. Browne (1884–1975), who, as mentioned earlier, edited David's posthumous *Geology of the Commonwealth of Australia* (David 1950; see Branagan 2005). Browne (1969, p. 560) envisaged the following sequence for the physiographical history of eastern Australia:

1. Elevation and dissection of a Cretaceous peneplain during Late Eocene–Early Oligocene (Kiandra Epoch).[37]

2. Alluviation of valleys followed by effusion of flood basalts in Oligocene.

3. Peneplanation followed by duricrusting: Miocene.[38]

---

[37]Kiandra, now a ghost town, was a gold-mining locality in the Snowy Mountains region of New South Wales

in the 1860s; it had auriferous gravels buried under basalt and was one of the first areas investigated by Andrews. The name stems from his work.

[38]Woolnough was a leading student of duricrust studies in Australia and coined the term (Woolnough 1928 for 1927). Duricrust is formed in semi-arid conditions when mineral solutions in a soil rise by capillary

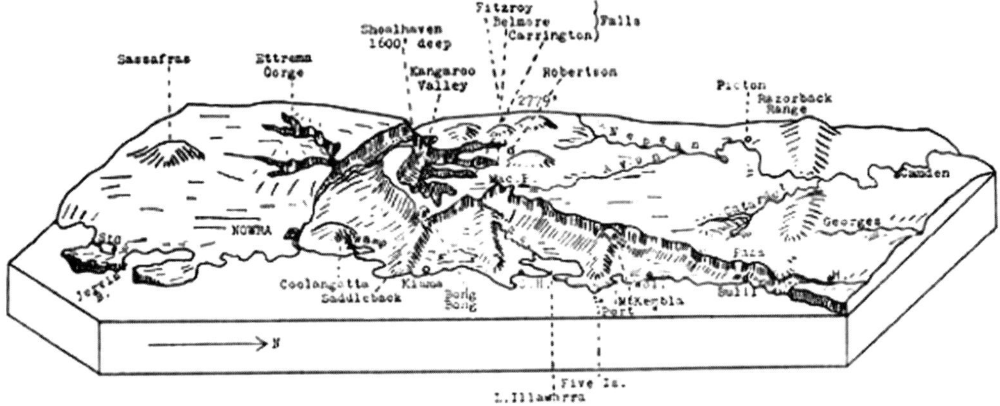

Fig. 17. Block diagram representing southern parts of the Sydney Basin, from the teaching handbook: Taylor & Taylor (1925, p. 12). Wol., Wollongong; S. H., Shellharbour; H., Helensburgh. The 'Illawarra Scarp' is shown running diagonally from the centre of the figure to the lower right. I thank D. Branagan for the loan of this publication.

4. Uplift at close of Miocene, followed by Pliocene stillstand (Macleay Epoch); excavation and alleviation of wide mature valleys and exhumation of basalt-filled valleys.
5. Differential and intermittent uplift in Late Pliocene–Early Pleistocene, initiating vigorous erosion (Kosciusko Epoch).

Thus, ideas such as those of Andrews (1934) and Taylor (1958b) were still extant in the late 1960s. Browne believed that excavation of the inner gorges of the mountains was accomplished during the Pleistocene, when rainfall was supposedly higher.[39]

But Browne was about the last geologist to attempt to apply Davisian ideas directly to the Sydney scene and more recent work has thrown the whole nice synthesis of Andrews, Taylor, etc., into disarray. Robert Young, of Wollongong University, working in 1970 near Nowra (about 150 km south of Sydney) in the southern part of the (geological) Sydney Basin (see Fig. 17), found evidence that was incompatible with the older ideas that invoked a relatively recent uplift to account for so many of the features of the geomorphology of the Sydney Basin. The older view was that near the close of the Tertiary an erosional surface (marked by a duricrust) had subsequently been elevated to form the inland plateau surface. But Young found well-developed duricrust on Permian sediments near Nowra that

were below the mountain scarp but generally similar to those on the plateau country. If the high- and low-level duricrusts were of the same age (which he regarded as mid-Tertiary), then the difference of elevation need *not* be owing to quite recent uplift.

Young's fieldwork confirmed that the Permian rocks of the southern part of the Sydney Basin dipped eastwards from the plateau country to the coast, and further north they were covered by almost horizontal Triassic strata, so that there was an unconformity between them. Thus, there was apparently an *ancient* (*pre*-Triassic) warp. Further, he suggested that the major uplift in the Nowra region must have been completed by the Oligocene (Young 1974). This did not imply that the Lapstone Monocline (in Triassic rocks) near Sydney was formed in the mid-Tertiary, but it was suggestive that there was a well-established inclined landform in the older rocks before the Triassic sediments were deposited; and the movements that gave rise to the monocline could have followed an earlier lineament.[40] In the southern part of the Sydney Basin, some of the older literature (e.g. Harper 1915) had supposed that the northward-dipping surface of the Nowra Sandstone (Permian), inland from Nowra – the so-called Yalwal Ramp (see Fig. 17: the surface in which the Ettrema Gorge is incised) – was the result of

---

action and evaporate during dry conditions, slowly depositing a 'hard crust' (silcrete, ferricrete, or calcrete) on the upper surface of the ground.

[39]Today, Australia is thought to have been abnormally arid during the Ice Ages.

---

[40]Young (1974, p. 285) figured an east–west section, north of Nowra, with the Narrabeen rocks following the line of the inclined Permian surface, but thinning out to the east, while the overlying Hawkesbury Sandstone was almost horizontal (but not figured as unconformable to the Narrabeen).

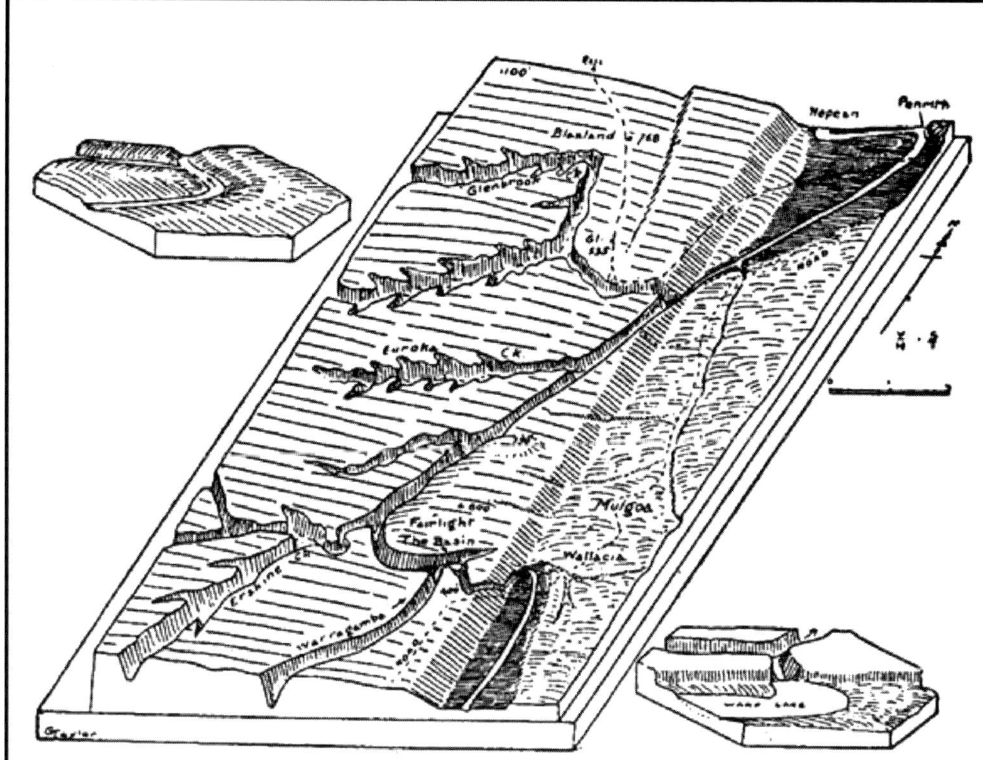

Block diagram of the Fairlight Gorge (Nepean River) and of the silted warp lakes at Wallacia and Penrith, looking north, that is, downstream. The small upper figure (looking west) shows the meander near Fairlight and the lower Warragamba before the warp. The small lower figure shows a later stage after a rapid joggle-up. The warp lake formed is later silted up to form the area now being trenched by the Nepean River.

Fig. 18. Block diagram of the river systems near the southern end of the Lapstone Monocline, showing the Blue Mountain Plateau and some of its canyons, the Lapstone Monocline and the Cumberland Plain (Wianamatta Stillstand); from Taylor (1958b, p. 89). The Basin =Norton's Basin. The monocline and its associated Kurrajong Fault are indicated at the top centre of the sketch. The figure does not indicate the rather complicated structure of the 'Lapstone Structural Complex'.

Pliocene–Pleistocene monoclinal warping, like the Nepean Ramp further north, see Fig. 3. But Young (1977) regarded it as an exhumed Permian (syndepositional) feature, and thus had nothing to do with anything such as the supposed late Kosciusko Uplift.[41]

---

[41]Leslie Frank Harper (1873–?) was a geologist with the NSW Geological Survey, with particular interests in economic geology. He served as Survey head from 1933 to 1937.

Further evidence in favour of quite early uplift became available following the K–Ar dating of numerous basalts in Victoria and New South Wales by Peter Wellman & Ian McDougall (1974a, b) of the Australian National University. In the Blue Mountains, basalts as at Mount Tomah were dated at 14.6 Ma and a little older near neighbouring Mount Wilson. Down south in the Shoalhaven area, they were substantially older, as for example at Caoura (46.1 Ma). This latter had been studied much earlier by Craft (1931), who had found a lava flow *in an old river valley* descending *westwards* from 700 m on the plateau to 550 m. This meant that the high ground and the valley must have been present at least as far back as the Eocene. So the Pliocene–Pleistocene Kosciusko Uplift hypothesis was fatally compromised. The old ideas of Andrews on the role of basalts in the tectonic scenario no longer held water in the light of their radiometric dating.

So far little mention has been made of the volcanic necks (diatremes) that are found in the Sydney area and in the Blue Mountains, but they too have presented problems for Kosciusko Uplift theorists. There are many of these diatremes scattered around the region, in addition to the larger outcrops of volcanics in the Blue Mountains, such as those at Mount Tomah, previously mentioned. Interestingly, one of the diatremes passes directly through the southern (faulted) extension of the Lapstone Monocline, at a place called Norton's Basin, not far from where the Warragamba River cuts through its gorge to join the Nepean, which then passes through the Nepean Gorge before the waters flow out onto the Cumberland Plain (see Fig. 4). The river courses are quite irregular there (see Fig. 18), but this may be attributable, at least in part, to the occurrence of the diatreme, which provides rock that is susceptible to greater weathering and erosion than is the Hawkesbury Sandstone through which it passes. The locality is marked in Taylor's diagram (Fig. 18). It is a 'nice question' whether the 'boathook bend' there is the result of disruption of the drainage pattern by Earth movements or the presence of the more easily weathered material of the diatreme. But either way, the Norton Basin diatreme is the only one that passes through the monocline. Moreover, it is the only one with an elongated (east–west) outcrop. Possibly, therefore, it is not a vertical structure.

It will be recalled that the gravels on the top and sides of the Lapstone Hill (and also to its south) and on the plains below were taken by Taylor and other authors as evidence of the earlier course of the Nepean, which formerly meandered over a plain, but was then disturbed by the geologically rather recent Kosciusko Uplift. For part of its course, the Nepean was able to maintain its path, thus forming the Nepean Gorge. But near Lapstone/ Glenbrook it supposedly got pushed off the rising ramp of the monocline and remained on the lower ground, while gravels were left 'high and dry' on the hilltop and hillside.[42]

But there are many problems with this older interpretation. Are all the gravels of the same age? How many watercourses do they represent? (Could there have been a single braided river system?) Do they necessarily predate the monocline (so that they were left 'high and dry' after its formation)? Could the gravels on the side of the scarp/ramp be reworked material? Perhaps not, as they sometimes appear as valley-fill and have also apparently participated in the folding process(es). But putting that point aside, how old is the monocline? And was it formed by upward movement of the crust or by downwarping of the area of the Cumberland Plain (the area that Taylor called a 'Stillstand')?

Today, the monocline is thought to be much older than Taylor *et al.* supposed. Using palaeomagnetic evidence, Paul Bishop (then at Monash University, Melbourne, but presently at Glasgow University) *et al.* (1982) published evidence that it was formed at least before the end of the Miocene. There is datable hematite in the Hawkesbury Sandstone that forms the Lapstone Monocline; and its crystals' magnetic orientations were compared with what might be expected from the known Apparent Polar Wander Path for Australia in the Tertiary. From this work, it appeared that the hematite was deposited within the sandstone *after* its folding; and as the hematite was found to be 15 ± 7 Ma, the monocline was manifestly formed well before the hypothesized Kosciusko Uplift. Therefore, the old theory of Andrews, Taylor *et al.* was once again invalidated.

Further, John Pickett (NSW Geological Survey) and Bishop (Pickett & Bishop 1992) have presumed that the Norton's Basin diatreme passes vertically through the inclined strata of the monocline, implying that it was not deformed by the Earth movements that created the fold (and associated faults) or ones that may have occurred subsequently. The intrusion therefore post-dated the monocline. But the formation of the diatreme brought material containing early Jurassic palynoflora to the surface (cf. Helby 1971). So the monocline could (in principle) date back to (Late?) Jurassic. Even if the structure is not so old as that, the Taylor *et al.* theory could still not be correct as regards the late timing of a 'Kosciusko Uplift' with concomitant formation of the monocline.[43]

---

[42]But, as previously indicated, the exact former course of the (proto-) Nepean is uncertain. We do not know for sure which gravels represent which particular river courses. Some of the gravels are presumably quite old as they are covered by silcrete.

[43]The diatremes of the Sydney area were for long thought to be Tertiary (Crawford *et al.* 1980, p. 322), but the Jurassic palynoflora suggests that they may be considerably older. The weathered condition of many of the materials within the volcanic necks has rendered their dating difficult, and their three-dimensional form is sometimes hard to

However, the vertical form of the diatreme, which was well mapped by Osborne (1920) as being elongated east–west in outcrop, was seemingly assumed by Pickett & Bishop (1992) rather than proved. Modern maps are broadly in agreement with Osborne's findings as regards the outcrop. The excellent map of the Sydney region by Willan (1925) also shows a distinctly elongated outcrop for the diatreme. It is the only vent in the district that has this form; *and* it is the only one that cuts right through the monocline. Further, it is close to one of the subsidiary faults associated with the southern end of the monocline. Another outcrop of diatreme material, only a little to the west and off the monocline, has a usual rounded outcrop. This state of affairs does, therefore, raise doubts about the Pickett & Bishop model. Perhaps the diatreme at Norton's Basin is, in fact, bent?

Without resolving that difficulty here, we should note Pickett & Bishop's (1992) suggested explanation of the 'gravel problem', namely that the gravels are not all of the same age, despite their having similar constituents. The authors offered a sequence of events illustrated by seven diagrams (see Fig. 19), according to which a substantial overburden of sediments (probably Jurassic: Fig. 19a) was hypothesized as having formerly covered the Wianamatta sediments (Triassic) but subsequently removed by erosion, an idea previously discussed but firmly rejected by Branagan (1983).[44] The present relationships (Fig. 19g) were thus brought into being over time, with gravels being deposited at different times on a surface of hard Hawkesbury Sandstone and soft Wianamatta Shales and postulated Jurassic(?) sediments. The latter were eroded away relatively rapidly, whereas the resistant Hawkesbury Sandstone was comparatively unaffected. It will be evident that, in this model, the Lapstone Monocline predated the gravels. This 'scenario' had the advantage, compared with the older theory, of not requiring an enormous amount of erosion – producing the huge Blue Mountain valleys – during the relatively brief Pliocene–Recent time span. But it also assumed a vertical form for the Norton's Basin diatreme and a 'lost'

---

ascertain. Also, there may have been multiple intrusions at any given vent. But almost certainly they are not Pliocene–Pleistocene. The microflora are mixed, the oldest components being Early Jurassic and the youngest Middle Triassic (Wianamatta) (Crawford *et al.* 1980, p. 322).

[44]Earlier studies of vitrinite reflectance on the coals of the Sydney Basin suggested a loss of perhaps, 1 km of overburden (but pressure could be effected laterally as well as by a vertical overburden.

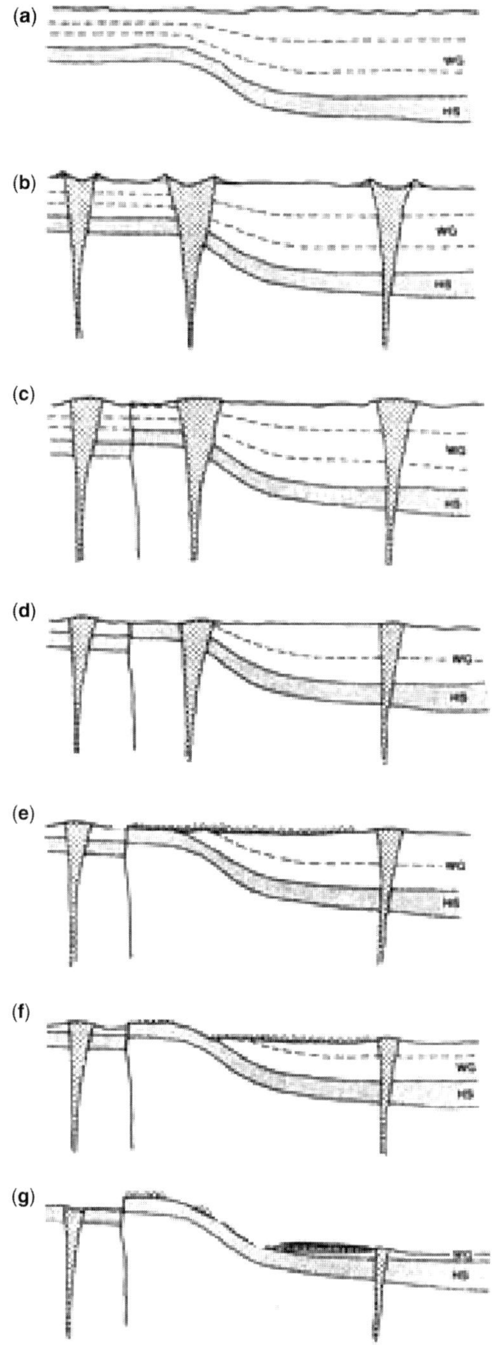

Fig. 19. Development (**a**–**g**) of the area around the Lapstone Monocline, according to Pickett & Bishop (1992, p. 24). WG, Wianamatta Group; HS, Hawkesbury Sandstone. Diatremes are marked with stipples. The small circles represent gravels. The middle diatreme is omitted by the authors from (f) and (g) so as not to obscure the other depicted relationships; but it is understood as still being present.

sequence, with Jurassic fossil traces only found in the palynoflora in the diatreme(s).

An alternative (although, in a way, complementary) model has been offered by David Branagan of Sydney University and his Iranian student Hamid Pedram (Branagan & Pedram 1990). Together they undertook a careful mapping of the Lapstone Monocline area (calling it the Lapstone Structural Complex) and interpreted it as a structure that had evolved over a long period and which was aligned along the course of an old approximately north–south lineament known to exist in the underlying basement rocks. (Extensions of the underlying lineament can be discerned in the pre-Permian rocks both north and south of the Sydney Basin.) They suggested that the Triassic sedimentation may have been controlled by basement deformation, and the Cumberland (or 'Sydney') Basin was initiated by tectonic activity in the basement rocks. Then 'after sedimentation ceased, uplift possibly accentuated topographic variation by adjustment along structural weaknesses echoing those in the basement' (Branagan & Pedram 1990, p. 34). These later movements were suggested to be associated with sea-floor spreading (in the region of the Tasman Sea) during the Cretaceous. Relative uplift of the Lapstone Complex continued in the Tertiary. As for the gravels, their history is uncertain in detail, as some of them could have been reworked down the monocline. Before that, compression (chiefly from the NE) could have given the Sydney Basin at least part of its present form.

The idea of basement control has recently been reaffirmed by Christopher Fergusson (2006) of Wollongong University, who, following Herbert (1989), 'recognises' some quite high-angle reverse faulting, parallel with the strike of the 'Monocline' along parts at least of its course. He suggests, further, that this faulting may be aligned with a 'pre-existing west-dipping thrust in the basement under Sydney' (Fergusson 2006, p. 48), which became reactivated during the Late Neogene under the influence of an east–west contractional regime, which might be associated with east–west convergence across the Australian–Pacific plate boundary (in far-away New Zealand). Such movement may even be on-going, as evidenced by the low-level seismic activity of the general area of the monocline.

Other 'post-Taylor' work has not confirmed the older understanding of the history of the Blue Mountains either, and (as might be expected) the whole problem has come to be construed in terms of the plate-tectonic paradigm (as just implied). Clifford Ollier (1978, p. 37), then at the University of New England in northern NSW, dismissed the idea of a relatively recent Kosciusko Uplift as 'largely a myth' and suggested (Ollier 1982) that the present eastern coastal shelf originated as the side of a rift valley with a 'chasmic fault', as Australia separated from other parts of Gondwanaland about 80 Ma ago. He contended that 'the landscape of eastern Australia ... [could not] be explained in terms of simple [Davisian] cycles of erosion' (Ollier 1982, p. 13). Also, he sketched two alternative models (see Fig. 20), favouring the second of the two, which was intended to display analogies with the situation for the African Rift Valley.

Ollier pointed out that the first alternative would not lead to a differentiation between the great escarpment observed parallel with the coastline in eastern Australia and the Great Divide. On the first of the two models, the Divide and the Escarpment would coincide, which was manifestly not the case. Ollier took the view that the large scarp that separates the eastern tablelands of Australia from the lowland plain(s) was a *unique* feature, which could not be accounted for in terms of Davisian cycle theory, but rather by the specific tectonic event associated with the Tasman Sea rifting.[45] His views were reiterated in Ollier (1995), although he supported in a general way Taylor's old idea of a general reversal of drainage: for Ollier, it involved a modification of an older river system in which the rivers of SE Australia formerly all tended to flow towards the NW. The rivers were reversed because of 'back-tilting' associated with rifting in the area of the present Tasman Sea, as per Figure 18. But it is not obvious (to me) that Ollier's model gives a good account of the structure of the Sydney region, which has its basin structure cutting across the line of orientation suggested by this model.

Gil Jones and John Veevers (1983), of Macquarie University, Sydney, also discussed a major change in tectonic regime at about the time of the Cenomanian (*c.* 95–90 Ma), following which the area of the eastern part of the Australian continent did not front a plate boundary (Pacifica), but a back-arc basin. Along with this model, they (like Taylor long before) envisaged a westward migration of the highland divide – but with the migration slowing over time, instead of culminating with a supposed Kosciusko Uplift.

An interesting series of articles published in the 1990s in the A[*ustralian*] G[*eological*] S[*urvey*]

---

[45]If there were such a cycle operating in SE Australia, it is certainly by no means complete with the formation of a new peneplain for the area; and, given the present rate of erosion, it is unlikely that the cycle will be completed before being disrupted by renewed tectonic activity of some kind.

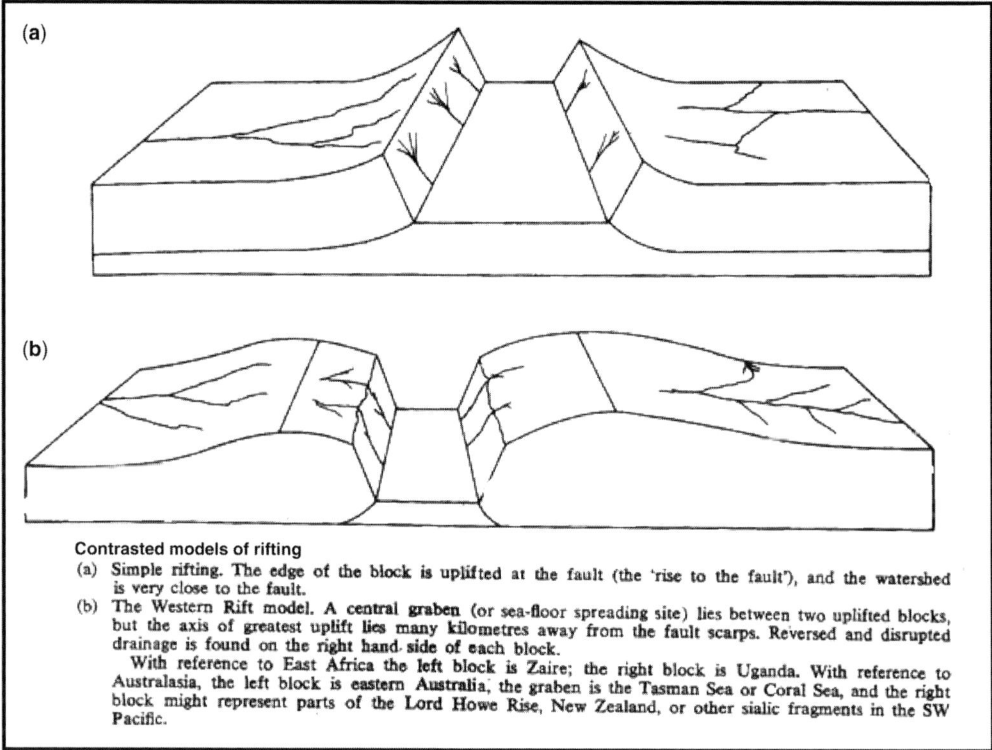

Fig. 20. Alternative models of rifting in the Tasman Sea area, according to Ollier (1982, p. 21).

O[rganization] Journal of Australian Geology and Geophysics revealed that interpretations were anything but settled at that time (see Ollier & Pain 1994, 1996; Bishop 1996; Nott 1996; Roach *et al.* 1996; Shu Li *et al.* 1996). But one can readily see the efforts being made to subsume the geomorphological and other evidence under the plate-tectonic paradigm.

Going back to proposals put forward by Ollier (1978, 1982),[46] Ollier and Colin Pain (of the Australian National University and AGSO, respectively) accepted in broad terms the idea of the old rivers flowing westwards from Pacifica onto 'Australia' before the rifting of Australia from Gondwanaland, approximately along the line of Australia's present eastern continental shelf. The rifting supposedly led to downwarping along the

Tasman Sea margin but was accompanied by the formation of the Great Divide between the Tasman Sea drainage system and that of the inland sediment basins. So the flow of the rivers east of this divide was reversed. Following the rifting and break-up of the supercontinent, sea-floor spreading developed in the Tasman Sea between Australia and the Lord Howe Island Rise. In a sense, the Ollier–Pain model harked back to the old idea of Taylor (1911a, 1958b): that there had formerly been a landmass to the east of Australia, and that our present mountain range migrated there from an earlier more easterly situation.

But Bishop (1996) objected that the Ollier–Pain model was only one of several on offer; and the idea that there had formerly been a mountain range well to the east of present Australia was by no means proven. 'Barbed drainage' and 'boathook bends' were not foolproof evidence of reversal of drainage, as in some cases they might be explained in terms of variation of rock type. Jonathan Nott (1996), then at the Australian National University, pointed out that there were two recognizable schools of thought on the question: those who supposed that individual drainage lines, and hence divides, have remained

---

[46]Ollier (1978, p. 43) acknowledged that the elevation of the Eastern Highlands under a tensional regime was not really understood (at that time), but he noted that there was a 'chain of swells' associated with African rift valleys and suggested that the Australian case might be analogous.

essentially the same from at least the mid-Tertiary; and those who held that the main divide had moved westwards with consequent reorganization of the drainage system. Griffith Taylor was said to be the founding-father of the latter school.

As mentioned earlier, Bishop (1982) had earlier made a useful distinction between two schools of thought: the 'geological' and the 'geomorphological'. The former interested themselves in the ancient directions of river flows by studying river courses that had been 'preserved' by lava flows, below which the structures of the river gravels could be studied, and the direction of water flow inferred. The latter interested themselves in such phenomena as 'boathook bends', and inferred drainage disruptions and Earth movements accordingly. Taylor belonged to that clan. Unfortunately, however, it lies beyond the scope of the present paper to trace out, or adjudicate, these differences of opinion – or even rather heated debates.[47]

But perhaps the computer can come to the rescue? For example, Peter Van der Beek, Anna Pulford and Jean Braun (Van der Beek *et al.* 2001) (of the Australian National University, Canberra) have given close attention to the Blue Mountain basalts and the land surfaces onto which they were extruded. It appears that they did not flow down pre-existing valleys, as in some other parts of eastern Australia, but were extruded onto fairly level surfaces during the Miocene. The large Blue Mountain canyons are attributed to headward erosion from knick-points, starting from the Late Cretaceous or Eocene. Thus, the *process* was as Taylor *et al.* envisaged; but the timing was quite different. The authors conducted computer modelling to endeavour to choose between different models, assuming: (1) erosion working on an initial structure similar to that envisaged by Pickett & Bishop (1992) (see Fig. 19a) but with an eastwardly inclined plateau, topped by soft Wianamatta Shales; (2) erosion of the same, but with the Wianamatta surface initially horizontal; (3) erosion working on a plateau that was elevated, or already high, in the Late Cretaceous–Early Tertiary (as envisaged by Branagan & Pedram 1990) on which was imposed an uplift along the line of the present main divide in the Eocene. The last model successfully 'produced' a drainage system that had rivers running along the foot of the ('electronic') Lapstone scarp – but it did not predict capture of 'western' rivers by those on the eastern side of the divide. Still, the model of Tertiary

uplift occurring parallel with the coastline, and along the line of the main divide, gained credibility with the help of the computer simulation.

Van der Beek *et al.* also sought to explain the history of the Sydney region in terms of plate-tectonic theory. As for so many issues in the history of geology, the problem was the *cause* of uplift. Like Ollier (and Pain, and most other geologists today), Van der Beek *et al.* assumed that eastern Australia is a rifted continental margin (*not* an analogue of California as Andrews had imagined!). But why should the separation of two land masses produce uplift on one of them; and why should the present-day stressfield in eastern Australia be compressive and directed (very) approximately east–west? Sediment loading of the Tasman Sea region might, it was suggested, have produced on-land upwards warping. Also, the timing of uplift, which seems to have coincided with the onset of volcanism in eastern New South Wales, may have been due to magmatic underplating accompanying the volcanism – a popular *deus ex machina* for many current geological problems. *This* is not the place to explore *that* issue. We may note, however, that uplift is required for the computer to model the river system development with any success. But one may remark that the general stress system in the Sydney Basin area is aligned approximately NE–SW to east–west (Enever & Clark 1997) or NE–SW (Hillis & Reynolds 2000, p. 919), in accordance with the axis of the basin, rather than east–west. This can hardly be accounted for by the system of forces that might be anticipated by Ollier's (1982) model (see Fig. 20).[48]

Taylor's 'River X' again and submarine topography

While Taylor initiated a discussion about drainage reversals and migration of uplands that is still exercising geomorphologists' minds today, his account of the development of the form of the coastline north of Sydney in terms of the meanders of a former northward-flowing 'River X' is not thought to hold good. For the offshore submarine topography of the area around Sydney reveals former waterways (now submarine valleys) running west–east from the line of the present coastline (Albani *et al.* 1988). Therefore the series of bays along the coastline of Sydney's 'Northern

---

[47]Ollier & Pain (1996, p. 329) complained that Bishop 'wallow[ed] in all the difficulties of reconstruction of ancient drainage' and went further to assert that 'the present is not the key to the past!'.

[48]The general question of the stress field in Australia is still controversial, given that it differs from what might be anticipated on the basis of a supposed northward 'drift' of the continent.

Beaches' suburbs seem to mark a set of quite short, formerly eastward-flowing streams, which were not connected to the waterway in the Pittwater inlet.

But the form and alignment of the Pittwater inlet (see Figs 4 and 14b) – evidently a drowned valley that now links up with the Broken Bay estuary – still appears as a difficulty. Running parallel to and inland from the northward-extending Pittwater Peninsula, it seems too large to have been cut by the small river (McCarr's Creek) that presently feeds it and its orientation seems anomalous. However, Alberto Albani of the University of New South Wales and co-workers (Albari & Johnson 1974; Albani *et al.* 1988) have shown, by echo sounding of the area, that Pittwater's submarine topography reveals a drainage line on the floor of the inlet, and that the ancestral river formerly made a sharp turn to the east before reaching the sea. A small ridge-line – now thought to be owing to a rather large dyke that indurated the sandstone into which it had been injected – stood in the way of the ancestral Pittwater stream and prevented its waters from flowing directly into Broken Bay. Thus, the ancestral Pittwater waterway made a right-turn and joined the Hawkesbury River at a spot that is presently out to sea. That is, the ancestral Pittwater stream and the Hawkesbury River formerly utilized the same outlet to the sea, but at a place different from either of the modern outlets. When sea levels rose after the Ice Age, the Pittwater stream was able to flow over the dyke/sandstone obstruction directly into Broken Bay and a new barrier/tombolo was created to the east by deposition of sand along the present Palm Beach at the northern end of the Pittwater Peninsula. So the former 'right-turn waterway' was abandoned. Thus, we have the south–north alignment of Pittwater seen today. The present waterway has not, it is thought, been created along the line of some now hidden fault, although there is north–south jointing in the area. Anyway, although the little McCarr's Creek has managed to excavate quite a substantial valley before it flows into Pittwater, one cannot imagine it cutting out the Pittwater estuary unaided with the sea at its present level – but presumably it was at a lower level during the Pleistocene.

So Taylor's hypothesis about a northward-flowing 'River X' has been supported to a minor extent by subsequent investigations of the ancestral river traces identified on the bottom of Pittwater. But there is no evidence that there was such a river flowing some considerable distance from the south – for example, from the Botany Bay or Sydney areas. And Taylor's hypothesized 'River X' supposedly had a meandering course that lay to the *east* of the palaeo-drainage system that underlies the modern Pittwater. Nevertheless, he

had an interesting idea – and it was one that can be traced back to the old ideas of Davisian geomorphology. Needless to say, Taylor had no data about the details of the submarine topography, as Pittwater contains much sediment that depth sounding with a line and weight could not have provided the information needed to develop the modern model. He was presumably unaware of the series of small generally north-dipping folds along the Avalon Peninsula (see Packham 1976, p. 4), which may account (at least in part) for the succession of headlands and beaches of Sydney's Northern Beaches (as the hard rocks do not maintain a constant elevation).

Taylor's ideas about the geomorphological history of Botany Bay have not stood the test of time either. As mentioned, he thought it represented a drowned meandering stream at the area of his 'Stillstand' that escaped uplift. But work by Albani & Rickwood (1998) on the bedrock topography of the area of the bay shows that there were two separate drainage 'trees' by which water flowed from the bay area to the sea via *different* outlets: one through the present mouth of the bay; and the other (to the south) that flowed across what is today a tombolo connecting the mainland to what was formerly an island at the southern entrance to the bay. But, again, Taylor had no means of knowing the details of the bedrock topography. Davisian theory, not backed up by an adequate empirical basis, was a rather weak instrument.

Concluding remarks

Geomorphology may operate on a very small scale: is this particular river pattern a manifestation of river capture? What is the reason for the different forms of Sydney Harbour and its neighbour Botany Bay? Why are there 'bottleneck' valleys in the Blue Mountains? This was the kind of work that Taylor and other Davisians sought to undertake.

Alternatively, much larger structures may be examined and attempts made to give an account of their histories. How and why, for example, was Australia's Great Dividing Range formed? Such questions involve both geomorphological understanding and the application of tectonic theory. This latter kind of work has been dubbed 'megageomorphology' (Gardner & Scoging 1983); and Bishop (1988) has considered its application to eastern Australia, showing that a wide variety of opinions had already been expressed up to that time. Something about some of these ideas has been said above and need not be repeated here. But a particular point made by Bishop is worth emphasis. That is, the old idea of a rapid uplift during the Pliocene–Pleistocene, *followed by remarkably rapid erosion*, as envisaged by the likes of Andrews and Taylor, is unacceptable. In fact, Bishop emphasizes, the rate of erosion in continental Australia during the Pliocene–Pleistocene has been very slow compared with that which took place during the same period in the northern hemisphere. 'Thus, apparently youthful forms which figured strongly in [explanations proposed in] the first three-quarters of this [the twentieth] century are actually very old forms which are changing very slowly' (Bishop 1988, p. 176). Indeed, it might be added that the too-ready application of northern hemisphere geomorphological ideas to Australia has not always been helpful.

In fact, the problems discussed in the present paper cannot be said to have been resolved to everyone's satisfaction at the time of writing. This is surprising in a way, as a preliminary glance at the geological map of the Sydney region seems to indicate a remarkably simple structure. But in detail, the region poses many tricky questions, only a small portion of which have been discussed in the present paper.

Be that as it may, the old Davisian cyclic theory, which Taylor strongly espoused for Australia, is now superseded, for various reasons such as have been outlined in the present paper. (In 1998, Paul Larson posted a query on the internet (http://main. amu.edu.pl/-sgp/gw/wmd/wmd.html) seeking information as to how many people accepted Davis's cyclic model. The response was overwhelmingly negative. In particular, it was not thought to be applicable to landscapes such as Australia's.) But a synthesis of geomorphological investigations with tectonics remains a necessity for understanding the geological history of any region. Such work continues. The 'old geomorphology' of Davis and his followers (in Australia and elsewhere) may be obsolete, but its memory lives on. And Taylor and Andrews were notable pioneers 'down under'. Geomorphology must be a partner with other branches of geoscience in attempts to elucidate the histories of different regions of the Earth.

I am much indebted to D. Branagan and R. Young for their careful reading of earlier drafts of this paper, their numerous suggestions for correction and improvement, and the provision of some relevant articles, maps and books. Any errors remaining should be attributed to me, not to them! A. Albani kindly provided first-hand information about his work on submarine topographies of the Sydney area and gave me copies of his publications on the topic. I also thank the National Library of Australia and the Basser Library of the Australian Academy of Science, Canberra, for granting access to papers of Taylor and Andrews, respectively; and the Archives of the University of Sydney allowed me to consult some Taylor materials in the Edgeworth David collection for the purposes of the present paper. Specifics are provided in the relevant figure captions. R. H. Grapes kindly assisted with the improvement of the figures.

References

ALBANI, A. D. & JOHNSON, B. D. 1974. The bedrock topography and origin of Broken Bay, N.S.W. *Journal of the Geological Society of Australia*, **21**, 209–214.

ALBANI, A. D. & RICKWOOD, P. C. 1998. *Botany Bay: bedrock topography and geological history.* Paper ?presented at the conference of the Geological Society of Australia: Environmental, Engineering and Hydrogeology Specialists Group on the Environmental Geology of Botany Bay.

ALBANI, A. D., TAYTON, J. W., RICKWOOD, P. C., GORDON, A. & HOFFMAN, J. G. 1988. Cainozoic morphology of the inner continental shelf near Sydney, N.S.W. *Journal and Proceedings Royal Society of New South Wales*, **121**, 11–28.

ANDREWS, E. C. 1903*a*. Notes on the geography of the Blue Mountains and Sydney District. *Proceedings of the Linnean Society of New South Wales*, **28**, 786–825 and plates.

ANDREWS, E. C. 1903*b* (1905 for 1900–1904). An outline of the Tertiary history of New England. *Records of the Geological Survey of New South Wales*, **7**, 140–216 and plates.

ANDREWS, E. C. 1905. *An Introduction to the Physical Geography of New South Wales.* William Brooks & Co., Sydney.

ANDREWS, E. C. 1910. Geographical unity of eastern Australia in Late and Post Tertiary time, with applications to biological problems. *Journal and Proceedings of the Royal Society of New South Wales for 1910*, **44**, 420–480.

ANDREWS, E. C. 1934. The origin of modern mountain ranges. With special reference to the Eastern ?Australian Highlands. *Journal and Proceedings of the Royal Society of New South Wales for 1933*, **67** (Part 2), 251–350.

ANDREWS, E. C. not dated. Autobiographical notes ?prepared during period 21/11/3 – (manuscript). Andrews Papers, Basser Library, Australian Academy of Science, Canberra, Box 1.

ANON., 1952. Ernest Clayton Andrews, 1870–1948. *Proceedings of the Linnean Society of New South Wales*, **77**, 98–103.

BISHOP, P. 1982. Stability or change: a review of ideas on ?ancient drainage in eastern New South Wales. *?Australian Geographer*, **32**, 219–230.

BISHOP, P. 1988. The eastern Highlands of Australia: the evolution of an intraplate highland belt. *Progress in Physical Geography*, **12**, 159–182.

BISHOP, P. 1996. Discussion: landscape evolution and tectonics in southeastern Australia (Ollier & Pain 1994). *AGSO Journal of Australian Geology and Geophysics*, **16**, 315–317.

BISHOP, P., HUNT, P. & SCHMIDT, P. W. 1982. Limits to the age of the Lapstone Monocline, N.S.W. – a palaeomagnetic study. *Journal of the Geological Society of Australia*, **29**, 319–326.

BRANAGAN, D. F. 1983. The Sydney Basin and its vanished sequence. *Journal of the Geological Society of Australia*, **30**, 75–84.

BRANAGAN, D. F. 2005. *T.?W.?Edgeworth David: A Life*. National Library of Australia, Canberra.

BRANAGAN, D. F. & PEDRAM, H. 1990. The Lapstone Complex, New South Wales. *Australian Journal of Earth Sciences*, **37**, 23–36.

BROWNE, W. R. 1969. Geomorphology. *In*: PACKHAM, G. H. (ed.) The geology of New South Wales. *Journal of the Geological Society of Australia*, **16**, 559–580.

CARNE, J. E. 1908. *Geology and Mineral Resources of the Western Coalfield [of the Sydney Basin]; with maps and sections*. Memoirs of the Geological Survey of New South Wales, Geology, **6**.

CHAMBERLIN, T. C. & SALISBURY, R. D. 1906. *Geology Earth History in Three Volumes Vol. III. Mesozoic, Cenozoic*. John Murray, London.

CLARKE, W. B. 1878. *The Sedimentary Formations of New South Wales*, 4th edn. Government Printer, Sydney.

CRAFT, F. A. 1928. The physiography of the Cox River Basin. *Proceedings of the Linnean Society of New South Wales*, **53**, 207–254 and plates.

CRAFT, F. A. 1931. The physiography of the Shoalhaven River Valley.?I.?Tallong–Bungonia. *Proceedings of the Linnean Society of New South Wales*, **56**, 99–132 and plates.

CRAFT, F. A. 1933. The coastal tablelands and streams of New South Wales. *Proceedings of the Linnean Society of New South Wales*, **29**, 437–460.

CRAWFORD, E., HERBERT, C., TAYLOR, G., HELBY, R., MORGAN, R. & FERGUSON, J. 1980. Diatremes of the Sydney Basin. *In*: HERBERT, C. & HELBY, R. (eds) *A Guide to the Sydney Basin*. Department of Mineral Resources and Geological Survey of New South Wales, Bulletin, **26**.

DAVID, T.?W.?E. 1897. Summary of our present knowledge of the structure and origin of the Blue Mountains of New South Wales. *Journal and Proceedings of the Royal Society of New South Wales for 1896*, **30**, 33–69 and plates.

DAVID, T.?W.?E. 1902. An important geological fault at Kurrajong Heights, N.?S.?Wales. *Journal and Proceedings of the Royal Society of New South Wales*, **36**, 359–370.

DAVID, T.?W.?E. 1912 for 1911. Notes on some of the?-chief tectonic lines of Australia [part of a ?Presidential Address]. *Journal and Proceedings of the Royal Society of New South Wales for 1911*, **45**,?4–60.

DAVID, T.?W.?E. 1950. *The Geology of the Commonwealth of Australia by the Late T.?W.?Edgeworth David; edited and much supplemented by W.?R.?Browne*. E. Arnold, London.

DAVIS, W. M. 1895. The development of certain English rivers. *Geographical Journal*, **5**, 127–146.

DAVIS, W. M. 1899. The geographical cycle. *Geographical Journal*, **14**, 481–504.

ENEVER, J. R. & CLARK, J. 1997. The current stress state in the Sydney Basin. *In*: BOYD, R. L. & ALLEN, K. (eds) *Proceedings of the Thirty First Newcastle Symposium on 'Advances in the Study of the Sydney Basin'*. Department of Geology, University of Newcastle, Newcastle, 91–98.

FERGUSSON, C. L. 2006. Review of structure and basement control of the Lapstone Structural Complex, Sydney Basin, eastern New South Wales. *In*: HUTTON, A. & GRIFFIN, J. (eds) *Proceedings of the Thirty-sixth Sydney Basin Symposium, Advances in the Study of the Sydney Basin*. School of Earth and Environmental Sciences, University of Wollongong, Wollongong, 45–50.

GARDNER, R. & SCOGING, H. (eds) 1983. *Mega-Geomorphology*. Clarendon Press, Oxford.

GOBERT, V. 1976. Nepean Cainozoic sediments and Mines Department Bore Core Library. *In*: *Excursion Guide 22B*. Sydney, 25th International Geological Congress.

HARPER, L. F. 1915. *Geology and Mineral Resources of the Southern Coalfield*. Memoirs of the Geological Survey of New South Wales, Geology, **7**.

HEDLEY, C. 1912. A study of marginal drainage. *Proceedings of the Linnean Society of New South Wales for the Year 1911*, **36**, 13–39 and plates.

HELBY, R. J. 1971. *Palynological examination of material in the volcanic breccia at Kedumba neck and Norton's Basin New South Wales*. Palynological Report, **1971/?15** (unpublished) GS1971/676). Cited in Jones & Clark (1991, pp.?186).

HERBERT, C. 1989. The Lapstone Monocline–Nepean Fault – a high angle reverse fault system. *In*: *Advances in the Study of the Sydney Basin: Proceedings of the 23rd Symposium*. Department of Geology, University of Newcastle, 179–186.

HILLIS, R. R. & REYNOLDS, S. D. 2000. The Australian stress map. *Journal of the Geological Society, London*, **157**, 915–921.

JENSEN, H. J. 1912 *for* 1911. The river gravels between Penrith and Windsor. *Journal and Proceedings of the Royal Society of New South Wales for 1911*, **45**, 249–257.

JONES, D. C. & CLARK, N. R. (eds). 1991. *Geology of the Penrith 1:100 000 Sheet 9030*. New South Wales Geological Survey, Sydney.

JONES, J. G. & VEEVERS, J. J. 1983. Mesozoic origins and?antecedents of Australia's Eastern Highlands. *Journal of the Geological Society of Australia*, **30**, 305–322.

JOSE, A. W., TAYLOR, T. G. & WOOLNOUGH, W. G. (edited by DAVID, T.?W.?E.). 1912. *New South Wales: Historical, Physiographical, and Economic*. Whitcombe & Tombs, Melbourne.

MUELLER, F. VON. 1874. *Observations on New Vegetable Fossils of the Auriferous Drifts*. Geological Survey of Victoria. J. Ferres, Government Printer, Melbourne.

NICHOLAS, F. & NICHOLAS, J. M. 1989. *Charles Darwin in Australia*. Cambridge University Press, Cambridge.

NOTT, J. F. 1996. Discussion: landscape evolution and tectonics in southeastern Australia (Ollier & Pain 1994). *AGSO Journal of Australian Geology and Geophysics*, **16**, 319–321.

OLDROYD, D. R. 1994. Griffith Taylor and his views on race, environment and settlement. *In*: BRANAGAN, D. F. & McNALLY, G. H. (eds) *Useful and Curious Geological Enquiries Beyond the World: Pacific–Asia Historical Themes*. International Commission on the History of Geological Sciences (INHIGEO), Springwood (NSW), 251–274.

OLLIER, C. D. 1978. Tectonics and geomorphology of the Eastern Highlands. *In*: DAVIES, J. L. & WILLIAMS, M.?A.?J. (eds) *Landform Evolution in Australasia*. ?Australian National University Press, Canberra, 5–47.

OLLIER, C. D. 1982. The Great Escarpment of Eastern Australia: tectonic and geomorphic significance. *Journal of the Geological Society of Australia*, **29**, 13–23.

OLLIER, C. D. 1995. Tectonics and landscape evolution in southeast Australia. *Geomorphology*, **12**, 37–44.

OLLIER, C. D. & PAIN, C. F. 1994. Landscape evolution and tectonics in southeastern Australia. *AGSO Journal of Australian Geology and Geophysics*, **15**, 335–345.

OLLIER, C. D. & PAIN, C. F. 1996. Reply: landscape evolution and tectonics in southeastern Australia (Ollier & Pain 1994). *AGSO Journal of Australian Geology and Geophysics*, **15**, 325–331.

OSBORNE, G. D. 1920. The volcanic neck at The Basin, Nepean River. *Journal and Proceedings of the Royal Society of New South Wales*, **54**, 113–145.

PACKHAM, G. H. (ed.) 1969. The geology of New South Wales. *Journal of the Geological Society of Australia*, **16**(1).

PACKHAM, G. H. 1976. *Excursion 16B: Narrabeen – Palm Beach Triassic Sedimentation*. Sydney, 25th International Geological Congress.

PICKETT, J. W. & BISHOP, P. 1992. Aspects of landscape evolution in the Lapstone Monocline area, New South Wales. *Australian Journal of Earth Sciences*, **39**, 21–28.

POWELL, J. M. 1979. Thomas Griffith Taylor (1880–1963). *Geographers; Biobibliographical Studies*, **3**, 141–153.

RAGGATT, H. G. 1959. Walter George Woolnough (1876–1958). *AAPG Bulletin*, **43**, 2035–2040.

ROACH, I. C., MCQUEEN, K. G. & TAYLOR, G. 1996. Discussion: landscape evolution and tectonics in southeastern Australia (Ollier & Pain 1994). *AGSO Journal of Australian Geology and Geophysics*, **16**, 309–313.

SANDERSON, M. 1988. *Griffith Taylor: Antarctic Scientist and Pioneer Geographer*. Carleton University Press, Ottawa.

SCOTT, H. I. 1977. *The Development of Landform Studies in Australia*. Artarmon (Sydney), Bellbird Publishing Service; and Saxon House (Teakfield), Farnborough (UK).

SHU LI, WEBB, J. A. & FINLAYSON, B. L. 1996. Discussion: landscape evolution and tectonics in southeastern Australia. *AGSO Journal of Australian Geology and Geophysics*, **16**, 323–324.

SÜSSMILCH, C. A. 1910 for 1909. Notes on the physiography of the Southern Tablelands of New South Wales for 1909. *Journal and Proceedings of the Royal Society of New South Wales*, **43**, 331–354 and plates.

SÜSSMILCH, C. A. 1911. *An Introduction to the Geology of New South Wales*. Government Printer, Sydney.

TAYLOR, T. G. 1899. Prof. David's lectures on physiography in 1899. University of Sydney, Taylor papers P163/1/1.

TAYLOR, T. G. 1911a. *Physiography of Eastern Australia*. Commonwealth Bureau of Meteorology, Melbourne, Bulletin No. 8. Government of the Commonwealth of Australia, Melbourne.

TAYLOR, T. G. 1911b. *Australia in its Physiographic and Economic Aspects*. Clarendon Press, Oxford.

TAYLOR, T. G. 1916. *A Geography of Australasia*, 3rd edn. Clarendon Press, Oxford.

TAYLOR, T. G. 1921. Mountain building I: the scenery around Sydney. *The Sydney Morning Herald*, 3 December.

TAYLOR, T. G. 1923a. The warped littoral around Sydney. *Journal and Proceedings of the Royal Society of New South Wales*, **57**, 58–79.

TAYLOR, T. G. 1923b. The Blue (Mountain) Plateau. *In*: *Pan-Pacific Science Congress, Australia, 1923. Sydney 23rd August–3rd September. Guide-Book to the Excursion to Blue Mountains, Jenolan Caves, and Lithgow*. Government Printer, Sydney, 17–29.

TAYLOR, T. G. 1927a. Coast structure. Cronulla to Baranjoey. *The Sydney Morning Herald*, February 1927 (Taylor papers, National Library of Australia, MS1003/4/489).

TAYLOR, T. G. 1927b. The topography of Australia. *The Official Year Book of the Commonwealth of Australia*, **20**, 75–86.

TAYLOR, T. G. 1958a. *Journeyman Taylor: The Education of a Scientist*. Robert Hale, London.

TAYLOR, T. G. 1958b. *Sydneyside Scenery and How it Came About*. Angus & Robertson Ltd, Sydney.

TAYLOR, T. G. not dated. *Some new physiographic ?techniques*. Unpublished manuscript. National Library of Australia, Taylor papers, MS 1003, Box No. 35.

TAYLOR, T. G. & TAYLOR, D. 1925. *The Geographical Laboratory: A Practical Handbook for a Two Year Course in Australian Universities with 32 Block Diagrams, Etc.* Sydney University Union, Sydney.

TWIDALE, C. R. & CAMPBELL, E. M. 2005. *Australian Landforms: Understanding a Low, Flat, Arid and Old Landscape*. Rosenberg Publishing, Dural (Sydney).

VAN DER BEEK, P., PULFORD, A. & BRAUN, J. 2001. Cenozoic landscape development in the Blue Mountains (SE Australia): lithological and tectonic controls on rifted margin topography. *Journal of Geology*, **109**, 35–56.

WALKER, P. H. 1960. *A Soil Survey of the County of ?Cumberland, Sydney Basin, New South Wales*. Bulletin Soil Survey Unit, **2**. New South Wales Department of ?Agriculture, Sydney.

WELLMAN, P. & MCDOUGALL, I. 1974a. Potassium–Argon dating on the Cainozoic volcanic rocks of eastern Australia. *Journal of the Geological Society of Australia*, **21**, 359–376.

WELLMAN, P & MCDOUGALL, I. 1974b. Potassium–argon dating on the Cainozoic volcanic rocks of New South Wales. *Journal of the Geological Society of ?Australia*, **21**, 247–272.

WILLAN, T. L. 1923. The geology of the Sydney District. *In*: *Pan-Pacific Science Congress, Australia, 1923. Sydney 23rd August–3rd September. Guide-Book to the Excursion to Blue Mountains, Jenolan Caves, and Lithgow*. Government Printer, Sydney, 22–26.

WILLAN, T. L. 1925. *Geological Map of the Sydney District*. Prepared under the direction of E.?C.?Andrews, BA. Department of Mines, New South Wales. Alfred James Kent, Government Printer, Sydney.

WOOLNOUGH, W. G. 1928 for 1927. Presidential Address: Part 1, The chemical criteria of peneplanation; Part 2, The duricrust of Australia. *Journal and Proceedings of the Royal Society of New South Wales for 1927*, **61**, 17–53.

WOOLNOUGH, W. G. & TAYLOR, T. G. 1907–1908. A?striking example of river capture in the coastal ?district of NSW. *Proceedings of the Linnean Society of New South Wales for the Year 1906*, **31**, 546–554 and plates.

YOUNG, R. W. 1974. A major error in south-east ?Australian landform studies. *In*: *Proceedings of the International Geographical Union Regional Conference and Eighth New Zealand Geography Conference Palmerston North*. Wellington, New Zealand Geographical Society, 283–287.

YOUNG, R. W. 1977. Landscape development in the ?Shoalhaven River catchment of southeastern New South Wales. *Zeitschrift für Geomorphologie*, **21**, 262–283.

YOUNG, R. W. 1978. The study of landform evolution in the Sydney region: a review. *Australian Geographer*, **14**, 71–93.

YOUNG, R. W. 2007. Reverend W.?B.?Clarke, 'the Father of Australian Geology', on the origin of valleys. ?*Australian Journal of Earth Sciences*, **84**, 127–234.

Early geological investigations of the Pleistocene Tamala Limestone, Western Australia

WOLF MAYER

Research School of Earth Sciences, D. A. Brown Building, Australian National University, Canberra, ACT 0200, Australia (e-mail: wolf.mayer@ems.anu.edu.au)

Abstract: The first geological studies of the Quaternary deposits, which crop out extensively along the coast of Western Australia, were carried out by members of English and French expeditions of discovery, between 1791 and 1836. The exploring parties included scholars with a background in geology, zoology and botany, as well as knowledgeable surgeons and sea captains with a strong interest in the natural sciences. Their collective work established the continuity, over vast distances, of a sequence of sedimentary rocks composed of quartz grains and shell debris, which today form the major part of the Tamala Limestone sequence. Their observations of the internal features of these rocks led some among them to develop views on the nature and origin of the cementing substance that bonds sand grains and shell debris in sedimentary layers and in concretions. There was disagreement among successive parties of visitors on the nature and origin of rhizoliths and other petrified woody matter in calcareous rocks. The finding of well-preserved sea shells in rocks now above sea level provided convincing evidence to investigators that the ocean had, in recent times, retreated from the land. The discovery of species of mollusc, known to be extinct in Europe, raised questions about an assumed worldwide extent of sedimentary sequences.

The mariners of the Dutch ship, the *Eendracht* (*Unity*), sailing from Holland to Batavia in the East Indies in 1616 were the first Europeans to sight the low-lying, semi-arid western coast of the southern continent that would soon become known as New Holland. Their commander, Dirk Hartog, piloted his ship into the calm waters of Shark Bay and landed on an island that now bears his name. As they ascended the slopes of the coastal cliffs, the visitors climbed over outcrops of distinctly layered, light-coloured sedimentary rocks that we recognize today as part of the Tamala Limestone. Because of its extensive exposures along the Western Australian coast, this geological unit was the most frequently encountered by early explorers and naturalists, and would, in future years, receive their greatest attention. However, neither Hartog nor later seafarers – among them the English explorer William Dampier, in 1688 and 1699; the Dutch navigator Willem de Vlamingh in 1696–1697; and the French naval lieutenant François de Saint-Allouarn in 1772, who visited and charted parts of this inhospitable coast – recorded details of its rock formations.

It was the arrival of the French scientific expedition led by Nicolas Baudin (1754–1803) in 1801 that marked the start of the geological investigation of the continent's western margin and of its coastal deposits. The French ships, the *Géographe* and the *Naturaliste*, which also visited parts of Australia's southern coast, the east coast of Tasmania and the new English settlement at Sydney, brought to Australia a number of well-qualified scientists. They included Louis Depuch (1774–1803) and Charles Bailly (1777–1844), former students, respectively, of the École des Mines and the École Polytechnique in Paris, and the first 'tertiary-educated' geologists to visit the southern continent. Also among them was the multi-talented François Péron (1775–1810), whose zoological studies in Paris had been guided by George Cuvier (1769–1832) (Péron 1804, p. 438), but who also had a good understanding of contemporary geological thought and practice (Mayer 2005).

The geological observations made by Depuch, who did not survive the voyage, are known to us from a number of reports he wrote to his captain, some of which Baudin copied into his personal journal (Bonnemains *et al.* 2001). Bailly, who was assigned to other duties soon after his return to France, also left a number of manuscripts recording his geological work in Australia. Péron and, later, Louis de Freycinet (1779–1842), who had served as sublieutenant on the *Naturaliste*, included extracts from the writings of both Depuch and Bailly in the official history of the voyage (Péron 1807; Péron & Freycinet 1816).

Freycinet himself returned to Australia's west coast, leading his own expedition in the *Uranie* in 1816. The mission's surgeon/naturalist Jean René Quoy (1790–1869) and, to a lesser extent, one of

From: GRAPES, R. H., OLDROYD, D. & GRIGELIS, A. (eds) *History of Geomorphology and Quaternary Geology.*
Geological Society, London, Special Publications, **301**, 279–293.
DOI: 10.1144/SP301.17   0305-8719/08/\$15.00 © The Geological Society of London 2008.

its naval officers, Louis Duperrey (1786–1865), carried out geological work during the ship's call at Shark Bay (Freycinet 1827).

In 1822 Phillip Parker King (1791–1856), commanding the *Bathurst*, visited several localities along the West Australian coast and made an extensive collection of specimens from outcrops now regarded as part of the Tamala Limestone (King 1826). These specimens were later described in England by the Irish geologist William Henry Fitton (1780–1861) (Fitton 1826). Outcrops of similar calcareous rocks and sediments, extending to the southern coast of Western Australia, were first noted by George Vancouver (1757–1798), who visited and named King George Sound in the *Discovery* and *Chatham* in 1791. The branch-like concretionary structures he discovered in these sedimentary layers became the subject of debate and controversy over the next several decades. The early contributions by French naturalists and scientists to the exploration and understanding of Australian geology have previously been discussed by Vallance (1975, 1981–1982) and Branagan (2002).

The Tamala Limestone – seen through contemporary eyes

The rock exposures along the continent's western coast that first attracted the attention of European explorers and naturalists are composed of what are now known to be Quaternary deposits cropping out for many hundreds of kilometres along the continent's western margin from Geographe Bay in the south to Shark Bay and beyond in the north (Fig. 1). According to modern views, they overlie a thick sequence of Phanerozoic sedimentary strata deposited along a rifted continental margin in the Perth and Carnarvon basins, and are separated from the Precambrian Yilgarn Craton to the east by the prominent scarp of the Darling Fault.

The coastal outcrops are predominantly composed of coarse- to medium-grained eolian calcarenites, a mixture of wind-blown shell fragments – derived from molluscs, calcareous algae, foraminifera and echinoderms – and quartz sand, often showing large-scale cross-bedding. They include some richly fossiliferous marine intercalations of shell beds and coral reefs, as well as calcrete and fossil soil horizons. The deposits range from lightly cemented sediments to more consolidated layers, some of which contain large numbers of rhizoliths, or fossil root structures, along with other petrified vegetable matter and globular calcareous concretions.

Formerly collectively referred to as the 'Coastal Limestone' (Teichert 1950), and as 'Tamala Eolianite' (Logan *et al.* 1970), more recent work has assigned most of these deposits to formal stratigraphic units. Playford *et al.* (1976) grouped the eolian calcarenite deposited in coastal dune and beach near-shore environments in the Carnarvon and the Perth basins with the Tamala Limestone, and gave separate formation status to its several marine intercalations. They assigned an age range of Pleistocene–Early Holocene to this coastal sequence. Both Murray-Wallace & Kimber (1989) and Kendrick *et al.* (1991) have treated the Tamala Limestone as a single formation that includes both eolian and marine members. Shell beds in the Tamala Limestone have been dated by Kendrick *et al.* (1991) as Middle and Late Pleistocene in age.

The sedimentary sequence cropping out along the Western Australian coast has not yet been fully differentiated into formal stratigraphic units over the entire extent of its outcrop. Scattered, undifferentiated occurrences of calcarenites, which are also found along the southern coast of Western Australia, have been informally referred to as Tamala Limestone equivalent.

On Rottnest Island (Fig. 1), Playford (1988, p. 25, 1997, p. 795) has established that the wind-blown deposits of the Tamala Limestone lie both under and over the reef of the Rottnest Limestone. The latter unit has been dated at about 130 ka (Late Pleistocene). He suggested that the part of the eolianite sequence that underlies the reef may have formed at about 135–140 ka. Both the eolian and the reef sequences at Rottnest Island are overlain by the Herschell Limestone of Holocene age. Recently, Playford (2006) has announced dates for the Tamala Limestone from its type section on the Zuytdorp Cliffs in the Carnarvon Basin of about 250 000 ka (Middle Pleistocene).

The effects of Quaternary glaciation and sea-level changes on the morphology of the Western Australian coast were first recorded by Teichert (1950), Churchill (1959) and Fairbridge (1958, 1961), among others. As Playford (1988, p. 38, 1997, p. 798) has noted from his studies on Rottnest Island, 'this evidence is in the form of elevated marine deposits, elevated shoreline platforms and notches, and subaerially formed features that now extend below sea level'. Based on the hight of the coral reef of the Rottnest Limestone above the adjoining shoreline platform, he concluded that sea level during the last interglacial stood about 3 m higher than at present. He also argues that sea level was as much as 2.4 m higher than today during the Holocene when the Herschell Limestone was deposited.

Kendrick *et al.* (1991, p. 433) conclude that 'the available evidence from Western Australia points to sea level being near its present position a number of

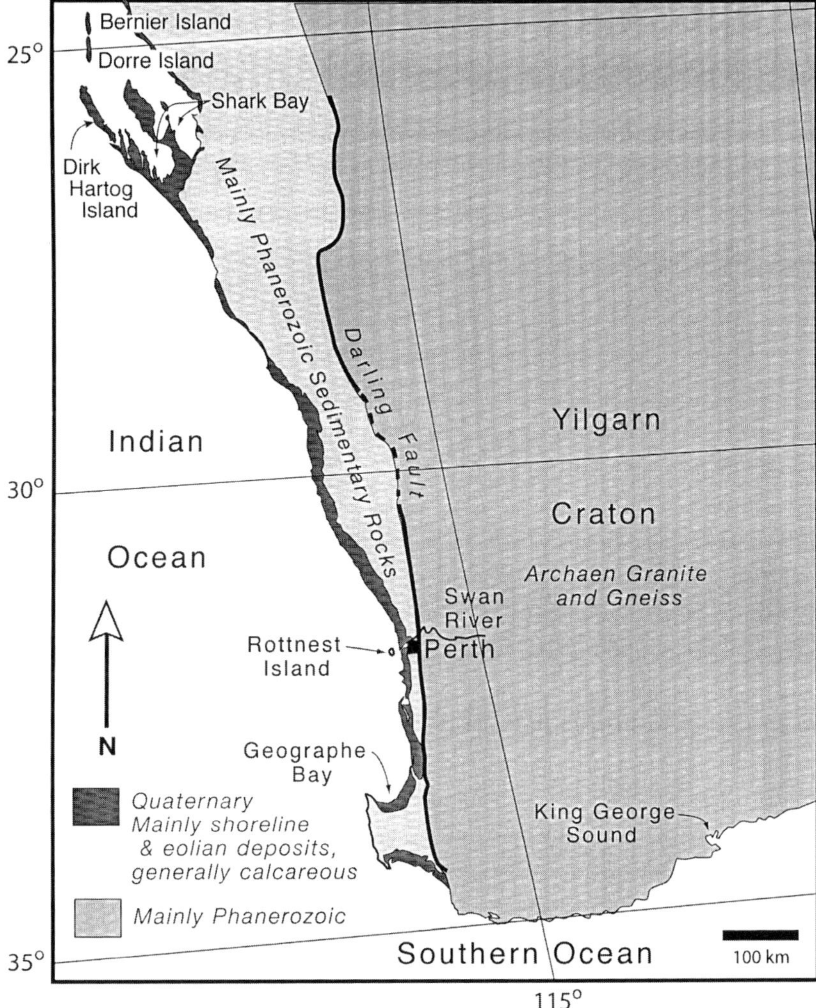

Fig. 1. Simplified geological map of the southwestern part of Western Australia. Based on Myers & Hocking (1988).

times during the Plio-Pleistocene'. They link the formation of various shell beds and coral reefs in the Tamala Limestone in the Perth Basin to former higher sea levels stands, near to its present height, during the 'Last' and 'Penultimate Interglacials' (oxygen isotope stages/marine isotope stages (OIS/MIS) 5 and 7), respectively. Although they concede that Shackleton's (1987) data suggest that sea level during OIS/MIS 7 may have been somewhat lower.

There is debate about the prevailing conditions that led to the formation of the very extensive eolian dunes which make up the greater part of the Tamala Limestone sequence. Playford (1997, p. 794) suggests that 'the Tamala Limestone

accumulated as coastal dunes, mostly during periods of lowered sea level during the Pleistocene and early Holocene, when wide areas of continental shelf were exposed, and carbonate productivity was high'. He states that the sea reached its lowest level of about −130 m at about 18 ka, when the coastline was located some 12 km west of Rottnest Island, and that the dune limestone around the coast of Rottnest Island probably extends to a depth of at least 70 m below sea level. In a contrary view, Hearty (2002) argues that the eolianites on Rottnest Island were formed mainly during two of the three highstand events over the last 125 000 ka (Chappell & Shackleton 1986); that is, the first and second interstadials of the last glacial cycle (MIS 5a and

5c, about 75–85 and 95–110 ka). Among the arguments in support of his view he expresses his doubts about voluminous carbonate production being possible in cool ocean waters near the edge of the continental shelf, and states his belief that rapid induration of the sand deposits would prevent long distance migration of carbonate dunes (Hearty 2002, p. 220).

There is, as yet, no conclusive evidence as to the relative effects of tectonic adjustments v. eustatic sea-level changes along the Western Australian coast. Playford (1997, p. 802) believes that sea-level changes during the Pleistocene 'that affected the mainland and the adjoining part of the continental shelf around Rottnest clearly resulted from global eustatic changes associated with waxing and waning of continental ice sheets'. However, as Holocene highstands of sea level on Rottnest 'have no generally accepted eustatic correlatives elsewhere in Australia or the world', he suggested 'that the alternative of a tectonic origin also needed to be considered' (Playford 1997, p. 803). Kendrick *et al.* (1991, p. 421) write of the difficulty of separating 'with confidence any tectonic/isostatic adjustments from eustatic changes for the central Perth Basin sequence'. While they cite evidence of significant tectonic activity during the Late Cenozoic in parts of the Carnarvon Basin, with uplift of some hundreds of metres, they believe that such effects on the Pleistocene deposits of the Perth Basin have amounted to possibly a few tens of metres. They conclude with the cautionary note that 'whatever the figure, a tectonic influence cannot be ruled out for much of the region' (Kendrick *et al.* 1991, p. 422).

It is clear that further work is needed before the respective effects of eustatic sea-level change and of tectonic activity on the deposition of these sediments and on the morphology of the coastal regions can be fully ascertained.

The naturalists of the various expeditions mentioned above, who visited the Western Australian coast, as well as Fitton at the time he described the Australian collections in England, were unfamiliar with the concepts of glaciation and, particularly, glacial eustasy and its effects on sea-level changes, concepts which scientists would not propose in a substantive form until later years (Agassiz 1840; Suess 1888).

First descriptions and classification of the coastal rocks

As Baudin's ships sailed along the continent's western coast, the naturalists aboard were struck by the dry, flat and monotonous appearance of the land, and by the extensive areas of sand that covered it. Where they came ashore they noted outcrops of a succession of horizontal layers (Fig. 2) of often fossiliferous, calcareous sandstone and – more rarely – pure quartz sandstone. Many of the individual beds were of a uniform thickness and continuous in outcrop. Some contained curious, hard, globular bodies of similar composition to that of the sandy, calcareous matter that enclosed them. None of the investigators referred to the presence of conspicuous cross-bedding – but that phenomenon had not then been recognized in the geological literature as being significant.

Following his examination of the coastal rocks cropping out at both Geographe Bay and at Shark Bay (Fig. 3), and having noted the appearance of the low cliffs between these localities, Péron concluded that this entire stretch of coastline must be of a similar make-up and origin (Péron 1801*a*, p. 3). His view was reinforced by the work of Bailly at Rottnest Island and along the banks of the Swan River who, at these localities, also recorded 'calcareous and sandy rocks arranged in horizontal bands' some of which contained fossils (Fig. 4) (Péron & Freycinet 1816, vol. 1, Part 2, pp. 178–184).

Depuch concluded that these extensive outcrops of calcareous strata represented part of the *secondaire* (Depuch 1801*a*, p. 4): one of the four divisions in the classification of the Earth's crustal rocks that was then widely accepted by geologists in Continental Europe. His former professor at the École des Mines, Déodat de Dolomieu (1750–1801) had, in his lectures, grouped crustal rocks into four major types or *sortes*: (1) Primitive or primordial; (2) overlying bedded sedimentary rocks deposited in the sea; (3) debris transported in rivers and by other means on the surface of the Earth (equivalent to modern day alluvium); and (4) volcanic rocks (Cordier 1796). Depuch chose to use *secondaire* in preference to Dolomieu's *seconde sorte*. Like the *Monti secondari* of Arduino (1760) in Italy and the *Flötzgebirge* of Werner's (1786) classification in Germany, the term referred to stratified sedimentary rocks, often containing fossils. The term Quaternary was not introduced until 1829 for marine and alluvial sediments overlying the Tertiary in the Seine Basin (Desnoyers 1829).

Like his countrymen some 15 years earlier, Quoy, when examining coastal outcrops at Shark Bay, observed layers of calcareous quartz sandstones, sometimes containing broken bivalve shells, as well as thin beds of a red- and yellow-coloured mixture of clay and quartz. He found the horizontal layers to range from friable to very hard and to contain various concretionary structures. Duperrey, who observed red sandstone in another part of Shark Bay, was probably the first

Fig. 2. Outcrop of Tamala Limestone showing horizontal and cross-bedded calcarenite at Shark Bay.

to document outcrops of what is now termed the Peron Sandstone (Fig. 3), also of Pleistocene age and underlying the Tamala Limestone (Freycinet 1827, vol. 1, Part 2, pp. 471–475).

After studying the reports of the Baudin and King expeditions, and with the benefit of his own examination of samples brought to England by King, Fitton acknowledged the extensive occurrence of calcareous strata along the continent's western coast. He also intimated their continuation to the southern coast of Western Australia, when he noted the similarity between specimens from Dirk Hartog Island in Shark Bay and those collected from Bald Head at King George Sound (see Fig. 6 later in this paper). Taking into account the abundance of concretionary nodules and fragments that had been found in individual beds, he collectively referred to this sedimentary sequence as 'recent calcareous breccia'. Fitton noted its 'resemblance to the very recent lime-stone, full of marine shells,

Fig. 3. The darker, reddish-coloured Peron Sandstone at the base of the cliff is overlain by the cream–buff coloured Tamala Limestone. Dune sands cover the land surface above. Eagle Bluff, Shark Bay.

which abounds on the shores of the Mediterranean, the West Indies, and several other parts of the world' (Fitton 1826, p. 18). As an example, he cited the work of Daubeney (1825, pp. 117–118) who also used the term 'calcareous breccia' to describe a fossiliferous limestone from Sicily, containing concretions and 'being of recent formation'.

Early views on stratification and cementation of the sedimentary rocks

Both Depuch and Péron carefully examined the sedimentary deposits of Bernier Island in Shark Bay (Depuch 1801*a*; Péron 1801*a*). They were struck by finding sand and shell debris on the

Fig. 4. Marine intercalation in the Tamala Limestone in an outcrop at Peppermint Grove on the Swan River.

beaches that in appearance and composition mirrored the make-up of the adjacent rock strata. Péron, who appears to have applied uniformitarian principles before the concept became fashionable, observed that the debris he found on the beach was made up of equal amounts of lime and sand, similar in proportion to that of many of the adjacent sedimentary layers, and that, consequently, the hardened and stratified rocks must have formed by the solidification of former beach deposits. He suggested that during calm conditions waves would deposit fine sand and calcareous debris on the beach, which would then mix to form a more or less regular layer. The longer this steady process continued, the greater would be the thickness of the bed formed. However, when rougher weather brought coarser sand to the beach, further mixing with the previously deposited, finer-grained material would cease and a new, separate layer would result (Péron 1801a, p. 9). Péron does not appear to have considered the influence of wind action in the movement of debris and its effect on the formation of the sedimentary layers.

To explain the process of cementation of detrital particles Péron drew on his observations of 'remarkable incrustations', which he had noted in many localities along the Australian coast and which he examined with attention in the deposits of the Tamala Limestone. His observations led him to the view that the cementing material that had produced these incrustations consisted of a mixture of very fine sand grains and finely divided calcareous matter, which, when mixed with water and left to harden, was not only capable of forming a concretionary crust around tree roots, twigs and even trunks, but also bound together pebbles, shells and other organic matter. In the process it transformed loose debris into a solid rock (Péron & Freycinet 1816, vol. 2, p. 169).

Péron likened the process of cementation in sandstone to that by which mortar is produced by the mixing of appropriate proportions of lime and quartz sand (Péron 1801a, p. 7). The finely crushed shelly material, he argued, helped by the heat of a blazing sun and pervasive humidity, would undergo a kind of chemical decomposition to produce lime, which, helped by constant moistening with sea water, would result in the bonding of the molecules of lime with those of sand. Péron further suggested that the proportion of the principal constituents, and the time elapsed since the start of the bonding process, would determine the hardness of the beds. Cuvier would later comment approvingly on Péron's views on the processes of cementation (1825, vol. 1, pp. 16–17).

As he was convinced that the sea had steadily retreated from the land in recent times, Péron concluded that the highest layers of a stratified sequence must be the hardest and, therefore, the oldest, while new beds were currently forming along a receding shoreline. His observations at Bernier Island seemed to confirm this when he found hard rocks, most probably calcrete horizons, at elevated positions in the island's interior (Péron 1801a, pp. 10–11; Péron & Freycinet 1816, vol. 2, pp. 166–167). It is interesting to note here that Lamarck (1802, 1964, p. 42) also advocated the deposition of sediment along some coasts by 'a general movement [of] marine water [which] brings land-derived sediments towards the shores', from which the sea slowly retreats. The piled up sediment will then be covered by plants and become consolidated. However, to compensate for the growth of a continent on one side of the ocean, Lamarck, in contrast to Péron's more limited vision, suggested that sea waters would erode the land on its opposite shores, thus maintaining a balance between areas covered by land and sea.

There was general agreement among Péron's colleagues, Depuch and Bailly, that the sand grains were held together by calcareous cement, with the implication that the solution of shell debris supplied the bonding substance. While Quoy was able to confirm the findings of his countrymen, he made the important point that cementation occurred both in strata containing abundant fossil shells as well as in pure quartz sandstones (Freycinet 1827, vol. 1, Part 2, p. 472). Fitton (1826, p. 22), with the benefit of later studies from different parts of the world to draw on, gave examples in support of Péron's view that linked the presence of shell debris in sediments to the production of cement to provide the means for lithification.

Calcareous concretions

The naturalists all noted that while there was some regularity in the general stratigraphic arrangement of the sedimentary layers, they were not homogeneous in their composition. Examining the strata at Bernier Island, Depuch was particularly intrigued by the discovery of an aggregate of calcareous nodules (Fig. 5) so tightly bound in a sandy and ochreous earth that they could not be extracted without breaking them.

He further noted that these pebbles had undergone a particular modification that manifested itself in each by the presence of banded and concentric layers (Depuch 1801a, p. 2). Péron, with his typical attention to detail, described these pebbles as having a globular form, made up of a large number of concentric zones, which are formed around a central core of sparkling, brownish

Fig. 5. Calcareous nodules from calcarenite beds of the Tamala Limestone cropping out at Shark Bay.

sandstone. 'These differing bands are barely a few millimetres thick and show pleasing variations in colour, from a dark-red to a clear yellow' (Péron & Freycinet 1816, vol. 1, p. 111).

Following the teachings of Dolomieu (Cordier 1796, pp. 29–30), who did not use the term conglomerate in his classification of sedimentary rocks, Depuch (1801a, p. 4) referred to these deposits as breccias and speculated that its pebbles might have formed prior to becoming embedded in their sandy matrix. When he discovered isolated nodules in the sandy soil, probable erosion remnants from the underlying strata, he wrote:

I have noticed debris of calcareous stones on some of the sandy plains [on Bernier Island], smoothed because of their natural weakness rather than by any friction they have been subjected to: the nature of these fragments, the manner in which they are arranged and associated with the sand, could make one think that one day they will form into a breccia of the kind I have been talking of.

(Freycinet 1827, vol. 1, Part 2, p. 475)

Similar concretionary bodies, varying from the size of a cherry stone to that of a fist, were found by Quoy in the strata of neighbouring Dirk Hartog Island. He noted that these deposits, containing such rounded pebble-like objects, had been taken for conglomerates. However, he maintained that 'those who believe that these globules have been transported deceive themselves; I attribute their formation to a kind of crystallisation'.

Fitton, who referred to these calcareous nodules as 'one of the most remarkable productions of New Holland', also favoured an *in situ* formation for the concretions. He believed, however, that their distinction from conglomerates of mechanical origin could, in many cases, prove difficult. To explain their formation, he drew an analogy with the processes operating in the manufacture of earthenware where 'a sort of chemical reaction produces, under certain circumstances, a new arrangement of the parts', a process that could 'probably be extended to those nodular concretions' and which allowed them to be 'considered as contemporaneous with the paste in which they are enveloped' (Fitton 1826, p. 33). Fitton (1826, pp. 30–31) disagreed with Péron's views that heat and humidity were necessary for the solution of calcareous matter to produce cement. Citing an example from Cornwall of calcareous sand becoming agglutinated into a stone, he suggested that the process could also operate in temperate climates.

'Tangled masses of stony branches'

The presence of root structures, or rhizoliths, and other incrusted parts of trees and shrubs found in several horizons of the Tamala Limestone both fascinated early naturalists by their strange appearance and challenged their powers of deduction in attempting to explain their origin. In the study of these natural phenomena the first geological explorers of the continent's western coast were, however, forestalled by even earlier visitors who had discovered examples of these curious structures in some of the southernmost outcrops of calcareous rocks.

In 1791, during a stay in King George Sound on the southern coast of Western Australia (Fig. 6), George Vancouver, in the company of his naturalist, Archibald Menzies (1754–1842), ascended the granite slopes of Bald Hill on Flinders Peninsula.

On its summit they were confronted by an expanse of white sand from which protruded a 'tangled mass of stony branches' (Fig. 7). The appearance of these tubular, cemented sand objects, often with empty, but sometimes sand-filled, central cavities, led these first investigators to explain their origin by comparing them to known natural occurrences. They saw in the arrangement and size of these features a close similarity to the branched corals they had observed growing upwards from the bed of the sea and proposed a like origin for these remains. Their faith in this interpretation was strengthened by the finding of 'sea shells in great abundance, some nearly in a perfect state still adhering to the coral'. 'These fields of coral', as Vancouver

referred to them in the account of his voyage (Vancouver 1798, vol. 1, pp. 48–50), provided a major attraction for naturalists of later expeditions, who felt compelled to visit the site (McNamara & Dodds 1986).

The year after Vancouver's discovery, the French vessels *Recherche* and *Espérance*, commanded by Bruny d'Entrecasteaux (1737–1793), failed in their attempt to enter King George Sound and sailed on to Esperance Bay, some 500 km further east. There, Claude Riche (1762–1797), the mission's zoologist and stand-in mineralogist, discovered petrified tree trunks in calcareous sand very similar to that at Bald Head (Duyker & Duyker 2001, p. 119). Péron later made use of his countryman's description of this discovery to lend support to his own ideas of the origin of incrustations of vegetable matter (Péron & Freycinet 1816, vol. 2, p. 169).

The expedition led by Matthew Flinders (1774–1814) in the *Investigator* visited King George Sound and Vancouver's 'fields of coral' in 1801. The ship's botanist, Robert Brown (1773–1858), wrote in his journal: 'the top and even sides of the

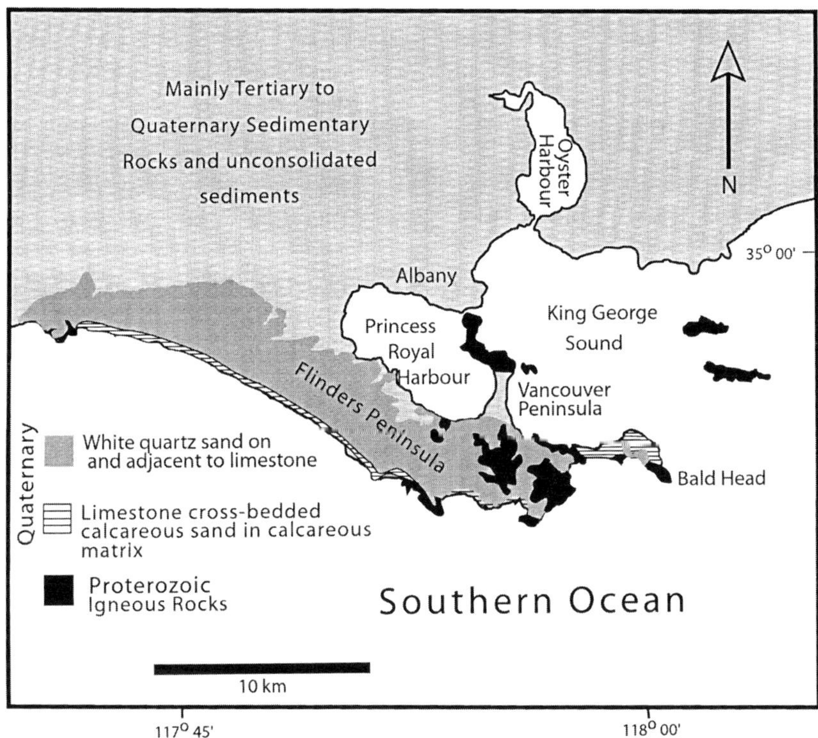

Fig. 6. Simplified geological map of the area surrounding King George Sound. Based on Muhling & Brakel (1985).

Fig. 7. Part of Vancouver's 'fields of coral': rhizoliths protruding through calcareous sands on the summit of Bald Head, King George Sound.

hill [are] formed principally of coral of some of which the ramifications were just visible'. He further noted that 'some of these coral rocks were hollowed out' displaying caverns, and that they consisted of calcareous sand in which univalve shells were embedded (Vallance *et al.* 2001, p. 92).

However, Brown was soon to have doubts about the origin of these branching forms. As Vallance *et al.* (2001, p. 92) have pointed out, the botanist, in an entry in his diary, questioned their coralline nature as 'all traces of their original structure have been obliterated' and suggested instead that they may represent petrified wood. Flinders appears not to have been privy to his botanist's change of mind. When referring to Vancouver's claim to have found corals high above sea level, he wrote in the account of his voyage that: '[t]his curious fact I was desirous to verify; and his [Vancouver's] description was proved to be correct' (Flinders 1814, vol. 1, p. 49). However, he went on to say that he also found 'two broken columns of stone three or four feet high, formed like stumps of trees ... but whether they were of coral, or wood now petrified ... I can not determine'. The belief therefore persisted among the scientific community that the stony branches emerging from the sand,

originated as coral, elevated more than 200 m above sea level.

Péron, who visited King George Sound during the Baudin expedition's return voyage in 1803, lamented the fact that, while spending 12 days there, none of the naturalists was able to go to the summit of Bald Head, the natural features of which he considered 'so invaluable to know' (Péron & Freycinet 1816, vol. 2, p. 176). Had he, perhaps in company with Bailly, succeeded in visiting this hill, he would almost certainly have recognized the branching structures as calcareous concretionary matter formed around tree roots and branches. Bailly, who had transferred to the *Géographe* to replace the mortally ill Depuch, now homeward bound on the *Naturaliste*, had 2 years earlier observed incrusted trees and roots, with their woody tissue still preserved, in the calcareous deposits cropping out along the Swan River (Péron & Freycinet 1816, vol 1, pp. 178–184; Bailly 1801, pp. 97–99). Péron himself gave a detailed explanation of the process leading to the incrustation of vegetable matter. A finely divided, wind-blown mixture of quartzose and calcareous material covers the lower twigs and branches of shrubs. As the calcareous envelope thickens, the woody tissue breaks down, leaving an empty tube equal in diameter to that of the branch that once occupied it. Finally, the cavity becomes filled with calcareous sand, which after some years is formed into sandstone. Only its plant-like form serves as a reminder of its former state (Péron & Freycinet 1816, vol. 2, p. 171).

Clearly familiar with Péron's work, Quoy gave a similar explanation for the formation of tubular structures of sand grains held together by a calcareous matrix, which he found in dunes at Shark Bay. He interpreted them as the 'moulds of twigs and roots' that had decayed to leave an empty space in the centre of now irregularly shaped cylinders (Freycinet 1827, vol. 1, Part 2, p. 472). In another part of the bay, on a small plain surrounded by sand-dunes, his colleague, Duperrey, discovered a large number of petrified tree trunks up to 1 m in height and still in their position of growth. He had no doubt that the woody tissue of these trunks, as well as that of their roots, had been replaced and transformed into a kind of sandstone (Freycinet 1827, vol. 1, Part 2, p. 473).

The failure of Péron to visit Bald Head, and his verbatim incorporation of Vancouver's account into his own work (Péron & Freycinet 1816, vol. 2, pp. 175–176), served to maintain the English navigator's explanation of these branched structures as corals elevated from the sea bed.

In 1822 members of the King expedition also made the now obligatory trek to the summit of Bald Head to examine the 'calcareous substance ... supposed to be coral'. An examination of the branching forms did not convince the knowledgeable King of the correctness of this supposition. He was aware of the work of the English botanist Clarke Abel (1780–1826) at the Cape of Good Hope (Abel 1818), who had discovered similar substances described by him as 'vegetables impregnated with carbonate of lime'. However, King reached the conclusion that occurrences at Bald Head represented 'neither coral, nor a petrified vegetable substance, but merely sand agglutinated by calcareous matter' (King 1826, vol. 1, pp. 12–13).

When examining samples of these enigmatical calcareous forms, Fitton was unable to reach a firm decision regarding their origin. He accepted that these 'irregular stem-like bodies with a rugged sandy surface, composed of sand and cemented by carbonate of lime, have some resemblance to the trunks or roots of trees'. However, in his view, the specimens did 'not really exhibit any traces of organic structure' and did 'so nearly resemble the irregular stalactitical concretions produced by the passage of calcareous or ferruginous solutions through sand that they are probably of the same origin'. 'But', he added, 'there is no reason to suppose that the trunks of trees, as well

as other foreign substances, may not be thus incrusted' (Fitton 1826, pp. 56–57). Elsewhere he concurred with the views of King that 'the ramified calcareous concretions, erroneously considered as corals by Vancouver and others' appear 'to be nothing more than a variety of recent limestone' (King 1826, vol. 2, p. 581).

At a meeting of the Geological Society in 1831, Archdeacon T. H. Scott (1783–1860) presented a short paper in which he related his geological observations along the Swan River during a visit to Western Australia (Scott 1834, p. 320). Clearly unaware of Bailly's earlier discovery of petrified roots and trees at the same localities, Scott described 'numerous concretions having the appearance of inclosing vegetable matter' and a 'sandstone that assume[d] the appearance of a thick forest, cut down about two or three feet from the surface'. Scott, in his talk, appears to have been somewhat ambiguous about the origin of these phenomena, as Charles Lyell (1797–1875), who had attended the meeting, felt prompted to express his own interpretation of the Reverend's musings in a letter, written in a form of shorthand, to Gideon Mantell (1790–1852) (Thackray 2003, p. 44):

[N]ear the shore blown calcareous sand ... including petrified trees or as some will have it stalactites – I rather think the sand & carb. of lime does form round trees & their branches, that there is a solution of carb. of lime by rain water acting on fine dust from shells comminuted & which containing animal matter give out carb. acid gas on putrefying.

Lyell's views closely paralleled those expressed earlier by Péron, and supported Bailly's identification of these concretionary structures as petrified remains of trees, shrubs and roots.

The uncertainty about the origin of these problematic petrifactions was finally resolved after a visit to Bald Head by Charles Darwin (1809–1882) in 1836. He recorded the following in his diary (Keynes 2003, pp. 412–413):

On the day I accompanied Captain Fitzroy to Bald Head [8 March, 1836] the place mentioned by so many navigators, where some imagined that they saw corals, and others that they saw petrified trees, standing in the position in which they had grown. According to our view, the beds have been formed by the wind having heaped up fine sand, composed of minute rounded particles of shells and corals, during which process branches and roots of trees, together with many land-shells [Bothriembryon melo, W. G. Kendrick, pers. commun. 2006] became enclosed. The whole then became consolidated by the percolation of calcareous matter; and the cylindrical cavities left by the decaying of the wood, were thus also filled up with a hard pseudo-stalactitical stone. The weather is now wearing away the softer parts, and in consequence the hard casts of the roots and branches of the trees project above the surface, and, in a singularly deceptive manner, resemble the stumps of a dead thicket.

Darwin seems to have been the first to have attributed the formation of the Tamala Limestone eolianites to wind action. He noted elsewhere that Fitton was 'inclined to attribute a concretionary origin to the branching bodies', but he countered this interpretation with his own observations of cylindrical stems in beds of sand in La Plata, 'which no doubt thus originated; but they differed much in appearance from those at Bald Head'. He also supported Péron's views 'with whose observations and opinions on the origin of the calcareous matter and branching casts, mine entirely accord' (Darwin 1842, pp. 147–148).

A current explanation of rhizolith formation in beds of the Tamala Limestone that crop out on Rottnest Island is that of Playford (1988, pp. 22–25), which is largely in accord with the views expressed by Péron and Darwin. In Playford's account, these structures:

were initiated by lime precipitation around roots of trees and shrubs that grew in the original dunes, the cavities left after decay of the roots being filled by clastic limestone and cement, although in some cases the wood of the root is replaced by calcium carbonate. This process can be seen to occur around living roots below modern dunes, where these have been exposed in blow-outs.

He adds that 'the rhizolith horizons commonly include conspicuous solution pipes – cylindrical bodies up to 0.5 m in diameter and originally several meters long'. They are believed to have formed around the tap roots of large trees (probably mainly eucalypts).

Sea-level changes

Péron wrote in his official history of the voyage to Australia (Péron & Freycinet 1816, vol. 2, pp. 165–166):

One of the greatest achievements of modern geological research and also one of its most indisputable, is the certain knowledge that, in the past, the level of the sea was higher than at the present time. At almost all places in the old and the new world is the proof of this phenomenon as numerous as it is evident. Only in *les Terres australes* was this still to be ascertained as, by virtue of its immense areal extent, it could have proved to be an important exception to the universality of the former domination of the ocean over the land.

Following his erroneous identification of corals at the top of Bald Hill, Vancouver (1798, vol. 1, p. 48) was the first to suggest that the land at this locality had been elevated above the ocean. From the good state of preservation of many of the 'corals' he concluded that the uplift was 'of moderate date' and added that 'I have seen coral in many places at a considerable distance from the sea; but in no other instance have I seen it [the land] so elevated'.

The Baudin expedition's two geologists, Depuch and Bailly, and in particular Péron, carefully observed the continent's coastal geology for evidence of changes in sea level. In his account of the voyage, Péron was able to state that at several localities, including in the calcarenites that now make up the Tamala Limestone, 'I . . . found invaluable remains, which are the indisputable witnesses of the revolutions of nature' (Péron & Freycinet 1816, v. 2, p. 164). He was referring specifically to the presence of fossil shells and other organic debris which he found in rocks and sediments above the present level of the sea.

Following their first landing on the shores of Geographe Bay, Depuch and Péron separately traversed a vast, almost flat, expanse of dunes. They noticed that some of the sands were partly cemented to form beds and contained marine organisms, particularly oyster shells (Depuch 1801*b*, pp. 3–4). In his account of this region Péron had no doubt that the land was formed by the retreat of the sea (Péron 1801*b*, p. 7). The young zoologist in his later description of older rock sequences on Maria Island in Tasmania demonstrated his adherence to the Neptunist views promoted by Werner (1786). Péron explained the formation of these rocks by a gradual diminishing of the oceans' waters, with exposure of the Primitive granite mountains, followed by the deposition in a shallower sea of the stratified and fossiliferous sedimentary Secondary rocks (Péron 1802, pp. 17–19; Plomley *et al.* 1990, pp. 19–20). Péron equally advocated a slow retreat of the sea, leading to the gradual emergence of the calcarenites along the western coast of Australia. His discovery of fossilized shells in elevated strata at localities from Tasmania in the south to Timor in the north, led him to conclude that the sea had withdrawn from the land over a distance of 2000 *milles* (3898 km) in that part of the world (Péron & Freycinet 1816, vol. 2, p. 182). While he sometimes wrote of upheavals and revolutions of nature, echoing the catastrophist ideas of Cuvier, his former teacher, he seems to suggest that, with respect to Australian coastlines, a lowering of sea level was the main cause of the emergence of land (Péron & Freycinet 1816, vol. 2, p. 166; Plomley *et al.* 1990, p. 18). He did not put forward specific reasons for the withdrawal of the oceans from the land.

On the evidence of horizontally bedded fossil-bearing rocks along the banks of the Swan River, Bailly agreed with the views of Péron. He found that the shells and concreted woody matter they contained had been little degraded, and concluded that all the land had, not long ago, been covered by the sea. He implied agreement with his colleague's views that the sea had retreated to expose these beds (Bailly 1801, p. 98; Péron & Freycinet,

1816, vol. 1, p. 179). Lamarck (1805, pp. 39–40), who summarized Péron's findings, agreed with him that the sea once covered elevated parts of the land. He proposed, however, that the cause of sea-level change was the slow migration of the Earth's equatorial bulge, which resulted in the progressive emergence of land that was formerly covered by the sea, and in the steady displacement of the ocean basins (Lamarck 1802, 1964).

The thoughts of Fitton were more concerned with the causes that resulted in land emerging above the level of the sea. He wrote that 'as *marine* [*sic*] shells are found in the cemented masses, at heights above the sea, to which no ordinary natural operation could have conveyed them, the elevation of these shells to their actual place, (if not that of the rock in which they are agglutinated,) must be referred to some other agency' (Fitton 1826, p. 27).

After a review of the literature, Fitton came to favour the 'heaving up of the land by a force from beneath', the most likely force being provided by earthquakes. He based this view on information provided from several sources after the Chilean Earthquake of 1820, which raised an extensive stretch of the coast by several feet and where there was evidence of beaches raised in earlier times to a height of 30 feet (10 m). Such evidence, he argued, 'concur[red] to demonstrate the existence of most powerful expansive forces within the Earth, and to testify their agency in producing the actual condition of its surface' (Fitton 1826, pp. 28–30).

'Living fossils'

An aspect of considerable interest to the naturalists of the Baudin expedition was the finding of living organisms that were known to them only in the fossil state in Europe. Péron's discovery of the shells of a living species of Trigonia on one of Australia's southern beaches led Lamarck (1804, pp. 352–353) to speculate that small marine organisms in particular may not really have become extinct but, instead, have undergone mutations resulting in some differences between the fossil and the living species. Both Péron, and later Duperrey, believed that some mollusc shells on beaches in Shark Bay were similar to those found in the fossilized state in adjacent beds (Péron & Freycinet 1816, vol. 1, p. 110; Freycinet 1827, vol. 1, Part. 2, p. 473). From an outcrop of Tamala Limestone in Shark Bay, Péron reported (Péron & Freycinet 1816, vol. 1, pp. 110–111):

The embedded shells in this mass of rocks are almost all univalve; they belong chiefly to the genus Natice of M. Lamarck, and have the closest relationship with the species Natice which is found living at the foot of these rocks. They have, without doubt, been petrified for many centuries, for, besides the difficulty of removing them from their matrix, owing to their close adhesion to it, one observes them again at more than 150 *pieds* [about 50 m] above the present sea level.

While the appearance of the two types of shells found by Péron was indeed similar, the fossil mollusc that Péron examined was probably the land snail *Bothriembryon onslowi* (Cox) (Kendrick pers. comm. 2006), while the living form was *Polinices (Conuber) conicus* (Lamarck) (McNamara & Dodds 1986, p. 27).

The German geologist Leopold von Buch (1774–1853) examined the Baudin expedition's rock collection, which now appears to be lost, and published his views on the significance of their discoveries. Referring to specimens collected by Bailly (1804, p. 7) at Shark Bay, he also alluded to the presence there of living molluscs that are extinct in Europe. This caused him to reflect on the possibility that, contrary to the views expressed by Werner, his former teacher, some sedimentation may have been restricted to local areas, rather than being uniform across the globe (Buch 1814, p. 235).

Conclusion

The extensive outcrops of what are now known to be Quaternary deposits at the margins of the Western Australia coast have provided successive parties of early European naturalists with a rich source of geological material for study. The examination of the composition and characteristics of its strata enabled them to test their knowledge, their ideas and theories – the summation of long-established scholarship in the old world – against some of the natural features in parts of a little explored continent. While they were able to apply their learning to explain many of the geological phenomena they observed in this distant land, they also made discoveries of a novel nature for which their European experience and training had not fully prepared them. The accounts of their observations and their attempts at a rational explanation of their discoveries in Australia created much interest and stimulated debate among scholars in Europe, and added to the sum of scientific knowledge.

After the establishment of Western Australia as a British Crown Colony in 1829, the geological exploration of this large area of the continent became mainly the task of colonial officials and interested members of the resident population. As many of the records of the first geological investigators were not readily available to this new breed of land-based explorers, the latter were often denied the benefit of familiarity with the earlier work of their predecessors. Scholars in more recent times have begun to re-examine the work of these first scientific observers and have recognized the historical value of these early contributions to the geological sciences.

The author is grateful to G. Baglione of the Muséum d'histoire naturelle at Le Havre for her help in providing copies of manuscripts relating to this study. I am greatly indebted to R. Barwick of the Department of Earth and Marine Sciences at the Australian National University, who prepared the two maps for this article. G. Kendrick of the Western Australian Museum generously gave of his time to introduce the author to species of molluscs referred to in the accounts of the early naturalists. P. Playford of the Geological Survey of Western Australia, K. McNamara of the Western Australian Museum and J. Chappell of the Research School of Earth Sciences at the Australian National University provided much helpful information. I also thank my referees for their constructive comments.

References

ABEL, C. 1818. *Narrative of a Journey in the Interior of China, and of a Voyage to and from that Country in the Years 1816 and 1817*. Longman, Hurst, Rees, Orme & Brown, London.

AGASSIZ, L. J. R. 1840. *Etudes sur les glaciers*. Jent & Gassman, Neuchâtel.

ARDUINO, G. 1760. Due lettere del Sig. Giovanni Arduino sopra varie sue osservazioni naturali. Al chiaris. Sig. Antonio Vallisnieri. *Nuovo raccolta d'opusculi scientfici e filologici*, **6**, 99–180.

BAILLY, J. C. 1801. *Observations minéralogiques et géologiques faites dans la Rivière des Cygnes*. MS, Service Hydrographique de la Marine, **1**, Carton 12, 97–99.

BAILLY, J. C. 1804. *Catalogue des substances minérales recueillies pendant le voyage de découvertes commandé par le Capitaine Baudin*. MS, Service Hydrographique de la Marine, Carton 22/23, 4–81.

BONNEMAINS, J., ARGENTIN, J.-M. & MARIN, M. 2001. *Mon voyage aux Terres Australes: Journal personnel du commandant Baudin*. Imprimerie Nationale, Paris.

BRANAGAN, D. 2002. Australian stratigraphy and palaeontology: the nineteenth century French contribution. *Comptes Rendus Palevol*, **1**, 657–662.

BUCH, L. VON. 1814. Einige Bemerkungen über die geognostische Constitution von Van Diemens Land. *Magazin für die Neuesten Entdeckungen in der Gesammten Naturkunde*, **6**, 234–240.

CHAPPELL, J. & SHACKLETON, N. J. 1986. Oxygen isotopes and sea level. *Nature*, **324**, 137–140.

CHURCHILL, D. M. 1959. Late Quaternary eustatic changes in the Swan River district. *Journal of the Royal Society of Western Australia*, **42**, 53–55.

CORDIER, L. 1796. *Extrait des leçons orales faites par Dolomieu sur les gisements de minéraux, au commencement de 1796, à l'École des mines de Paris. Rédigé par moi, alors que j'étais élève à l'École des Mines et obligé de justifier mon travail à chaque Professeur. Archives de l'Académie des Sciences, Paris,*

4J 16. Transcription at whttp:/www.musee.ensmp.fr/dolomieu:/www.musee.ensmp.fr/dolomieu.

CUVIER, G. 1825. *Recherches sur les ossemens fossiles*, 3rd edn, Voume 1. Dufour & d'Ocagne, Paris.

DARWIN, C. 1842. *The Structure and Distribution of Coral Reefs*. Smith, Elder, London.

DAUBENY, C. 1825. Sketch of the geology of Sicily. *Edinburgh Philosophical Journal*, **13**, 107–118.

DEPUCH, L. 1801a. *Observations géologiques sur celle des îles Stériles où nous sommes descendus le 9 messidor an IX* [=28 June 1801]. MS No. 08 033, Muséum d'histoire naturelle, Le Havre. Transcription by Jacqueline Bonnemains.

DEPUCH, L. 1801b. *Rapport fait au Commandant sur l'état géologique de la baie du Géographe, le 26 prairial an IX* [=15 June 1801]. MS No. 08 043, Muséum d'histoire naturelle, Le Havre. Transcription by Jacqueline Bonnemains.

DESNOYERS, J. 1829. Observations sur un ensemble des dépôts marins plus récents que les terrains tertiaires du bassin de la Seine, et constituent une formation géologique distincte: précédées d'une aperçu de la non-simultanéité des bassins tertiaires. *Annales des sciences naturelles*, **16**, 171–214, 402–491.

DUYKER, E. & DUYKER, M. (eds & translators). 2001. *Bruny d'Entrecasteaux: Voyage to Australia and the Pacific 1791–1793*. Melbourne University Press, Melbourne.

FAIRBRIDGE, R. W. 1958. Dating the latest movements of the Quaternary sea level. *Transactions of the New York Academy of Science*, Series 2, **20**, 471–482.

FAIRBRIDGE, R. W. 1961. Eustatic changes in sea level. *In*: *Physics and Chemistry of the Earth*, volume 4. Pergamon, London, 99–185.

FITTON, W. H. 1826. *An Account of some Geological Specimens from the Coasts of Australia*. W. Clowes, London.

FLINDERS, M. 1814. *A voyage to Terra Australis; undertaken for the purpose of completing the discovery of that vast country, and prosecuted in the years 1801, 1802 and 1803, in His Majesty's ship the* Investigator, *and subsequently in the armed vessel* Porpoise *and* Cumberland *schooner. With an account of the shipwreck of the* Porpoise, *arrival of the* Cumberland *at Mauritius, and imprisonment of the Commander during six years and a half in that island*. Volume 1, G. & W. Nicol, London.

FREYCINET, L. 1827. *Voyage autour du monde, entrepris par Ordre du Roi, exécuté sur les corvettes de S. M. l'Uranie et la Physicienne pendant les années 1817, 1818, 1819 et 1820*. Pillet Aîné, Paris.

HEARTY, P. J. 2002. Stratigraphy and timing of eolianite deposition on Rottnest Island, Western Australia. *Quaternary Research*, **60**, 211–222.

KENDRICK, W. G., WYRWOLL, K. H. & SZABO, B. J. 1991. Pliocene–Pleistocene coastal events and history along the western margin of Australia. *Quaternary Science Reviews*, **10**, 419–439.

KEYNES, R. D. 2003. *Charles Darwin's Beagle Diary*. Cambridge University Press, Cambridge.

KING, P. P. 1826. *Narrative of a Survey of the Intertropical and Western Coasts of Australia*. John Murray, London.

LAMARCK, J.-B. 1802. *Hydrogéologie eu recherches sur l'influence qu'ont les eaux sur la surface du globe terrestre; sur les causes de l'existence du bassin des mers, de son déplacement et de son transport successif sur les différens points de la surface de ce globe; enfin sur les changemens que les corps vivans exercent sur la nature et l'état de cette sufface.* Agasse & Maillard, Paris.

LAMARCK, J.-B. 1804. Une nouvelle espèce de Trigonie, et sur une nouvelle d'huitre, découvertes dans le voyage du Capitaine Baudin. *Annales du Muséum d'histoire naturelle*, **4**, 351–359.

LAMARCK, J.-B. 1805. Considérations sur quelques faits applicable à la théorie du globe, observés par M. Péron dans son voyage aux Terres australes, et sur quelques questions géologiques qui naissent de la connoissance de ces faits. *Annales du Muséum d'histoire naturelle*, **6**, 26–52.

LAMARCK, J.-B. 1964. *Hydrogeology* translated by A. V. Carozzi. University of Illinois Press, Urbana, IL.

LOGAN, B. W., READ, J. F. & DAVIES, G. R. 1970. History of carbonate sedimentation, Quaternary Epoch, Shark Bay, Western Australia. *AAPG, Memoir*, **13**, 38–84.

MCNAMARA, K. J. & DODDS, F. S. 1986. The early history of palaeontology in Western Australia: 1791–1899. *Earth Sciences History*, **5**, 24–38.

MAYER, W. 2005. Deux géologues français en Nouvelle-Hollande (Australie): Louis Depuch et Charles Bailly, membres de l'expédition Baudin (1801–1803). *Travaux du Comité français d'histoire de la géologie*, Series 3, **19**, 95–112.

MUHLING, P. C. & BRAKEL, A. T. 1985. Mount Barker – Albany, Western Australia. *Geological Survey of Western Australia 1:250 000 Geological Map and Explanatory Notes.* Sheets SI/50-11 and 50-15.

MURRAY-WALLACE, C. V. & KIMBER, R. W. L. 1989. Quaternary marine animostratigraphy – Perth Basin, Western Australia. *Australian Journal of Earth Sciences*, **36**, 553–568.

MYERS, J. S. & HOCKING, R. M. 1988. Geological map of Western Australia. *Geological Survey of Western Australia 1:2 500 000 Geological Map.*

PÉRON, F. 1801*a*. *Topographie générale de la petite île de Dorre.* MS No. 08 033, Muséum d'histoire naturelle, Le Havre. Transcription by Jacqueline Bonnemains. (Note that in his title Péron referred to Dorre Island instead of to neighbouring Bernier Island where he carried out his work.)

PÉRON, F. 1801*b*. *Topographie générale de la Baie du Géographe sur la côte Ouest.* MS No. 08 040, Muséum d'histoire naturelle, Le Havre.

PÉRON, F. 1802. *Topographie générale de l'île Maria sur la côte Orientale de la Terre de Diémen.* MS No. 18 043, Muséum d'histoire naturelle, Le Havre.

PÉRON, F. 1804. Sur le nouveau genre Pyrosoma. *Annales du Muséum d'histoire naturelle*, **4**, 437–446.

PÉRON, F. 1807. *Voyage de découvertes aux Terres Australes, exécuté par ordre de Sa Majesté l'Empereur et Roi, sur les corvettes le* Géographe *et le* Naturaliste *et la goëlette le* Casuarina *pendant les années 1800, 1801, 1802, 1803 et 1804.* Imprimerie Imperiale, Paris.

PÉRON, F. & FREYCINET, L. 1816. *Voyage de découvertes aux Terres Australes, exécuté par ordre de Sa Majesté l'Empereur et Roi, sur les corvettes le* Géographe *et le* Naturaliste *et la goëlette le* Casuarina *pendant les années 1800, 1801, 1802, 1803 et 1804.* Imprimerie Imperiale, Paris.

PLAYFORD, P. E. 1988. *Guidebook to the Geology of Rottnest Island. Excursion Guidebook No. 2.* Geological Society of Australia, Western Australian Division and the Geological Survey of Western Australia, Perth.

PLAYFORD, P. E. 1997. Geology and hydrology of Rottnest Island, Western Australia. *In*: VACHER, H. I. & QUINN, T. (eds) *Geology and Hydrology of Carbonate Islands.* Developments in Sedimentology series, **54**, 783–810.

PLAYFORD, P. E. 2006. The Tamala Limestone: its characteristics, origins and dangers. *Proceedings of the Royal Society of Western Australia*, **89**, 2–3.

PLAYFORD, P. E., COCKBAIN, A. E. & LOW, G. H. 1976. Geology of the Perth Basin, Western Australia. *Western Australian Geological Survey Bulletin*, **124**, 210.

PLOMLEY, B., CORNELL, C. & BANKS, M. 1990. *François Péron's natural history of Maria Island, Tasmania.* Records of the Queen Victoria Museum, **99**.

SCOTT, T. H. 1834. Geological remarks on the vicinity of Swan River and Isle Buâche [*sic*] or Garden Island, on the coast of Western Australia. *Proceedings of the Geological Society of London*, **1**, 320.

SHACKLETON, N. J. 1987. Oxygen isotopes, ice volume and sea level. *Quaternary Science Reviews*, **6**, 183–190.

SUESS, E. 1888. *Das Antlitz der Erde, II: Die Meere der Erde.* F. Tempsky, Vienna.

TEICHERT, C. 1950. Late Quaternary changes of sea level at Rottnest Island, Western Australia. *Royal Society of Victoria Proceedings*, **59**, 63–79.

THACKRAY, J. C. 2003. *To See the Fellows Fight: Eye Witness Accounts of Meetings of the Geological Society of London and its Club, 1822–1868.* British Society for the History of Science, Oxford, Monograph, **12**.

VALLANCE, T. G. 1975. Origins of Australian geology. *Proceedings of the Linnean Society of New South Wales*, **100**, 13–43.

VALLANCE, T. G. 1981–1982. Lamarck, Cuvier and Australian geology. *Histoire et Nature*, **19–20**, 133–136.

VALLANCE, T. G., MOORE, D. T. & GROVES, E. W. 2001. *Nature's Investigator: The Diary of Robert Brown in Australia, 1801–1805.* Australian Biological Resources Study, Canberra.

VANCOUVER, G. 1798. *A Voyage of Discovery to the North Pacific Ocean and Around the World*, Volume 1. G. G. & J. Robinson, London.

WERNER, A. G. 1786. Kurze Klassifikation und Beschreibung der verschiedenen Gebirgsarten. *In*: *Abhandlungen der Böhmischen Gesellschaft der Wissenschaften, auf das Jahr 1786.* Walterische Hofbuchhandlung, Prague, 272–297.

Sir Charles Cotton (1885–1970): international geomorphologist

RODNEY H. GRAPES

*Department of Earth and Environmental Sciences, Korea University,
Seoul, South Korea
(e-mail: grapes@korea.ac.kr)*

Abstract: Sir Charles Cotton (1885–1970), a New Zealander by birth, was Professor of Geology at Victoria University of Wellington, New Zealand, between 1921 and 1953. He produced a quartet of well-known textbooks, the most influential being *Geomorphology of New Zealand* first published in 1921, and a remarkable number of pioneering papers on a great variety of subjects in geomorphology. Essentially self-taught, much of Cotton's earlier work followed the ideas of W. M. Davis in terms of an explanatory description of landforms (structure, process, form), but he also emphasized the importance of climate change and tectonic movements in landscape-forming processes. His work was enhanced by the use of block diagrams to demonstrate progressive evolution of landscape features and his simple sketches, in particular, provided a clarity allowing people to see the land around them with new interest and understanding. His studies were never quantitative, and he remained sceptical about illusions of precision and accuracy in the new post-World War II trends and ideas in geomorphology. Cotton's range of interests was wide, but certain themes keep returning; in particular, shore processes and shoreline development and classification, the significance of faulting in all its forms, and the geomorphic history of the area of his Wellington home in New Zealand. Although Cotton's work became well known throughout the scientific world, he did not create a school of geomorphological thought. His international reputation came from his scientific papers and especially his books that captured the interest of generations of university students and citizens from all walks of life. Indeed, the honour of a knighthood in 1959 could well have come 30 years earlier when, at the age of 41, he had already gained such recognition. He made an outstanding contribution to our understanding of the evolution of New Zealand's landforms. Cotton's bibliography is included.

A widespread appreciation of geomorphology and landscape development both in New Zealand and internationally can be linked to the publications of Sir Charles Cotton (Fig. 1) – a man who was by nature shy and retiring, who took little part in public life, but was an exceptional scholar who brought international recognition to the developing science of geomorphology through his promotion of the rich and varied landforms of a tectonically active New Zealand. Cotton holds a special position in New Zealand science as he is one of the few who have been knighted solely for their scientific achievements, the citation of which succinctly sums up his contribution: 'For outstanding service as a geologist in New Zealand, particularly as Professor of Geology at Victoria University of Wellington, and for research and publications of international repute'. Cotton's enthusiasm for the New Zealand landscape was expressed as early as 1918: 'The shores of New Zealand are a veritable treasure house of beauty, rivalling in their charm the glaciers of the southern island, the geysers of the northern, and the mountains, lakes, and gorges of both. What a story it is that these fascinating coasts reveal!' His life work was devoted to

unravelling this story through word and illustration, and he achieved this almost single-handed.

Formative years

Born in Dunedin in the southern part of the South Island of New Zealand (see Fig. 2) on 24 February 1885, Charles Andrew Cotton spent much of his childhood aboard his father's sailing ship on trading routes between New Zealand, Australia, Mauritius and the Gilbert Islands, where he was taught by his mother, a Scottish school teacher. High school years at Christchurch (for location see Fig. 2) brought a more settled life, but here he tragically lost the sight of an eye because of a schoolboy prank with the pen. In 1904, Cotton entered Otago University, Dunedin, and the School of Mines where his interest focused on the technology of mining. He graduated with a Bachelor of Science degree in 1907 and, 2 weeks before obtaining his Master of Science degree with first class honours in geology, he was appointed Director of the Coromandel School of Mines in the North Island of New

From: GRAPES, R. H., OLDROYD, D. & GRIGELIS, A. (eds) *History of Geomorphology and Quaternary Geology.*
Geological Society, London, Special Publications, **301**, 295–313.
DOI: 10.1144/SP301.18   0305-8719/08/$15.00 © The Geological Society of London 2008.

Fig. 1. Sir Charles Cotton. Photograph taken on the occasion of his retirement in 1953 as Professor of Geology, Victoria University of Wellington, New Zealand (No. VUW100032 Photographer's Proof Series, held in the J. C. Beaglehole Room. Victoria University of Wellington Library).

Zealand (see Fig. 2) on 22 February 1908. Here he became involved in the petrology and mining aspects of the Coromandel goldfields, where large-scale cyanide-processing of quartz from gold-bearing reefs was being carried out. However, a desire to return to the academic world of the university prompted him to successfully apply for a position at Victoria University College (later Victoria University) in Wellington (for location see Fig. 2) and, at the age of 24, Cotton took up the position of Lecturer in Geology and Examiner in Physical Geography in April 1909. The same year, his first paper, on the 'Geology of Signal Hill, Dunedin', was published in the *Transactions of the New Zealand Institute* (later to become the *Transactions of the Royal Society of New Zealand*) together with two landscape sketches that were to become typical of Cotton's later work in a *New Zealand Geological Survey Bulletin* (Park 1909), and made while he was a student assistant during the fieldwork.

The Geology Department at Victoria University was a one-man department and the geology course required practical work, usually in the form of short field excursions during the weekend. Although the bedrock geology around Wellington is predominantly and monotonously of one rock type – highly deformed Mesozoic greywacke – the hilly physiography surrounding the large harbour and the rugged coastline facing Cook Strait between the North and South islands (see Fig. 2) offered a great deal of variety, and it was probably this, more than anything else, that led Cotton into the study of geomorphology. Many years later Cotton was to remark, 'Until I came to Wellington in 1909, my knowledge of geomorphology was absolutely nil. I had, however, discovered W. M. Davis and from 1909 to 1914 I read every word I could get hold of that he had written. This refers not so much to the early "classic" papers, which are criticised *ad nauseum* – favourably and unfavourably – but to the articles that were then coming from Davis's pen; I learned my geomorphology from its written words, more than from personal contacts' (quoted in Collins 1966).

Cotton worked in virtual academic and geographic isolation, drawing inspiration from the well-known North American physiographers J. K. Powell and G. K. Gilbert, as well as W. M. Davis. Davis had pioneered an approach to geomorphology that involved the study of landforms based on fieldwork, mapping and photography, leading to a systematic description and interpretation of the landscape. The approach typically involved explanations of the progressive development of landforms illustrated by sequential three-dimensional (block) diagrams proceeding from an initial stage, through youth, maturity to old age. With his geological background, Cotton was able to show that geomorphology could provide valuable evidence not only of the latest (Quaternary) geological history of an area, but also the importance of the composition and structure of the underlying rocks. His first geomorphological paper dealt with aspects of Wellington physiography, particularly the well-preserved coastal platforms (Cotton 1912), that is notable for the characteristic line drawings (Fig. 3) not dissimilar to the style of Davis. The paper introduced into this subdiscipline of New Zealand geology a new standard of description and explanation based on the Davisian 'generic explanatory method' of landscape interpretation. This was a paradigm that was to permeate all of Cotton's work, and it was reaffirmed by Davis personally when he visited New Zealand in 1914 as a member of the American research programme in the Pacific.

During these formative years Cotton was able to make similar studies in other parts of New Zealand,

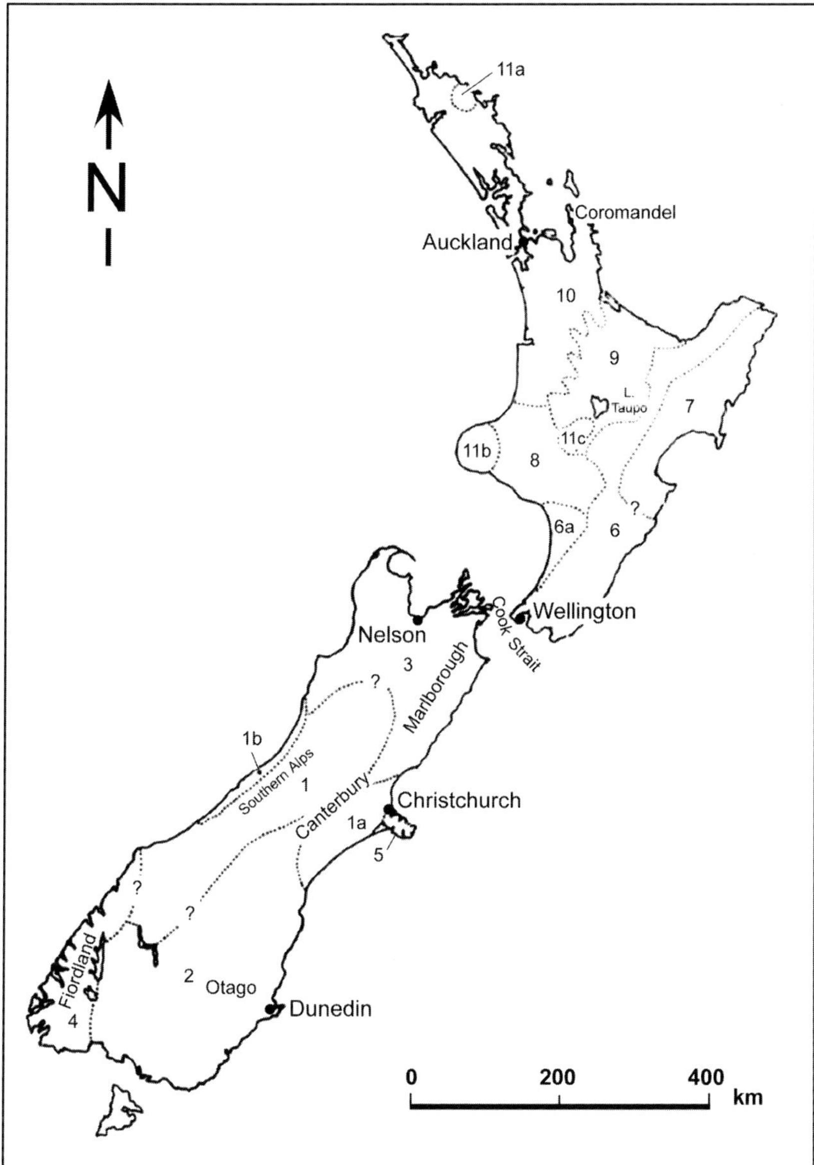

Fig. 2. Map of New Zealand showing localities mentioned in the text and geomorphic provinces delineated by Cotton in 1945. **1**. Axial range of the South Island (Alpine province including the Southern Alps) together with marginal moraine-covered area on the western side and associated alluvial plains of Canterbury (**1a**) and Westland (**1b**). Within the province are some tectonic basins and the boundary lines separating this province from provinces 2 and 3 are indefinite. **2**. Otago basin and range province. **3**. Nelson–Marlborough basin and range province transacted by numerous active faults. **4**. Fjordland province: a large terrane of crystalline rocks, heavily glaciated and fringed by fjords. **5**. Banks Peninsula province: a large dissected volcanic edifice. **6**. Wellington basin and range province transacted by numerous active NE-SW-trending faults. It extends northwards to include the axial ranges of the North Island and includes the Manawatu alluvial lowland (**6a**). **7**. Eastern maturely dissected province with complex structure. **8**. Western maturely dissected weak-rock province with simple structure and drainage pattern. **9**. Central volcanic plateau province containing young lava-built structures, lacustrine terraces and plains, and extensive areas covered by pumice and ignimbrite. **10**. Auckland basin and range province with relatively low average relief. **11a–c**. Young volcanoes and lava flows. There are also many small areas of the same nature in the Auckland province.

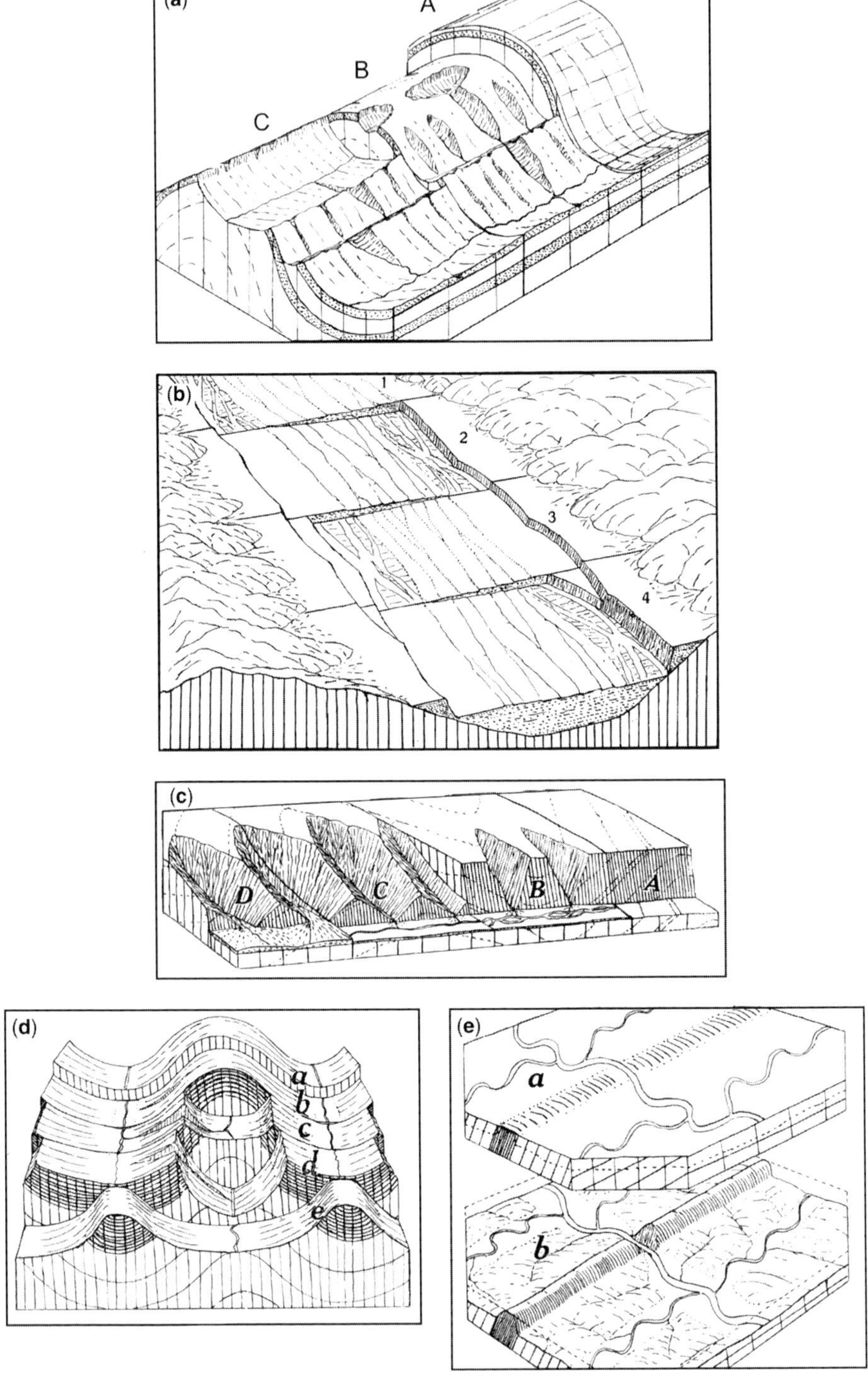

Fig. 3. Examples of block diagrams drawn by Cotton to illustrate sequential change of the landscape. (**a**) Development of subsequent drainage on folded rocks (fig. 78 in *Geomorphology* 1958). (**b**) Development of terrace slip-off across-valley slopes and their progressive destruction during successive stages (labelled 1–4) of terrace cutting (fig. 177 in *Landscape* 1957). (**c**) Dissection of a fault scarp: A, Initial form; B–D, sequential forms. Debris from

particularly Marlborough, north Canterbury, NW Nelson and north Otago (see Fig. 2 for locations), that were fundamental to developing his ideas on geomorphology. Davis had advised him to take up the study of block mountains, and from this work Cotton was to recognize and describe fault-block mountains, peneplains, fault scarps, orogenic and denudation episodes in a series of publications between 1913 and 1916. Perhaps the most important of these was his 1913 paper on the physiography of the middle Clarence Valley in Marlborough (see Fig. 2) in which he described the valley in terms of occupying a fault-angle depression between uplifted crustal blocks of the Inland and Seaward Kaikoura ranges. Several major stages of development were recognized: denudation of old basement highland; deposition of a thick sequence of mainly marine cover strata; orogenic uplift (the Kaikoura Orogeny); a cycle of erosion called the 'great denudation'; and a lesser regional uplift followed by renewed denudation that was still continuing. Many of the terms introduced in the paper quickly became part of the language of New Zealand geologists and the significance of vertical movement on the great faults traversing the area mapped by the Government Geologist Alexander McKay in the 1890s was clarified by Cotton's work. A paper on the structure and later geological history of New Zealand in 1916 more fully elaborated his ideas regarding the evolution of fault-block mountains and the idea that the landmass was a concourse of earth blocks of varying size and shape. Papers on block mountains and fossil denudation surfaces in the NW Nelson area published in 1916, enlarged to include examples from elsewhere in New Zealand in 1917, all supported by block diagrams, sketches and simplified maps encouraging the casual observer to see the landscape with a new interest and understanding. It was during these productive years that Cotton was elected Fellow of the Geological Society of London (F.G.S.) (1913) and in 1915, awarded a Doctorate in Science from the University of New Zealand.

Despite the honours and academic output, the war years were ones of austerity, and Cotton laboured on at the university with no support staff, no equipment and no money. By 1917 he was involved in giving a series of lectures to cadets and other officers of the Department of Agriculture, north of Wellington, a ride of some 2 h on his A.J.S. motorcycle. These trips provided the inspiration for a further paper in 1918 on coastal progradation along the west coast of the southern part of the North Island. The following year provided an opportunity for a longer trip together with Professor Noel Benson, Professor of Geology at Otago University, Dunedin, also riding a motorcycle, to see, photograph and sketch a variety of landscapes in Southland and Fiordland (see the locations in Fig. 2), many of which were used in his publications and signed 'C.A.C. 1918'.

Early in 1921, at the age of 36, Cotton was promoted to Professor of Geology at Victoria University and in the same year was elected a Fellow of the New Zealand Institute (F.N.Z.I.), later to become the Royal Society of New Zealand. At this point in his career he had published 29 papers and had clearly established himself as the undisputed leader of geomorphological studies in New Zealand. The year 1922 saw the publication of his first and most influential book, *Geomorphology of New Zealand*, bringing him almost immediate worldwide recognition as a leader in expounding the Davisian idea of landscape evolution in terms of structure, process and stage. The book was regarded as something new and dynamic, and it quickly became the standard textbook for students and working geologists alike. The impact of the book was largely determined by the fact that it was copiously illustrated with 442 figures, block diagrams, landscape sketches and half-tone pictures providing evidence of a remarkably compact diversity of physiographical features in New Zealand. The book was favourably reviewed by Davis (1923):

Cotton's book is the work of a young physiographic geologist [Davis was 85 at the time] who for the past fifteen years has had abundant observational experience in New Zealand, who has carried into his richly varied yet compact field a trained mind unusually competent in the rational study of landforms and a skilful hand exceptionally successful in reproducing landscapes in simple outlines, and who, when it comes to describing what he has seen, has decided unequivocally in favour of a modern, generic, explanatory method as against the old-fashioned and empirical method.

This last statement is hardly surprising given that the book closely followed the Davisian 'method' of explaining landscape development. The *Geomorphology of New Zealand* was to be republished seven times between 1924 and 1958, the third edition of 1942 being considerably revised, enlarged and its title changed to *Geomorphology: An Introduction to the Study of Landforms*.

In 1928 Cotton broadened his first-hand observational experience when he attended the

---

Fig. 3. (*Continued*) dissection of the fault scarp may be removed by the river in B, C; or accumulate as fans in D (fig. 173 in *Geomorphology* 1958). (**d**) Development of synclinal subsequent ridges and inversion relief (fig. 81 in *Geomorphology* 1958). (**e**) An even-crested ridge developed along the outcrop of a resistant rock layer preserving a remnant of a peneplain (flattened upper surface in a) at a later period (b) (fig.138 in *Geomorphology* 1958).

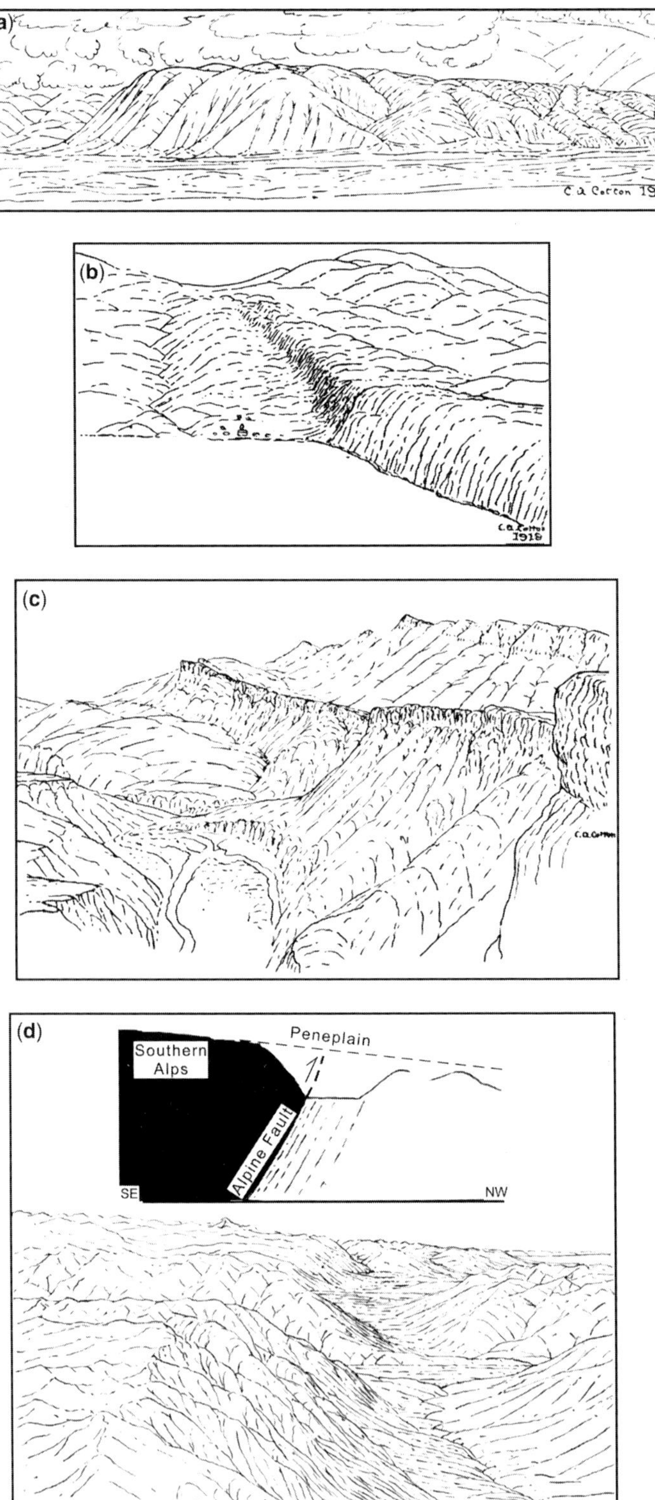

Fig. 4. Examples of sketches made by Cotton to illustrate geomorphic forms in New Zealand. (**a**) Small tilted block at Kurow, Otago, within a complex graben of the Waitaki Valley, Otago (drawn in 1916; reproduced as fig. 163 in

British Association meeting in Glasgow. Together with his wife, he took the opportunity of touring Britain and parts of the continent, visiting Norway and Spitzbergen on a German cruise ship where he observed and recorded by way of sketches and photographs glacial and fjord features, many of which were used to illustrate a later book.

The Depression and later university years

Despite his undoubted international reputation, Cotton's domain at Victoria University continued to be a one-man department regardless of the fact that by this time, geography was also his responsibility. Repeated requests for assistance had got nowhere, and it was only in 1930 that he was finally granted the use of a part-time assistant after volunteering money to help pay the salary. The person hired was Lester King, a brilliant student of Cotton's who had completed a thesis on uplifted beaches of Wellington's SE coast (King 1930) and who was to take up a position at Natal University, South Africa, in 1934, where he produced his monumental work, *Morphology of the Earth* in 1962.

The 1930s began with a reduction in government funds to the university, and by 1933 their reserves were all but exhausted. For funding to continue, the government requested that cuts be made and it was requested that the Department of Geology, conspicuous as the smallest, should be disestablished despite the fact that its professor was the most prolific producer of published work in the university. Accordingly, Cotton was given notice that his position would be terminated in March 1934. The other professors protested and Cotton attempted to justify continuance of geology at the university, arguing that it was of high academic, cultural, practical and aesthetic value, and that falling student numbers was largely the result of fee increases made when he was overseas in 1928–1929. The University Council were unpersuaded, but in the end Cotton was allowed to keep his position only after he volunteered to take a substantial cut in salary to almost half the normal professorial rate. He had to wait until 1941 before his full salary of £900 was reinstated. In recompense, the part-time position vacated by Lester King was replaced by a demonstrator to run laboratory classes.

These years were also lean of publications except for a number of review articles that were a measure of Cotton's international reputation. The new demonstrator, Maxwell Gage, who in 1966 was to become Professor of Geology at Canterbury University, Christchurch, found Cotton to be a rather remote figure, only appearing to give his scheduled lectures and leaving him alone to run the laboratory classes. The professor was busy at home writing research papers and the books that were to follow. Another trip to Europe in 1939 had to be cancelled because of the outbreak of war and, although Cotton was unfit for military service because of his sight impediment, he nevertheless joined the Home Guard. Although student numbers plummeted to only one in 1941, the war years were to mark a milestone in Cotton's creativity with the publication of three books: *Landscape* in 1941, *Climatic Accidents in Landscape Making* in 1942 and *Volcanoes and Landscape Forms* in 1944. These, together with the revised edition of *Geomorphology* in 1942, make up the quartet of books for which Cotton is known as one of the world's outstanding geomorphologists, In 1945 two further books, *The Earth Beneath: An Introduction to Geology for Readers in New Zealand* and *Living on a Planet*, introduced geomorphology to the wider public.

At the university Cotton's geomorphology lectures were well illustrated by a large selection of lantern slides and photographic projections. This managed to keep the students interested because his actual lectures tended to be rather 'dry', usually read verbatim from one of his textbooks with no words wasted. In many ways the lectures were a reflection of Cotton's writing, which, for the most part, consisted of short, incisive sentences, but often, to the frustration of readers, punctuated with unfamiliar technical terms such as *anteconsequent, resequent, insequent, cryopertuerbation, cryergic, feral*, etc. On the other hand, Cotton's use of block diagram sequences to convey the notion of evolutional change in the landscape (Fig. 3) and simple line drawings (Figs 4–6), allowed many to see and understand the landscape for the first time and this undoubtedly contributed to the popularity of the subject.

Fig. 4. (*Continued*) *Geomorphology* 1958). (**b**) Escarpments of homoclinal ridges developed on outcrops of limestone, Waipara Valley, Canterbury (reproduced as fig. 98 in *Landscape* 1957). (**c**) A fault-scarp controlling the western edge of Lake Taupo, Central Volcanic area of the North Island, trending inland in the southerly direction (drawn in 1919; reproduced as fig. 54 in *Landscape* 1957). (**d**) View together with key diagram (above) looking SW along the trace of the Alpine Fault at the base of the northern part of the Southern Alps, Westland, showing what Cotton interpreted to be a gently westward-dipping remnant peneplain surface across the top of the Alps that in the distance can be seen on both sides of the Alpine Fault (after fig. 328 in *Landscape* 1957).

During World War II coasts and beaches became the focus of intensive research because of the needs of amphibious landings, and this channelled geomorphology into a new and highly quantitative direction in the post-war years. Cotton was not part of this trend. While a new generation of graduate students were feeding computers with data on beach and river sediment grain sizes, etc., slope angles, wave dimensions and river flow rates, Cotton continued to publish in qualitative terms. His work was totally devoid of graphs, numerical estimates of geomorphological processes and rates. The exercise of gathering, organizing and analysing numbers never appealed to him, and he remained cynical about what he perceived to be illusions of precision and accuracy in the 'new geomorphology'. He considered that little was known of the absolute rate at which landscape changes, except that the rate must vary within wide limits dependent on such variables of climate and the relative hardness of the rocks being eroded. However, he considered that the mere description of a landscape was never an end in itself. He considered that a reasoned understanding of geomorphic processes was a necessary part of the equipment of a geologist for the interpretation of geological history. His approach was to define problems by looking at actual landscapes, at photographs and maps. His explanations of landscape development arose intuitively, were tested against geological data, and compared with his extensive personal observations, his comprehensive knowledge of the literature and his many overseas contacts. Thus, although Cotton's geomorphology

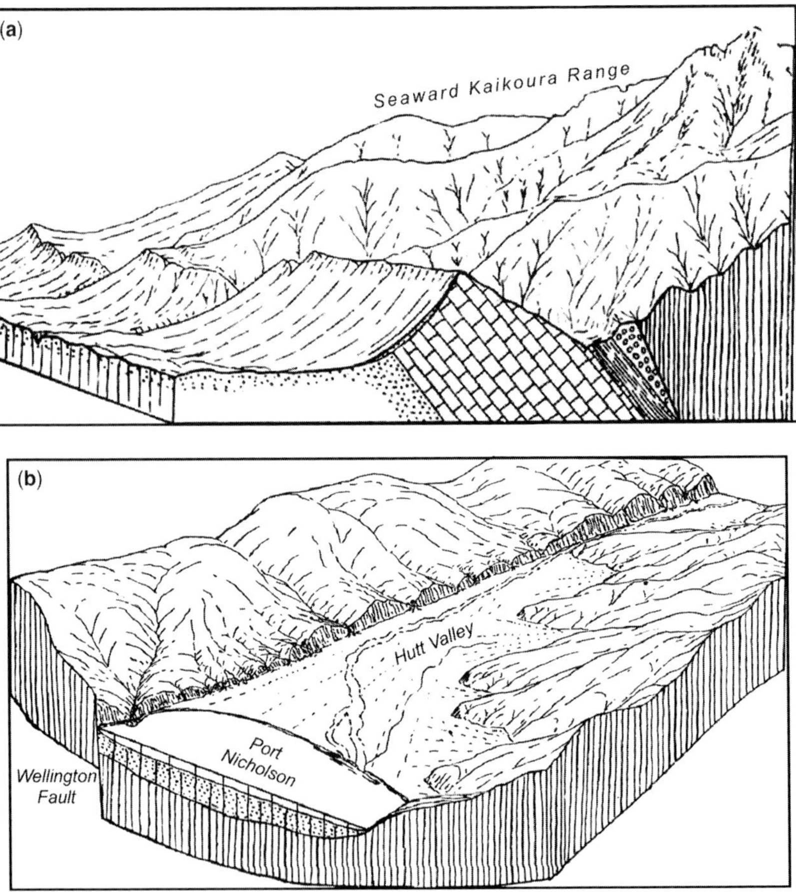

Fig. 5. Block diagrams drawn by Cotton illustrating geomorphic features in New Zealand. (**a**) Water gaps at D and L of superposed origin in a homoclinal ridge parallel to the front of the Seaward Kaikoura mountains, Marlborough (after fig. 123 in *Landscape* 1957). (**b**) A tectonic depression (Port Nicholson Depression; see Fig. 8) caused, in part, by movement on the Wellington Fault that has formed the alluvial–marine sediment-filled Hutt Valley and Wellington Harbour (Port Nicholson), Wellington (after fig. 176 in *Geomorphology* 1958).

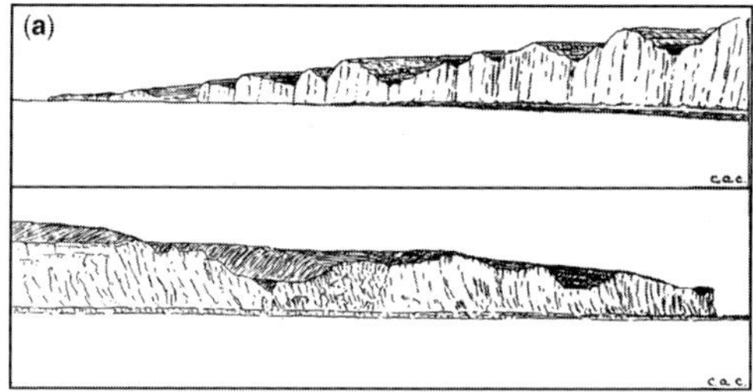

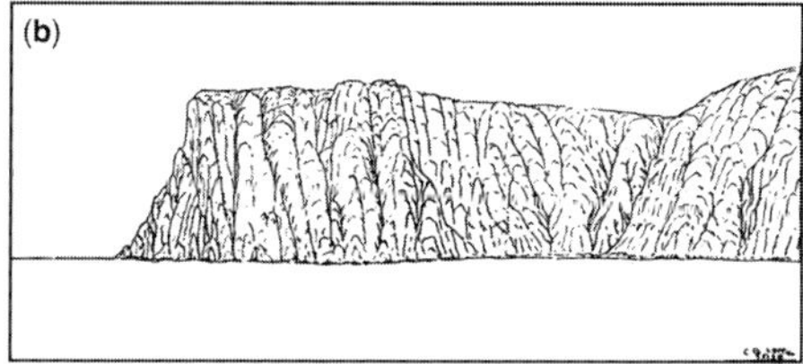

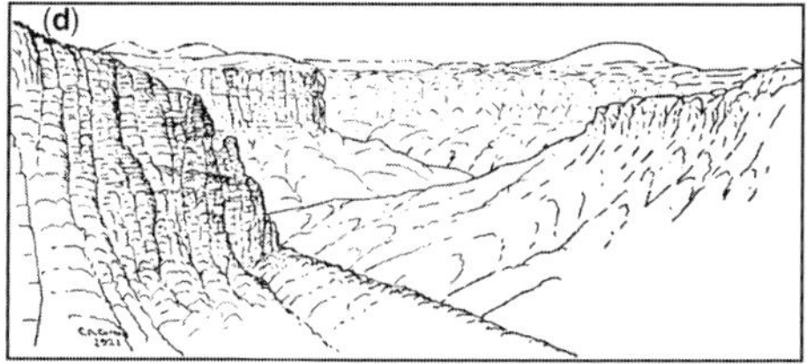

Fig. 6. Sketches made by Cotton of geomorphic features outside New Zealand. (**a**) Betrunked and incipiently rejuvenated valleys on the French (above) and English (below) coasts of the Straits of Dover (drawn from photographs, fig. 441 in *Geomorphology* 1958). (**b**) The plateau of Norway at North Cape drawn by Cotton in 1928 (fig. 221 in *Landscape* 1957). (**c**) Incised meander of the Meuse River, Fumay, northern France (drawn from a photograph, fig. 252 in *Landscape* 1957). (**d**) The valley of the Grose, Blue Mountains, New South Wales, Australia. The upper surface is the

remained qualitative, which is probabily the main reason for its lasting popularity, although it is also the reason why he never established a school of geomorphological thought.

By the late 1950s many geomorphologists had abandoned the Davisian 'cyclic' analysis of landforms, but Cotton continued to advocate its value, particularly as it facilitated a systematization of landforms that could be readily understood by those beginning geomorphic studies. In a postscript to the seventh edition of *Geomorphology* in 1958, he reiterated that a cycle may never complete the full-circle and that, although there may be apparent reversals in the order of evolutionary stages owing to their shortening or suppression, cyclic stages provided an unrivalled system of explanatory description of landforms with the cyclic concept providing a useful framework 'fitted with convenient pegs on which to hang discussions of important problems'.

By the end of World War II New Zealand had been covered by low-level (12 000 ft height) aerial photographs that provided a tremendous leap forward in mapping the country, and making it possible for the first time to observe the landscape directly from above and in stereoscopic vision. Unfortunately, Cotton, with only one good eye, was unable appreciate the 3-D image of the landscape from a pair of overlapping aerial photographs. He did, however, establish a working relationship with V. C. Browne associated with the New Zealand Public Works Department and with N. Z. Aerial Mapping who took Cotton on many flying trips around New Zealand in which he used an ex-US airforce handle-grip camera to take areal photographs while holding the camera outside the plane window. Many of these oblique aerial photographs were used by Cotton in later editions of his books and they provide an additional, dynamic dimension, complimentary to his line sketches, for understanding and recognizing landscapes and landscape-forming processes from another perspective.

A recurring theme

One interest that Cotton returned to several times was the geomorphic development of the rugged landscape of the Wellington area where he had began his studies in 1912 with the recognition of

---

Fig. 6. (*Continued*) plateau composed of hard Hawkesbury Sandstone that is being incised to form to escarpment cliffs (drawn in 1921, fig. 117 in *Landscape* 1957). (**e**) Needle karst forms (high-peaked hums), Kwangsi, China (drawn from a photograph, fig. 368 in *Landscape* 1957).

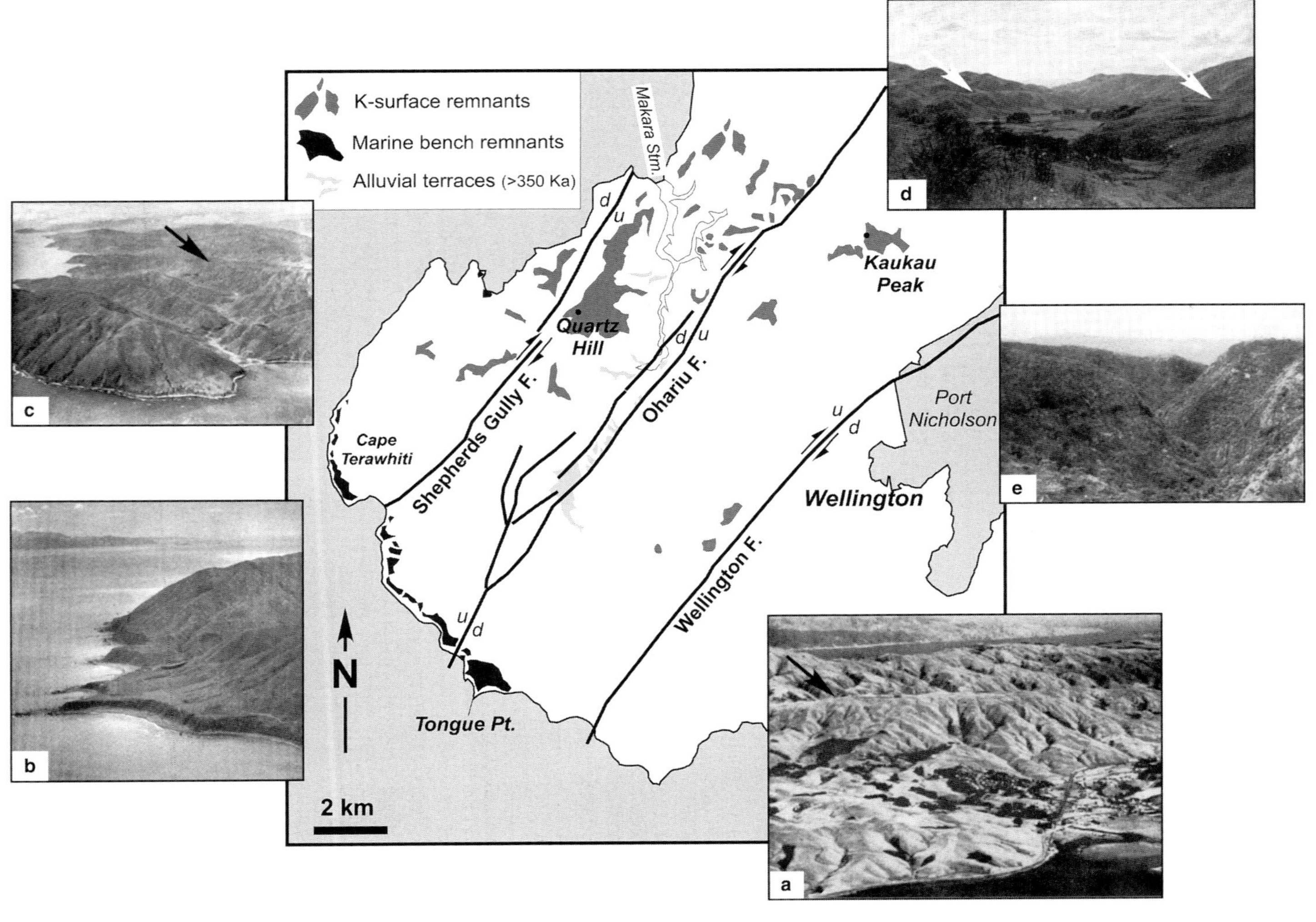

Fig. 7. Map of the Wellington Peninsula showing distribution of K-surface, marine benches and alluvial terraces described by Cotton in terms of three erosion cycles (see the text). The Makara Stream valley consists of Holocene alluvial gravels. Also shown are the Wellington, Ohariu and Shepherds Gully faults with upthrown (*u*), downthrown

cycles of erosion, regional tilting (warping) and active faulting (Fig. 7). In 1912 Cotton explained the geomorphic development of the Wellington Peninsula in terms of three Davisian erosion cycles corresponding to different base levels. In addition to the three main erosion events, Cotton also recognized other landform remnants, such as successive marine platforms and alluvial terraces, that he considered corresponded to other, short pauses or stillstands in the overall uplift of the peninsula. His deductions, in Cotton's own words, were based on 'hasty traverses and observations made from a distance', which is probably why his 37-page 1912 paper is called, 'Notes on Wellington physiography'. Nevertheless, the general conclusions reached have survived the test of time and were little modified in subsequent papers except for the dropping of any rigorous adherence to erosion cycles. Cotton recognized the oldest erosion feature in an otherwise hilly landscape as remnants of a mature low relief surface (a peneplain he termed the K (= key)-surface) with a base level of 244–275 m above present sea level (Fig. 7a). Outcrops of the surface are deeply weathered and commonly littered with lag boulders of greywacke and vein quartz, and the areas of low relief are bordered by precipitous slopes and ravines produced during the subsequent cycle. Cotton considered that this oldest erosion cycle (termed the *Kaukau cycle* after the highest peak preserving the K-surface, Fig. 7) reached a stage of late maturity with an overall elevation close to that of base level. Height differences of the K-surface remnants were attributed to later differential movement or warping of rocks of uniform composition (greywacke) between the NNE-trending faults that traverse the Wellington Peninsula (Fig. 7c). Today, Cotton's 'K-surface' is recognized as a key pre-late Quaternary marker surface (Ota *et al.* 1981; Begg & Mazengarb 1996), although with an uncertain age of between 0.4 and 4 Ma. One small outcrop of fossiliferous Early–Middle Pliocene (*c.* 3.6–5 Ma) marine mudstone overlying downfaulted

greywacke basement on the Ohariu Fault in the Makara Valley (Fig. 7) (Grant-Taylor & Hornibrook 1964) may represent part of the Tertiary sediments that once covered the K-surface.

The next main erosion cycle, the '*Tongue Point*' or '*Intermediate*' *cycle*, proposed by Cotton was recognized as the most prominent marine bench preserved on the Wellington coast at Tongue Point (Fig. 7b), indicating a base level of approximately 50 m above present sea level. Today, this marine bench is considered to represent the Last Interglacial highstand in sea level (oxygen isotope stage 5e) at approximately 125 000 BP. Inland, for example in the Makara Valley (Fig. 7), Cotton considered that there was coeval development of graded streams associated with the formation of wide flood plains (Fig 7d) and that the morphological stage reached by the Intermediate cycle was one of adolescence or early maturity in which divides were reduced to altitudes of 180–240 m by 'rounding' and slope grading.

The youngest geomorphic features on the Wellington Peninsula, representing what Cotton terms the '*Present cycle*' of erosion, are characterized by the steep lower slopes of valley sides and flood plains and incisement, in many cases with streams cutting steep gorges into the basement greywacke (Fig. 7e), indicating a youthful stage of development.

Later Cotton had doubts about the correlation of the marine benches with broad alluvial terraces inland, and considered the possibility that the latter were a more ancient feature and intermediate in height (and therefore age) between the proposed Kaukau and Tongue Point cycle base-level elevations. The alluvial gravel surfaces in the Makara Valley (Fig. 7d) are now considered to be Late Quaternary and more than 350 ka old. Also, an older, more dissected, marine beach remnant identified by Cotton landward of the Tongue Point bench at *c.* 75–100 m above sea level (B in Fig. 7) is now considered to represent the erosion surface created during a sea-level high at approximately 195 ka (Ota *et al.* 1981). Cotton's later, more detailed, work describing a well-developed marine bench exposed to the east of

---

Fig. 7. (*Continued*) (*d*) sides and sense of horizontal movement (arrows) indicated (after Ota *et al.* 1981; Begg & Mazengarb 1996). *Photographs.* (**a**) Looking SE across the Wellington Peninsula (Port Nicholson in the background), showing the remnant of the Pliocene–late Quaternary K-surface (arrowed; first recognized or earliest erosion cycle) in the vicinity of Quartz Hill (see the map for location). (**b**) View looking west along the south Wellington Coast showing the last Interglacial marine bench at Tongue Point (second erosion cycle) (see the map for location). (**c**) View looking east across the western part of the Wellington Peninsula with Cape Terawhiti in the foreground and Quartz Hill, with the K-surface (arrowed) in the distance (see the map for location). Note the height accordance (indicative of the first erosion cycle) of the NNE–SSW-trending parallel ridges that reflect the dominant drainage pattern controlled by regional strike of the greywacke basement rocks and main faults shown in the map. (**d**) View looking down the Makara Valley showing Late Quaternary alluvial terrace remnants (arrowed; tentatively correlated with the second erosion–deposition cycle) (see the map for location). (**e**) Stream gorge of the youngest, or third, erosion cycle incised into the inferred second cycle mature surface (upthrown side of the Wellington Fault). See the text.

Port Nicholson (Wellington Harbour) (Cotton 1921), and clearly correlated with the one he described in 1912 west of Wellington, make no mention of the Tongue Point erosion cycle. From height approximations of this and older, higher, marine bench remnants, he proposed an overall westward tilt towards Port Nicholson (termed the Port Nicholson depression) bounded on the western side by the Wellington Fault, and a similar, more gentle tilting westwards from Tongue Point. A section shown in Figure 8, constructed from localities and heights mentioned in Cotton's 1912, 1916 and 1921 papers, illustrates this. A detailed survey of the successive marine benches preserved along the Wellington coast with possible correlations, was only made 60 years later (Ota *et al.* 1981), and is in broad agreement with Cotton's work (Fig. 8). Of particular interest was the effect of uplift, maximum of 2.5 m, on the western side of Palliser Bay (Fig. 8) and westward tilting that affected the whole of the Wellington Peninsula caused by the 23 January 1855 M8 earthquake, described by Lyell (1868) from information supplied to him on changes in coastal elevation by New Zealand eyewitnesses. Cotton realized that the pattern of westward tilting caused by this earthquake event was the same as the much older marine benches, implying a long period of earthquake activity to explain the overall uplift of the Wellington area. In this respect, Cotton was following Lyell (1868) who stated that, 'The geologist has rarely enjoyed so good an opportunity as that afforded him by this convulsion in New Zealand, of observing one of the steps by which those great displacements of the rocks called "faults" may in the course of ages be brought about. The manner also in which the upward movement increased from north-west to south-east explains the manner in which beds may be made to dip more and more in a given direction by each successive shock' (p. 89).

In the case of the 1855 earthquake, uplift was caused by movement on the now-called Wairarapa Fault, as recorded by Lyell (1868), but Cotton was also interested in the geomorphic effects of uplift on the Wellington Fault, which he also described in 1912 in terms of an eroded fault scarp where it borders the western side of Wellington Harbour (Fig. 9) and the presence of truncated valleys where they intersect the fault. In 1954 he described stream alignments and 'trailing' tributary streams along the active trace (called a *cicatrice*) and, more importantly, reported dextral displacement of one stream and two ridges by approximately 60 m (Fig. 9). These offsets were intepreted as being the result of cumulative displacement of a number of earthquakes, as were the dextrally trailing tributaries the result of many small

earthquake lateral offsets developed on the SE side of the fault (Fig. 9). These were the first observations of lateral displacement of Quaternary geomorphic features on any of the major active faults in New Zealand, all of which are now known to have had a much greater lateral (dextral) than vertical movement, a feature first noted by Cotton for movement on the Wellington Fault.

Active retirement and honours

In 1953, at the age of 65, Cotton retired from the Chair of Geology at Victoria University after 32 years. He was elected Emeritus Professor and was awarded an Honorary LLD from the University of New Zealand the following year, when the university marked his retirement with the publication of a book, *New Zealand Geomorphology*, containing reprints of his selected papers from between 1912 and 1925, a decade that was one of amazingly fruitful activity in New Zealand geological research led by Cotton's statements of new ideas on the later geological history of the country. An 'Appreciation' at the beginning of the book contributed by one of the leading geomorphologists, Henri Baulig in Strassbourg, paid tribute to this work pointing out that it 'demonstrated the fundamental validity of the Davisian evolutionary concept and its durable fruitfulness'. Throughout his long career, Cotton received many awards and honours. He received the highest distinctions of New Zealand science, the Hector Medal in 1927 and the Hutton Medal in 1947 from the Royal Society of New Zealand. Overseas recognition came in the form of the Victoria Medal of the Royal Geographical Society of London in 1951, and the Dumont Medal of the Geological Society of Belgium in 1954. In the Queen's Birthday Honours of 1959, Cotton was made Knight Commander of the Order of the British Empire (K.B.E.). With characteristic modesty, together with mild humour, he replied to one congratulatory letter: 'I do not know whether I am very much better off as Sir Charles: but I do know my wife gets a great deal of satisfaction out of being Lady Cotton'.

In retirement, Cotton continued to publish from his home address on a multitude of topics – coastal geomorphology, climate change, sea-level change, landscapes formed by fault movement, Pacific islands, erosion in tropical and temperate climates (see the Bibliography). The papers developed both original ideas as well as revived old ones in a new context that gave due prominence to the post-Davisian views of Albrecht and Walther Penck, Lester King, as well as others. As evident from his bibliography, Cotton retained his mental alertness, and was well aware of the changes in

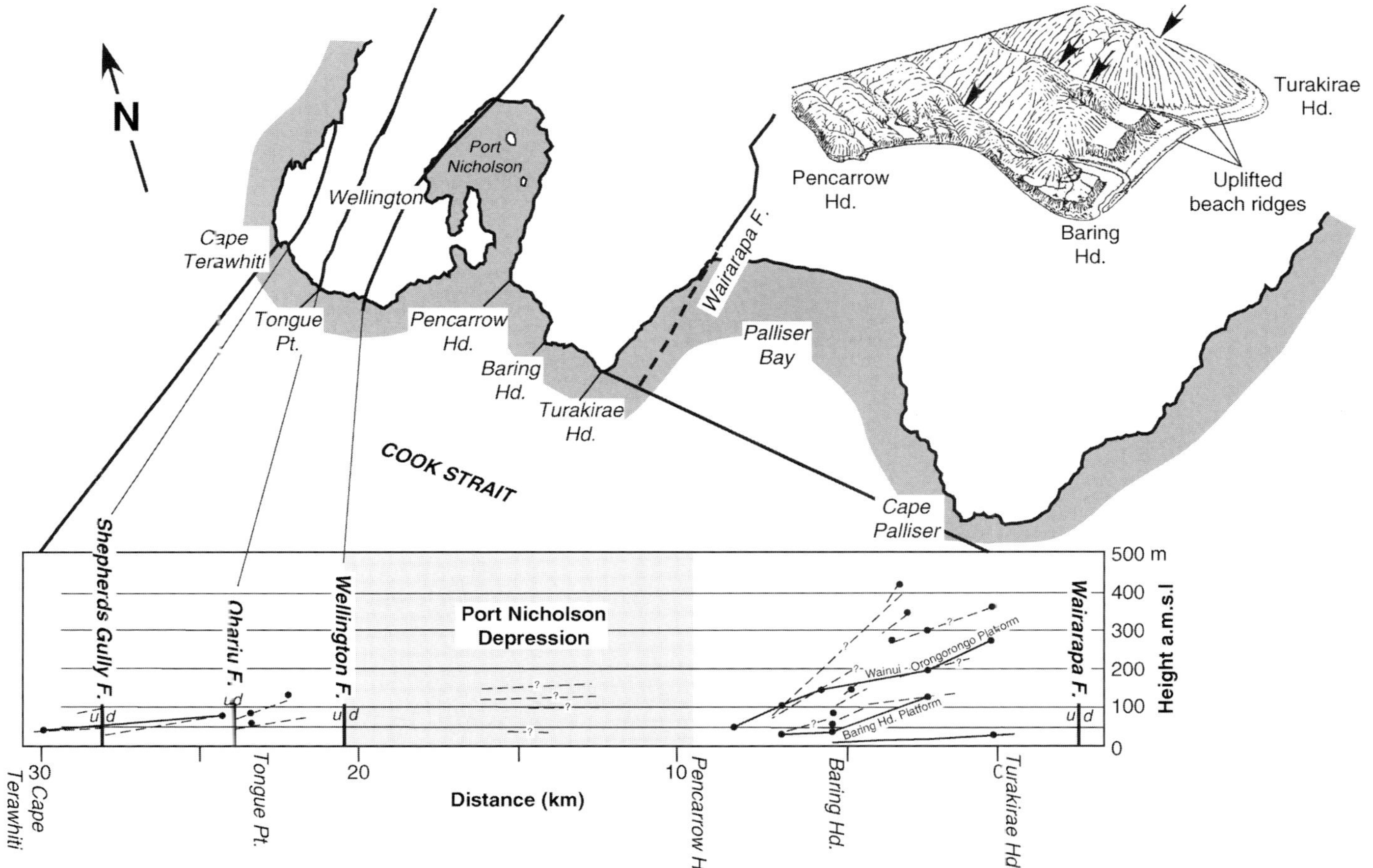

Fig. 8. Height distribution of marine benches along the south Wellington coast projected onto a vertical section striking 300° between Turakirae Head and Cape Terawhiti as shown on the map of the southern part of the North Island of New Zealand. Solid circles and connecting solid lines are heights and correlations given by Cotton (1912, 1916, 1921). For comparison, thin dashed lines that are inferred marine bench correlations (with question marks) determined from surveyed heights mapped by Ota et al. (1981) are also shown. The block diagram (after fig. 3 of Cotton 1921) shows the marine benches and earthquake-uplifted beach lines, west of Port Nicholson. A block diagram of the form of the Port Nicholson depression is shown in Figure 5b.

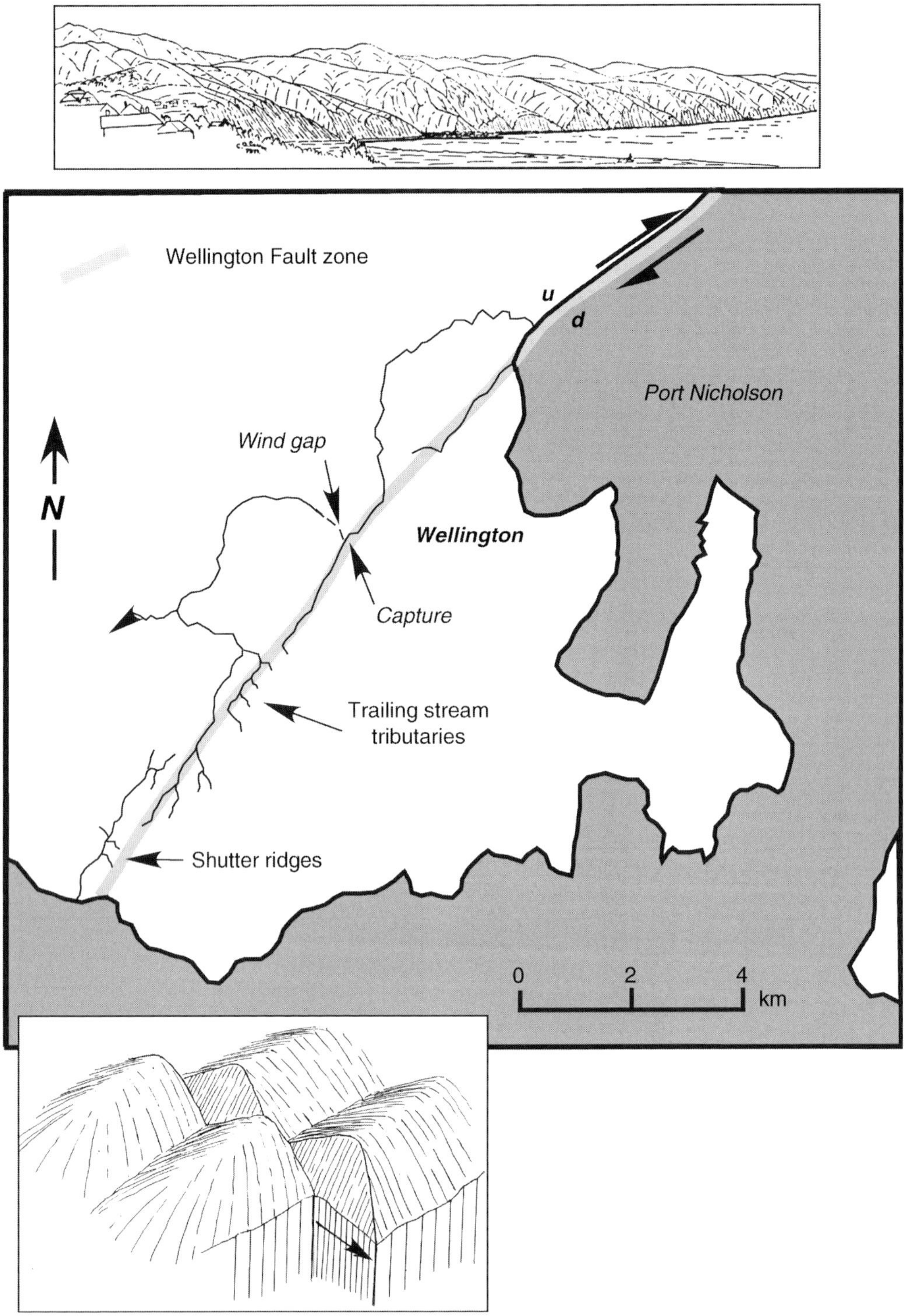

Fig. 9. Map showing the surface trace of the southern part of the Wellington Fault (after fig. 2 of Cotton 1951), showing stream alignments, trailing stream tributaries and the location of dextrally displaced (shutter) ridges illustrated in the block diagram below (fig. 4. of Cotton 1951). The illustration above shows the eroded scarp of the Wellington Fault that forms the western side of Port Nicholson. The position of the active fault trace lies offshore. See the text.

outlook, methods and purpose in geomorphology that had been underway since the 1950s, until his death on 29 June 1970, aged 85. Out of a total of 185 research papers, only five are co-authored, two of which were published posthumously. A collection of 16 of his most important papers on coastal geomorphology, which had been annotated by Cotton just before his death, were published as the book *Bold Coasts* (Collins 1974).

Illustrative skills

While Cotton's life's work was devoted to unravelling the story of landscape form and development through word and illustration, his landscape sketches are particularly distinctive and possess a simplicity of style that reflects his skill as a landscape artist. They are one of the most endearing aspects of Cotton's work, a clarification that clears the landscape of vegetation and buildings to expose the underlying structure of the land, and in so doing capture a moment of geological time. A letter from Japan commemorating his 80th birthday pays particular tribute to his artistic ability: '... we [i.e. Japanese geomorphologists] especially entertain friendly sentiment for his sketches, which sometimes remind us of pictures drawn in Chinese ink – of the Southern China School' – i.e. monochrome ink painting. In addition to Cotton's original sketches, he also made a large number from photographs or from sketches drawn by overseas geomorphologists (see Fig. 6). In the 33 year period between 1911 and 1944, Cotton produced 422 pen and ink drawings, and, in total, his line drawings amount to almost 600. They stand together with his writing as a tribute to the unequalled lifetime work of one man.

Final comment

Cotton pointed out on numerous occasions that one must bear in mind that no feature of the Earth's surface is a finished product; 'the agencies that effect changes of form are everywhere at work; every part of the surface is even now undergoing change, and its future forms will differ from the present as the present differs from the past' (Cotton 1947; Preface to 2nd edition of *Landscape as Developed by Processes of Normal Erosion*). His life's work was an attempt to document these changes and trace their evolution in terms of a systematized cycle of erosion, fluvial, glacial, marine, arid and karstic, in relation to the Davisian paradigm of structure, process and time that defines the landscape. Field observation was the core of his geomorphological investigations and his qualitative, largely deductive, conclusions. However, like many of his contemporaries, Cotton also emphasized the 'process' aspect in introducing the

effects of climate and sea-level change, and tectonic forces, particularly those involving uplift and faulting, i.e. the interaction of exogenic and endogenic forces advanced by Walther Penck. Indeed, Cotton could not have been in a better place to develop these ideas but in New Zealand, a tectonically active country where the relationship between geologically recent Earth movements, denudation, erosion and associated deposition is particularly obvious and ongoing.

I would like to thank M. Johnston (Nelson, New Zealand) and R. Twidale (University of Adelaide, Australia) for their constructive comments on an earlier draft of the manuscript.

Bibliography of Sir Charles Cotton

Books (date of first editions)

1922. *Geomorphology of New Zealand: Part 1 – Systematic: An Introduction to the Study of Landforms.* Dominion Museum, Wellington.
1941. *Landscape as Developed by Processes of Normal Erosion.* Cambridge University Press, Cambridge.
1942. *Climatic Accidents in Landscape Making.* Whitcombe and Tombs, Christchurch.
1944. *Volcanoes as Landscape Forms.* Whitcombe and Tombs, Christchurch.
1945. *Earth Beneath: An Introduction to Geology for Readers in New Zealand.* Whitcombe and Tombs, Christchurch.
1945. *Living on a Planet.* Whitcombe and Tombs, Christchurch.
1955. *New Zealand Geomorphology* (reprints of selected papers 1912–1925). New Zealand University Press, Wellington.

Journal articles

1909. Geology of Signal Hill, Dunedin. *Transactions of the New Zealand Institute,* **41,** 111–126.
1912. Typical sections showing the junction of the Amuri limestone and Weka Pass Stone at Weka Pass. (Abs). *Transactions of the New Zealand Institute,* **44,** 84–85.
1912. Notes on Wellington physiography. *Transactions of the New Zealand Institute,* **44,** 245–265.
Recent and sub-Recent movements of uplift and subsidence near Wellington, New Zealand. *Scottish Geographic Magazine,* **28,** 306–312.
1913. The physiography of the Middle Clarence Valley, New Zealand. *Geographic Journal,* **42,** 225–246.

The Tuamarina Valley: A note on the Quaternary history of the Marlborough Sounds district. *Transactions of the New Zealand Institute*, **45**, 316–322.

1914. On the relations of the Great Marlborough Conglomerate to the underlying formations in the Middle Clarence valley, New Zealand. *Journal of Geology*, **22**, 346–363.
Preliminary note on the uplifted east coast of Marlborough. *Transactions of the New Zealand Institute*, **46**, 286–294.
Supplementary notes on Wellington physiography. *Transactions of the New Zealand Institute*, **46**, 294–298.
Outline of the physiography of Wellington. *In*: Handbook for Scientific Visitors. NZ Science Congress Committee, Wellington, 20–21.
The Wellington fault-scarp. *In*: Handbook for Scientific Visitors. NZ Science Congress Committee, Wellington, 21.
Raised beaches at Wellington. *In*: Handbook for Scientific Visitors, NZ Science Congress Committee, Wellington, 21–22.

1916. The structure and later geological history of New Zealand. *Geological Magazine*, **53**, 243–249.
Block mountains and a 'fossil' denudation plain in northern Nelson. *Transactions of the New Zealand Institute*, **48**, 59–75.
Fault coasts in New Zealand. *Geographic Review*, **1**, 20–47.

1917. The fossil plains of North Otago. *Transactions of the New Zealand Institute*, **49**, 429–432.
Block Mountains in New Zealand. *American Journal of Science*, **44**, 249–293.

1918. River terraces in New Zealand. *New Zealand Journal of Science and Technology*, **1**, 145–152.
The geomorphology of the coastal district of south-western Wellington. *Transactions of the New Zealand Institute*, **50**, 212–222.
Mountains. *New Zealand Journal of Science and Technology*, **1**, 280–285.
Conditions of deposition of the continental shelf and slope. *Journal of Geology*, **26**, 135–160.
The outline of New Zealand. *Geographic Review*, **6**, 320–340.

1919. Problems presented by the Notocene beds of central Otago. *New Zealand Journal of Science and Technology*, **2**, 69–72.
Rough Ridge, Otago, and its splintered fault scarp. *Transactions of the New Zealand Institute*, **51**, 281–285.
Geomorphology. (Letter to Editor.) *Science*, **49**, 423–424.

1921. For how long will Wellington escape destruction by earthquake? *New Zealand Journal of Science and Technology*, **3**, 220–231.
Origin of the Upper Waitaki depression. (Letter.) *New Zealand Journal of Science and Technology*, **4**, 207–208.
Report of physiographic features of Australasia Committee: Advances in physiography of New Zealand in the last ten years. *Reports of the Australasian Association for the Advancement of Science*, **15**, 335–342.
The warped land surface on the southeastern side of the Port Nicholson depression. *Transactions of the New Zealand Institute*, **53**, 131–143.

1922. The geomorphology of Wellington. *In*: New Zealand Nature Notes. Government Printer, Wellington, 3–10.
Glaciers and glaciation if the Mount Cook district. *In*: New Zealand Nature Notes. Government Printer, Wellington, 46–48.

1925. Evidence of Late Tertiary or post-Tertiary orogeny in New Zealand. Gedenkboek Verbeek, *Verhenigin Geologie-Mijnbouw, Genootschap Nederland en Kolonien Geologische Series*, **8**, 89–110.

1926. [Recent Papers on] Geomorphology in New Zealand, 1921–1924. *Reports of the Australasian Association for the Advancement of Science*, **17**, 50–56.

1936. Geomorphology of Wellington. *In*: Handbook for New Zealand. Australia and New Zealand Association for the Advancement of Science, Wellington, 20–21.

1938. The Haldon Hills problem. *Journal of Geomorphology*, **1**, 187–198.
Some peneplanations in Otago, Canterbury, and the North Island of New Zealand. *New Zealand Journal of Science and Technology*, **20B**, 1–8.

1939. Lateral planation in New Zealand. *New Zealand Journal of Science and Technology*, **B20**, 227–232.
Glacial spur-truncation in New Zealand. *Journal of Geomorphology*, **2**, 70–72.

1940. Classification and correlation of river terraces. *Journal of Geomorphology*, **3**, 27–37.
Taylor's hypothesis of headward glacial erosion. *Journal of Geomorphology*, **3**, 346–352.
Glacial motion: A review of theories. *New Zealand Journal of Science and Technology*, **B21**, 281–286.

1941. The longitudinal profiles of glaciated valleys. *Journal of Geology*, **49**, 113–128.

Notes on two transverse-profile geomorphic problems. *Transactions of the Royal Society of New Zealand*, **71**, 1–6.

Moraines and outwash of the Tasman Glacier. *Transactions of the Royal Society of New Zealand*, **71**, 204–207.

The shoulders of glacial troughs. *Geological Magazine*, **78**, 81–96.

Basal remnants of truncated spurs in glaciated troughs. *Journal of Geomorphology*, **4**, 65–70.

Some volcanic landforms in New Zealand. *Journal of Geomorphology*, **4**, 297–306.

1942. Shorelines of transverse deformation. *Journal of Geomorphology*, **5**, 45–58.

1943. Oahu Valley sculpture: A composite review. *Geological Magazine*, **80**, 237–243.

1944. Tropical weathering and landforms: a correction. *Geological Magazine*, **81**, 141–142.

1945. The significance of terraces due to climatic oscillation. *Geological Magazine*, **82**, 10–16.

Geomorphic provinces in New Zealand. *New Zealand Geographer*, **1**, 40–47.

Ruapehu in 1945. *New Zealand Science Review*, **3**, 3–4.

1946. The 1945 eruption of Ruapehu. *The Geographical Journal*, **107**, 140–143.

1947. The Alpine Fault of the South Island of New Zealand from the air. *Transactions of the Royal Society of New Zealand*, **76**, 369–372.

Revival of major faulting in New Zealand. *Geological Magazine*, **84**, 79–88.

1948. The landscape of the Acheron Valley system. (Note.) *New Zealand Geographer*, **4**, 197.

Recent volcanic activity in New Zealand. (Note.) *Geographic Journal*, **112**, 258.

The Hamner Plain and the Hope Fault. *New Zealand Journal of Science and Technology*, **B29**, 10–17.

A note on tectonic and structural landforms. *New Zealand Journal of Science and Technology*, **B29**, 96.

Query as to the tempo of Australian denudation. *Geological Magazine*, **85**, 54–56.

Volcano eustatism, positive or negative? *Geological Magazine*, **85**, 201–202.

The buckling theory of orogeny, island festoons and foredeeps. *New Zealand Science Review*, **6**, 8–12.

The present-day status of coral reef theories. *New Zealand Science Review*, **6**, 111–113.

1949. Plunging cliffs, Lyttelton Harbour. *New Zealand Geographer*, **5**, 130–136.

Rejuvenation of the Awatere Fault cicatrice. *Transactions of the Royal Society of New Zealand*. **77**, 273–274.

Recent volcanic activity in New Zealand. *Geographic Journal*, **112**, 258.

Earth birth – a flashback to primordial chaos. *New Zealand Science Review*, **7**, 120–122.

A review of tectonic relief in Australia. *Journal of Geology*, **57**, 280–296.

1950. Tectonic scarps and fault valleys. *Bulletin of the Geological Society of America*, **61**, 717–758.

Quelques Aspectes de Relief de Failles. *Revue Geomorphologie Dynamique*, **1**, 226–234.

Tectonic relief. *Annual Association of American Geographers*, **40**, 181–187.

The cause of Ice Ages – is the sun a variable star. *New Zealand Science Review*, **8**, 31–33.

Climatic significance of terraces. (Note.) *New Zealand Geographer*, **6**, 85–87.

Axes of warping in the New Zealand seismic region. *Geological Magazine*, **87**, 360–368.

1951. Redeposition theory of sedimentation. *New Zealand Journal of Science and Technology*, **B32**, 19–25.

Post-Hokonui Orogeny, erosion, and planation. *New Zealand Journal of Science and Technology*, **B33**, 173–178.

Fault valleys and shutter ridges at Wellington. *New Zealand Geographer*, **7**, 62–68.

Seacliffs of Banks Peninsula and Wellington: Some criteria for coastal classification. *New Zealand Geographer*, **7**, 103–120.

Atlantic gulfs, estuaries and cliffs. *Geological Magazine*, **88**, 113–128.

Accidents and interruptions in the cycle of marine erosion. *Geographic Journal*, **117**, 343–349.

Une Côte de déformation transverse à Wellington (Nouvelle-Zélande). *Revue Géomorphologie Dynamique*, **2**, 97–109.

1952. The Wellington coast: An essay in coastal classification. *New Zealand Geographer*, **8**, 46–62.

Cyclic reseaction of headlands by marine erosion. *Geological Magazine*, **89**, 221–225.

The erosional grading of convex and concave slopes. *Geographic Journal*, **118**, 197–204.

1953. Submarine canyon hypotheses. *New Zealand Science Review*, **11**, 110–113.

Tectonic relief: With illustrations from New Zealand. *Geographic Journal*, **119**, 213–222.

1954. Submergence in the lower Wairau Valley. *New Zealand Journal of Science and Technology*, **35**, 364–369.
Terrestrial carbon. *New Zealand Science Review*, **12**, 54–55.
Earth's core and its mantle. *New Zealand Science Review*, **12**, 67–71.
Notocenozoic: The New Zealand Cretaceo-Tertiary (Sixth Hudson Lecture). *New Zealand Science Review*, **12**, 105–112.
Tests of a German non-cyclic theory and classification of coasts. *Geographic Journal*, **120**, 353–361.
Deductive morphology and the genetic classification of coasts. *Science Monthly*, **78**, 163–181.

1955. Review of the Notocenozoic or Cretaceo-Tertiary of New Zealand. *Transactions of the Royal Society of New Zealand*, **82**, 1071–1122.
Ten thousand years of accelerated erosion. *Geographic Journal*, **121**, 241–244.
Peneplanation and pediplanation. *Bulletin of the Geological Society of America*, **66**, 1213–1214.
The theory of secular marine planation. *American Journal of Science*, **253**, 580–589.
Aspects géomorphologiques de la flexure continentale. *Annales Societe Géologique de Belgique*, **78**, B403–418.
& TE PUNGA, M. T. Solifluxion and periglacially modified landforms at Wellington, New Zealand. *Transactions of the Royal Society of New Zealand*, **82**, 1001–1031.
& TE PUNGA, M. T. Fossil gullies in the Wellington landscape. *New Zealand Geographer*, **11**, 72–75.

1956. Coastal history of southern Westland and northern Fiordland. *Transactions of the Royal Society of New Zealand*, **83**, 483–488.
Hawke's Bay coastal types. *Transactions of the Royal Society of New Zealand*, **83**, 687–694.
Geomechanics of New Zealand mountain-building. *New Zealand Journal of Science and Technology*, **38**, 187–200.
The question of polar wanderings. *New Zealand Science Review*, **14**, 92–93.
Rias *sensu stricto* and *sensu lato*. *Geographic Journal*, **122**, 360–364.
–, Baulig, H. & Wallace, W. H. Geomorphology, geomorphography, geomorphogeny, and geography. *New Zealand Geographer*, **12**, 72–75.
& Tricart, J. Discussion à propos de l'analyse d'un ouvrage. *Revue Geomorphologie Dynamique*, **7**, 148–149.

1957. Tectonic features in a coastal setting at Wellington. *Transactions of the Royal Society of New Zealand*, **84**, 761–790.
Pleistocene shorelines on the compound coast of Otago. *New Zealand Journal of Science and Technology*, **38**, 750–762.
An example of transcurrent-drift tectonics. *New Zealand Journal of Science and Technology*, **38**, 939–942.
The Pukurua tectonic corridor, Wellington. *New Zealand Geographer*, **13**, 117–124.
A slow tempo of denudation claimed for Eastern Australia. *Geological Magazine*, **94**, 341–342.
Geomorphic evidence of major structures associated with transcurrent faults in New Zealand. *Revue Géographie Physique Géologie Dynamique*, 16–30.
Geomorphic contrasts in central New Zealand. *Tijdschriftt Kon Nederllande Aardrijksk Generalle*, **74**, 248–253.

1958. Dissection and redissection of the Wellington landscape. *Transactions of the Royal Society of New Zealand*, **85**, 409–425.
Evidence of late Pliocene warmth in New Zealand. *New Zealand Journal of Geology and Geophysics*, **1**, 470–473.
Alternating Pleistocene morphogenetic systems. *Geological Magazine*, **95**, 125–136.
The rim of the Pacific. *Geographic Journal*, **124**, 223–231.
Eustatic river terracing complicated by seaward downflexure. *Transactions of the Edinburgh Geological Society*, **17**, 165–178.
Fine-textured erosional relief in New Zealand. *Zeischrift für Geomorphogie*, **2**, 187–210.

1959. Notocenozoic; Notopleistocene. *In*: FLEMING, C. A. (ed.) *Lexique Stratigraphique International* (Oceane), Fasicule 4. New Zealand. Centre Nationale de la Rescherche Scientifique; for Stratigraphic Commission of International Geological Congress, Paris, Sce, **6**, 262–263.

1960. Hutt Valley fault scarps. *New Zealand Journal of Geology and Geophysics*, **3**, 218–221.
The origin and history of central-Andean relief: divergent views. *Geographic Journal*, **126**, 476–478.
Levelling by rock-floor robbing in Brazil. *Journal of Geololgy*, **68**, 573–574.

1961. The last glacial age. *New Zealand Science Review*, **19**, 58–60.
Growing mountains and infantile islands of the western Pacific. *Geographic Journal*, **127**, 209–211.

The theory of savanna planation. *Geography*, **46**, 89–101.

1962. Low seal levels in the Late Pleistocene. *Transactions of the Royal Society of New Zealand (Geology)*, **1**, 249–252.

Plains and inselbergs of the humid tropics. *Transactions of the Royal Society of New Zealand (Geology)*, **1**, 269–277.

The origin of New Zealand feral (fine-textured) relief. *New Zealand Journal of Geology and Geophysics*, **5**, 269–270.

The volcano-tectonic theory of block faulting no longer tenable. *New Zealand Science Review*, **20**, 48–49.

La signification de la fossilisation d'un relief de microdissection sous le head en Novelle Zélande. *Rev. Geomorphol. Dyn.*, **10**, 30–32.

Dating recent mountain growth by fossil pollens. *Tuatara* (Journal of the Biological Society, Victoria University of Wellington, New Zealand), **10**, 5–12.

Holocene marine benches in South America and elsewhere. *Geologia Colombiana*, **3**, 3–15.

1963. The question of high Pleistocene shorelines. *Transactions of the Royal Society of New Zealand (Geology)*, **2**, 51–61.

Development of fine-textured landscape relief in temperature pluvial climates. *New Zealand Journal of Geology and Geophysics*, **6**, 528–533.

A new theory of the sculpture of middle-latitude landscapes. *New Zealand Journal of Geology and Geophysics*, **6**, 769–774.

The rate of down-wasting of land surfaces. *Tuatara* (Journal of the Biological Society, Victoria University of Wellington, New Zealand), **11**, 26–27.

Did the Murrumbidgee Aggradations take place in Glacial ages? *Australian Journal of Science*, **26**, 54.

Levels of planation of marine benches. *Zeitschrift für Geomorphogie*, **7**, 97–111.

1964. The control of drainage density. *New Zealand Journal of Geology and Geophysics*, **7**, 348–352.

The cryergic-fluvial valley form. *New Zealand Journal of Geology and Geophysics*, **7**, 918–919.

1966. The continental shelf. *New Zealand Journal of Geology and Geophysics*, **9**, 105–110.

Antarctic scablands. *New Zealand Journal of Geology and Geophysics*, **9**, 130–132.

1968. Plunging cliffs and Pleistocene coastal cliffing in the southern hemisphere. *Mélanges de Géographie, 1. Géographie Physique et Géographie Humaine*, 37–59.

1969. The pedestals of oceanic volcanic islands. *Bulletin of the Geological Society of America*, **80**, 749–760.

Skeleton islands of New Zealand and elsewhere. *Earth Science Journal (Waikato Geological Society)*, **3**, 61–76.

The Turakirae coastal plain. (Letter.) *New Zealand Journal of Geology and Geophysics*, **12**, 838–841.

Marine cliffing according to Darwin's theory. *Transactions of the Royal Society of New Zealand (Geology)*, **6**, 187–208.

1970. Pedestals of oceanic volcanic islands: reply. *Bulletin of the Geological Society of America*, **81**, 1605.

1972. The duration of the Pleistocene. *Journal of the Royal Society of New Zealand*, **2**, 541–544.

References

BEGG, J. G. & MAZENGARB, C. 1996. *Geology of the Wellington Area*. Institute of Geological and Nuclear Sciences Geological Map 1:50 000, 1 sheet. New Zealand Institute of Geological and Nuclear Sciences, Lower Hutt.

COLLINS, B. W. 1966. Sir Charles Cotton (80th Birthday Issue). *New Zealand Journal of Geology and Geophysics*, **9**.

COLLINS, B. W. (ed.). 1974. *Bold Coasts* (Annotated reprints of selected papers on coastal geomorphology 1916–1969). A. H. and A. W.Reed, Wellington.

DAVIS, W. M. 1923. C. A. Cotton. Geomorphology of New Zealand. Part 1 – Systematic: An Introduction to the Study of Land-forms. *Geographical Review*, **13**, 321–322.

GRANT-TAYLOR, T. L. & HORNIBROOK, N. DE. B. 1964. The Makara faulted outlier and the age of Cook Strait. *New Zealand Journal of Geology and Geophysics*, **7**, 299–313.

KING, L. C. 1930. Raised Beaches and other features of the south-east coast of the North Island of New Zealand. *Transactions of the New Zealand Institute*, **61**, 498–523.

KING, L. C. 1962. *Morphology of the Earth*. Oliver & Boyd, Edinburgh.

LYELL, C. 1868. *Principles of Geology*, Volume 2, 10th edn. John Murray, London, 82–89.

OTA, Y., WILLIAMS, D. N. & BERRYMAN, K. 1981. *Late Quaternary Tectonic Map of New Zealand*. Sheets Q27, R27 and R28, Wellington, Scale 1:50 000 with notes. New Zealand Department of Scientific and Industrial Research, Lower Hutt.

PARK, J. 1909. *The Geology of the Queenstown Subdivision*. New Zealand Geological Survey, Bulletin, **7**.

George Leslie Adkin (1888–1964): glaciation and earth movements in the Tararua Range, North Island, New Zealand

MARTIN S. BROOK

Geography Programme, School of People, Environment and Planning, Private Bag 11-222, Massey University, Palmerston North, New Zealand (e-mail: m.s.brook@massey.ac.nz)

Abstract: In the northern hemisphere the broad extent of glaciation had been mostly accepted by the start of the twentieth century, but in New Zealand at that time even the general picture of glaciation was uncertain. Julius von Haast had established that in the South Island the glaciers had extended out considerably from the Southern Alps, but the extent of glaciation in the North Island was unknown. Two rival viewpoints were put forward in 1909. James Park thought that there was a widespread ice sheet, which covered the South Island, Cook Strait and much of the North Island; Patrick Marshall thought that the extent of ice was much less, and argued that much of Park's evidence was spurious. The argument was 'somewhat resolved' by a young amateur geologist, Leslie Adkin, who showed that glaciation in the North Island was at most modest, and largely confined to the uppermost part of the main axial ranges. Adkin was a farmer and had no university education, but he published nearly 40 articles in scientific journals on topics as varied as Maori archaeology, and glacial and tectonic geomorphology. This article examines the evidence he adduced for the occurrence of a limited Pleistocene glaciation in the Tararua Range in the south of the North Island, and considers the role of the amateur in New Zealand geology.

In the northern hemisphere the general geographical extent of glaciation during the Pleistocene was accepted by most geologists by the beginning of the twentieth century, but there was considerable interest in knowing how far this was reflected in the southern hemisphere, especially in a small island country like New Zealand. Although geology was well established in Europe in the nineteenth century, few scientists in New Zealand in the second half of the century were specialized in the subject, with geological progress at that time being due to men with a combination of botany, zoology, geology and the physical sciences. This situation continued well into the twentieth century. Early geological work in New Zealand was shaped by geologists working at the Geological Survey (established 1865), along with the University of Otago, the Otago School of Mines, the Canterbury Museum and University, and Victoria University College, Wellington. The German émigré geologist Julius von Haast (1822–1887) was the first to establish that in New Zealand's South Island, glaciers had extended considerably out from the Southern Alps, onto the adjacent plains (Fig. 1). Julius von Haast (1822–1887) studied geology and mineralogy at the University of the Rhine, before arriving in New Zealand in 1858. He made a topographical and geological survey of the West Coast of the South Island for the Nelson Provincial Government, and was then appointed Provincial Geologist for Canterbury. Haast's Canterbury explorations (1861–1868) included journeys to the Rangitata, Ashburton and Rakaia headwaters, to the glaciers around Mt Cook, and to the West Coast south to Franz Josef Glacier (which he named).

This period of New Zealand geology, and science in general, was dominated by the 'New Zealand Society', originally set up by a group of enthusiastic Wellington citizens in 1851 (Fleming 1987). The 'New Zealand Institute' was established in 1867, as a colonial academy of science, bringing together the existing local scientific or 'philosophical' societies. These members made up a significant part of New Zealand science, and members who undertook scientific studies mostly did so as a gentlemanly or spare-time pursuit. They made important contributions to knowledge, however, and for many years outnumbered the professional or salaried scientists. In an age when the distinction between amateur and professional was less important than it appears now, and in a small isolated country, the influence of individuals and the interplay between personalities were clearly marked (Galbreath 1998). James Hector (1834–1907), the inaugural Director of the Geological Survey of New Zealand, was appointed manager of the New Zealand Institute, and editor of its *Transactions*. The period 1867–1903 in New Zealand science has been described by Fleming (1987) as one of 'Hector hegemony'. The period from 1903 to 1926 has become known as the 'Thomson Period' in

From: GRAPES, R. H., OLDROYD, D. & GRIGELIS, A. (eds) *History of Geomorphology and Quaternary Geology.* Geological Society, London, Special Publications, **301**, 315–328.
DOI: 10.1144/SP301.19   0305-8719/08/$15.00 © The Geological Society of London 2008.

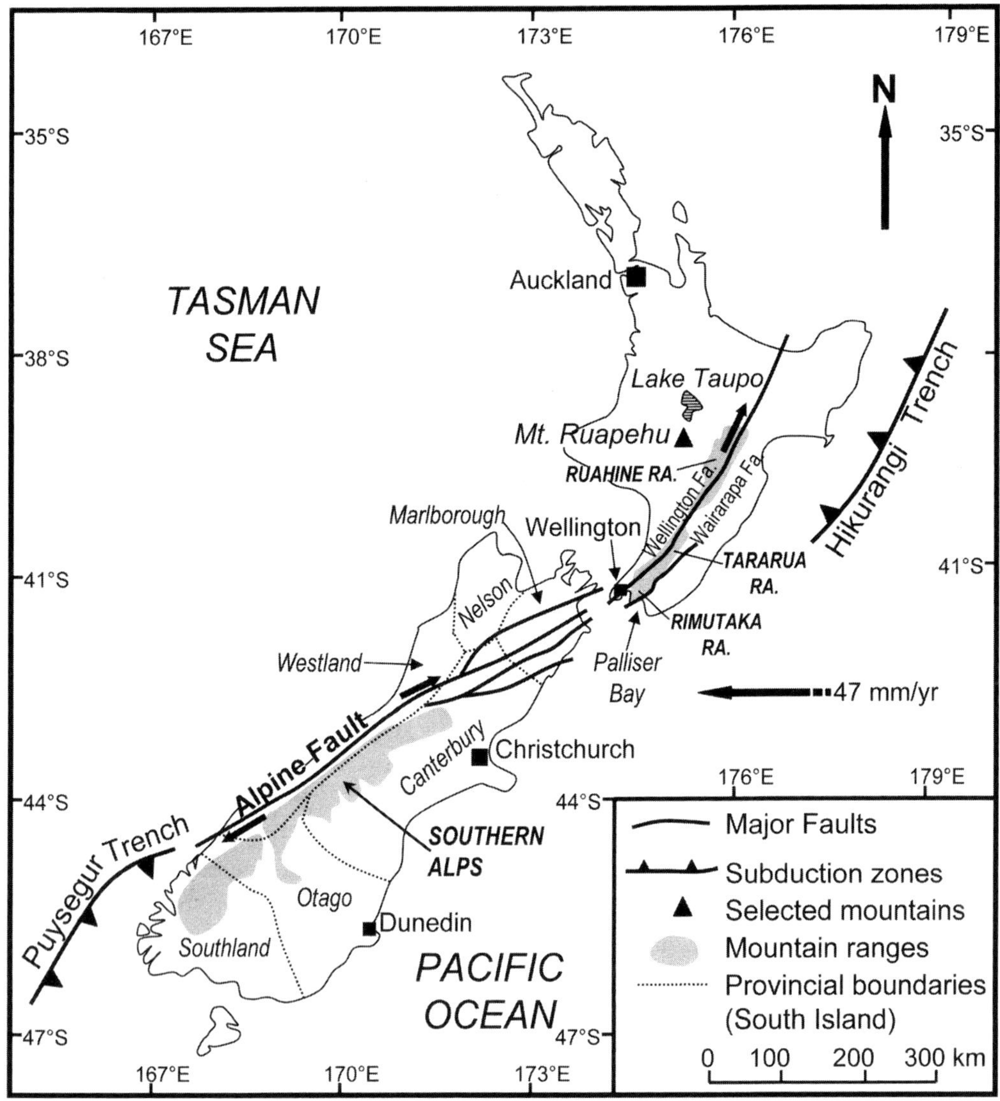

Fig. 1. Map of New Zealand showing geological/topographic features and localities mentioned in the text.

New Zealand, owing to the influence of George Malcolm Thomson (1848–1933), and his son, Allan (Galbreath 2002), amateur and professional scholar, respectively. For G. M. Thomson, a Member of Parliament for Dunedin North, the scientific study of the natural world was a passion, not a profession, and he well represented the older school of natural historians and gentleman scientists in New Zealand. His son Allan, by contrast, represented a new generation of university-trained professionals, being a skilled palaeontologist, trained at the Otago School of Mines (Galbreath 2002). The Thomsons made a formidable parliamentary–civil servant partnership to promote the cause of New Zealand science in government (Hoare 1984).

As well as professional geologists, there were in the early twentieth century a significant number of amateur geologists (who generally attended regular meetings at regional branches of the New Zealand Institute with their professional counterparts). John Hardcastle (1847–1927) was one such amateur. (Hardcastle spent much of his life working for the *Timaru Herald* newspaper, and as an amateur scientist published four important geological papers, two dealing with the origin of loess. He recognized that loess was normally produced during glacial periods, and that deposits of this material at Timaru (on the SE coast of the South Island) indicated several periods of cold-climate conditions. It was the first time that such inferences had been made, and it was many decades before the concept that presence of loess indicated former glaciation was accepted internationally.) He recognized that dust deposits SE of the Southern Alps were loess, formed during glacial periods (Hardcastle 1891). Alexander McKay (1875) had also noted that extensive sedimentary deposits east of the Southern Alps were moraines, produced during past glacial activity. However, the extent of glaciation in the North Island was uncertain in the early twentieth century.

In 1909 two rival viewpoints were put forward by geologists from two kindred institutions (Hocken 2003). James Park (1857–1946), at the Otago School of Mines and Professor of Mining at the University of Otago, thought that there was a former widespread ice sheet that covered the South Island, Cook Strait (the seaway between the North and South Islands) and much of the North Island (Park 1909a). Park joined the New Zealand Geological Survey in 1878 as a field assistant, later becoming a mining geologist. New Zealand was only partially explored at that time and his duties led him into remote areas such as King Country (the broad extent of hill country that lies to the west of Lake Taupo and Ruapehu), NW Nelson and NW Otago. His well-documented *Geology of New Zealand* (Park 1910), published when he was Director at Otago School of Mines, is considered by many as a milestone in New Zealand geology. Park's reasoning for the ice sheet was partly based on the subdued islands around the Wellington coastline (Fig. 1), which he thought were roches moutonnées (Park 1909b); and lahar deposits near the 2757 m active volcano Mt Ruapehu in the central part of the North Island (Fig. 1), which he assumed were of glacial origin (Park 1909c). Park also suggested that andesite blocks at the head of a catchment south of Mt Ruapehu had been transported over a drainage divide by ice (Park 1909d).

In contrast, Patrick Marshall (1869–1950), of the University of Otago, believed that the extent of ice in the North Island was much less, and argued that most of the evidence put forward by Park was spurious (Watters 2006). Marshall was appointed lecturer in geology at Otago University in 1901 and promoted to professor of geology and mineralogy in 1908. He accompanied the Philosophical Institute of Canterbury's 1907 expedition to the sub-Antarctic islands and worked on regional bulletins for the New Zealand Geological Survey. Scientific disputes with James Park, Director of the Otago School of Mines, may have been a factor in his decision in 1916 to resign from the university. Marshall also wrote *Geology of New Zealand* (Marshall 1912). In particular, Marshall (1910) pointed out that the lack of *roches moutonnées*, glacial till and striated boulders near the coasts of both the North and South Islands indicated that Pleistocene glaciers did not reach the coast. Marshall (1910, p. 345) was also sceptical of the existence of the striated erratic boulders identified as such by Park (1909c) south of Mt Ruapehu: 'the blocks are not markedly angular and no striated boulders have yet been found in the deposits' (Marshall 1910, p. 345). Be that as it may, despite visiting the glaciers and moraines on Mt Ruapehu, neither of these early geologists ventured into the axial ranges of the North Island to examine further the possibility of Pleistocene glaciation. The axial ranges are composed of grey sandstone and mudstone, referred to in New Zealand as greywacke. They are of Mesozoic age. Tertiary sedimentary rocks formerly covered the greywackes but have now been largely stripped from the ranges.

The axial ranges of the North Island are over 2000 m high in the Ruahine Range, SE of Ruapehu (Fig. 1), although no evidence of glaciation has ever been reported there. Further south along the axial ranges, in the southern part of the island, the dominant axial range is the Tararua, which reaches a height of 1571 m at Mitre. It was not until early in the twentieth century that indications that the range might formerly have been glaciated were first observed by G. Leslie Adkin (1888–1964). However, since he made his first observations in Park Valley (Fig. 1) in the centre (and highest) part of the range, the conclusiveness of his original observations and arguments (Adkin 1911a) has remained equivocal. Since his work, the majority of researchers (e.g. Willett 1950; Stevens 1974; Moffat 1990; Pillans *et al.* 1993) have accepted Late Quaternary glaciation of the Tararua Range, but some scepticism remains among modern geologists (e.g. Shepherd 1987).

G. L. Adkin: background

Leslie Adkin (Fig. 2) was born in Wellington on 26 July 1888, the first of seven children of William George Adkin, a draper, and his wife, Annie, née Denton. His mother's family were keen naturalists and photographers, and purchased 100 acres of land on the Horowhenua coastal plain to the north of Wellington, adjacent to the Tararua Range (Fig. 3). During his 2 years as a boarder at Wellington College in 1903–1904 Adkin developed an enthusiasm for collecting plants and rocks, and learned to process his own photographs. When he returned to work on the family farm he

Fig. 2. Self-portrait of Leslie Adkin, taken in September 1913 on his farm, Horowhenua coastal plain at Levin, looking east. To Adkin's right are some of the high peaks of the Tararua Range: Waiopehu (1094 m) and Twin Peak (1097 m). (Source: Leslie Adkin/Museum of New Zealand Te Papa Tongarewa (A.005940)).

began a diary, which he kept to within a month of his death. It was an account of farm, community and family life, and it is also a record of his travels and scientific observations (Dreaver 2006). In addition to his daily diary and published work, Adkin also built up a collection of geological notebooks, of personal records of geological observations, with an emphasis on the landscape of the lower North Island, within about 50 km of Wellington (the 46 geological notebooks, each indexed by Adkin, are held in the library of the Institute of Geological and Nuclear Sciences, Lower Hutt, Wellington). Whilst his future geological adversary Charles Cotton was Lecturer in Geology at Victoria University College, Adkin carried out geological and geomorphological investigations in the vicinity of the family farm in his spare time, notably during the evenings and on Sundays. (Charles Cotton (1885–1970) trained at the University of Otago. In 1909 he became the first lecturer in geology at Victoria College (now Victoria University of Wellington). Cotton was inspired by the writings of American physiographers such as G. K. Gilbert and, in particular, W. M. Davis (Stevens 2006) (see Grapes 2008).) He also extended his education by reading issues of the *Transactions of the New Zealand Institute*, the leading scientific journal in New Zealand at that time.

Adkin's first foray into geology and geomorphology was in the Tararua Range, and the adjacent coastal plain on which Levin and the wider

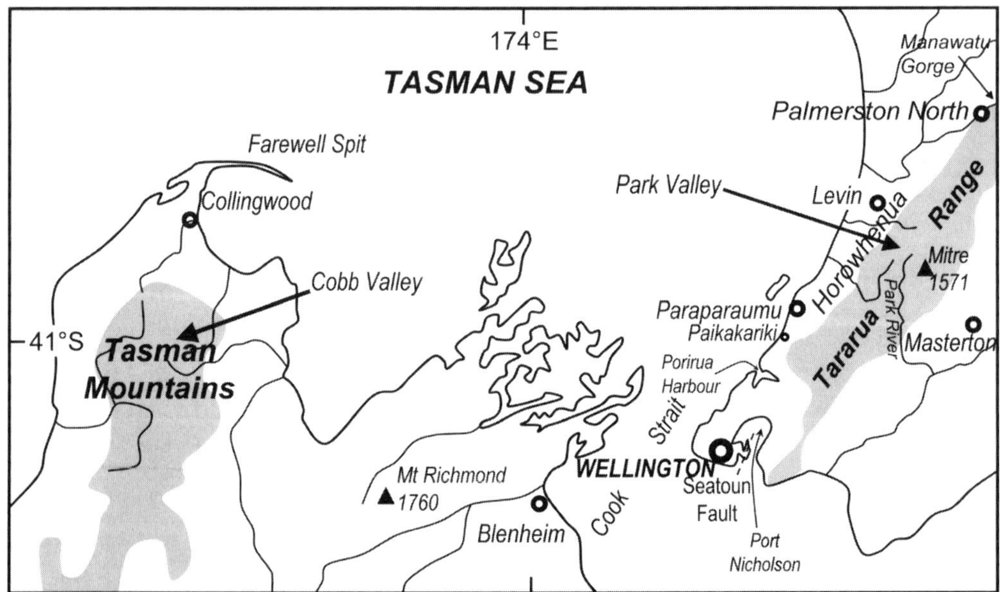

Fig. 3. Map showing locations in southern North Island and northern South Island, New Zealand, mentioned in the text. Note the similar latitude (about 25 km apart) of Cobb Valley and Park Valley.

Horowhenua County are situated (Adkin 1911*b*). His grandfather, George Denton, had introduced him to the Wellington Philosophical Society, and there he delivered a paper about the Horowhenua coastal plain to one of its meetings in 1910, having been given only 2 days' notice to prepare his presentation.

Adkin and glaciation in the Tararua Range

The Tararua Range forms a backdrop to the Horowhenua coastal plain (Fig. 3). In 1909, although some of the highest peaks in the range had been surveyed, no Europeans are known to have traversed the range in either direction. In February 1909 Adkin and an older friend, Len Lancaster, set out on a west-to-east crossing of the range from Levin to Masterton, and it was on this journey that Adkin noted that the heads of some of the valleys near the crest of the range appeared U-shaped and therefore had presumably been excavated by glaciers, although this hypothesis contradicted the prevailing view of the time. Adkin named the deep, U-shaped valley in the centre of the range, Park Valley, after James Park, whose work he had read (Fig. 4). He also noted that the local Maori, the Muaupoko, spoke of a 'lost lake' in the hills and Adkin thought that this could have been a moraine-dammed lake.

Adkin, Lancaster and another colleague, Harry Thompson, made another trip in March 1911 seeking further evidence for glaciation in Park Valley. As they reached Arete Peak at the valley head on 12 March 1911, they noted small uphill-facing ('anti-slope') scarps, which suggested the de-buttressing effect of glacier recession just below the ridge tops. He wrote: 'there were two more trenches (containing the tarns) striking N and S and paralleled [*sic*] terrace like ridges – the whole seeming to indicate faults and slipping of the hill side on a large scale containing tarns, which represent slope failures' (Adkin, Diary, 12 March 1911). However, it was the topography of Park Valley, to the SW, that most intrigued Adkin:

the topography of the upper portions of this valley is undoubtedly of glacial origins. The head of the valley, which is U-shaped, is a glacial cirque, the bounding precipices of which were preserved on the southwest face of Arete Peak, half-way up. The continuity of the precipices is broken by three large U-shaped hanging valleys. The largest of these lies on the south side of Arete Peak and has a small cirque at its head.

(Adkin, Diary, 12 March 1911)

Adkin's observations in Park Valley were recorded in a field sketch (Fig. 5). He also recorded that spurs had been truncated, and that there was an abrupt transition from U-shaped to V-shaped cross-profiles 3.2 km from the valley head, presumably marking the down-valley extent of glaciation. He also

Fig. 4. Park Valley, Tararua Range, photographed from the southern slopes of Arete. Note the broad, U-shaped valley bottom, indicative of glacial erosion (source: M. S. Brook, January 2006).

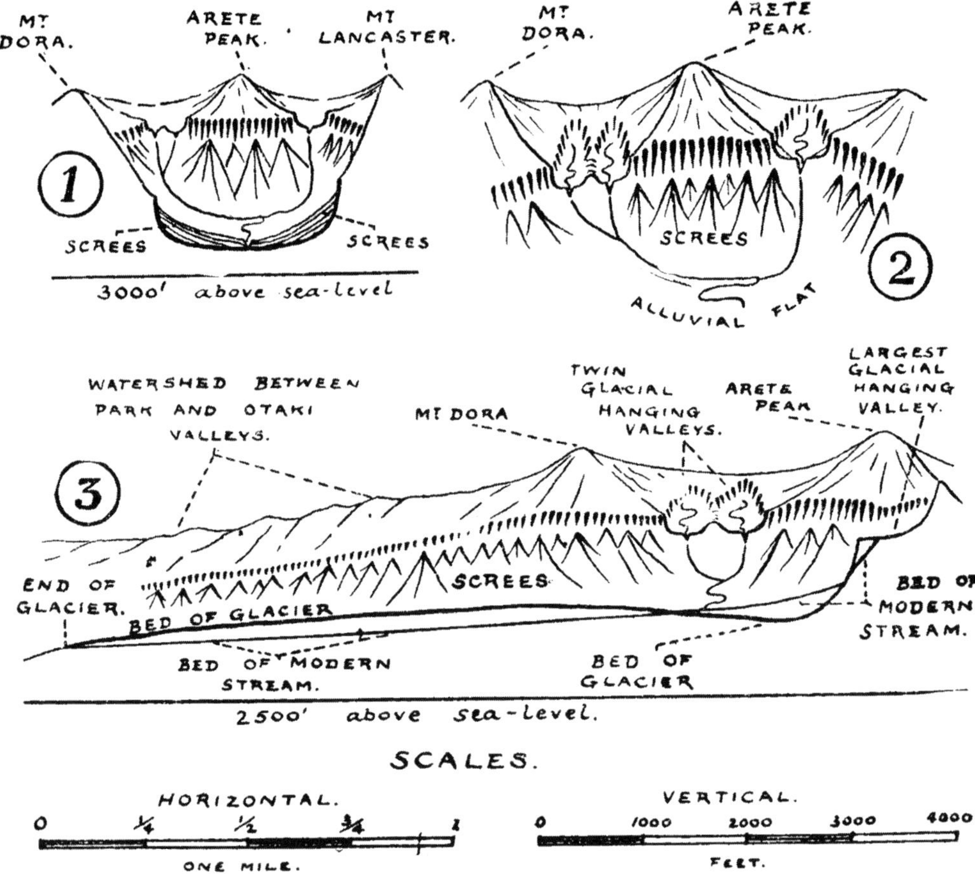

Fig. 5. Field sketch of the long-profile of Park Valley (from Adkin 1911*a*). Note the three 'cirques' (glacial hanging valleys in '2' and '3') identified by Adkin, which form tributaries of the main valley.

noticed, however, that other glacial landforms, such as striae, roches moutonnées or a terminal moraine, were not present in the area that had supposedly been glaciated, according to the apparent form of the valley (Adkin 1911*a*). Intriguingly, although Adkin depicted the presence of a ridge (Figs 6–8) between the two westernmost cirques (Adkin termed these 'U-shaped hanging valleys'), he did not consider that it might have been a lateral or medial moraine. If he had investigated this elongate ridge further and recognized its glacial origin, this might have convinced his later detractors. Adkin (1911*a*) noted further evidence of glaciation in the form of U-shaped valleys and cirques in four other valleys in the central Tararua Range – Waiohine, Arete Stream, Dorset Creek and South Mitre Stream – although none of these are as spectacularly shaped as Park Valley (Adkin 1911*a*).

On his return from the Park Valley, Adkin worked in his spare time on a paper focused on glaciation and the long-term evolution of the Tararua Range, and, at the invitation of the Wellington Philosophical Society, he presented his findings at its meeting on 6 September 1911. Unbeknown to Adkin, however, Cotton was promoting a different hypothesis about the development of the Tararua Range, which was presented the following month to the Philosophical Society (Cotton 1912). Adkin was thus challenging the theories that: (1) the Tararua Range had not been glaciated (e.g. Marshall 1910); and (2) that the topography of the Tararua Range was a northern continuation of the 'geanticlinal' (i.e. arising from the inversion of a geosyncline) structure of the South Island (Hutton 1885; Park 1910). This theory contended that the axial ranges of both islands were part of a geosyncline that had been subsequently folded upwards into a large anticlinal structure (e.g. Park 1910).

Cotton's theory of the geological development of the range was simpler, namely that the valleys

Fig. 6. Adjoining 'hanging valleys' on the western side of Park Valley, Tararua Range. The 'hanging valleys' are separated by a debris ridge, although the 'hanging valley' on the left (west) of the debris ridge is little more than a small cirque basin (source: M. S. Brook).

Fig. 7. The elongate debris ridge on the western side of Park Valley, separating the two 'hanging valleys' identified by Adkin (1911*b*), photographed in 2005 from the eastern side of Park Valley. Although not discussed by Adkin, the ridge with its smooth, even surface and truncated down-valley end is now identified as a moraine (source: M. S. Brook).

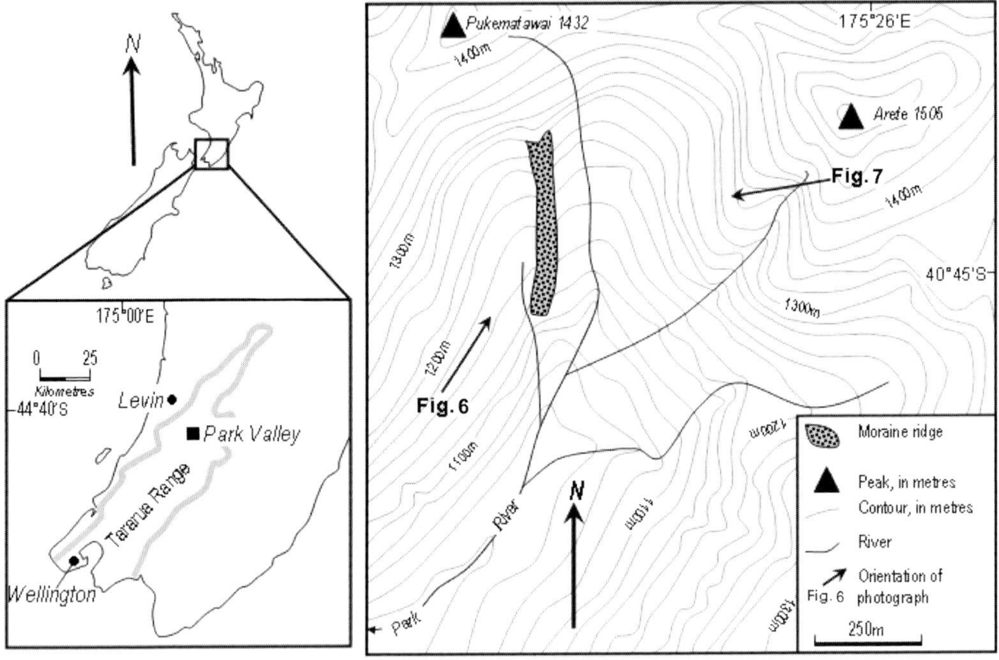

Fig. 8. Map of Park Valley showing location of the moraine in the Park Valley and the location of Figures 6 and 7.

were formed by 'adjustment to structure', whereby the rocks had been uplifted and had formed a peneplain, but were also split by longitudinal faults, forming alternating bands of resistant and weak rocks, the weaker bands having become incised into valleys. This theory largely followed ideas put forward by William Morris Davis (1850–1934) in the United States, by whom Cotton was heavily influenced. Davis's (1909) pioneering approach, involving the systematic description and interpretation of landforms through precise field observation, was applied to the New Zealand landscape by Cotton, especially using sequential diagrams. Cotton (1912, p. 264) referred to the Tararua Range as having an 'Appalachian-type' topography, although he cautioned that the analogy with the Appalachian Mountains should not be taken too far, as early cycles of planation were not completed. Adkin's 'anticlinorium theory' was somewhat more complex, and he concluded that the range was formed by east–west compression in the Earth's crust, which had buckled the range into a large anticline, traversed by a series of active faults. The Wairarapa Fault, one of several large NNE–SSW-trending faults bordering the eastern edge of the Rimutaka–Tararua ranges (Fig. 1), was thought by Adkin to be further evidence of this compression. This fault

had ruptured in 1855 during a M8.0+ earthquake that both uplifted and horizontally offset the ground by up to 3.8 and 17.5 m, respectively. As an amateur scientist living in provincial New Zealand, Adkin had limited access to overseas scientific literature. However, this meant that he was not necessarily influenced by preconceived ideas, and was able to develop his own thoughts based on his own judgement and careful field observations.

Adkin's presentation in September 1911 was not met with universal approval, in particular by Cotton (Dreaver 1997). Adkin noted that Cotton said that 'he considered the evidence of former glaciation inconclusive; that the photographs should show the glacial features; that there had been one period of folding in the Jurassic times and none since; that my [Adkin's] anticlinorium hypothesis was improbable as there had been insufficient time for base levelling [a Davisian concept] several times over' (Adkin, Diary, 7 September 1911). Furthermore, the Geological Survey palaeontologist, J. Allan Thomson (1881–1928), thought that 'the supposed glacial topography was more likely due to stream erosion ... [as] the evidence of the former was inadequate as there was no striae, roches moutonnées or moraines' (Adkin, Diary, 7 September 1911). Nevertheless, and perhaps more importantly, Adkin did win support from the Director of the Geological Survey, Percy Gates Morgan (1867–1927), who said he 'considered my case as regards former glaciation and generally backed me up' (Adkin, Diary, 7 September, 1911).

It must have been daunting for the amateur, Adkin, to present ideas that were contrary to those of some of the leading geologists in Wellington's scientific community. In fact, it was but the first of several heated debates between Adkin and Cotton on the geology and geomorphology of the lower North Island (Adkin 1911*b*, 1919; Cotton 1912, 1918). Despite the opposition, however, Adkin's work on glaciation was published soon after (Adkin 1911*a*). The central area of the Tararuas had not been previously geologically mapped, so Adkin (1911*a*) included a detailed sketch of the long profile of Park Valley (see Fig. 5), as well as a basic geomorphological map (Fig. 9). However, a further paper outlining his hypothesis of the long-term geological development of the Tararuas was rejected by the editor, Charles Chilton, of the *Transactions of the New Zealand Institute* in 1913. The grounds for the rejection were thought by Adkin to be that his paper ran counter to the Davisian 'adjustment to structure' theory preferred by Cotton (Dreaver 1997, p. 81). Although it is difficult to ascertain whether this was the case, the editor did remark that Adkin's

manuscript was 'based on too slender evidence, and . . . [the theory was] erroneous' (Dreaver 1997, p. 81).

Work on glaciation in the Tararuas since Adkin

Opinions of the nature and extent of glaciation in the Tararuas and the North Island have ranged from Park (1909*a*), who proposed that ice streams may have reached as low as the Wairarapa Valley and Cook Strait, to those of Marshall (1910) who claimed that few, if any, of the North Island mountains were glaciated during the Pleistocene. Speight (1939), Willett (1950) and Stevens (1974) accepted Adkin's (1911*a*) contention that glaciers may have existed in the Tararua Range. However, more recently, Mike Shepherd, a geomorphologist at Massey University in Palmerston North, rejected Adkin's hypothesis of glaciation, proposing that the U-shaped valleys were caused by periglacial processes. Shepherd (1987) thought that the centre of the Tararua Range represented an 'old' landscape and that active scree slopes indicated periglacial processes throughout the Pleistocene would have been sufficient to steepen slopes, and to develop a U-shaped

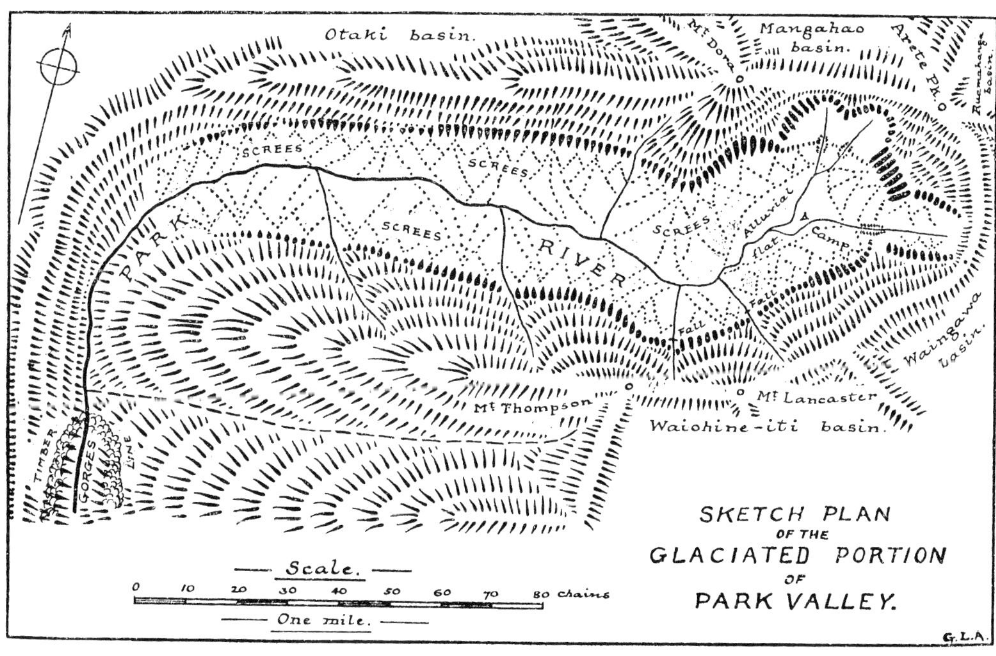

Fig. 9. Adkin's geomorphological map of Park Valley (Adkin 1911*a*).

valley profile. However, Shepherd (1987) made no mention of the elongate debris ridge (Figs 5–7) at the head of Park Valley, which was first documented by Moffat (1990). This ridge has since been excavated by Brook & Crow (2008) who have demonstrated that it is a lateral moraine. Topographic analyses of the Tararua Range (Brook & Brock 2005; Brook *et al.* 2005) have compared valley cross-profiles with data from other glacial valleys (e.g. Li *et al.* 2001) in Tian Shan (China), New Zealand's Southern Alps (Fig. 1), Patagonia and the Beartooth Mountains in Montana. These analyses confirm that the Park Valley has been glacially eroded, thus vindicating Adkin's (1911*a*) theory. Furthermore, a palaeo-glacier reconstruction of Park Valley (Brook *et al.* 2005), and published evidence for valley glaciation in Cobb Valley (Shulmeister *et al.* 2005), NW Nelson, 25 km in latitude to the south (Fig. 3), further support Adkin's (1911*a*) hypotheses. Following Shulmeister *et al.* (2005), a cooling of 4 °C would lower the glacier equilibrium line altitude by approximately 670 m, which is substantially below the crest of the Tararua Range.

Subsequent geological work by Adkin

In addition to his Tararua work, Adkin undertook other geological investigations, including a study of the geomorphic evolution of the Horowhenua coastal plain, the coastal lowland to the west of the Tararua Range (Adkin 1919) (Fig. 3). This was in response to a paper by Cotton (1918) that had rejected Adkin's (1911*b*) work on the geological development and Late Quaternary evolution of the Horowhenua area. Cotton (1918, p. 220) wrote: 'the account is somewhat difficult to follow and his conclusion ... is an extremely doubtful one'. In his reply to Cotton, Adkin (1919, p. 108) discussed 'the main points raised by the dissension of opinion between Mr Cotton (1918) and myself'. Adkin was forthright in this paper, noting that 'Cotton's views ... appear to result from misconception arising from his observations'; and 'one of the chief causes of Cotton's divergent opinion ... is the greater complexity of the lowland than he at present recognises' (Adkin 1919, p. 113). Adkin's paper contained detailed diagrams, including a conceptual model of the formation of the coastal plain. The model comprised a seven-stage schema of how the Ohau River, draining the western slopes of the Tararua Range, evolved, including aggradation and dissection of the floodplain, forming terraces and progradation of the coast. Despite

his critique of Adkin's (1911*b*) earlier work, Cotton did not respond to Adkin's (1919) thoroughly prepared counter-analysis through the journal literature.

Adkin's interest in geology after a 6-year hiatus in publication (although he still made geological observations) appears to have been rekindled by the Horowhenua coastal plain dispute with Cotton. Indeed, he locked horns in the literature with Cotton again, who had published two papers that referred to the Horowhenua coastal area (Cotton 1912, 1918), including a study of river captures north of Levin (Adkin 1920) and a study of the geomorphic development of Porirua Harbour (Adkin 1921*a*) (Fig. 3). Adkin (1920) was particularly critical of Cotton's (1918) 'adjustment to structure' theory of the Tararua Range evolution. Adkin (1920, p. 184) concluded that Cotton's (1918) theory that valleys in the Tararuas are aligned along weak bands of rock, caused by ancient folding and subsequent faulting, 'appears inadequate when confronted by the ... transverse ridges and certain arresting characteristics of the hydrography ... [so that] there are reasons for the belief that orogenic folding and uplift ... is the most satisfactory explanation'. Unlike Cotton (1918), Adkin clearly thought the Tararua Range was still being uplifted. In 1921 he hypothesized that the 1855 Wellington Earthquake had uplifted the shoreline around Porirua Harbour, 20 km to the north of Wellington (Adkin 1921*a*), coeval with the well-known uplift that had occurred along the Wellington waterfront (Stevens 1974). As highlighted by Adkin (1921*a*, p. 146), Cotton's (1912, p. 257) contrasting view of the Porirua inlet, focusing 2 km inland from the coast was that: 'a downward movement of 30 ft or 40 ft subsequent to the general movement of elevation of the Wellington Peninsula had occurred'. Adkin (1921*a*, p. 257) also drew attention to Cotton's (1912, p. 257) statement on the Porirua coastline that 'there appears to have been little or no movement either up or down in 1855. Raised rock platforms similar to those at Wellington are not found'. Adkin (1921*a*, p. 146, emphasis added) boldly stated that 'I shall be able to show, however, that raised shore-platforms of wave-planed rock *do* occur along a very considerable part of the Porirua shore-line'. In mapping the recently uplifted shore platforms, he noted that they 'form one of its most conspicuous features'. He also quoted an elderly local person who had memories of the pre- and post-1855 Porirua shoreline, thus demonstrating that at least 1 m of uplift had occurred.

In the early 1920s Adkin expanded his interests with a paper on engineering geology, in relation to a hydro-electric scheme in the Tararua Range (Adkin 1921*b*). He also conducted a study of the origin and evolution of the Manawatu Gorge (Adkin 1930) (Fig. 3), formed by incision of an antecedent river that crosses the axial ranges of the North Island near Palmerston North (the idea of which was, in this case, Davisian).

In 1946, at the age of 58, Adkin was appointed a full-time palaeontologist's assistant at the New Zealand Geological Survey in Wellington. He had approached a friend at the Lands and Survey Department, who then contacted the Director of the Geological Survey, Montague Ongley, and this led to an invitation to Adkin to visit the director to discuss employment opportunities. Adkin was interviewed by Ongley and offered the post a few weeks later (Dreaver 1997). He was finally taking up the work for which he had groomed himself over his lifetime. While with the Geological Survey, he presented two papers at the 1947 Royal Society Congress in Wellington. One of them was on mounds in Horowhenua around 75 cm high, which he interpreted as surficial evidence of artesian springs (Adkin 1948). The other resurrected the topic that had been rejected in 1913 – the long-term geomorphic development of the Tararua Range (Adkin 1949). Of Cotton's (1918) theory, Adkin (1949, p. 261) wrote: 'it has become the current fashion to ascribe the major physiographic trends (ridges and drainage lines) to adjustment to structure. Among the adherents ... reasoning in a circle seems to have occurred'. Adkin (1949, p. 272) concluded that 'the deformation hypothesis herein outlined provides an adequate explanation of the form and origin of the ... Tararua–Rimutaka Range'. In his diary on 22 May 1947 he wrote that: 'Professor Cotton got up to annihilate me but found he had to agree to much of my thesis ... 36 years later it seems that I have vindicated my thesis'.

Adkin's 'anticlinorium theory', was essentially that a cover of Tertiary sediments, now largely removed, had been draped over a rising anticlinal ridge of greywacke rocks, formed by east–west compression across the North Island, the idea of which has stood the test of time. Furthermore, Adkin (1949) had recognized the asymmetric nature of the structure of the Tararuas, due to the ranges being north-west tilted (and folded and faulted) blocks, so that the 'anticlinal' structure is more apparent than real. Cotton (1956) eventually accepted Adkin's work and graciously acknowledged that his own 'theories ... , [which were] always consistently opposed by Adkin (1919, p. 291), are now discredited, at any rate in their widest application; and a hypothesis of tectonic origin of many features, including most of those with NNE elongation, is here adopted' (Cotton 1956, p. 763).

Adkin's (1949) paper failed to refer to the work of Harold Wellman (1948) on the geological evolution of the Tararua Range. (Harold Wellman (1909–1999), born in England, and educated in New Zealand at Canterbury University College and Victoria University College, Wellington (MSc 1941, DSc 1956), joined the permanent professional staff of the Geological Survey of New Zealand in 1938. Between 1937 and 1952 Wellman published 36 papers and reports, his most famous discovery, the Alpine Fault, was a landmark for New Zealand geology (Grapes 2006). Wellman was a very influential geologist and professor at Wellington. He made his name by the mapping of the Alpine Fault in the South Island and recognizing that it had a dextral strike-slip of 480 km. See Nathan (2005).) Nevertheless, it must have been gratifying for Adkin to find his theories regarding east–west crustal compression leading to formation and uplift of the range were finally accepted. Cotton's work had drawn heavily on ideas proposed by Davis (1899) in landscapes in the United States. But Cotton's (1912, 1918) hypothesis that the Tararua Range summits were remnants of an eroded peneplain, with valleys aligned along weak bands of bedrock ('adjustment to structure'), similar to the Appalachians is now rejected. Today, it is believed that the Tararua Range, part of the axial ranges of New Zealand's North Island, is part of the on-land sector of the Hikurangi forearc, forming above the subducting Pacific plate (Neef 1999). The landscape is actively evolving, with uplift rates in this part of the forearc recently estimated at $1.0-1.6$ mm year^{-1} (Litchfield & Berryman 2006).

Adkin (1951) made a further study of the southernmost part of the Horowhenua coastal plain at Paekakariki (Fig. 3), adding sedimentological data from new road-cutting sections. This study reiterated his disagreement (Adkin 1919) with Cotton's hypothesis on the geological development of the Horowhenua lowland (Cotton 1912, 1918). Adkin (1953) analysed the neotectonics of the lower North Island, outlining various landforms resulting from active strike-slip and normal faulting. The study exemplified the wide variety of journal literature and textbooks in the Geological Survey that he now had access to, as well as his own increasing experience as a researcher. Indeed, Adkin (1953, p. 503) referred to overseas work, introducing phrases to describe the geomorphology around fault traces, and he also put tectonically-active New Zealand into a global context: 'New Zealand is a tectonically active region sharing this state with some others, notably, Western United States, Japan, and Alaska; the distribution is circum-Pacific and is evidence of Pleistocene–Recent diastrophic stress in this segment of the globe' (Adkin 1953, p. 501).

In 1956 Adkin had a study published concerning the origin of the depressions that form the Port Nicholson basin (Wellington Harbour) and the Porirua Harbour (Fig. 3). Adkin (1951, 1953) had previously mapped the Porirua end of this depression and speculated as to its origin. He analysed the elevation of Cotton's (1926, 1956) 'K-surface' (see Grapes 2008, p. 301). This is a peneplain that can be observed on a number of summits around the outskirts of the Wellington city area, and Cotton (1926, 1956) analysed its morphogenesis, concluding that the surface represents a planation surface that was subsequently buried, then uplifted, before being exhumed and (is currently being) dissected (see Grapes 2008). Adkin (1953) recognized that the elevation of Cotton's (1926) 'K-surface' was around 230 m above the adjacent Port Nicholson–Porirua depression to the east. Wellman (1955) had previously maintained that the faults in the Wellington city region were mostly strike-slip in nature. Adkin (1956) had a different view: that marked subsidence had occurred on what were thought to be predominantly strike-slip faults. His discovery and description of the normal nature of movement on the Seatoun Fault (Fig. 3) was taken as evidence of this. It had been a typical Adkin study, resulting from happening upon a fault scarplet while out walking with his wife. He then undertook detailed analysis in the field, consulted the relevant literature and concluded with an idea that challenged generally accepted theory (e.g. Wellman 1955), before becoming an established concept (e.g. Stevens 1974).

Cotton's (1956) review article on the tectonic history of the Wellington region drew heavily on the empirical observations published over the years by Adkin, and was generally accepting of Adkin's conclusions. However, Cotton (1956, p. 768) selectively quoted Adkin's (1951) study as indicative that Adkin's views on the tectonics of the area were that motion on the Wellington Fault (Fig. 1) is simply normal, rather than strike-slip in nature. In reality, Adkin's (1956) later study clearly outlined his interpretation that the Wellington Fault (and other faults in the region) have undergone both normal and strike-slip motion (Adkin 1956, p. 273).

Conclusions

Adkin's retirement from the New Zealand Geological Survey in 1955 allowed him to focus on Maori archaeology, his other great interest, and he held seats on the councils of both the New Zealand Archaeological Association and the Polynesian Society. He was an archaeologist of some reputation (he was often sent artefacts by members of the public) and was invited by the Dominion Museum to join an expedition to Palliser Bay, Wellington, in June 1959. Adkin died in 1964 leaving a significant legacy. He was a meticulous collector and recorder of information, and his co-authored *Bibliography of New Zealand Geology to 1950* (Adkin & Collins 1967, published posthumously) contained 5310 references. Perhaps a fitting tribute to his work is the Institute of Geological and Nuclear Sciences (IGNS) *Bibliography of Wellington Geology* (Begg *et al.* 1994), where 37 pages of the 77-page report contain lists of the contents of Adkin's 46 geological field notebooks. Adkin was both an inductive and deductive researcher, sometimes generating new theories, and often testing others' established hypotheses. He was also a pioneer in his ideas about glaciation in the Tararua Range, whilst his ideas on crustal compression and mountain building as far back as 1909 were ahead of their time.

In Britain, Torrens (2006) has highlighted the role of the 'amateur' in the development of geology, noting that such people can have great advantages, often not being constrained by paradigms. He concludes that the often significant contributions made by such geologists should encourage the idea that the 'amateur' geologist is not 'second rate'. I posit that this idea should similarly be applied to New Zealand geology in general and Adkin in particular.

I am indebted to A. Dreaver and S. Nathan for discussions and advice. The staff at the Alexander Turnbull Library, within the National Library of New Zealand, assisted with access to the Adkin Collection, and M. Shepherd is thanked for introducing me to Tararua glaciation and Adkin's early work. This research benefited from Massey University funding (MURF Grant 05/2059). M. Johnston and D. R. Oldroyd are thanked for assistance on drafts of this paper.

References

ADKIN, G. L. & DIARIES. Held at Alexander Turnbull Library, National Library of New Zealand, Wellington (micro-MS-0600).

ADKIN, G. L. 1911*a*. The discovery and extent of former glaciation in the Tararua Ranges, North Island, New Zealand. *Transactions of the New Zealand Institute*, **44**, 308–316.

ADKIN, G. L. 1911*b*. The post-Tertiary geological history of the Ohau River and of the adjacent coastal plains, Horowhenua County, North Island. *Transactions of the New Zealand Institute*, **43**, 496–520.

ADKIN, G. L. 1919. Further notes on the Horowhenua coastal plain and the associated physiographic form.

Transactions of the New Zealand Institute, **51**, 108–118.

ADKIN, G. L. 1920. Examples of drainage readjustment on the Tararua western foothills. *Transactions of the New Zealand Institute*, **52**, 183–191.

ADKIN, G. L. 1921*a*. Porirua Harbour: a study of its shoreline and other physiographic features. *Transactions of the New Zealand Institute*, **53**, 144–156.

ADKIN, G. L. 1921*b*. Mangahao hydro-electric scheme. *New Zealand Journal of Science and Technology*, **4**, 1–4.

ADKIN, G. L. 1930. The origin of the Manawatu Gorge. *New Zealand Journal of Science and Technology*, **11**, 353–356.

ADKIN, G. L. 1948. On the occurrence of natural artesian springs in the Horowhenua district. *New Zealand Journal of Science and Technology*, **B29**, 266–272.

ADKIN, G. L. 1949. The Tararua Range as a unit of the geological structure of New Zealand. *Transactions of the Royal Society of New Zealand*, **77**, 260–272.

ADKIN, G. L. 1951. Geology of the Paekakariki area of the coastal lowland of western Wellington. *Transactions of the Royal Society of New Zealand*, **79**, 157–176.

ADKIN, G. L. 1953. Late Pleistocene and recent faulting in the Otaki–Porirua and Dalefield–Waipoua districts of South Wellington, North Island, New Zealand. *Transactions of the Royal Society of New Zealand*, **81**, 501–519.

ADKIN, G. L. 1956. The Seatoun Fault, Miramar Peninsula, Wellington. *Transactions of the Royal Society of New Zealand*, **84**, 273–277.

ADKIN, G. L. & COLLINS, B. W. 1967. A bibliography of New Zealand geology to 1950. *New Zealand Geological Survey, Bulletin*, **65**, 1–244.

BEGG, J. G., MAZENGARB, C. & ZUCCHETTO, R. G. 1994. *A Bibliography of Wellington Geology*. Institute of Geological and Nuclear Sciences, Lower Hutt, New Zealand.

BROOK, M. S. & BROCK, B. W. 2005. Valley morphology and glaciation in the Tararua Range, southern North Island, New Zealand. *New Zealand Journal of Geology and Geophysics*, **48**, 717–724.

BROOK, M. S. & CROW, T. V. H. 2008. A debris ridge in Park Valley, Tararua Range, New Zealand, as evidence for Pleistocene glaciation. *New Zealand Journal of Geology and Geophysics*, **51**, 23–28.

BROOK, M. S., PURDIE, H. L. & CROW, T. V. H. 2005. Valley cross-profile morphology and glaciation in Park Valley, Tararua Range, New Zealand. *Journal of the Royal Society of New Zealand*, **35**, 399–407.

COTTON, C. A. 1912. Notes on Wellington physiography. *Transactions of the New Zealand Institute*, **44**, 245–265.

COTTON, C. A. 1918. The geomorphology of the coastal district of southwestern Wellington. *Transactions of the New Zealand Institute*, **50**, 212–222.

COTTON, C. A. 1926. *The Geomorphology of New Zealand: An Introduction to the Study of Landforms*. Dominion Museum, Wellington.

COTTON, C. A. 1956. Tectonic features in a coastal setting at Wellington. *Transactions of the New Zealand Institute*, **84**, 761–790.

DAVIS, W. M. 1899. The geographical cyle. *Geographical Journal*, **14**, 481–504.

DAVIS, W. M. 1909. *Geographical Essays*. Ginn & Co., Boston (reprinted by Dover Publications, New York, 1954).

DREAVER, A. 1997. *An Eye for Country*. Victoria University Press, Wellington.

DREAVER, A. 2006. Adkin, George Leslie 1888–1964. *Dictionary of New Zealand Biography* (http://www.dnzb.govt.nz/), updated 7 April 2006.

FLEMING, C. A. 1987. Science, settlers, and scholars: the centennial history of the Royal Society of New Zealand. *Royal Society of New Zealand Bulletin*, **25**, 1–353.

GALBREATH, R. 1998. *DSIR: Making Science Work for New Zealand*. Victoria University Press, Wellington.

GALBREATH, R. 2002. *Scholars and Gentlemen Both: G. M. and Allan Thomson in New Zealand Science and Education*. Royal Society of New Zealand, Wellington.

GRAPES, R. H. 2006. Wellman, Harold William 1909–1999. *Dictionary of New Zealand Biography* (http://www.dnzb.govt.nz/), updated 7 April 2006.

GRAPES, R. H. 2008. Sir Charles Cotton (1885–1970): International geomorphologist. *In*: GRAPES, R. H., OLDROYD, D. R. & GRIGELIS, A. (eds) *History of Geomorphology and Quaternary Geology*. Geological Society, London, Special Publications, **301**, 291–308.

HARDCASTLE, J. 1891. On the Timaru loess as a climate register. *Transactions of the New Zealand Institute*, **23**, 324–332.

HOARE, M. E. 1984. The Board of Science and Art 1913–1930: a precursor to the DSIR. *In*: HOARE, M. E. & BELL, E. L. G. (eds) *In Search of New Zealand's Scientific Heritage. Royal Society of New Zealand Bulletin*, **21**, 25–48.

HOCKEN, A. G. 2003. *Geology at the University of Otago: The First 100 years*. Geological Society of New Zealand, Miscellaneous Publication, **115**.

HUTTON, F. W. 1885. A sketch of the geology of New Zealand. *Quarterly Journal of the Geological Society, London*, **41**, 191–220.

LI, Y., LIU, G. & CUI, Z. 2001. Glacial valley cross-profile morphology, Tian Shan Mountains, China. *Geomorphology*, **38**, 153–166.

LITCHFIELD, N. & BERRYMAN, K. 2006. Relations between postglacial fluvial incision rates and uplift rates in the North Island, New Zealand. *Journal of Geophysical Research*, **111**, F02007, doi:10.1029/2005JF000374.

MCKAY, T. 1875. The glacial period of New Zealand. *Transactions of the New Zealand Institute*, **7**, 447–417.

MARSHALL, P. 1910. The glaciation of New Zealand. *Transactions of the New Zealand Institute*, **42**, 334–348.

MARSHALL, P. 1912. *Geology of New Zealand*. Government Printer, Wellington.

MOFFAT, J. G. 1990. *Surficial geology of the Park Valley, Northern Tararua Range, New Zealand*. BSc (Hons) thesis, Victoria University of Wellington.

NATHAN, S. 2005. *Harold Wellman: A Man Who Moved New Zealand*. Victoria University Press, Wellington.

NEEF, G. 1999. Neogene development of the onland part of the forearc in northern Wairarapa, North Island, New Zealand: a synthesis. *New Zealand Journal of Geology and Geophysics*, **42**, 113–135.

PARK, J. 1909*a*. Some evidences of glaciation on the shores of Cook Strait and Golden Bay. *Transactions of the New Zealand Institute*, **42**, 585–588.

PARK, J. 1909*b*. The great ice age of New Zealand. *Transactions of the New Zealand Institute*, **42**, 589–612.

PARK, J. 1909*c*. Further Notes on the Glaciation of the North Island of New Zealand. *Transactions of the New Zealand Institute*, **42**, 580–584.

PARK, J. 1909*d*. On the glacial till in the Hautapu Valley, Rangitikei, Wellington. *Transactions of the New Zealand Institute*, **42**, 575–580.

PARK, J. 1910. *The Geology of New Zealand.* Whitcombe and Tombs, Christchurch.

PILLANS, B., MCGLONE, M., PALMER, A., MILDENHALL, D., ALLOWAY, B. & BERGER, G. 1993. The Last Glacial Maximum in central and southern North Island, New Zealand: a palaeoenvironmental reconstruction using the Kawakawa Tephra Formation as a chronostratigraphic marker. *Palaeogeography, Palaeoclimatology, Palaeoecology*, **101**, 283–304.

SHEPHERD, M. J. 1987. Glaciation of the Tararuas: fact or fiction? *Proceedings of the 14th New Zealand Geography Conference, Palmerston North*, New Zealand Geographical Society, Aukland, 114–115.

SHULMEISTER, J., FINK, D. & AUGUSTINUS, P. C. 2005. A cosmogenic nuclide chronology of the last glacial transition in north-west Nelson, New Zealand – insights into Southern Hemisphere climate forcing during the last deglaciation. *Earth and Planetary Science Letters*, **233**, 455–466.

SPEIGHT, R. 1939. Presidential address 'Some Aspects of Glaciation'. *Report of Australian and New Zealand Association for Advancement of Science*, **24**, 49–71.

STEVENS, G. R. 1974. *Rugged Landscape: The Geology of Central New Zealand.* Reed, Wellington.

STEVENS, G. R. 2006. Cotton, Charles Andrew 1885–1970. *Dictionary of New Zealand Biography* (http://www.dnzb.govt.nz/), updated 7 April 2006.

TORRENS, H. S. 2006. Notes on 'The Amateaur' in the development of British geology. *Proceedings of the Geologists' Association*, **117**, 1–8.

WATTERS, W. A. 2006. Marshall, Patrick 1869–1950. *Dictionary of New Zealand Biography* (http://www.dnzb.govt.nz/), updated 7 April 2006.

WELLMAN, H. W. 1948. Tararua Range summit height accordance. *New Zealand Journal of Science & Technology*, **B30**, 123–127.

WELLMAN, H. W. 1955. New Zealand Quaternary Tectonics. *Geologische Rundschau*, **43**, 248–257.

WILLETT, R. W. 1950. The New Zealand Pleistocene snowline, climatic conditions and suggested biological effects. *New Zealand Journal of Science and Technology*, **32**, 18–48.

Index